Recent Advances in the Behavior of Liquids in Honor of Prof. Dr. William Acree Jr.

Recent Advances in the Behavior of Liquids in Honor of Prof. Dr. William Acree Jr.

Guest Editors

William E. Acree, Jr.
Juan Ortega Saavedra

MDPI

Basel • Beijing • Wuhan • Barcelona • Belgrade • Novi Sad • Cluj • Manchester

Guest Editors
William E. Acree, Jr.
Department of Chemistry
University of North Texas
Denton, TX
USA

Juan Ortega Saavedra
Division of Thermal
Engineering and
Instrumentation
University of Las Palmas de
Gran Canaria
Las Palmas de Gran Canaria
Canary Islands
Spain

Editorial Office
MDPI AG
Grosspeteranlage 5
4052 Basel, Switzerland

This is a reprint of the Special Issue, published open access by the journal *Liquids* (ISSN 2673-8015), freely accessible at: https://www.mdpi.com/journal/liquids/special_issues/70AAVLE51A.

For citation purposes, cite each article independently as indicated on the article page online and as indicated below:

Lastname, A.A.; Lastname, B.B. Article Title. *Journal Name* **Year**, *Volume Number*, Page Range.

ISBN 978-3-7258-4191-2 (Hbk)
ISBN 978-3-7258-4192-9 (PDF)
https://doi.org/10.3390/books978-3-7258-4192-9

Contents

About the Editors

William E. Acree, Jr.

William E. Acree, Jr. received his bachelor's degree (1975), master's degree (1977), and doctorate (1981) in chemistry from the University of Missouri at Rolla. He taught at Kent State University from September 1982 to August 1988 before moving to the University of North Texas, where he is currently a professor in the Chemistry Department. His research interests include spectrofluorometric probe studies, thermodynamic properties of non-electrolyte solutions, solid–liquid equilibrium in organic non-electrolyte solutions, and developing linear free energy relationships to describe mathematically solute transfer processes of chemical importance. To date, his research has resulted in the publication of more than 1,070 peer-reviewed research articles, one research monograph titled "Thermodynamic Properties of Non-electrolyte Solutions", six volumes in the IUPAC-NIST Solubility Data Series, one Liquid's Special Issue Reprint, as well as several encyclopedia articles, book chapters, and educational articles in the Journal of Chemical Education.

Juan Ortega Saavedra

Juan Ortega Saavedra graduated in industrial engineering (1976) and received his PhD in 1979, he was the first doctor from the University of Las Palmas de Gran Canaria (Canary Islands, Spain), where he worked as a Thermodynamics professor at the High School of Engineering. He directed a research team working on the phase equilibrium of solutions with young PhD students. He has contributed to numerous papers and published almost 300 research articles, including a Liquid's Special Issue Reprint, several book chapters, divulgation articles, and contributions to databases. Currently, he is Professor Emeritus of the University, where he teaches and conducts his research, which is now directed towards modeling.

Preface

Many important biochemical and industrial processes occur in aqueous and liquid organic solvent media. Significant experimental and theoretical studies have been performed in recent years to better understand the role of molecular interactions in affecting chemical reaction rates and product yields, chemical equilibrium, partition coefficients, and physical properties of liquid solutions. This volume is the second part of an earlier Special Issue titled Modeling of Liquids Behavior: Experiments, Theory and Simulations, and honors Professor William Acree Jr. for his extensive and outstanding research accomplishments in thermodynamic and physicochemical properties of non-electrolyte solutions.

The comprehensive review article and eighteen original research articles published in the Special Issue are authored by leading experts in their respective fields and report experimental: (a) diffusion coefficients for anti-tuberculosis drugs in cyclodextrin-containing aqueous solutions; (b) spectroscopic probe studies in deep eutectic solvents composed of lanthanide salts and urea; (c) preferential solvation and apparent thermodynamic properties for phenothiazine in two binary aqueous-alcoholic solvent mixtures; (d) solubility data for solutes dissolved in water, organic mono-solvents, and mixtures; and (e) vapor pressures and enthalpies of vaporization of 5α-androstane, 5α-cholestane, three triazolium ionic liquids, and two cinnamate derivatives with notable flavor and fragrance characteristics. Predictive expressions based on quantitative structure–property relationships (QSPRs), atom fragment additivity approaches, machine learning models, and quantum mechanical methods are provided for estimating: (a) enthalpies and free energies of solvation; (b) partition coefficients for solute transfer into membranes and organic solvents from both water and the gas phase; and (c) Abraham model solute descriptor values. A study involving molecular dynamic simulations explored the influence of the structure of the cations of a series of aprotic ([CnC1Im][TFSI] and [CnCnIm][TFSI]) and protic ([HCnIm][TFSI]) alkylimidazolium bistrilimides on the domain structure of their liquid phase. Measured experimental data, combined with semi-theoretical models and molecular simulations, have provided valuable insight into intermolecular and intramolecular hydrogen-bond formation, as well as the formation of solute–solvent association, preferential solvation in mixed solvents, and the polarity of novel ionic liquids, renewable ester solvents prepared dimer carboxylic acids, and deep eutectic solvent systems. Information gained from such studies has resulted in improved and more efficient industrial manufacturing processes.

The wide and diverse range of topics covered in this Special Issue illustrates that important research is still being conducted into aqueous and non-electrolyte solutions. Major scientific discoveries are the products of more accurate experimental measurements, state-of-the-art computational methods, and better scientific instrumentation. This Special Issue will be a valuable resource for researchers conducting experimental and computational studies in liquid solutions.

William E. Acree, Jr. and Juan Ortega Saavedra
Guest Editors

liquids

MDPI

Review

On the Solute-Induced Structure-Making/Breaking Phenomena: Myths, Verities, and Misuses in Solvation Thermodynamics

Ariel A. Chialvo

Independent Researcher, Knoxville, TN 37922-3108, USA; ovlaich@gmail.com

Abstract: We review the statistical mechanic foundations of the *fundamental structure-making/breaking functions*, leading to the rigorous description of the solute-induced perturbation of the solvent environment for the understanding of the solvation process of any species regardless of the type and nature of the solute–solvent interactions. Then, we highlight how these functions are linked to unambiguous thermodynamic responses resulting from changes in state conditions, composition, and solute–solvent intermolecular interaction asymmetries. Finally, we identify and illustrate the pitfalls behind the use of surrogate approaches to structure-making/breaking markers, including those based on Jones–Dole's B-coefficient and Hepler's isobaric-thermal expansivity, while highlighting their ambiguities and lack of consistency and the sources of misinterpretations.

Keywords: solute-induced microstructural perturbation; fundamental structure-making/breaking functions; aqueous electrolyte solutions; solution non-idealities; solute–solvent intermolecular asymmetries; Hepler's criterion; Jones–Dole's B-coefficient-based criterion; preferential solvation function; Krichevskii parameter; second osmotic virial coefficients

1. Introduction

Since the early days of the molecular modeling of fluid mixtures, the main effort has focused on drawing formal (albeit rigorous) connections between the microstructural features of the system of interest and their macroscopic manifestation as thermodynamic properties. This approach follows logically from a pair of fundamental facts, namely: (a) the microstructure of a system arises from the presence of interparticle interactions, where an ideal gas represents the complete absence of them so that non-zero (statistical mechanic) pair correlation functions reflect the deviations of the actual pair distribution functions from the (ideal gas) uniform distribution counterpart triggered by the strength of the interactions; and (b) the microstructure of the mixture stems from the differences (e.g., asymmetries) between the unlike-pair interactions, giving rise to contrasting pair correlation functions [1]. Consequently, the macroscopic manifestations of these microscopic behaviors become described by the thermodynamic residual properties as deviations from those of the ideal gas model in the (a)-case and by the thermodynamic excess properties as deviation from a precisely-defined ideal solution reference in the (b)-case [2,3]. These two fundamental concepts become the blueprint for building the rigorous microscopic-to-macroscopic understanding of solvation processes, one based on the analysis of the solute-induced perturbation of the surrounding environment, leading to the unambiguous definition of the fundamental structure-making/breaking functions [4].

In fact, the isobaric–isothermal solute-induced microstructural distortion described as a local solvent-density perturbation (direct effect) caused by the intermolecular interaction asymmetry between the strength of the original solvent–solvent and the resulting solute–solvent interactions and propagating across the system up to a (correlation length) distance from the perturbing center (indirect effect) [5,6] leads to its macroscopic interpretation in terms of thermodynamics of solution non-ideality [7–14]. Thus, a great deal of effort in solution thermodynamics has been placed on the development of experimental methods to extract the structural information of aqueous electrolyte and non-electrolyte solutions from scattering experiments, such as X-ray (including XANES and EXAFS) [15–22], neutron diffraction with isotope substitution (NDIS) [20,23–26], hybrid empirical microstructural refinement techniques [27–33], and the molecular simulation-based interpretation of NDIS raw data [34–40], frequently complemented with infrared [41,42], Raman [43–47], NMR [48,49], and femtosecond pump-probe spectroscopies [50,51].

This research effort has often led to loose interpretations of the solvation behavior of solutes as either structure-makers or structure-breakers [52–56], followed by the more pretentious kosmotrope (order-maker) and chaotrope (disorder-maker) descriptors [57], usually used as an interchangeable alternative notation for structure-makers and structure-breakers, respectively [57–62], i.e., an unfortunate circumstance that introduced additional ambiguities and confusion into the reported interpretation of the solvation behavior of solutes. In fact, this sentiment was vented earlier as "*Concepts like 'structure-breaking' and 'structure-promoting' solutes appear almost in every discussion of the physicochemical properties of aqueous solutions... Yet in spite of the great number of articles written on this subject, the very basic questions of how to define the 'structure of water' and which experiment should be done to measure the extent of 'structural changes in the solvent' were left unanswered. Almost invariably, these concepts slip into the discussion in a somewhat qualitative manner, i.e., it is assumed that they are already understood intuitively and no further clarification or elaboration is needed.*" [63].

The common source of ambiguity around the concept of structure-maker/breaker solutes (or kosmotropes/chaotropes, for that matter) has been the lack of an explicit identification of the structure-making/breaking (kosmotrope/chaotrope) signature, and consequently, the absence of any definite formal link between the microstructural effect of the solute on the surrounding solvent environment—the solute-induced local density perturbation—and the corresponding macroscopic response whose magnitude can be measured experimentally [54,64]. Moreover, the alluded restructuring process highlights a few essential requirements to avoid the frequently used vague language while facilitating the interpretation of solvation phenomena, e.g., the need for (i) a meaningful unambiguous account for the microstructural perturbation of the environment around any species, (ii) the development of precisely defined descriptors of the solvation phenomena, and (iii) the identification of the corresponding macroscopic responses as thermodynamic signatures of the species solvation leading to the solution of thermodynamic non-ideality.

Based on the discussed scenario, the main goal of this manuscript is twofold: first, review the fundamentals underlying the rigorous and unambiguous interpretation of the solute-induced structure-making/breaking phenomena indicated in the title, and second, identify the myths, verities, and misuses of the structure-making/breaking markers in solvation thermodynamics, as emphasized in the subtitle of this manuscript. Given the scope of the current manuscript, and to keep the discussion on focus, this review mainly comprises our own developments on the matter while being illustrated with the analysis of data from numerous sources from the literature and aims to raise awareness on what can and cannot be construed from frequently invoked markers of structure-making/breaking phenomena.

For that purpose, we start reviewing the statistical mechanic foundations underlying the solute-induced distortion of the solvent environment in Section 2, where we discuss the interplay of two length scales in the solvation process associated with the microstructural perturbation and its propagation, leading to the explicit unambiguous and natural definition of the *fundamental structure-making/breaking functions*. In the process, we show how these functions become rigorously linked to thermodynamic responses, including the isochoric–isothermal pressure change upon solvation, the osmotic second virial coefficient, and a number of frequently invoked thermodynamic preferential solvation parameters. In Section 3, we discuss the responses of the *fundamental structure-making/breaking functions* to changes in state conditions, composition, and solute–solvent intermolecular interaction asymmetries. Then, in Section 4, we argue and illustrate the pitfalls behind the use of surrogate approaches, including those based on Jones–Dole's B coefficient and Hepler's isobaric-thermal expansivity, while highlighting their ambiguities and lack of consistency and fulfillment of required constraints, as well as identifying the sources of misinterpretations.

2. Rigorous Definition of Solute-Induced Microstructural Perturbation of the Solvent Environment

As indicated in the introduction, we must fulfill a few essential requirements in the microscopic characterization of the solvation process, including the unambiguous and precise description of the perturbation of the solvent environment upon solute solvation at the prevailing state conditions of the solution, and the rigorous linking of the microscopic behavior to the corresponding macroscopic manifestations leading to the experimental measurements [4]. For that purpose, we review the statistical mechanical-based foundations for the *fundamental structure-making/breaking* concept as a thermodynamic state function, link it to the pressure perturbation counterparts and preferential solvation phenomena, and discuss its attributes, including the universality of the definition.

2.1. Fundamental Split between Short- and Long-Range Contributions to the Solvation Behavior of Solutes

The solvation of any solute (electrolyte or otherwise) entails a disruption of the system microstructure resulting from the solute–solvent intermolecular interaction asymmetry, described formally and unambiguously as a finite local density perturbation (i.e., a short-range or direct effect) whose propagation across the system (i.e., a longer-range or indirect effect) is felt up to a distance given by the correlation length of the environment and scales with its isothermal compressibility [65]. Note that the local density perturbation associated with the solvation process remains finite even under special thermodynamic conditions, such as in the solvation of a solute at infinite dilution in a solvent at its critical conditions where the correlation length becomes divergent [12,66]. More importantly, the alluded split is rigorously described by the Ornstein–Zernike equation [67], and therefore, the local density perturbation becomes naturally linked to the direct correlation functions, while the ensuing propagation becomes linked to the indirect correlation functions [5,68,69]. This feature provides the means to unravel the rigorous connections between the species-induced microstructural perturbation in a solution, its propagation across the system, and the driving force underlying the solvation of any species at any state condition or composition. This formal characterization becomes more powerful through their reinterpretation according to the species' molecular-based *fundamental structure-making/breaking functions* [4].

2.2. Definition and Attributes of the Fundamental Structure-Making/Breaking Functions $\mathcal{S}_{\alpha\beta}$

To dissipate the controversy underlying the meaning of the structure-making/breaking concept, we must (i) identify the precise link between the microstructural impact of the solute species on its surrounding environment, i.e., the induced local density perturbation and the ensuing macroscopic manifestation to be ultimately measured, (ii) provide a direct path for the calculation of the magnitude of the microstructural perturbation of the system according to an explicit, general, and rigorous molecular-based thermodynamic formalism, and (iii) be able to apply the concept equally and consistently to any solute–solvent intermolecular interaction asymmetry regardless of either the state conditions, number of components, or system composition [4].

The magnitude of the local density perturbation of the j-solvent environment around the i-solute can be rigorously measured in terms of Kirkwood–Buff [70] total correlation function integrals (TCFI) as follows [4,71]:

$$\begin{aligned} \mathcal{S}_{ij}(T,P,x_i) &= \rho x_j (G_{ij} - G_{jj}) \\ &= (\mathcal{N}_{ij} - \mathcal{N}_{jj}) \end{aligned} \tag{1}$$

where $\mathcal{N}_{ij}(T,P,x_i) = \rho x_j G_{ij}$ describes the absolute average number of j-solvent molecules around any central i-solute at the prevailing state conditions and system composition. Note that the TCFIs in Equation (1), $G_{\alpha\beta} = 4\pi \int_0^\infty h_{\alpha\beta}(r_{\alpha\beta}) r_{\alpha\beta}^2 dr_{\alpha\beta}$, are volume integrals over the pair correlation function $h(r_{\alpha\beta}) = \left\langle g_{\alpha\beta}^\infty(r_{\alpha\beta}, \omega_\alpha, \omega_\beta) \right\rangle_{\omega_\alpha \omega_\beta} - 1$, where $g_{\alpha\beta}^\infty(r_{\alpha\beta}, \omega_\alpha, \omega_\beta)$ depicts the spatial pair distribution function as the statistical mechanical microstructural descriptor, $\langle \cdots \rangle_{\omega_\alpha \omega_\beta}$ represents an orientational average, ω_α describes the orientation of the α-species, and $r_{\alpha\beta}$ defines the distance between the centers of mass of the two species. In other words, $\mathcal{S}_{ij}(T, P, x_i)$ characterizes the average number of j-solvent molecules around the central i-solute in excess (deficit) to that around any j-solvent and comprises the key microscopic information required to assess the magnitude of the i-induced microstructural perturbation of the j-solvent environment, i.e.,

$$G_{jj}(T, P, x_i) \xrightarrow[j\text{-solvent perturbation}]{i\text{-solute solvation}} G_{ij}(T, P, x_i) \tag{2}$$

where the arrow highlights the microstructural evolution of the solvent environment from the original TCFI, $G_{jj}(T, P, x_i)$, to the final TCFI, $G_{ij}(T, P, x_i)$. This microstructural perturbation becomes characterized by three possible outcomes according to the sign of the difference $(G_{ij} - G_{jj})_{TPx}$ as follows:

$$(G_{ij} - G_{jj})_{TPx} \Rightarrow \begin{cases} G_{ij} > G_{jj} \rightarrow \mathcal{S}_{ij} > 0 \; \textit{structure-maker} \\ G_{ij} \cong G_{jj} \rightarrow \mathcal{S}_{ij} \cong 0 \; \textit{neither maker-nor breaker} \\ G_{ij} < G_{jj} \rightarrow \mathcal{S}_{ij} < 0 \; \textit{structure-breaker} \end{cases} \tag{3}$$

Therefore, according to Equations (1)–(3), the solute-induced solvent perturbation becomes unambiguously described as a strengthening of the solvent environment around the i-solute when $G_{ij} > G_{jj}$, i.e., $\mathcal{S}_{ij}(T, P, x_i) > 0$, as the i-solute behaves as a structure-maker species. Conversely, the solvent perturbation weakens the solvent environment around the i-solute when $G_{ij} < G_{jj}$, i.e., $\mathcal{S}_{ij}(T, P, x_i) < 0$, as the i-solute exhibits a structure-breaking effect. As we will discuss below, the j-solvent environment might remain unperturbed despite the presence of the i-solute as $(G_{jj} \cong G_{ij})_{TPx}$, i.e., $\mathcal{S}_{ij}(T, P, x_i) \cong 0$. Note that the unperturbed condition represents also the signature of a special case of Lewis–Randall solution ideality because $G_{jj}(T, P, x_i) = G_{ij}(T, P, x_i) \neq 0$ describes the scenario where the i-solute behaves as the j-solvent, as well as the more general solution ideality condition $\Delta_{ij}(T, P, x_j) \equiv (G_{ii} + G_{jj} - 2G_{ij}) = 0$, leading to $(G_{ii} - G_{ij})_{TPx} = -(G_{jj} - G_{ij})_{TPx} \lesseqgtr 0$ [72,73].

At this point, we fulfilled the first requirement (i), and now we must provide a direct path for the calculation of the magnitude of the microstructural perturbation described by the three conditions in Equation (3), i.e., requirement (ii). For that purpose, we have invoked Kirkwood–Buff fluctuation formalism [70] and derived the following expressions for the general type of (electrolyte or non-electrolyte) i-solute at any system composition including infinite dilution [1,4]:

$$\mathcal{S}_{ij}(T, P, x_j) = 1 - \rho\hat{v}_i/\nu D \Rightarrow \begin{cases} \rho\hat{v}_i < \nu D \rightarrow \mathcal{S}_{ij} > 0 \; \textit{structure-maker} \\ \rho\hat{v}_i \cong \nu D \rightarrow \mathcal{S}_{ij} \cong 0 \; \textit{neither maker-nor breaker} \\ \rho\hat{v}_i > \nu D \rightarrow \mathcal{S}_{ij} < 0 \; \textit{structure-breaker} \end{cases} \tag{4}$$

where ν is the stoichiometric coefficient of the electrolyte i-solute, $\hat{v}_i$ denotes its partial molar volume, ρ is the molar density of the solution, and $D = \left[1 + (\partial \ln \gamma_j / \partial \ln x_j)_{TP}\right]$ describes the material stability coefficient in terms of the composition slope of the activity coefficient of the j-solvent [74]. In other words, given the pair of volumetric properties $(\hat{v}_i, \rho)$ and the stability coefficient D, Equation (4) describes the thermodynamic response to the solute-induced effects on the microstructure of the fluid mixture and rigorously and unambiguously reveals the possible behaviors of the i-solute without requiring any explicit microstructural information of the system.

Finally, requirement (iii) is automatically satisfied given the absence of any constraint on the type of interactions, nature of the species, and number of components in the solution underlying the generality of the Kirkwood–Buff formalism. In fact, a universal and fundamentally based structure-making/breaking function (aka criterion or marker) must apply equally to species such as a non-interacting ideal gas *i*-solute [14,75], any *i*-species behaving identically to the *j*-solvent [12], and either weakly or strongly interacting *i*-electrolyte solutes, regardless of the type and description of the solute–solvent interactions. Therefore, the structure-making/breaking criterion must be able to identify the null solute-induced effect when the solute–solvent interaction asymmetry crosses the transition boundary between the solute-induced structure-making and the structure-breaking perturbations. Note that this transition condition might portray two significantly different scenarios, namely: (i) one characterized as a special cases of Lewis–Randall solution ideality, either when the interacting solute species behaves just as another solvent species and therefore keeps the solvent environment unperturbed [76], or (ii) when a non-interacting ideal gas species (IG_i) is solvated by a real *j*-solvent along the thermodynamic path $\rho kT\kappa(T,P,x_j) = 1$ [14,71]. In fact, from (4) and the fact that $\hat{v}_i(T,P,x_j)^{G_{ii}=G_{ij}=G^o_{jj}} = \nu v^o_i(T,P) = \nu\rho^{-1}$ with $G^o_{jj} = kT\kappa^o_j - v^o_j$, we find that $\mathcal{S}_{ij}^{G_{ii}=G_{ij}=G^o_{jj}}(T,P,x_j) = 1 - \rho v = 0$. Likewise, because $\hat{v}_i^{IG\_i}(T,P,x_j) = kT\kappa$ [14] it follows immediately that $\mathcal{S}_{ij}^{IG\_i}(T,P,x_j) = 1 - \rho kT\kappa = 0$, i.e., when the actual compressibility of the system becomes equal to that of the system behaving as an ideal gas environment, $\kappa(T,P,x_j) = \kappa^{IG} = (\rho kT)^{-1}$ (e.g., see Figure 6 and Appendix A of Ref. [71]).

2.3. *Link between $\mathcal{S}_{\alpha\beta}$, the Isochoric-Isothermal Pressure Perturbation upon Solvation and Krichevskii's Parameter*

The isobaric–isothermal change of residual (R) partial molar Gibbs free energy for the solvation of a ν-dissociable *i*-solute in a *j*-solvent can be expressed as follows,

$$\begin{aligned} \mu_i^R(T,P,x_i) - \nu\mu_j^R(T,P,x_i) &= \nu kT\ln\left(\hat{\phi}_i/\hat{\phi}_j\right)_{TP} \\ &= \int_0^\rho (\partial P/\partial x_i)_{T\rho}\left(d\rho/\rho^2\right) \end{aligned} \tag{5}$$

where $\ln\hat{\phi}_\alpha(T,P,x_\alpha)$ denotes the partial molar fugacity coefficient of the α-species in solution at the prevailing state conditions and composition, while the isochoric-isothermal pressure perturbation $(\partial P/\partial x_i)_{T,\rho}$ becomes the thermodynamic response to the microstructural reorganization of the system upon solvation, which in turn can be expressed in terms of the species partial molar volumes and the isothermal compressibility of the system as follows [4,12],

$$(\partial P/\partial x_i)_{T\rho} = \rho\left(\hat{v}_i - \nu\hat{v}_j\right)/\kappa \tag{6}$$

Equations (5) and (6) provide the macroscopic-to-microscopic rigorous link between the solvation-induced pressure perturbation and the resulting microstructural response of the system,

$$(\partial P/\partial x_i)_{T\rho} = \nu D\kappa^{-1}\left[\rho_j\left(G_{jj} - G_{ij}\right) - \rho_{diss}\left(G_{ii} - G_{ij}\right)\right] \tag{7}$$

where we have invoked the following identities,

$$\begin{aligned} \rho\hat{v}_{diss}(T,P,x_i) &= D\left[1 + \rho_j\left(G_{jj} - G_{ij}\right)\right] \\ &= \rho\hat{v}_i/\nu \end{aligned} \tag{8}$$

With the diffusional stability coefficient $D = 1 + (\partial\ln\gamma_i/\partial\ln x_i)_{TP}$ [74], and

$$\rho\hat{v}_j(T,P,x_i) = D\left[1 + \rho_{diss}\left(G_{ii} - G_{ij}\right)\right] \tag{9}$$

As well as,

$$\kappa(T,P,x_i) = (D/kT\rho)\left[1 + \rho\left(x_jG_{jj} + x_{diss}G_{ii}\right) + \rho^2x_{diss}x_j\left(G_{jj}G_{ii} - G_{ij}^2\right)\right] \tag{10}$$

Which define the partial molar volumes of the species, and the corresponding isothermal compressibility of the system with $x_{diss} = \nu x_i$ and $\hat{v}_i = \nu \hat{v}_{diss}$, according to the inversion of the Kirkwood-Buff formalism [1].

From Equations (1) and (6) we immediately find the explicit connection between the pressure perturbation upon solvation and the microstructural responses, i.e.,

$$(\partial P/\partial x_i)_{T\rho} = \nu D \kappa^{-1}(\mathcal{S}_{ji} - \mathcal{S}_{ij})_{TPx} \quad (11)$$

where the phase stability requires $\kappa(T,P,x_i) > 0$ and $D(T,P,x_i) > 0$ [74] so that,

$$D(T,P,x_i) = \left(1 - x_{diss}\mathcal{S}_{ij} - \nu x_j \mathcal{S}_{ji}\right)^{-1} \quad (12)$$

Leading finally to,

$$(\partial P/\partial x_i)_{T\rho} = \nu(\mathcal{S}_{ji} - \mathcal{S}_{ij})_{TPx} / \left[\kappa(1 - x_{diss}\mathcal{S}_{ij} - \nu x_j \mathcal{S}_{ji})\right]_{TPx} \quad (13)$$

Note that the pressure perturbation upon solvation $(\partial P/\partial x_i)_{T,\rho}$ in a stable phase is always finite regardless of the state condition and composition, even though the right-hand-side of Equation (11) involves potentially divergent compressibility-driven contributions [12,77]. In fact, for the binary system comprising an infinitely dilute *i*-solute at the critical point of the pure *j*-solvent, the partial molar volume of the solute at infinite dilution will diverge at the same rate as the isothermal compressibility of the pure solvent, yet,

$$\begin{aligned} \lim_{T\rho \to critical} (\partial P/\partial x_i)^{\infty}_{T\rho} &= \lim_{T\rho \to critical} \left(\rho_j^o \hat{v}_i^{\infty} - \nu\right)/\kappa_j^o \\ &= \mathcal{A}_{Kr} \to finite \end{aligned} \quad (14)$$

where the so-called Krichevskii parameter $\mathcal{A}_{Kr}$ [77] becomes defined in terms of the limiting behavior of the structure-making/breaking function of the *i*-solute, i.e., [78]

$$\mathcal{A}_{Kr} = -\nu \lim_{T,\rho_j^o \to critical} \left(\mathcal{S}_{ij}^{\infty}/\kappa_j^o\right) \quad (15)$$

The finiteness of the alluded limiting function becomes more obvious after recalling that $\lim_{T,\rho_j^o \to critical} \left(\mathcal{S}_{ij}^{\infty}/\kappa_j^o\right) = \mathcal{A}_{Kr}^{IG\_i} \lim_{T,\rho_j^o \to critical} \mathcal{S}_{ij}^{\infty}(SR)$, with $\mathcal{A}_{Kr}^{IG\_i} = kT_c\rho_c$ where SR denotes the short-range or solvation contribution to the property [79].

2.4. Link between $\mathcal{S}_{\alpha\beta}$ and the Second Osmotic Virial Coefficient of the Solute

The second osmotic virial coefficients derive from the isothermal solute-density expansion of the solute activity along different thermodynamic Θ-paths including $\Theta \equiv (\mu_j, P, \rho_j)$ (see e.g., Figure 4 and discussed in section SI-A of the Supplementary Information of Ref. [71]). According to the Lewis-Randall (LR) solution ideality reference, the structure-making/breaking function $\mathcal{S}_{ij}^{\infty}(T,P)$ can be rigorously linked to the three distinctive types of second osmotic virial coefficients [71]. In fact, after invoking the limiting composition slope $(\partial \ln \gamma_i^{LR}/\partial x_i)^{\infty}_{TP} = -\rho_j^o\left(G_{jj}^o + G_{ii}^{\infty} - 2G_{ij}^{\infty}\right)_{TP}$ [80], and recalling the definition of the isobaric–isothermal second osmotic virial coefficient, $B_i'(T,P) = -0.5\left(G_{ii}^{\infty} - G_{ij}^{\infty}\right)$, it has been demonstrated that

$$\mathcal{S}_{ij}^{\infty}(T,P) = \left(\partial \ln \gamma_i^{LR}/\partial x_i\right)^{\infty}_{TP} - 2\rho_j^o B_i'(T,P) \quad (16)$$

Likewise, from the definition of the isothermal iso $-\mu_j$ osmotic second virial coefficient, $B_i^*(T,P) = -0.5G_{ii}^{\infty}$ [81], follows immediately that,

$$\left(\partial \ln \gamma_i^{LR}/\partial x_i\right)_{TP}^{\infty} = \rho_j^o\left(kT\kappa_j^o - v_j^o\right) + 2\rho_j^o\left(B_i^* - \hat{v}_i^{\infty}\right) \tag{17}$$

And therefore,

$$\begin{aligned} \mathcal{S}_{ij}^{\infty}(T,P) &= -2\rho_j^o B_i^{*,LR-IS} + 2\rho_j^o\left(B_i^* - \hat{v}_i^{\infty}\right) - 2\rho_j^o B_i' \\ &= 2\rho_j^o\left(B_i^{*,E} - B' - \hat{v}_i^{\infty}\right) \\ &= 2\rho_j^o\left(B_i^* - B' - \hat{v}_i^{\infty}\right) - \mathcal{S}_{ij}^{\infty,IG\_i} \end{aligned} \tag{18}$$

where

$$B_i^{*,LR-IS}(T,P) = 0.5\left(v_j^o - kT\kappa_j^o\right) = 0.5v_j^o\mathcal{S}_{ij}^{\infty,IG\_i} \tag{19}$$

And,

$$\begin{aligned} B_i^{*,E}(T,P) &= B_i^* - B_i^{*,LR-IS} \\ &= B_i^* - 0.5v_j^o\mathcal{S}_{ij}^{\infty,IG\_i} \end{aligned} \tag{20}$$

Moreover, from the definition of the isochoric-isothermal second osmotic virial coefficient [81], and subsequent interpretation in terms of Kirkwood-Buff integrals $B''(T,P) = 0.5\left[G_{ij}^{\infty,2}/\left(v_j^o + G_{jj}^o\right) - G_{ii}^{\infty}\right]$ [75], follows that [71],

$$\left(\mathcal{S}_{ij}^{\infty} - 1\right)^2 = \rho_j^o kT\kappa_j^o\left[1 + \rho_j^o B_i'' - \left(\partial \ln \gamma_i^{LR}/\partial x_i\right)_{TP}^{\infty}\right] \tag{21}$$

We must highlight that the derived $\mathcal{S}_{ij}^{\infty} = \Im\left(B_i^{\otimes}\right)$ relationships are all rigorous expressions involving neither approximations nor constraints on the type of solute, nature of the solvent, and magnitude of the intermolecular interactions asymmetries. However, for *i*-solutes whose non-idealities follows the $\ln \gamma_i(T,P,x_i) \sim (1-x_i)^2$ functionality, as in the case of gases in aqueous and organic solutions, [75,82] the limiting composition slope becomes $\left(\partial \ln \gamma_i^{LR}/\partial x_i\right)_{TP}^{\infty} = -2\ln \gamma_i^{LR}(T,P)$. Consequently, the *fundamental structure-making/breaking function* and the three osmotic second virial coefficients become related as follows,

$$\mathcal{S}_{ij}^{\infty}(T,P) = -\left(\ln \gamma_i^{LR} + \rho_j^o B_i^{*,E}\right)_{TP} \tag{22}$$

$$\mathcal{S}_{ij}^{\infty}(T,P) = -2\left(\ln \gamma_i^{LR} + \rho_j^o B_i'\right)_{TP} \tag{23}$$

$$\left(\mathcal{S}_{ij}^{\infty} - 1\right)^2 = \rho_j^o kT\kappa_j^o\left[1 + \rho_j^o B_i'' + 2\ln \gamma_i^{LR}\right] \tag{24}$$

While we have illustrated the links for a non-electrolyte solute, similar connections can be drawn for a dissociative solute as described explicitly for electrolyte species in Ref. [83].

2.5. Preferential Solvation Phenomenon as a Balance among Structure-Making/Breaking Effects

Preferential solvation $\mathcal{PS}$ usually refers to the effect of a third species (*aka* cosolvent) on the solubility of a solute, a phenomenon of great interest for the understanding of the thermodynamics underlying the solubility enhancement of low-solubility species in highly compressible environments [10,84–87], and the manipulation of the relative volatility of species to improve separations [88,89]. The behavior of these systems has been frequently interpreted by a diversity of thermodynamic preferential interaction parameters leading to a number of non-equivalent microscopic markers for the preferential solvation phenomenon [90,91]. The common denominator underlying this scenario is again the lack of an explicit microscopic definition of the underlying $\mathcal{PS}$ marker as previously highlighted, i.e., "*how do we measure the $\mathcal{PS}$ of a given solute in a given mixture?*" leading to "*what are the* *molecular reasons that cause a solute to prefer one component over the other and, hence, alter the composition in its local environment?"* [92].

To confront this issue, we have recently introduced a *universal preferential solvation* (state) function, $\mathcal{PS}(T,P,x_j) = \rho_o\left(G_{ik}^{\infty} - G_{ij}^{\infty}\right)$ [91], as unique microscopic measure of how much different the j-solvent environment around the i-solute is in comparison to the k-solvent environment counterpart, where $\rho_o = \rho_j + \rho_k$. Thus, $\mathcal{PS}(T,P,x_j)$ accounts unambiguously for the solvation of an infinitely dilute i-solute species in a mixed-solvent (i.e., a j-solvent plus a k-cosolvent) environment at fixed state conditions in terms of the solute-induced microstructural perturbations of the solvent and cosolvent environments resulting from the species intermolecular interaction asymmetries. This solute-induced differential microstructural perturbation, caused by the disparity of affinities between the i-solute and each of the solvents species, becomes more revealing when written explicitly written in terms of Kirkwood-Buff integrals [70] as follows,

$$\begin{aligned} \mathcal{PS}(T,P,x_j) &= \rho_o\left(G_{ik}^{\infty} - G_{ij}^{\infty}\right) \\ &= N\left[\left(\mathcal{N}_{ik}^{res,\infty}/N_k\right) - \left(\mathcal{N}_{ij}^{res,\infty}/N_j\right)\right] \end{aligned} \tag{25}$$

where $\rho_\beta = \rho_o x_\beta$, and $\mathcal{N}_{\alpha\beta}^{res,\infty} \equiv \rho_\beta G_{\alpha\beta}^{\infty}$ represents the average number of β-solvent molecules around the infinitely dilute α-solute in either excess or deficit to that of the $\alpha\beta$-ideal gas uniform distribution. From Equation (25) follows immediately that $\mathcal{PS}(T,P,x_j) = 0$ for two precisely-defined scenarios; namely: (a) $\left(G_{ik}^{\infty} = G_{ij}^{\infty}\right)_{TPx} \neq 0$ for any interacting i-solute species at infinite dilution, where the j-solvent and the k-cosolvent behave identically as a single pure solvent leading to $G_{jj} = G_{kk} = G_{jk} = \left(kT\kappa_j^o - v_j^o\right)$, and (b) $\left(G_{ik}^{\infty} = G_{ij}^{\infty}\right)_{TPx}^{IG\_i} = 0$ for the solvation of a non-interacting i-solute (ideal gas) species at infinite dilution in a mixed $(j+k)$-real solvent.

On the one hand, note that $\mathcal{PS}(T,P,x_j) = 0$ implies that $\left(\mathcal{N}_{ik}^{res,\infty}/\mathcal{N}_{ij}^{res,\infty}\right) = (N_k/N_j)$, in other words, $\mathcal{PS}(T,P,x_j)$ describes the average deviation of the mixed-solvent distribution around the i-solute at infinite dilution from that of the corresponding uniform bulk-phase value. On the other hand, $\mathcal{PS}(T,P,x_j) \neq 0$ highlights the microstructural response of the system to the solute–solvent and solute-cosolvent interaction asymmetries, leading to the differential affinities of the solute with the solvent, whose macroscopic responses translate into thermodynamic excess properties describing the solution thermodynamic non-idealities [74]. Moreover, given the fact that the ternary system under consideration becomes fully characterized either by five Kirkwood-Buff integrals (G_{ij}^{∞}, G_{ik}^{∞}, G_{jk}, G_{jj}, G_{kk}), or four *fundamental structure-making/breaking functions*, ($\mathcal{S}_{ij}^{\infty}$, $\mathcal{S}_{ik}^{\infty}$, $\mathcal{S}_{jk}$, $\mathcal{S}_{kj}$), where $\mathcal{S}_{kj}\left(\mathcal{S}_{ij}^{\infty}\right)$ and $\mathcal{S}_{jk}\left(\mathcal{S}_{ik}^{\infty}\right)$ with $\mathcal{S}_{kj} \neq \mathcal{S}_{jk}$. In other words, this ternary system becomes fully described by two independent *fundamental structure-making/breaking functions,* and consequently, the interpretation of preferential solvation phenomena based on $\mathcal{PS}(T,P,x_j)$, Equation (25), becomes entirely described in terms of a linear combination of four fundamental structure-making/breaking functions $\mathcal{S}_{\alpha\beta}$ as follows [82,91],

$$\mathcal{PS}(T,P,x_j) = \left(\mathcal{S}_{ik}^{\infty} - \mathcal{S}_{jk}\right)/x_k - \left(\mathcal{S}_{ij}^{\infty} - \mathcal{S}_{kj}\right)/x_j \tag{26}$$

We must emphasize that Equations (25) and (26) are exact statistical mechanical expressions for the alluded ternary system regardless of the type or nature of the species involved, while Equation (26) explicitly identifies the formal contribution of the relevant fundamental structure-making/breaking functions. Moreover, this expression becomes rigorously connected to the thermodynamic preferential solvation parameter in the isothermal–isobaric closed system $\Gamma_{i\alpha}^{(x_i\to 0)}(x_j) \equiv (\partial\mu_i^{\ominus}/\partial\mu_\alpha)_{TP}$ (often identified as $\nu_{i\alpha}$, i.e., in Refs. ([93,94]) for $\alpha = (j,k)$, as follows:

$$\begin{aligned}\Gamma_{ij}^{(x_i\to 0)}(x_j) &\equiv (\partial\mu_i^{r,\infty}/\partial\mu_j)_{TP} \\ &= \left[\partial\ln(z\hat{\phi}_i)^{\infty}/\partial\ln x_j\right]_{TP}/\left[1+\left(\partial\ln\gamma_j^{x}/\partial\ln x_j\right)_{TP}\right] \\ &= x_j\mathcal{PS}(T,P,x_j)\end{aligned} \tag{27}$$

where $\mu_i^{r,\infty}(T,P,x_j) = kT\lim_{\rho_i\to 0}\ln(z\hat{\phi}_i)_{T,P,\rho_i,x_j}$. Thus, by invoking the Gibbs–Duhem relation for the $(j+k)$-binary mixture, we find the complementary thermodynamic link,

$$\begin{aligned}\Gamma_{ik}^{(x_i\to 0)}(x_j) &\equiv (\partial\mu_i^{r,\infty}/\partial\mu_k)_{TP} \\ &= x_k\mathcal{PS}(T,P,x_j)\end{aligned} \tag{28}$$

indicating that the isothermal–isobaric closed system $\Gamma_{i\alpha}^{(x_i\to 0)}(x_j) \equiv (\partial\mu_i^{\ominus}/\partial\mu_\alpha)_{TP}$ comprises formally similar behaviors for the diffusible α-species, resulting from the combined i-solute-induced perturbations of the j-solvent and k-cosolvent, as described by a linear combination of the four structure-making/breaking functions $\mathcal{S}_{\alpha\beta}$.

3. Microstructural Responses to Intermolecular Interaction Asymmetries, Environmental State Conditions, and Composition

Because the solvation process comprises the solute-induced effect on the solution microstructure at the prevailing state conditions and composition, it becomes instructive to evaluate how the solute–solvent molecular asymmetry, as well as individual perturbations of pressure, temperature, and composition, can modify the solvent microstructure around a solute species.

3.1. Response to Solute–Solvent Molecular Asymmetry

As a state function, the behavior of the *fundamental structure-making/breaking function* $\mathcal{S}_{ij}(T,P,x_j)$, Equation (1) of a stable fluid mixture becomes a continuous function of the magnitude of the solute–solvent interaction asymmetry, as well as the state conditions and system composition. Typically, the idea of structure-making/breaking behavior has been applied to solutes at infinite dilution [57,95], and therefore, we review first this limiting behavior for a binary fluid mixture, i.e., $\mathcal{S}_{ij}^{\infty} = \mathcal{S}_{ij}(T,P,x_j\to 1)$. For the sake of generality, we start with Equation (1), written as follows:

$$\mathcal{S}_{ij}^{\infty}(T,P) = 1 - \rho_j^{o}\bar{v}_i^{\infty}/v \tag{29}$$

whose isochoric-isothermal pressure perturbation upon solvation becomes [4]

$$\mathcal{S}_{ij}^{\infty} = -\left(\kappa_j^{o}/v\right)(\partial P/\partial x_i)_{T\rho_j^{o}}^{\infty} \tag{30}$$

Equations (29) and (30) comprise two remarkable attributes: (*a*) the first indicates that the structure-making/breaking functions scale with the partial molar volume of the solute at infinite dilution, and consequently, with the magnitude of the solute–solvent intermolecular interaction asymmetry, and (*b*) the second provides an alternative avenue to experimentally measure the degree of dissimilarity between the solute and the solvent, i.e., the source of thermodynamic non-ideality of the resulting mixture.

In fact, Equation (30) indicates that an i-solute behaves as a *structure-maker* when $(\partial P/\partial x_i)_{T\rho_j^{o}}^{\infty} < 0$, i.e., conventionally known as non-volatile [77] or attractive solutes [7]. Conversely, an i-solute behaves as a *structure-breaker* if $(\partial P/\partial x_i)_{T\rho_j^{o}}^{\infty} > 0$, i.e., characterized as volatile [77] or weakly-attractive and repulsive [7] in the jargon of supercritical fluid solutions [12,85]. Obviously, nature presents a wide and continuous range of solute–solvent interaction asymmetries in binary fluid solutions. For instance, for a particular real *j*-solvent, we might consider a non-interaction (ideal gas) *i*-solute, IG_i, characterized by $\nu = 1$ so that [96]

$$\hat{v}_i^{\infty,IG\_i}(T,P) = kT\kappa_j^o > 0 \tag{31}$$

$$\mathcal{S}_{ij}^{\infty,IG\_i}(T,P) - 1 - \kappa_j^o kT\rho_j^o \lesseqqgtr 0 \tag{32}$$

and a finite $(\partial P/\partial x_i)_{T\rho_j^o}^{\infty,IG\_i} \lesseqqgtr 0$. By cranking up the solute–solvent interactions, such as that of an inert gas solute, B_i, the binary system can be characterized by the second virial coefficients $B_{ii}(T)$ and $B_{ij}(T)$ while keeping the solvent-solvent interactions unchanged, i.e., [4]

$$\hat{v}_i^{\infty,B\_i}(T,P) = kT\kappa_j^o + 2B_{ij} \tag{33}$$

and

$$\mathcal{S}_{ij}^{\infty,B\_i}(T,P) = \mathcal{S}_{ij}^{\infty,IG\_i}(T,P) - 2\rho_j^o B_{ij}(T) \tag{34}$$

This expression indicates that $\mathcal{S}_{ij}^{\infty,B\_i}(T,P) > \mathcal{S}_{ij}^{\infty,IG\_i}(T,P) > 0$ when the unlike-pair interaction second virial coefficient $B_{ij}(T) < 0$, and consequently, the B_i solute induces an additional microstructural perturbation on the surrounding solvent, over that of the ideal gas solute, and becomes a *structure-making* solute. Otherwise, when $B_{ij}(T) > 0$, there are two possible alternatives: either $0 < B_{ij}(T) < 0.5v_j^o\mathcal{S}_{ij}^{\infty,IG\_i}$ giving rise to a *structure-maker* with $0 < \mathcal{S}_{ij}^{\infty,B\_i} < \mathcal{S}_{ij}^{\infty,IG\_i}$ or $0 < B_{ij}(T) > 0.5v_j^o\mathcal{S}_{ij}^{\infty,IG\_i}$ resulting in a *structure-breaker*. Note that at a specific (T,P) state condition at which $B_{ij}(T) = -0.5G_{jj}^o(T,P)$, Equation (34) indicates that the B_i solute does not perturb the microstructure of the *j*-solvent.

Should we keep increasing the magnitude solute–solvent interactions, up to the point that the solute–solvent interactions become equal to the corresponding solvent-solvent interactions, then we would have reached the null solute–solvent asymmetry, i.e., $G_{jj}^o(T,P) = G_{ij}^\infty(T,P) \;\rightarrow\; \mathcal{S}_{ij}^\infty(T,P) = 0$, and the *i*-solute would not perturb the surrounding solvent environment. Not surprisingly, the most frequent scenario comprises systems with solute–solvent interactions different (either weaker or stronger) from the solvent–solvent interactions. When $G_{ij}^\infty(T,P) < G_{jj}^o(T,P)$, such as in the case of non-polar gases in water, we expect $\hat{v}_i^\infty(T,P) > 0$ and $(\partial P/\partial x_i)_{T\rho_j^o}^\infty > 0$ so that $\mathcal{S}_{ij}^\infty(T,P) < 0$, i.e., the *i*-solute behaves as a *structure-breaker* species. Conversely, when $G_{ij}^\infty(T,P) > G_{jj}^o(T,P)$, such as in the case of strongly interactive solutes including strong electrolytes, we observe $\hat{v}_i^\infty(T,P) - v_j^o(T,P) < 0$ and $(\partial P/\partial x_i)_{T\rho_j^o}^\infty < 0$ so that $\mathcal{S}_{ij}^\infty(T,P) > 0$, i.e., the *i*-solute becomes a *structure-maker*. Note, however, that the inequality $\hat{v}_i^\infty(T,P) - v_j^o(T,P) < 0$ can be satisfied by $\hat{v}_i^\infty(T,P) \lesseqgtr 0$, i.e., by either $\hat{v}_i^\infty(T,P) < 0$ or $0 < \hat{v}_i^\infty(T,P) < v_j^o(T,P)$, as illustrated in Figure 1 below, where we provide a schematic with the discussed relationships among $\mathcal{S}_{ij}^\infty(T,P)$, $\hat{v}_i^\infty(T,P)$, and $(\partial P/\partial x_i)_{T\rho}^\infty = -\mathcal{S}_{ij}^\infty/\kappa_j^o$ while highlighting the relevant boundaries.

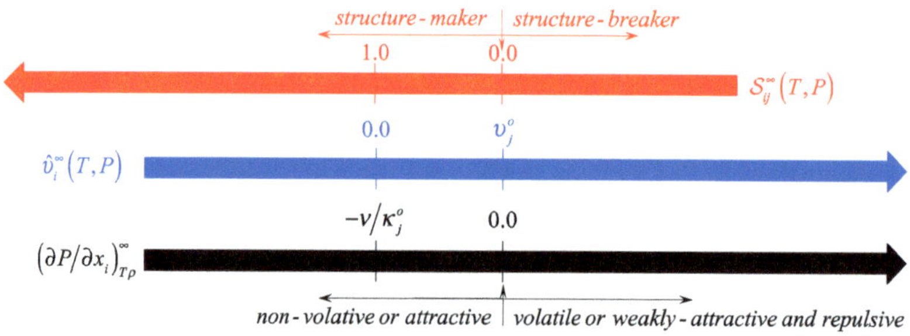

Figure 1. Relationships among $\mathcal{S}_{ij}^\infty(T,P)$, $\hat{v}_i^\infty(T,P)$, and $(\partial P/\partial x_i)_{T\rho}^\infty$, including some relevant boundaries.

3.2. Response to Changes in State Conditions

From Equation (29), we immediately have the following temperature and pressure derivatives:

$$\begin{aligned}\left(\partial \mathcal{S}_{ij}^{\infty}/\partial T\right)_P &= \rho_j^o \hat{v}_i^{\infty}\left(\beta_{Tj}^o - \beta_{Ti}^{\infty}\right)/\nu \\ &= \left(1-\mathcal{S}_{ij}^{\infty}\right)\left(\beta_{Tj}^o - \beta_{Ti}^{\infty}\right)\end{aligned} \tag{35}$$

and

$$\begin{aligned}\left(\partial \mathcal{S}_{ij}^{\infty}/\partial P\right)_T &= \rho_j^o \hat{v}_i^{\infty}\left(\kappa_i^{\infty} - \kappa_j^o\right)/\nu \\ &= \left(\mathcal{S}_{ij}^{\infty}-1\right)\left(\kappa_j^o - \kappa_i^{\infty}\right)\end{aligned} \tag{36}$$

where $\kappa_\alpha^{\oplus} = -(\partial \ln v_\alpha^{\oplus}/\partial P)_T$ and $\beta_{T\alpha}^{\oplus} = (\partial \ln v_\alpha^{\oplus}/\partial T)_P$ identify the isothermal compressibility and isobaric thermal expansivity of the species α at the limiting composition condition $\oplus$, i.e., either pure component $\oplus = o$ or infinite dilution $\oplus = \infty$. Consequently, the magnitude and sign of the $\mathcal{S}_{ij}^{\infty}(T,P)$ response to changes in either temperature or pressure depends on the product of two differences, i.e., either $\left(1-\mathcal{S}_{ij}^{\infty}\right)\left(\beta_{Tj}^o - \beta_{Ti}^{\infty}\right)$ or $\left(\mathcal{S}_{ij}^{\infty}-1\right)\left(\kappa_j^o - \kappa_i^{\infty}\right)$, respectively. In fact, for a structure-making (SM) solute, $\mathcal{S}_{ij}^{\infty}(T,P) > 0$, characterized by the following condition:

$$\nu > \rho_j^o|\hat{v}_i^{\infty}| \rightarrow |\hat{v}_i^{\infty}(T,P)| < \nu v_j^o(T,P) \tag{37}$$

There are four distinctive responses to changes in temperature,

$$\begin{aligned}&\left(\partial \mathcal{S}_{ij}^{\infty}/\partial T\right)_P^{SM} = \left(1-\mathcal{S}_{ij}^{\infty}\right)\left(\beta_{Tj}^o - \beta_{Ti}^{\infty}\right) \rightarrow \left\{\begin{array}{l} >0 \ if \ \beta_{Tj}^o > \beta_{Ti}^{\infty} \\ <0 \ if \ \beta_{Tj}^o < \beta_{Ti}^{\infty}\end{array}\right\} for \ 0 < \mathcal{S}_{ij}^{\infty} < 1 \\ &\left(\partial \mathcal{S}_{ij}^{\infty}/\partial T\right)_P^{SM} = \left(1-\mathcal{S}_{ij}^{\infty}\right)\left(\beta_{Tj}^o - \beta_{Ti}^{\infty}\right) \rightarrow \left\{\begin{array}{l} >0 \ if \ \beta_{Tj}^o < \beta_{Ti}^{\infty} \\ <0 \ if \ \beta_{Tj}^o > \beta_{Ti}^{\infty}\end{array}\right\} for \ \mathcal{S}_{ij}^{\infty} > 1\end{aligned} \tag{38}$$

And another four distinctive responses to changes in pressure,

$$\begin{aligned}&\left(\partial \mathcal{S}_{ij}^{\infty}/\partial P\right)_T^{SM} = \left(\mathcal{S}_{ij}^{\infty}-1\right)\left(\kappa_j^o - \kappa_i^{\infty}\right) \rightarrow \left\{\begin{array}{l} >0 \ if \ \kappa_j^o < \kappa_i^{\infty} \\ <0 \ if \ \kappa_j^o > \kappa_i^{\infty}\end{array}\right\} for \ 0 < \mathcal{S}_{ij}^{\infty} < 1 \\ &\left(\partial \mathcal{S}_{ij}^{\infty}/\partial P\right)_T^{SM} = \left(\mathcal{S}_{ij}^{\infty}-1\right)\left(\kappa_j^o - \kappa_i^{\infty}\right) \rightarrow \left\{\begin{array}{l} >0 \ if \ \kappa_j^o > \kappa_i^{\infty} \\ <0 \ if \ \kappa_j^o < \kappa_i^{\infty}\end{array}\right\} for \ \mathcal{S}_{ij}^{\infty} > 1\end{aligned} \tag{39}$$

Otherwise, for a structure-breaking (SB) solute, $\mathcal{S}_{ij}^{\infty}(T,P) < 0$, is characterized by the following conditions:

$$\nu < \rho_j^o \hat{v}_i^{\infty} \rightarrow \hat{v}_i^{\infty}(T,P) > \nu v_j^o(T,P) > 0 \tag{40}$$

There are only two possible responses to changes in temperature

$$\left(\partial \mathcal{S}_{ij}^{\infty}/\partial T\right)_P^{SB} = \left(1-\mathcal{S}_{ij}^{\infty}\right)\left(\beta_{Tj}^o - \beta_{Ti}^{\infty}\right) \rightarrow \left\{\begin{array}{l} >0 \ if \ \beta_{Tj}^o > \beta_{Ti}^{\infty} \\ <0 \ if \ \beta_{Tj}^o < \beta_{Ti}^{\infty}\end{array}\right. \tag{41}$$

and another two possible responses to changes in pressure

$$\left(\partial \mathcal{S}_{ij}^{\infty}/\partial P\right)_T^{SB} = \left(\mathcal{S}_{ij}^{\infty}-1\right)\left(\kappa_j^o - \kappa_i^{\infty}\right) \rightarrow \left\{\begin{array}{l} >0 \ if \ \kappa_j^o < \kappa_i^{\infty} \\ <0 \ if \ \kappa_j^o > \kappa_i^{\infty}\end{array}\right. \tag{42}$$

In summary, Equations (38)–(42) indicate contrasting solvent microstructural responses to changes in system pressure and temperature regardless of the structure-making/breaking nature of the i-solute. In fact, according to the thermodynamic relation $\beta_T = \kappa(\partial P/\partial T)_\rho$ with $(\partial P/\partial T)_\rho > 0$, the microstructural response to either a pressure increase or temperature decrease accentuates both the intrinsic nature of the original structure-making when $\mathcal{S}_{ij}^{\infty}(T,P) > 1$ and structure-breaking solutes whenever the isothermal compressibility and the isobaric thermal expansivity of the pure solvent are both larger than those corresponding to the infinite dilute solute. Otherwise, when $0 < \mathcal{S}_{ij}^{\infty}(T,P) < 1$, the same changes in state conditions will weaken the intrinsic nature of the original structure-making and structure-breaking solutes whenever the isothermal compressibility and isobaric thermal expansivity of the pure solvent are both smaller than those of the infinite dilute solute.

Note that, according to the analysis of Section 3.1, we immediately find that

$$\mathcal{S}_{ij}^{\infty,G_{jj}^{o}=G_{ij}^{\infty}=G_{ii}^{o}}(T,P) = \left(\partial \mathcal{S}_{ij}^{\infty,G_{jj}^{o}=G_{ij}^{\infty}=G_{ii}^{o}}/\partial T\right)_P = \left(\partial \mathcal{S}_{ij}^{\infty,IG_{jj}^{o}=G_{ij}^{\infty}=G_{ii}^{o}}/\partial P\right)_T = 0 \quad (43)$$

For which $G_{jj}^{o}(T,P) = G_{ij}^{\infty}(T,P) = G_{ii}^{\infty}(T,P) = kT\kappa_j^o - v_j^o$, i.e., when a solute behaves identically as a solvent species, the structure-making/breaking function and its responses to (T,P)-perturbations are all zero. Moreover, for the non-interacting ideal gas solute, for which we have $\mathcal{S}_{ij}^{\infty,IG\_i}(T,P) = 1 - \kappa_j^o kT\rho_j^o$, we have

$$\left(\partial \mathcal{S}_{ij}^{\infty,IG\_i}/\partial T\right)_P = -\kappa_j^o kT\rho_j^o\left(\beta_{Tj}^o - \beta_{Ti}^{\infty,IG\_i}\right) \rightarrow \begin{cases} <0 \ if \ \beta_{Tj}^o > \beta_{Ti}^{\infty,IG\_i} \\ >0 \ if \ \beta_{Tj}^o < \beta_{Ti}^{\infty,IG\_i} \end{cases} \quad (44)$$

$$\left(\partial \mathcal{S}_{ij}^{\infty,IG\_i}/\partial P\right)_T = \kappa_j^o kT\rho_j^o\left(\kappa_j^o - \kappa_i^{\infty,IG\_i}\right) \rightarrow \begin{cases} >0 \ if \ \kappa_j^o > \kappa_i^{\infty,IG\_i} \\ <0 \ if \ \kappa_j^o < \kappa_i^{\infty,IG\_i} \end{cases} \quad (45)$$

The sign and magnitude of the responses to (T,P)-perturbations caused by a non-interacting solute in real solvents depend on how different β_{Ti}^{∞} and κ_i^{∞} are from its ideal gas values. Note that the authors have erroneously assumed in Ref. [4] that $\beta_{Ti}^{\infty,IG\_i} = T^{-1}$ and $\kappa_i^{\infty,IG\_i} = P^{-1}$, i.e., equivalent to assume that $\left(\partial \ln \kappa_i^{IG\_i}/\partial T\right)_P^{\infty} = \left(\partial \ln \kappa_i^{IG\_i}/\partial P\right)_T^{\infty} = 0$.

3.3. Response to Changes of State Conditions at Finite Composition

For the binary system forming a thermodynamic stable phase at finite compositions, i.e., exhibiting a finite D-coefficient, we deal with two *fundamental structure-making/breaking functions*, i.e., $\mathcal{S}_{ij}(x_j) \neq \mathcal{S}_{ji}(x_j)$, whose isothermal–isobaric composition dependence has been derived elsewhere [1]

$$\mathcal{S}_{ij}(T,P,x_j) = 1 - \rho\hat{v}_i/v\left[1 + (\partial \ln \gamma_j/\partial \ln x_j)_{TP}\right] \quad (46)$$

and

$$\mathcal{S}_{ji}(T,P,x_j) = 1 - \rho\hat{v}_j/\left[1 + (\partial \ln \gamma_j/\partial \ln x_j)_{TP}\right] \quad (47)$$

Therefore,

$$(\partial \mathcal{S}_{ij}/\partial T)_{Px} = (1 - \mathcal{S}_{ij})\left[\beta_T - \beta_{Ti} + (\partial \ln D/\partial T)_{Px}\right] \quad (48)$$

with $\beta_T = (\partial \ln v/\partial T)_P$, $\beta_{T\alpha} = (\partial \ln \hat{v}_\alpha/\partial T)_P$, while $\hat{v}_\alpha$ identifies the partial molar volume of the α-species, and

$$(\partial \mathcal{S}_{ji}/\partial T)_{Px} = (1 - \mathcal{S}_{ji})\left[\beta_T - \beta_{Tj} + (\partial \ln D/\partial T)_{Px}\right] \quad (49)$$

with $D = \left[1 + (\partial \ln \gamma_j / \partial \ln x_j)_{TP}\right]$ so that [74]

$$(\partial \ln D/\partial T)_{Px} = -\left(kT^2 D\right)^{-1}\left(\partial \hat{h}_j/\partial \ln x_j\right)_{Px} \tag{50}$$

where $\hat{h}_j(T,P,x_j)$ denotes the partial molar enthalpy of the j-species at the prevailing state conditions and composition.

Likewise,

$$(\partial \mathcal{S}_{ij}/\partial P)_{Tx} = (\mathcal{S}_{ij} - 1)[\kappa - \kappa_i + (\partial \ln D/\partial P)_{Tx}] \tag{51}$$

where $\kappa = -(\partial \ln v/\partial P)_T$, $\kappa_\alpha = -(\partial \ln \hat{v}_\alpha/\partial P)_T$, and

$$(\partial \mathcal{S}_{ji}/\partial P)_{Tx} = (\mathcal{S}_{ji} - 1)[\kappa - \kappa_j + (\partial \ln D/\partial P)_{Tx}] \tag{52}$$

with

$$(\partial \ln D/\partial P)_{Tx} = (kTD)^{-1}(\partial \hat{v}_j/\partial \ln x_j)_{Px} \tag{53}$$

while $\hat{v}_j(T,P,x_j)$ denotes the partial molar volume of the j-species at the prevailing state conditions and composition.

As expected, these expressions converge into the corresponding ones derived for the case of the i-species becoming infinitely dilute in the j-species. In fact, considering that $\lim_{x_j \to 1} D = 1$, the limiting behavior of Equation (46) becomes Equation (29) while $\lim_{x_j \to 1} \mathcal{S}_{ji} = 1$. Similarly, Equations (48) and (51) become identical to Equations (35) and (36) given that $\lim_{x_j \to 1} \beta_T = \beta_{Tj}^o$, $\lim_{x_j \to 1} \kappa = \kappa_j^o$, $\lim_{x_j \to 1} \beta_{Ti} = \beta_{Ti}^\infty$, and $\lim_{x_j \to 1} \kappa_i = \kappa_i^\infty$, while Equations (49) and (52) become null because $\lim_{x_j \to 1} \mathcal{S}_{ji} = 1$. Consequently, as for the case of the infinite dilution, the magnitude and sign of the $\mathcal{S}_{\alpha\beta}(T,P,x_\beta)$ response to changes in either temperature or pressure depend on two factors, $(1 - \mathcal{S}_{\alpha\beta})[\beta_T - \beta_{T\alpha} + (\partial \ln D/\partial T)_{Px}]$ and $(\mathcal{S}_{\alpha\beta} - 1)[\kappa - \kappa_\alpha + (\partial \ln D/\partial P)_{Tx}]$, and give rise to either two sets of four possible outcomes for $\mathcal{S}_{\alpha\beta}(T,P,x_\beta) > 0$ or two sets for $\mathcal{S}_{\alpha\beta}(T,P,x_\beta) < 0$. For example, for $\mathcal{S}_{\alpha\beta}(T,P,x_\beta) > 0$, we have that $v v_\beta^o D > \left|\hat{v}_\alpha^\infty(T,P,x_\beta)\right| > 0$ so that

$$\begin{aligned}
(\partial \mathcal{S}_{\alpha\beta}/\partial T)_{Px}^{SM} &\to \left\{ \begin{array}{l} > 0 \ if \ [\beta_T - \beta_{T\alpha} + (\partial \ln D/\partial T)_{Px}] > 0 \\ < 0 \ if \ [\beta_T - \beta_{T\alpha} + (\partial \ln D/\partial T)_{Px}] < 0 \end{array} \right\} for\ 0 < \mathcal{S}_{\alpha\beta} < 1 \\
(\partial \mathcal{S}_{\alpha\beta}/\partial T)_{Px}^{SM} &\to \left\{ \begin{array}{l} > 0 \ if \ [\beta_T - \beta_{T\alpha} + (\partial \ln D/\partial T)_{Px}] < 0 \\ < 0 \ if \ [\beta_T - \beta_{T\alpha} + (\partial \ln D/\partial T)_{Px}] > 0 \end{array} \right\} for\ \mathcal{S}_{\alpha\beta} > 1
\end{aligned} \tag{54}$$

$$\begin{aligned}
(\partial \mathcal{S}_{\alpha\beta}/\partial P)_{Tx}^{SM} &\to \left\{ \begin{array}{l} < 0 \ if [\kappa - \kappa_\alpha + (\partial \ln D/\partial P)_{Tx}] > 0 \\ > 0 \ if \ [\kappa - \kappa_\alpha + (\partial \ln D/\partial P)_{Tx}] < 0 \end{array} \right\} for\ 0 < \mathcal{S}_{\alpha\beta} < 1 \\
(\partial \mathcal{S}_{\alpha\beta}/\partial P)_{Tx}^{SM} &\to \left\{ \begin{array}{l} > 0 \ if \ [\kappa - \kappa_\alpha + (\partial \ln D/\partial P)_{Tx}] > 0 \\ < 0 \ if \ [\kappa - \kappa_\alpha + (\partial \ln D/\partial P)_{Tx}] < 0 \end{array} \right\} for\ \mathcal{S}_{\alpha\beta} > 1
\end{aligned} \tag{55}$$

Otherwise, for $\mathcal{S}_{\alpha\beta}(T,P,x_\beta) < 0$ we have that $\hat{v}_\alpha^\infty(T,P,x_\beta) > v v_\beta^o D > 0$ so that there are only two possible responses to isobaric changes in temperature

$$(\partial \mathcal{S}_{\alpha\beta}/\partial T)_P^{SB} \to \left\{ \begin{array}{l} > 0 \ if \ [\beta_T - \beta_{T\alpha} + (\partial \ln D/\partial T)_{Px}] > 0 \\ < 0 \ if \ [\beta_T - \beta_{T\alpha} + (\partial \ln D/\partial T)_{Px}] < 0 \end{array} \right. \tag{56}$$

and another two possible responses to changes in pressure

$$(\partial \mathcal{S}_{\alpha\beta}/\partial P)_T^{SB} \to \left\{ \begin{array}{l} > 0 \ if \ [\kappa - \kappa_\alpha + (\partial \ln D/\partial P)_{Tx}] < 0 \\ < 0 \ if \ [\kappa - \kappa_\alpha + (\partial \ln D/\partial P)_{Tx}] > 0 \end{array} \right. \tag{57}$$

Note that, according to the analysis of Section 3.1, we immediately find that

$$\mathcal{S}_{\alpha\beta}^{G_{\beta\beta}=G_{\alpha\beta}=G_{\alpha\alpha}}(T,P,x_\alpha) = \left(\partial \mathcal{S}_{\alpha\beta}^{G_{\beta\beta}=G_{\alpha\beta}=G_{\alpha\alpha}}/\partial T\right)_{Px} = \left(\partial \mathcal{S}_{\alpha\beta}^{G_{\beta\beta}=G_{\alpha\beta}=G_{\alpha\alpha}}/\partial P\right)_T = 0 \quad (58)$$

for which $G_{\beta\beta}(T,P,x_\alpha) = G_{\alpha\beta}(T,P,x_\alpha) = G_{\alpha\alpha}(T,P,x_\alpha) = kT\kappa_\beta^o - v_\beta^o$, i.e., when a solute behaves identically as a solvent species, the *fundamental structure-making/breaking function* and its responses to (T,P,x_α)-perturbations are all zero. Moreover, for the system comprising a non-interacting finite composition ideal gas solute in a real j-solvent, there is only one structure-making/breaking function, $\mathcal{S}_{ij}^{IG\_i}(T,P,x_j)$ (see Appendix A), i.e.,

$$\mathcal{S}_{ij}^{IG\_i}(T,P,x_j) = -\rho_j G_{jj} \quad (59)$$

And therefore,

$$v(T,P,x_j) = \hat{v}_j\left(1 - x_i \mathcal{S}_{ij}^{IG\_i}\right) \quad (60)$$

whose isobaric temperature derivative results in the following expression:

$$\left(\partial \mathcal{S}_{ij}^{IG\_i}/\partial T\right)_P = \left(1-\mathcal{S}_{ij}^{IG\_i}\right)\left(\beta_{Tj} - \beta_{Ti}^{IG\_i}\right) \to \begin{cases} \left.\begin{matrix} >0 \; if \; \beta_{Tj} > \beta_{Ti}^{IG\_i} \\ <0 \; if \; \beta_{Tj} < \beta_{Ti}^{IG\_i} \end{matrix}\right\} for \; \mathcal{S}_{ij}^{IG\_i} < 1 \\ \left.\begin{matrix} <0 \; if \; \beta_{Tj} > \beta_{Ti}^{IG\_i} \\ >0 \; if \; \beta_{Tj} < \beta_{Ti}^{IG\_i} \end{matrix}\right\} for \; \mathcal{S}_{ij}^{IG\_i} > 1 \end{cases} \quad (61)$$

Moreover, for the isothermal pressure derivative, we find

$$\left(\partial \mathcal{S}_{ij}^{IG\_i}/\partial P\right)_T = \left(1-\mathcal{S}_{ij}^{IG\_i}\right)\left(\kappa_i^{IG\_i} - \kappa_j\right) \to \begin{cases} \left.\begin{matrix} >0 \; if \; \kappa_i^{IG\_i} > \kappa_j \\ <0 \; if \; \kappa_i^{IG\_i} > \kappa_j \end{matrix}\right\} for \; \mathcal{S}_{ij}^{IG\_i} < 1 \\ \left.\begin{matrix} <0 \; if \; \kappa_i^{IG\_i} > \kappa_j \\ >0 \; if \; \kappa_i^{IG\_i} > \kappa_j \end{matrix}\right\} for \; \mathcal{S}_{ij}^{IG\_i} > 1 \end{cases} \quad (62)$$

Note that (61)–(62) converge into (44)–(45) when the i-solute becomes infinitely dilute in the j-solvent.

4. Pitfalls from Surrogate Experimental Techniques

Researchers have often invoked surrogate experimental techniques to investigate the solvent environment around a solute and searched for correlations between the observed thermodynamic manifestations and the corresponding responses from different microscopic probes hoping to gain some understanding of the effect of a solute on the microstructure of the solution. Unfortunately, these research tools frequently provide a rather limited (i.e., short-ranged and/or orientational) view of the solute-induced effects, might invoke solvent-specific (e.g., water or other hydrogen-bonding species) approaches, or would depend on other markers to conjecture microscopic-to-macroscopic correspondence while overlooking that correspondence does not necessarily means cause-effect connection [97,98].

In what follows, we review two examples of these surrogate approaches, i.e., the conjectured markers associated with the behavior of the B-coefficient in Jones–Dole's equation [99] together with its isobaric temperature derivative, and the isobaric-thermal expansivity of the solute in solution [95], whose misguided use has perpetuated their involvement as structure-making/breaking markers as if they had already been proven adequate for this purpose.

4.1. Myths and Verities about the Use of Jones–Dole's B- and $(\partial B/\partial T)_P$-Coefficients as Structure Markers and Their Connection to the Fundamental Structure-Making/Breaking Function S_{ij}^{∞}

The empirical Jones–Dole's equation for the relative viscosity of the dilute solution reads as follows:

$$\begin{aligned} \eta_r(T,P,c_i)-1 &= Ac_i^{0.5}+Bc_i \text{ as } c_i \to 0 \\ &\cong \ln \eta_r(T,P,c_i) \end{aligned} \tag{63}$$

where $\eta_r = \eta/\eta_j^o$ with $\eta_j^o \equiv \eta(c_i = 0)$, c_i denotes the molar concentration of the i-solute, and the coefficients A and B account for the direct ion-ion and the (ion) solute–solvent interactions, respectively [99]. If we assume that Jones–Dole's equation provides an accurate representation of the shear viscosity of a dilute solution, the B-coefficient becomes determined from the limiting isothermal–isobaric composition slope, i.e.,

$$\begin{aligned} B &= \lim_{c_i \to 0}\left[\left(\eta_r - 1 - Ac_i^{0.5}\right)/c_i\right] \\ &\cong \lim_{c_i \to 0}\left[\left(\ln \eta_r - Ac_i^{0.5}\right)/c_i\right] \end{aligned} \tag{64}$$

leading to $B \gtrless 0$. Because the B-coefficient is the coefficient of the linear concentration term in Jones–Dole's equation, this coefficient has been associated with the solvent-mediated solute–solute interactions, and therefore, with the conjectured $B > 0$ for a structure-maker species and $B < 0$ for a structure-breaker species [53,100–102], later supplemented with the alternative isobaric temperature derivative $(\partial B/\partial T)_P \gtrless 0$ [57,103–110].

Despite early warnings, e.g., "*This type of coefficient is therefore probably related to the disturbance of the structure which is present in such liquids...No general theory has yet been developed for the B-coefficient but important qualitative explanations have been advanced in terms of ion-solvent interactions...Interesting as these various correlations are, it should be emphasized that they are, as yet, qualitative in nature and a quantitative theory for the B-coefficient still awaits development*" [111] that would have prompted legitimate questions, as identified below, there was instead a proliferation of rather convoluted claims involving unproven connections between the signs of either the B-coefficient or its temperature derivative $(\partial B/\partial T)_P$ and the solute-induced perturbation of the solvent microstructure, as if the alluded links were either self-evident or already established and validated. In other words, instead of addressing some genuine issues including whether Jones–Dole's B-coefficient contains any embedded microstructural information, whether we could assign any definite foundation to the widely used assumption about the sign of the B-coefficient and its structure-making/breaking trend, and whether we could find a rational justification for the use of the sign of $(\partial B/\partial T)_P$ as a structure-making/breaking marker, unfortunately, the literature has been flooded with conjectures born on dicey narratives, including "*A positive B-coefficient indicates that the ions tend to order the solvent structure and increase the viscosity of the solution while a negative B-coefficient indicates disordering and a decrease of viscosity*" [102] *and* "*the values of temperature coefficient of B, i.e.,* (dB/dT) *for the solute can provide direct evidence regarding their structure-making or breaking effect in solution. The values of* (dB/dT) *are positive for hydrophilic structure-breaking groups. For hydrophobic structure-making groups, the value of* (dB/dT) *is negative*" [112], and a number of variations, as quoted in the Supporting Information of Ref. [113].

In a series of recent publications [4,71,113,114], we have addressed the issues identified above to test the validity and consistency of the outcomes from the application of the B-coefficient and/or its temperature derivative $(\partial B/\partial T)_P$ as structure-making/breaking markers. While all outcomes pointed to the lack of uniqueness of these markers, as clearly illustrated in Table 1, where we observe that the same $[B, (\partial B/\partial T)_P]$ sign-pair combination could represent opposite structure-making/breaking parameter $S_{ij}^{\infty}(T,P)$ and, consequently, confirm the lack of uniqueness of the $[B, (\partial B/\partial T)_P]$ sign-pair combination as a structure-making/breaking marker.

Table 1. Representative systems illustrating the eight $[B, (\partial B/\partial T)_P]$ sign-pair combinations and their resulting structure-making/breaking parameter $\mathcal{S}_{ij}^{\infty}(T,P)$ for infinitely dilute solutes at ambient conditions.

Mixture	B	$(\partial B/\partial T)_P$	$\mathcal{S}_{ij}^{\infty}(T,P)$	Ref.
$NaCl + H_2O$	>0	>0	>0 (maker)	[104]
$Me_4NBr + H_2O$	>0	>0	<0 (breaker)	[104]
$Ni(NO_3)_2 + H_2O$	>0	<0	>0 (maker)	[115]
$Et_4NBr + H_2O$	>0	<0	<0 (breaker)	[104]
$KBr + H_2O$	<0	>0	>0 (maker)	[116]
$KI + H_2O$	<0	>0	<0 (breaker)	[104]
$IG\_i + H_2O$ [(a)]	<0	<0	>0 (maker)	[113]
[Emim][NTf2] + EG	<0	<0	<0 (breaker)	[117]

[(a)] Supercritical conditions where $\kappa_j^{o,IG} < \kappa_j^o$.

To trace back the microscopic origin for such behavior according to explicit statistical mechanics arguments, we have interpreted the plausible links between the behavior of the B-coefficient (and/or its temperature derivative $(\partial B/\partial T)_P$) and the (electrolyte or non-electrolyte) solute-induced perturbation of the solvent environment. For that purpose, we invoked Feakins et al.'s [118,119] rigorous expression for the B-coefficient, derived according to the transition state theory [120], as follows:

$$B = \beta v_j^o \left(\Delta\mu_i^{\natural,\infty} - \nu\Delta\mu_j^{\natural,o}\right) - \left(\hat{v}_i^{\infty} - \nu v_j^o\right) \tag{65}$$

for the infinitely dilute i-solute where $\nu = \nu^+ + \nu^-$ with $\nu = 1$ for non-electrolyte solutes. $\Delta\mu_\alpha^{\natural,\oplus}$ and $\hat{v}_\alpha^{\oplus}$ denote the molar Gibbs free energy of activation of viscous flow and the corresponding partial molar volume for the α-species, respectively, at the $\oplus$-composition condition, i.e., either infinite dilution or pure component. Equation (65) contributes two terms to the B-coefficient, namely the change of Gibbs free energy of activation for the viscous flow, $\left(\Delta\mu_i^{\natural,\infty} - \nu\Delta\mu_j^{\natural,o}\right)$ and the volumetric change when ν-solvent molecules are alchemically mutated into a solute during the formation of an i-solute at infinite dilution in an otherwise pure j-solvent, $\left(\hat{v}_i^{\infty} - \nu v_j^o\right)$. This volumetric term provides direct access to the microstructural perturbation during the solvation process since, according to the rigorous definition of the structure-making/breaking parameter $\mathcal{S}_{ij}^{\infty}(T,P)$, as described by Equation (4), $\left(\hat{v}_i^{\infty} - \nu v_j^o\right) = -\nu v_j^o \mathcal{S}_{ij}^{\infty}$ and Equation (65) becomes

$$B = v_j^o\left[\nu\mathcal{S}_{ij}^{\infty} + \beta\left(\Delta\mu_i^{\natural,\infty} - \nu\Delta\mu_j^{\natural,o}\right)\right] \tag{66}$$

Equation (66) warrants the following important observations: (i) the B-coefficient must be defined by the limiting condition of infinite dilution, vide supra Equation (64), to be consistent with its TS-based interpretation, and (ii) the TS-based interpretation clearly illustrates that the B-coefficient results from the combination of the solute-induced perturbation of the solvent structure and the corresponding induced difference between the Gibbs free energy of activation for the viscous flow of the solute and the solvent. Consequently, Equation (66) not only confirms that the B-coefficient comprises some information about the solute-induced effect on the solvent microstructure but also suggests that the sign of the B-coefficient cannot unambiguously describe the structure-making/breaking nature of the dilute i-solute, contrary to the widespread practice in the literature [53,57,100,102,112,121–128].

To illustrate the lack of uniqueness of the B-coefficient and/or its temperature derivative $(\partial B/\partial T)_P$ as markers for the structure-making/breaking behavior, we first test the behavior of a prototypical solute (see Equation (15) of [113]) and then introduce a novel statistical mechanics proof. (Equation (15) of [114]). For example, in Ref. [113], we have determined the conditions at which the *B*-coefficient for the non-interacting ideal gas solute (*IG_i*) might predict either a structure-making, $\mathcal{S}_{ij}^{\infty,IG\_i} > 0$, a structure-breaking, $\mathcal{S}_{ij}^{\infty,IG\_i} < 0$, or an unperturbed microstructure, $\mathcal{S}_{ij}^{\infty,IG\_i} = 0$, as illustrated below.

This straightforward and rigorous thermodynamic analysis illustrates the lack of uniqueness of the *B*-coefficient as a structure-making/breaking marker, in that either $B^{IG\_i} < 0$ or $B^{IG\_i} > 0$ could represent a structure-making, $\mathcal{S}_{ij}^{\infty,IG\_i} > 0$, and simultaneously a structure-breaking, $\mathcal{S}_{ij}^{\infty,IG\_i} < 0$, behavior depending on the state conditions of the pure solvent (see Table 1).

$$B^{IG\_i} \rightarrow \begin{cases} \mathcal{S}_{ij}^{\infty,IG\_i} > 0 \ \rightarrow \begin{cases} B^{IG\_i} > 0 \ \rightarrow \begin{cases} \Delta\mu_j^{\natural,o} > 0 \text{ if } \mathcal{S}_{ij}^{\infty,IG\_i} > \rho_j^o B^{IG\_i} \\ \Delta\mu_j^{\natural,o} < 0 \text{ otherwise} \end{cases} \\ B^{IG\_i} < 0 \ \rightarrow \ \Delta\mu_j^{\natural,o} > 0 \end{cases} \\ \mathcal{S}_{ij}^{\infty,IG\_i} = 0 \rightarrow \left\{ \begin{array}{l} B^{IG\_i} > 0 \ \rightarrow \ \Delta\mu_j^{\natural,o} < 0 \\ B^{IG\_i} < 0 \ \rightarrow \ \Delta\mu_j^{\natural,o} > 0 \end{array} \right\} \leftarrow \kappa_j^o = kT\rho_j^o \\ \mathcal{S}_{ij}^{\infty,IG\_i} < 0 \ \rightarrow \begin{cases} B^{IG\_i} > 0 \ \rightarrow \ \Delta\mu_j^{\natural,o} < 0 \\ B^{IG\_i} < 0 \ \rightarrow \begin{cases} \Delta\mu_j^{\natural,o} > 0 \text{ if } \left|\mathcal{S}_{ij}^{\infty,IG\_i}\right| < \rho_j^o \left|B^{IG\_i}\right| \\ \Delta\mu_j^{\natural,o} < 0 \text{ otherwise} \end{cases} \end{cases} \end{cases} \quad (67)$$

Moreover, for the non-interacting ideal gas solute, the system comprises $\nu = 1$, $\Delta\mu_i^{\natural,\infty,IG\_i} = 0$ [120], and $\mathcal{S}_{ij}^{\infty,IG\_i} = 1 - kT\rho_j^o\kappa_j^o$ [4] so that its B^{IG_i}-coefficient would read as follows:

$$\begin{aligned} B^{IG\_i} &= v_j^o\left(\mathcal{S}_{ij}^{\infty,IG\_i} - \beta\Delta\mu_j^{\natural,o}\right) \\ &= v_j^o\left[1 - \beta\Delta\mu_j^{\natural,o} - \left(\kappa_j^o/\kappa_j^{IG}\right)\right] \end{aligned} \quad (68)$$

with $\beta\Delta\mu_j^{\natural,o} = \ln\left(\eta_j^o v_j^o/h\mathcal{N}\right)$ [120], where η_j^o and v_j^o denote the shear viscosity and molar volume of the pure *j*-solvent, while h and $\mathcal{N}$ identify Planck and Avogadro constants, respectively. Therefore, from Equation (68), we immediately find that

$$B^{IG\_i} = 0 \rightarrow \begin{cases} \beta\Delta\mu_j^{\natural,o} + kT\rho_j^o\kappa_j^o = 1 \ \rightarrow \ \mathcal{S}_{ij}^{\infty,IG\_i} = \beta\Delta\mu_j^{\natural,o} \neq 0 \\ \beta\Delta\mu_j^{\natural,o} = 0 \rightarrow \mathcal{S}_{ij}^{\infty,IG\_i} = 0 \end{cases} \quad (69)$$

emphasizing that the condition of the null solute-induced effect on the viscosity of the dilute solution, $B^{IG\_i} = 0$, represents two contrasting structure-making/breaking scenarios. The first one, the top line of Equation (69), indicates a non-zero solute-induced effect on the structure of a real solvent, $\mathcal{S}_{ij}^{\infty,IG\_i} \neq 0$, whose sign depends on the $\left(\eta_j^o v_j^o/h\mathcal{N}\right) \lessgtr 1$ condition. The second one, the bottom line of Equation (69), predicts an unperturbed structure of the solvent $\beta\Delta\mu_j^{\natural,o} = 0$ or $\left(\eta_j^o v_j^o/h\mathcal{N}\right) = 1$, a condition highly unlikely to occur in real solvents, as illustrated in Table 2 for a series of condensed solvents.

Table 2. Representative systems illustrating the $[B, (\partial B/\partial T)_P]$ sign-pair combinations and their resulting structure-making/breaking parameter $\mathcal{S}_{ij}^{\infty}(T,P)$ for and infinitely dilute ideal gas solute in a variety of solvents mostly at ambient conditions (according to tabulated data from Ref. [129]).

j-Solvent	$\mathcal{S}_{ij}^{\infty,IG\_i}(T,P)$	$\left(\partial B^{IG\_i}/\partial T\right)_P$	B^{IG_i}	$\beta\Delta\mu_j^{\natural,o}$
ambient water	>0	>0	<0	3.7
supercritical water	<0	<0	<0	14.2
subcritical water	>0	<0	<0	8.1
dimethyl sulfoxide	>0	>0	<0	5.8
N-N dimethyl formamide	>0	>0	<0	5.0
Tetrahydrofuran	>0	>0	<0	4.6
NN-Dimethylacetamide	>0	>0	<0	5.4
acetonitrile	>0	>0	<0	3.9
ethylene glycol	>0	>0	<0	7.3
1,4 dioxane	>0	>0	<0	5.5
methanol	>0	>0	<0	4.0
ethanol	>0	>0	<0	5.1
1-propanol	>0	>0	<0	5.9
t-butanol	>0	>0	<0	6.9
trifluoroethanol	>0	>0	<0	5.7
acetone	>0	>0	<0	4.0
propylene carbonate???	>0	>0	<0	6.3
alfa-butyrolactonate	>0	>0	<0	5.8
formamide	>0	>0	<0	5.8
n-methyl formamide	>0	>0	<0	5.5
tetramethylurea	>0	>0	<0	6.1
n-methylpyrrolidinone	>0	>0	<0	5.9
pyridine	>0	>0	<0	5.7
benzonitrile	>0	>0	<0	5.7
nitromethane	>0	>0	<0	4.7
nitrobenzene	>0	>0	<0	6.1
chloroform	>0	>0	<0	4.7
1-1 dichloroethane	>0	>0	<0	4.6
1-2 dichloroethane	>0	>0	<0	5.0
tetramethylenesulfone	>0	>0	<0	7.8
tetrahydrothiophene	>0	>0	<0	5.3
trimethylphosphate	>0	>0	<0	6.4

According to the isobaric temperature derivative of Equation (66), Ref. [113] discusses the ability of the $(\partial B/\partial T)_P$ coefficient to discriminate between structure-making and structure-breaking solutes, under the four categories introduced by Tsangaris and Martin [103]. The resulting expression is

$$\begin{aligned}(\partial B/\partial T)_P &= \beta_{T,j}^{o}B + v_j^{o}\left[\nu\left(\partial \mathcal{S}_{ij}^{\infty}/\partial T\right)_P - \left(\Delta\hat{h}_i^{\natural,\infty} - \nu\Delta h_j^{\natural,o}\right)/kT^2\right] \\ &= \beta_{T,j}^{o}B + v_j^{o}\left[\nu\left(1-\mathcal{S}_{ij}^{\infty}\right)\left(\beta_{T,j}^{o} - \beta_{T,i}^{\infty}\right) - \left(\Delta\hat{h}_i^{\natural,\infty} - \nu\Delta h_j^{\natural,o}\right)/kT^2\right]\end{aligned} \quad (70)$$

where the second line of Equation (70) highlights the replacement of $\left(\partial\mathcal{S}_{ij}^{\infty}/\partial T\right)_P$ with its corresponding explicit form previously derived [4]. Equation (70) indicates that the behavior (i.e., magnitude and sign) of $(\partial B/\partial T)_P$ depends on those of B, $\mathcal{S}_{ij}^{\infty}$, and $\left(\partial\mathcal{S}_{ij}^{\infty}/\partial T\right)_P$, as well as the differences $\left(\Delta\mu_i^{\natural,\infty} - \nu\Delta\mu_j^{\natural,o}\right)$, $\left(\Delta\hat{h}_i^{\natural,\infty} - \nu\Delta h_j^{\natural,o}\right)$, and $\left(\beta_{T,j}^{o} - \beta_{T,i}^{\infty}\right)$ giving rise to sixteen possible structure-making/breaking outcomes, as discussed in the Supporting Information of Ref. [113].

In brief, there are eight possible $[B,(\partial B/\partial T)_P]$ sign-pair combinations that could lead to a structure-making outcome $\mathcal{S}_{ij}^{\infty} > 0$, with four exhibiting $\left(\partial\mathcal{S}_{ij}^{\infty}/\partial T\right)_P > 0$ and another four exhibiting $\left(\partial\mathcal{S}_{ij}^{\infty}/\partial T\right)_P < 0$, while the remaining eight combinations could result in a structure-breaking outcome $\mathcal{S}_{ij}^{\infty} < 0$, with four exhibiting $\left(\partial\mathcal{S}_{ij}^{\infty}/\partial T\right)_P > 0$ and another four exhibiting $\left(\partial\mathcal{S}_{ij}^{\infty}/\partial T\right)_P < 0$. In other words, the sign of $(\partial B/\partial T)_P$ as a structure-making/breaking marker cannot distinguish between structure-making ($\mathcal{S}_{ij}^{\infty} > 0$) solutes and structure-breaking ($\mathcal{S}_{ij}^{\infty} < 0$). In summary, neither the $(\partial B/\partial T)_P$ alone nor the $[B,(\partial B/\partial T)_P]$ sign-pair combination can provide a unique one-to-one correspondence between their signs and the structure-making/breaking nature of the *i*-solute described here by the molecular-based parameter $\mathcal{S}_{ij}^{\infty}$, rendering these criteria deceptive markers of structure-making/breaking trends despite their widespread use in the literature, as illustrated in Table 3 below.

Table 3. Experimentally based structure-making/breaking parameter $\mathcal{S}_{i\text{-}H_2O}^{\infty}(T,P)$ for aqueous infinitely dilute solutes at ambient conditions in comparison with predictions based on Hepler's isobaric-thermal expansivity marker as well as Jones–Dole's B-coefficient and $(\partial B/\partial T)_P$ derivative criteria.

i-Solute	$\mathcal{S}_{i\text{-}H_2O}^{\infty}(T,P)$	$(\partial B/\partial T)_P$	B	$-T\left(\partial^2\bar{v}_i^{\infty}/\partial T^2\right)_P$	Ref.
water [(a)]	unperturbed	0	0	maker	[4]
urea	breaker	>0	>0	breaker [130]	[71,131,132]
N-methyl urea	breaker	>0	>0	breaker [133]	[71,132,134]
(1,3)-dimethyl urea	breaker	>0	>0	maker [130]	[71,132,134]
creatine	breaker	<0	>0	maker	[135]
creatinine	breaker	<0	>0	maker	[135]
nicotinic acid	breaker	<0	>0	maker	[136]
l-ascorbic acid	breaker	>0	>0	breaker	[71,136,137]
glycine	breaker	>0	>0	breaker	[138]
alanine	breaker	>0	>0	breaker	[138]
DTAB	breaker	>0	<0	breaker	[138]
caffeine	breaker	>0	>0	maker	[139]
theophilline	breaker	>0	>0	maker	[139]
theobromine	breaker	>0	>0	maker	[139]
l-serine	breaker	<0	>0	maker	[109]
l-arginine	breaker	<0	>0	maker	[109]
choline-biotinate	breaker	<0	>0	maker	[140]
choline-nicotinate	breaker	<0	>0	maker	[140]
choline-ascorbate	breaker	>0	>0	maker	[140]

Table 3. *Cont.*

i-Solute	$\mathcal{S}^{\infty}_{i\text{-}H_2O}(T,P)$	$(\partial B/\partial T)_P$	B	$-T\left(\partial^2 \bar{v}_i^{\infty}/\partial T^2\right)_P$	Ref.
$LiCy$	breaker	<0	>0	breaker	[141,142]
$NaCy$	breaker	~0	>0	maker	[141,142]
KCy	breaker	>0	>0	maker	[141,142]
$CaCl_2$	maker	>0	>0	breaker	[143,144]
$CdCl_2$	maker	<0	>0	breaker	[145,146]
$NiCl_2$	maker	>0	<0	breaker	[146,147]
NH_4NO_3	breaker	>0	<0	breaker	[148,149]
$MgCl_2$	maker	<0	>0	breaker	[146,150]

[(a)] solute identical to the solvent.

Tables 2 and 3 highlight the fact that, despite the lack of uniqueness, the structure-making/breaking marker based on the signs of $[B, (\partial B/\partial T)_P]$ have been regularly invoked in combination with, or as an alternative to, the isobaric expansivity-based criterion to be discussed in Section 4.3 [107,109,110,137,139,151–163].

4.2. *Sources of Ambiguities Underlying the Use of Jones–Dole's B- and $(\partial B/\partial T)_P$-Coefficients as Structure Markers*

Having illustrated the lack of uniqueness of the B-, $(\partial B/\partial T)_P$-, or any $[B, (\partial B/\partial T)_P]$ sign-pair combination as structure-making/breaking markers, the last piece of the puzzle to be solved was the origin of such behavior. For that purpose, Ref. [114] presented an alternative, yet equivalent, representation of Feakins' TST approach by rewriting the first term of Equation (65) as follows:

$$\beta v_j^o\left(\Delta\mu_i^{\natural,\infty} - \nu\Delta\mu_j^{\natural,o}\right)_{TP} = \beta v_j^o \nu \int_0^P \left(v_j^o \mathcal{S}_{ij}^{\infty} - v_j^{\natural,o}\mathcal{S}_{ij}^{\natural,\infty}\right) dP' \tag{71}$$

which, after invoking the mean-value theorem for integrals [164], provided a more informative version of the link between the B-coefficient and the microstructure of the solvent around the solute as described by the *fundamental structure-making/breaking function* $\mathcal{S}_{ij}^{\infty}$,

$$B(T,P) = \nu z_j^o \overline{\left(v_j^o \mathcal{S}_{ij}^{\infty} - v_j^{\natural,o}\mathcal{S}_{ij}^{\natural,\infty}\right)} - \nu v_j^o \mathcal{S}_{ij}^{\infty} \tag{72}$$

where $z_j^o = \beta P v_j^o$ identifies the compressibility factor of the pure j-solvent at the prevailing (T, P) state conditions, and the horizontal bar on the RHS of Equation (72) describes the mean value of the function between parenthesis within the $[0, P]$ range.

Equation (72) highlights the interplay between microstructural perturbations at two different levels, the ground state and the transition state, where the perturbation associated with the activated state for the viscous flow is incorporated as a pressure integral, which defines the activation molar Gibbs free energy. Consequently, Feakins' TST-based analysis of the B-coefficient describes a combination of disparate structure-making/breaking quantities, i.e., ground state versus transition state, which becomes the most obvious and direct theoretical attributes giving rise to the unavoidable drawback of the B-coefficient as a structure-making/breaking marker, beyond the ones previously discussed in detail elsewhere [78,113]. Likewise, under the same approach, the isobaric temperature derivative of Equation (72) led to

$$\begin{aligned}(\partial B/\partial T)_P = \; & \nu z_j^o \left\{ \beta_{T,j}^{o,R} \overline{\left(v_j^o \mathcal{S}_{ij}^{\infty} - v_j^{\natural,o} \mathcal{S}_{ij}^{\natural,\infty} \right)} + \overline{\left[\partial \left(v_j^o \mathcal{S}_{ij}^{\infty} - v_j^{\natural,o} \mathcal{S}_{ij}^{\natural,\infty} \right) / \partial T \right]_P} \right\} - \\ & \nu \left[\partial \left(v_j^o \mathcal{S}_{ij}^{\infty} \right) / \partial T \right]_P \end{aligned} \tag{73}$$

With the isobaric–isothermal residual thermal expansivity $\beta_{T,j}^{o,R} \equiv \left(\beta_{T,j}^{o} - \beta_{T,j}^{o,IG} \right)_{TP}$, where the isobaric temperature derivatives $\left(\partial \mathcal{S}_{ij}^{\infty} / \partial T \right)_P$ and $\left[\partial \left(v_j^o \mathcal{S}_{ij}^{\infty} \right) / \partial T \right]_P$ are simple functions of $\mathcal{S}_{ij}^{\infty}(T, P)$ by the exact relations [4]

$$\left(\partial \mathcal{S}_{ij}^{\infty} / \partial T \right)_P = \left(1 - \mathcal{S}_{ij}^{\infty} \right) \left(\beta_{T,j}^{o} - \beta_{T,i}^{\infty} \right) \tag{74}$$

and

$$\begin{aligned}\left[\partial \left(v_j^o \mathcal{S}_{ij}^{\infty} \right) / \partial T \right]_P & = v_j^o \left(\beta_{T,j}^{o} - \beta_{T,i}^{\infty} \right) + v_j^o \beta_{T,i}^{\infty} \mathcal{S}_{ij}^{\infty} \\ & = v_j^o \beta_{T,j}^{o} - \hat{v}_i^{\infty} \beta_{T,i}^{\infty} \end{aligned} \tag{75}$$

In other words, as in the case of the *B*-coefficient, $(\partial B/\partial T)_P$ involves solute-induced microstructural perturbations, and their temperature derivative, at not only the ground- but also the activated-state conditions. Consequently, neither the *B*-coefficient, $(\partial B/\partial T)_P$, nor any of its sign-pair combinations can unequivocally describe the actual solute-induced perturbation of the solvent microstructure at the ground state conditions, a conclusion that applies to any value of the stoichiometric coefficient ν.

4.3. Frequent Pitfalls in the Application of Feakins' TST Approach to the Study of B-Coefficients for Electrolytes

A persisting oversight in the application of Feakins' TST approach to the *B*-coefficients involving electrolyte systems is the omission of the stoichiometric coefficient $\nu \geq 2$ [165] in Equation (66) and related expressions associated with the transition-state properties. This omission amounts to using $\nu = 1$ in Equation (66) and leads to the corruption of the reported transition state thermodynamic quantities and their interpretation, given that it introduces significant errors in the values of the molar Gibbs free energy of activation for aqueous electrolytes involving $\nu = 2$ [109,141,166–174] depending on the magnitude of the $\left(v_j^o / \hat{v}_i^{\infty} \right)$ ratio and becoming even larger with increasing ν [175–179]. As a matter of fact, according to the tabulated data from Ref. [124], it becomes evident the significantly large error introduced when assuming $\nu = 1$ rather than the correct $2 \leq \nu \leq 5$ as illustrated in Table 4, which translates into $-57 \lesssim \; \delta \, \Delta\mu_i^{\natural,\infty}(\%) \lesssim 53$. In other words, the erroneous estimations of the molar Gibbs free energy of activation when assuming $\nu = 1$ are carried over to the representation of the *i*-solute-induced perturbation of the *j*-solvent microstructure and its corresponding (T, P)-derivatives, i.e.,

$$\overline{\left[\left(G_{ij}^{\natural,\infty} - G_{jj}^{\natural,o} \right)_{TP}^{\nu \geq 2} \right]} \neq \overline{\left[\left(G_{ij}^{\natural,\infty} - G_{jj}^{\natural,o} \right)_{TP}^{\nu = 1} \right]} \tag{76}$$

$$\overline{\left[\partial \left(G_{ij}^{\natural,\infty} - G_{jj}^{\natural,o} \right)^{\nu \geq 2} / \partial T \right]_P} \neq \overline{\left[\partial \left(G_{ij}^{\natural,\infty} - G_{jj}^{\natural,o} \right)^{\nu = 1} / \partial T \right]_P} \tag{77}$$

$$\overline{\left[\partial \left(G_{ij}^{\natural,\infty} - G_{jj}^{\natural,o} \right)^{\nu \geq 2} / \partial P \right]_T} \neq \overline{\left[\partial \left(G_{ij}^{\natural,\infty} - G_{jj}^{\natural,o} \right)^{\nu = 1} / \partial P \right]_T} \tag{78}$$

Table 4. Magnitude of the error in the calculated Gibbs free energy of activation when wrongly assuming $\nu = 1$ in Equation (66) instead of the correct $2 \leq \nu \leq 5$.

i-Solute	$\Delta\mu_i^{\natural,\infty}$ (kJ/mol)	$\delta\Delta\mu_i^{\natural,\infty}$ (%)
HCl	24.5	−12
NCl	26.7	−16
$RbCl$	12.6	38
$LiCl$	35.1	−24
$CsCl$	11.7	53
Et_4NCl	90.9	−14
Pr_4NCl	172.7	−26
NaF	39.5	−33
NaI	19.9	8
$NaOH$	41.4	−35
$MgCl_2$	76.0	−41
$CuCl_2$	73.9	−42
$CaCl_2$	62.6	−34
$NiCl_2$	74.8	−41
$CrCl_3$	130.3	−54
$NdCl_3$	113.8	−51
$LuCl_3$	108.3	−50
$ThCl_4$	152.6	−57

4.4. Myths and Verities about Hepler's Isobaric-Thermal Expansivity as Structure-Making/Breaking Marker

Another surrogate tool frequently invoked for the assessment of the structure-making/breaking ability of a solute has been proposed by Hepler [95], who suggested a marker based on the conjecture that "*since increasing pressure would also break up the bulky aggregates, the same reasoning suggests that the heat capacity of pure water should decrease with increasing pressure*" and that this behavior could be extended to species in solution at infinite dilution. Hepler's criterion is based on the sign of the isothermal pressure dependence of its partial molar heat capacity $\left(\partial \hat{C}_{Pi}^{\infty}/\partial P\right)_T$, or its more easily accessible Maxwell equivalent expression $-T\left(\partial^2 \hat{v}_i^{\infty}/\partial T^2\right)_P$. More specifically, this criterion states that whenever $-T\left(\partial^2 \hat{v}_i^{\infty}/\partial T^2\right)_P < 0$ or $\left(\partial^2 \hat{v}_i^{\infty}/\partial T^2\right)_P > 0$, the *i*-solute behaves as a structure-maker, and conversely, if $-T\left(\partial^2 \hat{v}_i^{\infty}/\partial T^2\right)_P > 0$ or $\left(\partial^2 \hat{v}_i^{\infty}/\partial T^2\right)_P < 0$, the *i*-solute behaves as a structure-breaker.

Unfortunately, as for the case of the *B*-coefficients, the lack of an explicit unambiguous cause-effect connection between either $\left(\partial \hat{C}_{Pi}^{\infty}/\partial P\right)_T$ or $-T\left(\partial^2 \hat{v}_i^{\infty}/\partial T^2\right)_P$ and the solute-induced perturbation of the solvent microstructure [1,4,71,78] makes any attempt to determine either the latter from volumetric data or the former from pressure perturbation calorimetry pointless [180–183]. Moreover, this so-called Hepler's criterion immediately fails the compliance of two simple constraints associated with the behavior of a system comprising either the largest or the null solute–solvent intermolecular asymmetry.

For the largest solute–solvent intermolecular asymmetry we have the *IG_i* solute, characterized by $\hat{v}_i^{\infty,IG\_i}(T,P) = \nu kT\kappa_j^o$ [96,184], leading to the exact expression for its structure-making/breaking function [4]

$$\mathcal{S}_{ij}^{\infty,IG\_i}(T,P) = 1 - kT\rho_j^o\kappa_j^o \tag{79}$$

Consequently, from Equation (4), we expect to observe

$$\mathcal{S}_{ij}^{\infty,IG\_i}(T,P)\begin{cases} >0 \rightarrow structure\ maker \rightarrow \kappa_i^{IG\_i} > \kappa_j^o \\ =0 \rightarrow unperturbed\ structure \rightarrow \kappa_i^{IG\_i} = \kappa_j^o \\ <0 \rightarrow structure\ breaker \rightarrow \kappa_i^{IG\_i} < \kappa_j^o \end{cases} \quad (80)$$

The significance of Equation (80) can be summarized as follows: (i) at ambient conditions, the IG_i in a real j-solvent exhibits a structure-making behavior, (ii) the microstructure of the pure j-solvent remains unperturbed along the states where the solvent behavior follows the $\kappa_j^o = \kappa_j^{o,IG} = \left[kT\rho_j^o\right]^{-1}$ condition, and (iii) after crossing this boundary, the ideal gas solute becomes a structure-breaker (e.g., see Figures 8 and 9 of Ref. [14] for the solvent water and the Lennard-Jones fluid, respectively).

Hepler's criterion cannot account for the behavior described by Equation (80), in particular, the inability of the isobaric-thermal expansivity-based criterion to cross the $\mathcal{S}_{\beta\alpha}^{\infty,IG\_\beta}(T,P) = 0$ line, as described by $-T\left(\partial^2 \bar{v}_\beta^{\infty,IG\_\beta}/\partial T^2\right)_P = 0$ in Hepler's marker. In fact, it has been shown that $\left(\partial^2 \bar{v}_i^{\infty,IG\_i}/\partial T^2\right)_P^{\kappa_i^o=\kappa_i^{IG\_i}} \neq 0$; in other words, Hepler's criterion cannot describe the structure-making-to-breaking transition for the ideal gas solute IG_i in a real j-solvent along the $\left(\kappa_j^o = \kappa_j^{o,IG}\right)$-line. Moreover, for the null solute–solvent intermolecular asymmetry, as characterized by an i-solute behaving as a j-solvent and regardless of the definition of structure-making/breaking, the outcome must be an unperturbed solvent structure depicted by $\mathcal{S}_{ij}^{\infty,G_{jj}^o=G_{ii}^\infty=G_{ij}^\infty\neq 0}(T,P) = 0$ in the schematic Equation (3). While the physical representation for the null solute–solvent intermolecular asymmetry is characterized by $\left(G_{jj}^o + G_{ii}^\infty - 2G_{ij}^\infty\right)_{TP} = 0$ [72], its simplest and most advantageous case occurs when the Kirkwood-Buff integrals obey the $\left(G_{jj}^o = G_{ii}^\infty = G_{ij}^\infty = kT\kappa_j^o - v_j^o \neq 0\right)_{TP}$ condition. In other words, this condition is not only the microstructural signature of a pure j-solvent [96] but also the prototypical fingerprint for the microstructural transition between structure-making and structure-breaking processes driven by the solute–solvent intermolecular interaction asymmetry. This general and rigorous condition for the structure-making/breaking transition provides another opportunity to test the validity of the thermal expansivity-based criterion, which in Hepler's terms becomes described by $-T(\partial^2 \bar{v}_i^\infty/\partial T^2)_P = -T\left(\partial^2 v_j^o/\partial T^2\right)_P$.

It becomes immediately obvious that, given the condition $T\left(\partial^2 v_j^o/\partial T^2\right)_P > 0$ for pure j-solvents at normal (around ambient), these fluids would describe the boundary $T\left(\partial^2 v_j^o/\partial T^2\right)_P = 0$ and its crossing only if they meet at least two rather unlikely properties, namely an isobaric thermal expansivity β_{Tj}^o independent of the state conditions, i.e., $\left(\partial \ln v_j^o/\partial T\right)_P \neq \mathcal{F}(T,P)$, and an isobaric-linear temperature-dependent isothermal compressibility, i.e., $\kappa_j^o(T,P) \equiv \mathcal{G}(P) + \mathcal{K}(P)T$. We are not aware of any real pure solvent following this bizarre behavior, and consequently, we must conclude that Hepler's criterion cannot possibly describe and/or predict the transition between structure-making and structure-breaking events involving common solvents (in particular water which was the target of Hepler's analysis). In summary, according to the above discussion and the evidence in Table 3, Hepler's conjectured marker cannot account for the solute-induced perturbation of the solvent microstructure of infinitely dilute solution comprising any solute–solvent intermolecular asymmetries, i.e., either the smallest (zero, for the special case of $LR - IS$) or the largest (for the IG_i solute).

5. Concluding Remarks and Outlook

The usefulness of the *structure-making/breaking* concept has been previously challenged due to its lack of uniqueness [64], i.e., forged by the absence of a precise definition for what is meant by structure, and how to measure its perturbation. In addressing this challenge, we have argued that any criterion or marker for the *structure-making/breaking* ability of a solute must represent a state function and apply to any type and magnitude of solute–solvent molecular interaction asymmetry and should comply with some fundamental molecular thermodynamic constraints. More importantly, we have emphasized that the uniqueness does not (and cannot) depend on the criterion or probe used for the test of the type of solute-induced perturbation of the solvent microstructure.

On the one hand, we have reviewed the approaches to rigorously describe the ability of a solute species to perturb the solvent environment, to assign a precise statistical mechanics meaning to the *structure-making/breaking* event, and to rigorously link it to its thermodynamic responses, including quantitative measures of solution thermodynamic non-ideality which can be assessed experimentally. On the other hand, we have lent true molecular thermodynamic support for the testability, and eventual falsifiability, of a pair of popular conjectured structure-making/breaking markers whose validity has never been established but assumed in the current literature.

Supported by rigorous statistical thermodynamics, we discussed the proper interpretation of the role played by the species-induced microstructural perturbations of the solvent, their response to environmental state conditions and composition, as well as our ability to probe the system and test all conjectures in the development of a causative explanation for a variety of solvation phenomena. In other words, "*...if a theory is testable, then it implies that events of a certain kind cannot happen; and so, it asserts something about reality. This is why we demand that the more conjectural a theory is, the higher should be its degree of testability.*" [185].

In closing, we must stress that often times, authors attempt to add value to their experimental data by introducing hand-waving interpretations of the trends in the observed thermodynamic properties based on umbrella concepts of structure-making/breaking as if, by placing names to unsupported or conjectured ideas, they provide additional understanding. In fact, in the words of Richard Feynman, [186] we should always recognize "*... the difference between knowing the name of something and knowing something*".

Funding: This research received no external funding.

Data Availability Statement: All data are available from the cited references.

Conflicts of Interest: The author declares no conflicts of interest.

Appendix A. Structure-Making/Breaking Function for a Non-Interacting Finite Composition Ideal Gas in a Real Solvent

Starting with the isothermal–isobaric composition evolution of the system microstructure, i.e., [80]

$$(\rho\Delta_{ij})_{TPx_i} = -\left(\partial \ln \gamma_i^{LR}/\partial x_i\right)_{TPx_i} / \left[x_j + x_j\left(\partial \ln \gamma_i^{LR}/\partial \ln x_i\right)_{TPx_i}\right] \quad \text{(A1)}$$

where for the general case $\Delta_{ij}(T,P,x_i) = \left(G_{jj} + G_{ii} - 2G_{ij}\right)_{TPx_i}$, though here, $\Delta_{ij}^{IG\_i} = G_{jj}(T,P,x_i)$.

Likewise, from the Kirkwood–Buff expressions for the partial molar volumes [70], we obtain the following equations for the real j-species and the ideal gas i-species [14]

$$\hat{v}_j(T,P,x_i) = kTv/\left(1 + x_i x_j \rho G_{jj}\right)_{TPx_i} \quad \text{(A2)}$$

$$\hat{v}_i(T,P,x_i) = v\left(1 + x_j \rho G_{jj}\right)_{TPx_i} / \left(1 + x_i x_j \rho G_{jj}\right)_{TPx_i} \quad \text{(A3)}$$

with the isobaric-thermal expansivity $\beta_T^{IG\_i} = (\partial P/\partial T)_\rho \kappa^{IG\_i}$, i.e.,

$$\kappa^{IG\_i}(T,P,x_i) = \left(1+\rho x_j G_{jj}\right)/\left[kT\rho\left(1+\rho x_i x_j G_{jj}\right)\right] \quad \text{(A4)}$$

$$\beta_T^{IG\_i}(T,P,x_i) = \left(1+\rho x_j G_{jj}\right)(\partial P/\partial T)_\rho/\left[kT\rho\left(1+\rho x_i x_j G_{jj}\right)\right] \quad \text{(A5)}$$

From Equations (A2)–(A4), we immediately find

$$\hat{v}_i(T,P,x_i) = kT\kappa^{IG\_i}(T,P,x_i) \quad \text{(A6)}$$

$$\hat{v}_j(T,P,x_i) = kT\kappa^{IG\_i}(T,P,x_i)/\left(1+\rho x_j G_{jj}\right) \quad \text{(A7)}$$

And from Equations (3) and (4), we determine the composition dependence of $\mathcal{S}_{ij}^{IG\_i}(T,P,x_i)$, i.e.,

$$\begin{aligned}\mathcal{S}_{ij}^{IG\_i}(T,P,x_i) &= -\rho x_j G_{jj} \\ &= \left(\partial \ln \gamma_i^{LR,IG\_i}/\partial x_i\right)_{TP}/\left[1+\left(\partial \ln \gamma_i^{LR,IG\_i}/\partial \ln x_i\right)_{TP}\right]\end{aligned} \quad \text{(A8)}$$

Consequently, from (A8) and Equation (6) follows that

$$(\partial P/\partial x_i)_{T\rho}^{IG\_i} = \left[kT\rho\mathcal{S}_{ij}^{IG\_i}/\left(\mathcal{S}_{ij}^{IG\_i}-1\right)\right]_{TPx_i} \quad \text{(A9)}$$

with the system isothermal compressibility $\kappa^{IG\_i}(T,P,x_i)$ written accordingly

$$\kappa^{IG\_i}(T,P,x_i) = \left(1-\mathcal{S}_{ij}^{IG\_i}\right)/\left[kT\rho\left(1-x_i\mathcal{S}_{ij}^{IG\_i}\right)\right] \quad \text{(A10)}$$

Moreover,

$$v(T,P,x_j) = \hat{v}_j\left(1-x_i\mathcal{S}_{ij}^{IG\_i}\right) \quad \text{(A11)}$$

so that its isobaric temperature derivative becomes

$$(\partial v/\partial T)_{Px} = \hat{v}_j\beta_{Tj}\left(1-x_i\mathcal{S}_{ij}^{IG\_i}\right) - x_i\hat{v}_j\left(\partial\mathcal{S}_{ij}^{IG\_i}/\partial T\right)_{Px} \quad \text{(A12)}$$

with $\beta_{T\alpha} = (\partial \ln \hat{v}_\alpha/\partial T)_{Px}$, whose rearrangement leads to

$$\left(\partial\mathcal{S}_{ij}^{IG\_i}/\partial T\right)_{Px} = \left(1-\mathcal{S}_{ij}^{IG\_i}\right)\left(\beta_{Tj}-\beta_{Ti}^{IG\_i}\right) \quad \text{(A13)}$$

Likewise, the isothermal pressure derivative results in the following expression

$$\left(\partial\mathcal{S}_{ij}^{IG\_i}/\partial P\right)_{Tx} = \left(\mathcal{S}_{ij}^{IG\_i}-1\right)\left(\kappa_j-\kappa_i^{IG\_i}\right) \quad \text{(A14)}$$

with $\kappa_\alpha = -(\partial \ln \hat{v}_\alpha/\partial P)_{Tx}$.

References

1. Chialvo, A.A. On the Elusive Links between Solution Microstructure, Dynamics, and Solvation Thermodynamics: Demystifying the Path through a Bridge over Troubled Conjectures and Misinterpretations. *J. Phys. Chem. B* **2023**, *127*, 10792–10813. [CrossRef] [PubMed]
2. Abbott, M.M.; Nass, K.K. Equations of State and Classical Solution Thermodynamics—Survey of the Connections. *ACS Symp. Ser.* **1986**, *300*, 2–40.
3. Chialvo, A.A. Accurate Calculation of Excess Thermal, Infinite Dilution, and Related Properties of Liquid Mixtures Via Molecular-Based Simulation. *Fluid Phase Equilibria* **1993**, *83*, 23–32. [CrossRef]
4. Chialvo, A.A. On the Solute-Induced Structure-Making/Breaking Effect: Rigorous Links among Microscopic Behavior, Solvation Properties, and Solution Non-Ideality. *J. Phys. Chem. B* **2019**, *123*, 2930–2947. [CrossRef]

5. Chialvo, A.A.; Cummings, P.T. Solute-induced Effects on the Structure and the Thermodynamics of Infinitely Dilute Mixtures. *Aiche J.* **1994**, *40*, 1558–1573. [CrossRef]
6. Levelt Sengers, J.M.H. Supercritical Fluids: Their Properties and Applications. In *Supercritical Fluids: Fundamentals and Applications*; Kiran, E., Debenedetti Pablo, G., Peters Cor, J., Eds.; Kluwer Academic Publishers: Dordrecht, The Netherlands, 2000; pp. 1–29.
7. Debenedetti, P.G.; Mohamed, R.S. Attractive, Weakly Attractive and Repulsive Near-Critical Systems. *J. Chem. Phys.* **1989**, *90*, 4528–4536. [CrossRef]
8. Petsche, I.B.; Debenedetti, P.G. Influence of Solute Solvent Asymmetry upon the Behavior of Dilute Supercritical Mixtures. *J. Phys. Chem.* **1991**, *95*, 386–399. [CrossRef]
9. O'Connell, J.P.; Liu, H.Q. Thermodynamic modelling of near-critical solutions. *Fluid Phase Equilibria* **1998**, *144*, 1–12. [CrossRef]
10. Mazo, R.M. Salting out near the critical point. *J. Phys. Chem. B* **2007**, *111*, 7288–7290. [CrossRef] [PubMed]
11. Gee, M.B.; Smith, P.E. Kirkwood-Buff theory of molecular and protein association, aggregation, and cellular crowding. *J. Chem. Phys.* **2009**, *131*, 165101. [CrossRef]
12. Chialvo, A.A. Solvation Phenomena in Dilute Solutions: Formal, Experimental Evidence, and Modeling Implications. In *Fluctuation Theory of Solutions: Applications in Chemistry, Chemical Engineering and Biophysics*, 1st ed.; Matteoli, E., O'Connell, J.P., Smith, P.E., Eds.; CRC Press: Boca Raton, FL, USA, 2013; pp. 191–224.
13. Pallewela, G.N.; Smith, P.E. Preferential Solvation in Binary and Ternary Mixtures. *J. Phys. Chem. B* **2015**, *119*, 15706–15717. [CrossRef] [PubMed]
14. Chialvo, A.A. On the Solvation Thermodynamics Involving Species with Large Intermolecular Asymmetries: A Rigorous Molecular-Based Approach to Simple Systems with Unconventionally Complex Behaviors. *J. Phys. Chem. B* **2020**, *124*, 7879–7896. [CrossRef] [PubMed]
15. Enderby, J.E. Neutron and X-ray Scattering from Aqueous Solutions. *Proc. R. Soc. Lond.* **1975**, *345*, 107–117.
16. Nishikawa, K.; Iijima, T. Small-Angle X-rays Scattering Study of Fluctuations in Ethanol and Water Mixtures. *J. Phys. Chem.* **1993**, *97*, 10824–10828. [CrossRef]
17. Yamanaka, K.; Yamagami, M.; Takamuku, T.; Yamaguchi, T.; Wakita, H. X-Ray-Diffraction Study on Aqueous Lithium-Chloride Solution in the Temperature-Range 138-373-K. *J. Phys. Chem.* **1993**, *97*, 10835–10839. [CrossRef]
18. Fulton, J.L.; Heald, S.M.; Badyal, Y.S.; Simonson, J.M. Understanding the Effects of Concentration on the Solvation Structure of Ca^{+2} in Aqueous Solution. I. The Perspective on Local Structure from EXAFS and XANES. *J. Phys. Chem. A* **2003**, *107*, 4688–4696. [CrossRef]
19. D'Angelo, P.; Roscioni, O.M.; Chillemi, G.; Della Longa, S.; Benfatto, M. Detection of second hydration shells in ionic solutions by XANES: Computed spectra for Ni2+ in water based on molecular dynamics. *J. Am. Chem. Soc.* **2006**, *128*, 1853–1858. [CrossRef]
20. Ansell, S.; Barnes, A.C.; Mason, P.E.; Neilson, G.W.; Ramos, S. X-ray and neutron scattering studies of the hydration structure of alkali ions in concentrated aqueous solutions. *Biophys. Chem.* **2006**, *124*, 171–179. [CrossRef]
21. Antalek, M.; Pace, E.; Hedman, B.; Hodgson, K.O.; Chillemi, G.; Benfatto, M.; Sarangi, R.; Frank, P. Solvation structure of the halides from x-ray absorption spectroscopy. *J. Chem. Phys.* **2016**, *145*, 044318. [CrossRef]
22. Zitolo, A.; Chillemi, G.; D'Angelo, P. X-ray Absorption Study of the Solvation Structure of Cu2+ in Methanol and Dimethyl Sulfoxide. *Inorg. Chem.* **2012**, *51*, 8827–8833. [CrossRef]
23. Enderby, J.E. Ion Solvation via Neutron Scattering. *Chem. Soc. Rev.* **1995**, *24*, 159–168. [CrossRef]
24. Badyal, Y.S.; Barnes, A.C.; Cuello, G.J.; Simonson, J.M. Understanding the effects of concentration on the solvation structure of Ca2+ in aqueous solutions. II: Insights into longer range order from neutron diffraction isotope substitution. *J. Phys. Chem. A* **2004**, *108*, 11819–11827. [CrossRef]
25. Mancinelli, R.; Botti, A.; Bruni, F.; Ricci, M.A.; Soper, A.K. Hydration of sodium, potassium, and chloride ions in solution and the concept of structure maker/breaker. *J. Phys. Chem. B* **2007**, *111*, 13570–13577. [CrossRef]
26. Bruni, F.; Imberti, S.; Mancinelli, R.; Ricci, M.A. Aqueous solutions of divalent chlorides: Ions hydration shell and water structure. *J. Chem. Phys.* **2012**, *136*, 064520. [CrossRef] [PubMed]
27. Pusztai, L.; McGreevy, R.L. MCGR: An inverse method for deriving the pair correlation function from the structure factor. *Physica* **1997**, *234*, 357–358. [CrossRef]
28. Pusztai, L. Partial pair correlation functions of liquid water. *Phys. Rev. B* **1999**, *60*, 11851–11854. [CrossRef]
29. Soper, A.K. Tests of the empirical potential structure refinement method and a new method of application to neutron diffraction data on water. *Mol. Phys.* **2001**, *99*, 1503–1516. [CrossRef]
30. Soper, A.K. Joint structure refinement of x-ray and neutron diffraction data on disordered materials: Application to liquid water. *J. Phys. Condens. Matter* **2007**, *19*, 335206. [CrossRef]
31. Harsanyi, I.; Pusztai, L. Hydration structure in concentrated aqueous lithium chloride solutions: A reverse Monte Carlo based combination of molecular dynamics simulations and diffraction data. *J. Chem. Phys.* **2012**, *137*, 204503–204509. [CrossRef]
32. Mile, V.; Gereben, O.; Kohara, S.; Pusztai, L. On the Structure of Aqueous Cesium Fluoride and Cesium Iodide Solutions: Diffraction Experiments, Molecular Dynamics Simulations, and Reverse Monte Carlo Modeling. *J. Phys. Chem. B* **2012**, *116*, 9758–9767. [CrossRef]
33. Pethes, I.; Pusztai, L. Reverse Monte Carlo modeling of liquid water with the explicit use of the SPC/E interatomic potential. *J. Chem. Phys.* **2017**, *146*, 064506. [CrossRef]

34. Chialvo, A.A.; Simonson, J.M. The Structure of Concentrated $NiCl_2$ Aqueous Solutions. What is Molecular Simulation Revealing about the Neutron Scattering Methodologies? *Mol. Phys.* **2002**, *100*, 2307–2315. [CrossRef]
35. Chialvo, A.A.; Simonson, J.M. The Structure of $CaCl_2$ Aqueous Solutions over a Wide Range of Concentrations. Interpretation of Difraction Experiments Via Molecular Simulation. *J. Chem. Phys.* **2003**, *119*, 8052–8061. [CrossRef]
36. Mason, P.E.; Neilson, G.W.; Dempsey, C.E.; Brady, J.W. Neutron diffraction and simulation studies of $CsNO_3$ and Cs_2CO_3 solutions. *J. Am. Chem. Soc.* **2006**, *128*, 15136–15144. [CrossRef] [PubMed]
37. Pluharova, E.; Fischer, H.E.; Mason, P.E.; Jungwirth, P. Hydration of the chloride ion in concentrated aqueous solutions using neutron scattering and molecular dynamics. *Mol. Phys.* **2014**, *112*, 1230–1240. [CrossRef]
38. Chialvo, A.A.; Vlcek, L. NO_3– Coordination in Aqueous Solutions by $^{15}N/^{14}N$ and ^{18}O/natO Isotopic Substitution: What Can We Learn from Molecular Simulation? *J. Phys. Chem. B* **2015**, *119*, 519–531. [CrossRef]
39. Kohagen, M.; Pluhařová, E.; Mason, P.E.; Jungwirth, P. Exploring Ion–Ion Interactions in Aqueous Solutions by a Combination of Molecular Dynamics and Neutron Scattering. *J. Phys. Chem. Lett.* **2015**, *6*, 1563–1567. [CrossRef]
40. Chialvo, A.A.; Vlcek, L. "Thought experiments" as dry-runs for "tough experiments": Novel approaches to the hydration behavior of oxyanions. *Pure Appl. Chem.* **2016**, *88*, 163–176. [CrossRef]
41. Max, J.J.; Chapados, C. Infrared spectroscopy of aqueous ionic salt mixtures at low concentrations: Ion pairing in water. *J. Chem. Phys.* **2007**, *127*, 114509. [CrossRef]
42. Nickolov, Z.S.; Miller, J.D. Water structure in aqueous solutions of alkali halide salts: FTIR spectroscopy of the OD stretching band. *J. Colloid Interface Sci.* **2005**, *287*, 572–580. [CrossRef] [PubMed]
43. Dillon, S.R.; Dougherty, R.C. Raman studies of the solution structure of univalent electrolytes in water. *J. Phys. Chem. A* **2002**, *106*, 7647–7650. [CrossRef]
44. Smith, J.D.; Saykally, R.J.; Geissler, P.L. The effects of dissolved halide anions on hydrogen bonding in liquid water. *J. Am. Chem. Soc.* **2007**, *129*, 13847–13856. [CrossRef]
45. Rinne, K.F.; Gekle, S.; Netz, R.R. Ion-Specific Solvation Water Dynamics: Single Water versus Collective Water Effects. *J. Phys. Chem. A* **2014**, *118*, 11667–11677. [CrossRef] [PubMed]
46. Shalit, A.; Ahmed, S.; Savolainen, J.; Hamm, P. Terahertz echoes reveal the inhomogeneity of aqueous salt solutions. *Nat. Chem.* **2017**, *9*, 273–278. [CrossRef] [PubMed]
47. Balos, V.; Imoto, S.; Netz, R.R.; Bonn, M.; Bonthuis, D.J.; Nagata, Y.; Hunger, J. Macroscopic conductivity of aqueous electrolyte solutions scales with ultrafast microscopic ion motions. *Nat. Commun.* **2020**, *11*, 1611. [CrossRef] [PubMed]
48. Kim, J.S.; Wu, Z.; Morrow, A.R.; Yethiraj, A.; Yethiraj, A. Self-Diffusion and Viscosity in Electrolyte Solutions. *J. Phys. Chem. B* **2012**, *116*, 12007–12013. [CrossRef]
49. Ma, K.; Zhao, L. The Opposite Effect of Metal Ions on Short-/Long-Range Water Structure: A Multiple Characterization Study. *Int. J. Mol. Sci.* **2016**, *17*, 602. [CrossRef]
50. Tielrooij, K.J.; Garcia-Araez, N.; Bonn, M.; Bakker, H.J. Cooperativity in Ion Hydration. *Science* **2010**, *328*, 1006–1009. [CrossRef]
51. Choi, J.H.; Cho, M. Ion aggregation in high salt solutions. VI. Spectral graph analysis of chaotropic ion aggregates. *J. Chem. Phys.* **2016**, *145*, 174501. [CrossRef]
52. Bernal, J.D.; Fowler, R.H. A Theory of Water and Ionic Solution, with Particular Reference to Hydrogen and Hydroxyl Ions. *J. Chem. Phys.* **1933**, *1*, 515–548. [CrossRef]
53. Frank, H.S.; Evans, M.W. Free Volume and Entropy in Condensed Systems.3. Entropy in Binary Liquid Mixtures—Partial Molal Entropy in Dilute Solutions—Structure and Thermodynamics in Aqueous Electrolytes. *J. Chem. Phys.* **1945**, *13*, 507–532. [CrossRef]
54. Ben-Naim, A. Structure-Breaking and Structure-Promoting Processes in Aqueous-Solutions. *J. Phys. Chem.* **1975**, *79*, 1268–1274. [CrossRef]
55. Collins, K.D. Sticky Ions in Biological-Systems. *Proc. Natl. Acad. Sci. USA* **1995**, *92*, 5553–5557. [CrossRef]
56. Waluyo, I.; Nordlund, D.; Bergmann, U.; Schlesinger, D.; Pettersson, L.G.M.; Nilsson, A. A different view of structure-making and structure-breaking in alkali halide aqueous solutions through X-ray absorption spectroscopy. *J. Chem. Phys.* **2014**, *140*, 244506. [CrossRef]
57. Marcus, Y. Effect of Ions on the Structure of Water: Structure Making and Breaking. *Chem. Rev.* **2009**, *109*, 1346–1370. [CrossRef] [PubMed]
58. Russo, D. The impact of kosmotropes and chaotropes on bulk and hydration shell water dynamics in a model peptide solution. *Chem. Phys.* **2008**, *345*, 200–211. [CrossRef]
59. Vranes, M.; Tot, A.; Papovic, S.; Panic, J.; Gadzuric, S. Is choline kosmotrope or chaotrope? *J. Chem. Thermodyn.* **2018**, *124*, 65–73. [CrossRef]
60. Ben Ishai, P.; Mamontov, E.; Nickels, J.D.; Sokolov, A.P. Influence of Ions on Water Diffusion—A Neutron Scattering Study. *J. Phys. Chem. B* **2013**, *117*, 7724–7728. [CrossRef]
61. Bonetti, M.; Nakamae, S.; Roger, M.; Guenoun, P. Huge Seebeck coefficients in nonaqueous electrolytes. *J. Chem. Phys.* **2011**, *134*, 114513. [CrossRef]
62. Chialvo, A.A.; Vlcek, L. Toward the understanding of hydration phenomena in aqueous electrolytes from the interplay of theory, molecular simulation, and experiment. *Fluid Phase Equilibria* **2016**, *407*, 84–104. [CrossRef]

63. Ben-Naim, A. Solubility, Hydrophobic Interactions and Structural Changes in the Solvent. In *Chemistry and Physics of Aqueous Gas Solutions*; Adams, W.A., Ed.; Electrochemical Society: Princeton, NJ, USA, 1975; pp. 8–18.
64. Ball, P.; Hallsworth, J.E. Water structure and chaotropicity: Their uses, abuses and biological implications. *Phys. Chem. Chem. Phys.* **2015**, *17*, 8297–8305. [CrossRef] [PubMed]
65. Fisher, M.E. The theory of equilibrium critical phenomena. *Rep. Prog. Phys.* **1967**, *30*, 615–730. [CrossRef]
66. Debenedetti, P.G.; Chialvo, A.A. Solute-Solute Correlations in Infinitely Dilute Supercritical Mixtures. *J. Chem. Phys.* **1992**, *97*, 504–507. [CrossRef]
67. Gray, C.G.; Gubbins, K.E. *Theory of Molecular Fluids*; Oxford University Press: Oxford, UK, 1985; Volume 1.
68. Chialvo, A.A.; Cummings, P.T.; Simonson, J.M.; Mesmer, R.E. Solvation in High-Temperature Electrolyte Solutions. II. Some Formal Results. *J. Chem. Phys.* **1999**, *110*, 1075–1086. [CrossRef]
69. Chialvo, A.A.; Kusalik, P.G.; Cummings, P.T.; Simonson, J.M.; Mesmer, R.E. Molecular approach to high-temperature solvation. Formal, integral equation and experimental results. *J. Phys. Condens. Matter* **2000**, *12*, 3585–3593. [CrossRef]
70. Kirkwood, J.G.; Buff, F.P. The Statistical Mechanical Theory of Solution. I. *J. Chem. Phys.* **1951**, *19*, 774–777. [CrossRef]
71. Chialvo, A.A.; Crisalle, O.D. Osmolyte-induced Effects on the Hydration Behavior and the Osmotic Second Virial Coefficients of Alkyl-Substituted Urea Derivatives. Critical assessment of their structure-making/breaking behavior. *J. Phys. Chem. B* **2021**, *125*, 6231–6243. [CrossRef]
72. Mazo, R.M. Statistical Mechanical Theory of Solutions. *J. Chem. Phys.* **1958**, *29*, 1122–1128. [CrossRef]
73. Ben-Naim, A. *Molecular Theory of Solutions*; Oxford University Press: Oxford, UK, 2006.
74. O'Connell, J.P.; Haile, J.M. *Thermodynamics: Fundamentals for Applications*; Cambridge University Press: New York, NY, USA, 2005.
75. Chialvo, A.A.; Crisalle, O.D. On the behavior of the osmotic second virial coefficients of gases in aqueous solutions: Rigorous results, accurate approximations, and experimental evidence. *J. Chem. Phys.* **2019**, *150*, 124503. [CrossRef]
76. Chialvo, A.A. Determination of Excess Gibbs Free Energy from Computer Simulation Via the Single Charging-integral Approach. I. Theory. *J. Chem. Phys.* **1990**, *92*, 673–679. [CrossRef]
77. Levelt Sengers, J.M.H. Solubility Near the Solvent's Critical Point. *J. Supercrit. Fluids* **1991**, *4*, 215–222. [CrossRef]
78. Chialvo, A.A.; Crisalle, O.D. Solute-Induced Perturbation of the Solvent Microstructure in Aqueous Electrolyte Solutions: Some Uses and Misuses of Structure Making/Breaking Criteria. *Liquids* **2022**, *2*, 106–130. [CrossRef]
79. Chialvo, A.A.; Crisalle, O.D. On the Molecular-Based Thermodynamics of Dilute Solutions Along Orthobaric Conditions Concise review of the fundamental microscopic constraints, rigorous formal results, and macroscopic modeling implications. *J. Mol. Liq.* **2023**, *390*, 122901. [CrossRef]
80. Chialvo, A.A. Alternative Approach to Modeling Excess Gibbs Free Energy in Terms of Kirkwood-Buff Integrals. In *Advances in Thermodynamics*; Matteoli, E., Mansoori, G.A., Eds.; Taylor & Francis: New York, NY, USA, 1990; Volume 2, pp. 131–173.
81. Koga, K.; Holten, V.; Widom, B. Deriving Second Osmotic Virial Coefficients from Equations of State and from Experiment. *J. Phys. Chem. B* **2015**, *119*, 13391–13397. [CrossRef] [PubMed]
82. Chialvo, A.A.; Crisalle, O.D. Gas Solubility and Preferential Solvation Phenomena in Mixed-Solvents; Rigorous relations between microscopic behavior, solvation properties, and solution non-ideality. *Fluid Phase Equilibria* **2024**, *581*, 114081. [CrossRef]
83. Chialvo, A.A. Molecular-Based Description of the Osmotic Second Virial Coefficients of Electrolytes: Rigorous Formal Links to Solute−Solvent Interaction Asymmetry, Virial Expansion Paths, and Experimental Evidence. *J. Phys. Chem. B* **2022**, *126*, 4339–4353. [CrossRef]
84. Schmitt, W.J.; Reid, R.C. The Use of Entrainers in Modifying the Solubility of Phenanthrene and Benzoic Acid in Supercritical Carbon Dioxide and Ethane. *Fluid Phase Equilibria* **1986**, *32*, 77–99. [CrossRef]
85. Chialvo, A.A. Solute-Solute and Solute-Solvent Correlations in Dilute Near-Critical Ternary Mixtures: Mixed Solute and Entrainer Effects. *J. Phys. Chem.* **1993**, *97*, 2740–2744. [CrossRef]
86. Gutkowski, K.I.; Fernandez-Prini, R.; Japas, M.L. Solubility of solids in near-critical conditions: Effect of a third component. *J. Phys. Chem. B* **2008**, *112*, 5671–5679. [CrossRef]
87. Shimizu, S.; Abbott, S. How Entrainers Enhance Solubility in Supercritical Carbon Dioxide. *J. Phys. Chem. B* **2016**, *120*, 3713–3723. [CrossRef]
88. Lei, Z.G.; Wang, H.Y.; Zhou, R.Q.; Duan, Z.T. Influence of salt added to solvent on extractive distillation. *Chem. Eng. J.* **2002**, *87*, 149–156. [CrossRef]
89. Zhao, J.; Dong, C.C.; Li, C.X.; Meng, H.; Wang, Z.H. Isobaric vapor-liquid equilibria for ethanol-water system containing different ionic liquids at atmospheric pressure. *Fluid Phase Equilibria* **2006**, *242*, 147–153. [CrossRef]
90. Smith, P.E. Equilibrium dialysis data and the relationships between preferential interaction parameters for biological systems in terms of Kirkwood-Buff integrals. *J. Phys. Chem. B* **2006**, *110*, 2862–2868. [CrossRef]
91. Chialvo, A.A. Preferential Solvation Phenomena as Solute-induced Structure-Making/Breaking Processes: Linking thermodynamic preferential interaction parameters to fundamental structure making/breaking functions. *J. Phys. Chem. B* **2024**, *128*, 5228–5245. [CrossRef]
92. Ben-Naim, A. Theory of Preferential Solvation of Non-Electrolytes. *Cell Biophys.* **1988**, *12*, 255–269. [CrossRef]
93. Shimizu, S.; Matubayasi, N. Preferential hydration of proteins: A Kirkwood-Buff approach. *Chem. Phys. Lett.* **2006**, *420*, 518–522. [CrossRef]

94. Smiatek, J. Aqueous ionic liquids and their effects on protein structures: An overview on recent theoretical and experimental results. *J. Phys. Condens. Matter* **2017**, *29*, 233001. [CrossRef]
95. Hepler, L.G. Thermal Expansion and Structure in Water and Aqueous Solutions. *Can. J. Chem.* **1969**, *47*, 4613–4617. [CrossRef]
96. Chialvo, A.A. Gas Solubility in Dilute Solutions: A Novel Molecular Thermodynamic Perspective. *J. Chem. Phys.* **2018**, *148*, 174502. [CrossRef]
97. Jani, A.; Malfait, B.; Morineau, D. On the coupling between ionic conduction and dipolar relaxation in deep eutectic solvents: Influence of hydration and glassy dynamics. *J. Chem. Phys.* **2021**, *154*, 164508. [CrossRef]
98. Balos, V.; Kaliannan, N.K.; Elgabarty, H.; Wolf, M.; Kuhne, T.D.; Sajadi, M. Time-resolved terahertz-Raman spectroscopy reveals that cations and anions distinctly modify intermolecular interactions of water. *Nat. Chem.* **2022**, *14*, 1031–1104. [CrossRef]
99. Jones, G.; Dole, M. The viscosity of aqueous solutions of strong electrolytes with special reference to barium chloride. *J. Am. Chem. Soc.* **1929**, *51*, 2950–2964. [CrossRef]
100. Gurney, R.W. *Ionic Processes in Solution*; McGraw Hill: New York, NY, USA, 1953.
101. Kaminsky, M. Ion-Solvent Interaction and the Viscosity of Strong Electrolyte Solutions. *Discuss. Faraday Soc.* **1957**, *24*, 171–179. [CrossRef]
102. Nightingale, E.R. Phenomenological Theory of Ion Solvation: Effective Radii of Hydrated Ions. *J. Phys. Chem.* **1959**, *63*, 1381–1387. [CrossRef]
103. Tsangaris, J.M.; Martin, R.B. Viscosities of Aqueous Solutions of Dipolar Ions. *Arch. Biochem. Biophys.* **1965**, *112*, 267–272. [CrossRef]
104. Out, D.J.P.; Los, J.M. Viscosity of aqueous solutions of univalent electrolytes from 5 to 95 °C. *J. Solut. Chem.* **1980**, *9*, 19–35. [CrossRef]
105. Hertz, H.G.; Maurer, R.; Killie, S. An NMR-Study of Dynamics in Aqueous Strong Acid Solutions. 1. The B′ Coefficients. *Z. Phys. Chem. Int. J. Res. Phys. Chem. Chem. Phys.* **1991**, *172*, 157–183.
106. Jenkins, H.D.B.; Marcus, Y. Viscosity B-coefficients of ions in solution. *Chem. Rev.* **1995**, *95*, 2695–2724. [CrossRef]
107. Zhu, C.Y.; Ren, X.F.; Ma, Y.G. Densities and Viscosities of Amino Acid plus Xylitol plus Water Solutions at 293.15 <= T/K <= 323.15. *J. Chem. Eng. Data* **2017**, *62*, 477–490.
108. Gupta, J.; Nain, A.K. Physicochemical study of solute-solute and solute-solvent interactions of homologous series of alpha-amino acids in aqueous-isoniazid solutions at temperatures from 293.15 to 318.15 K. *J. Mol. Liq.* **2019**, *278*, 262–278. [CrossRef]
109. Rani, R.; Kumar, A.; Sharma, T.; Sharma, T.; Bamezai, R.K. Volumetric, acoustic and transport properties of ternary solutions of L-serine and L-arginine in aqueous solutions of thiamine hydrochloride at different temperatures. *J. Chem. Thermodyn.* **2019**, *135*, 260–277. [CrossRef]
110. Sawhney, N.; Kumar, M.; Sandarve; Sharma, P.; Sharma, A.K.; Sharma, M. Structure-making behaviour of L-arginine in aqueous solution of drug ketorolac tromethamine: Volumetric, compressibility and viscometric studies. *Phys. Chem. Liq.* **2019**, *57*, 184–203. [CrossRef]
111. Stokes, R.H.; Mills, R. *Viscosity of Electrolytes and Related Properties*; Pergamon Press: Oxford, UK, 1965.
112. Dhondge, S.S.; Zodape, S.P.; Parwate, D.V. Volumetric and viscometric studies of some drugs in aqueous solutions at different temperatures. *J. Chem. Thermodyn.* **2012**, *48*, 207–212. [CrossRef]
113. Chialvo, A.A.; Crisalle, O.D. Can Jones-Dole's B-coefficient Be a Consistent Structure Making/Breaking Marker? Rigorous molecular-based analysis and critical assessment of its marker uniqueness. *J. Phys. Chem. B* **2021**, *125*, 12028–12041. [CrossRef]
114. Chialvo, A.A.; Crisalle, O.D. On the Transition-State theory approach to the Jones-Dole's viscosity B-coefficient: A novel molecular-based interpretation, assessment of its implications, and experimental evidence. *J. Mol. Liq.* **2023**, *379*, 121548. [CrossRef]
115. Abdulagatov, I.M.; Zeinalova, A.B.; Azizov, N.D. Viscosity of aqueous $Ni(NO_3)_2$ solutions at temperatures from (297 to 475) K and at pressures up to 30 MPa and concentration between (0.050 and 2.246) $mol{\cdot}kg^{-1}$. *J. Chem. Thermodyn.* **2006**, *38*, 179–189. [CrossRef]
116. Abdulagatov, I.; Azizov, N. Experimental Study of the Effect of Temperature, Pressure, and Concentration on the Viscosity of Aqueous KBr Solutions. *J. Solut. Chem.* **2008**, *37*, 3–26. [CrossRef]
117. Liu, Q.; Dai, H.; Chi, H.; Shi, K.; Zheng, Q.; Qi, Y. The density and dynamic viscosity for dilute solutions of [Emim][NTf2] [Bmim][NTf2], and [Bmmim][NTf2] in ethylene glycol. *J. Mol. Liq.* **2023**, *371*, 121080. [CrossRef]
118. Feakins, D.; Freemantle, D.J.; Lawrence, K.G. Transition-State Treatment of Relative Viscosity of Electrolyte Solutions: Applications to Aqueous, Non-Aqueous and Metanol+Water Systems. *J. Chem. Soc. Faraday Trans. I* **1974**, *70*, 795–806. [CrossRef]
119. Feakins, D.; Waghorne, W.E.; Lawrence, K.G. The Viscosity and Structure of Solutions. 1. A New Theory of the Jones-Dole B-coefficient and the Related Activation Parameters—Applications to Aqueous Solutions. *J. Chem. Soc. Faraday Trans. I* **1986**, *82*, 563–568. [CrossRef]
120. Glasstone, S.; Laidler, K.J.; Eyring, H. *The Theory of Rate Processes: The Kinetics of Chemical Reactions, Viscosity, Diffusion and Electrochemical Phenomena*; McGraw-Hill Book Company, Inc.: Chicago, IL, USA, 1941.
121. Robinson, R.A.; Stokes, R.H. *Electrolyte Solutions*, 2nd (revised) ed.; Butterworths: London, UK, 1970.
122. Conway, B.E. *Ionic Hydration in Chemistry and Biophysics*; Elsevier: Amsterdam, The Netherlands, 1981; Volume 12.
123. Franks, F. *Water: A Matrix of Life*; Royal Society of Chemistry: London, UK, 2000.

124. Waghorne, W.E. Viscosities of electrolyte solutions. *Philos. Trans. R. Soc. Lond. Ser. A-Math. Phys. Eng. Sci.* **2001**, *359*, 1529–1543. [CrossRef]
125. Hammadi, A. Electrical Conductance, Density, and Viscosity in Mixtures of Alkali-Metal Halides and Glycerol. *Int. J. Thermophys.* **2004**, *25*, 89–111. [CrossRef]
126. Jindal, R.; Singla, M.; Kumar, H. Transport behavior of aliphatic amino acids glycine/L-alanine/L-valine and hydroxyl amino acids L-serine/L-threonine in aqueous trilithium citrate solutions at different temperatures. *J. Mol. Liq.* **2015**, *206*, 343–349. [CrossRef]
127. Gadzuric, S.; Tot, A.; Armakovic, S.; Armakovic, S.; Panic, J.; Jovic, B.; Vranes, M. Uncommon structure making/breaking behaviour of cholinium taurate in water. *J. Chem. Thermodyn.* **2017**, *107*, 58–64. [CrossRef]
128. Zafarani-Moattar, M.T.; Shekaari, H.; Mostafavi, H.; Jafari, P. Thermodynamic and transport properties of aqueous solutions containing cholinium L-alaninate and polyethylene glycol dimethyl ether 250: Evaluation of solute-solvent interactions and phase separation. *J. Chem. Thermodyn.* **2019**, *132*, 9–22. [CrossRef]
129. Marcus, Y. *Ions in Solution and Their Solvation*; John Wiley and Sons: Hoboken, NJ, USA, 2015.
130. Batchelor, J.D.; Olteanu, A.; Tripathy, A.; Pielak, G.J. Impact of protein denaturants and stabilizers on water structure. *J. Am. Chem. Soc.* **2004**, *126*, 1958–1961. [CrossRef]
131. Hood, G.R. The viscosity of urea in aqueous solution. *Phys. A J. Gen. Appl. Phys.* **1933**, *4*, 211–214. [CrossRef]
132. Herskovits, T.T.; Kelly, T.M. Viscosity Studies of Aqueous Solutions of Alcohols, ureas, and Amides. *J. Phys. Chem.* **1973**, *77*, 381–388. [CrossRef]
133. Egorov, G.I.; Makarov, D.M. Densities and Volumetric Properties of Aqueous Solutions of {Water (1) + N-Methylurea (2)} Mixtures at Temperatures of 274.15-333.15 K and at Pressures up to 100 MPa. *J. Chem. Eng. Data* **2017**, *62*, 4383–4394. [CrossRef]
134. Singh, M.; Kumar, A. Hydrophobic interactions of methylureas in aqueous solutions estimated with density, molal volume, viscosity and surface tension from 293.15 to 303.15 K. *J. Solut. Chem.* **2006**, *35*, 567–582. [CrossRef]
135. Panic, J.; Vranes, M.; Tot, A.; Ostojic, S.; Gadzuric, S. The organisation of water around creatine and creatinine molecules. *J. Chem. Thermodyn.* **2019**, *128*, 103–109. [CrossRef]
136. Richu; Kumar, A. A Comprehensive Study on Molecular Interactions of l-Ascorbic Acid/Nicotinic Acid in Aqueous [BMIm]Br at Varying Temperatures and Compositions: Spectroscopic and Thermodynamic Insights. *J. Chem. Eng. Data* **2021**, *66*, 3859–3880. [CrossRef]
137. Sharma, R.; Thakur, R.C.; Kumar, H. Study of viscometric properties of L-ascorbic acid in binary aqueous mixtures of D-glucose and D-fructose. *Phys. Chem. Liq.* **2018**, *57*, 139–150. [CrossRef]
138. Hossain, M.F.; Biswas, T.K.; Islam, M.N.; Huque, M.E. Volumetric and viscometric studies on dodecyltrimethylammonium bromide in aqueous and in aqueous amino acid solutions in premicellar region. *Monatshefte Chem.* **2010**, *141*, 1297–1308. [CrossRef]
139. Banipal, T.S.; Beni, A.; Kaur, N.; Banipal, P.K. Volumetric, Viscometric and Spectroscopic Approach to Study the Solvation Behavior of Xanthine Drugs in Aqueous Solutions of NaCl at T = 288.15-318.15 K and at p = 101.325 kPa. *J. Chem. Eng. Data* **2017**, *62*, 20–34. [CrossRef]
140. Vranes, M.; Tot, A.; Jankovic, N.; Gadzuric, S. What is the taste of vitamin-based ionic liquids? *J. Mol. Liq.* **2019**, *276*, 902–909. [CrossRef]
141. Rudan-Tasic, D.; Klofutar, C.; Horvat, J. Viscosity of aqueous solutions of some alkali cyclohexylsulfamates at 25.0 degrees C. *Food Chem.* **2004**, *86*, 161–167. [CrossRef]
142. Horvat, J.; Bester-Rogac, M.; Klofutar, C.; Rudan-Tasic, D. Viscosity of aqueous solutions of lithium, sodium, potassium, rubidium and caesium cyclohexylsulfamates from 293.15 to 323.15 k. *J. Solut. Chem.* **2008**, *37*, 1329–1342. [CrossRef]
143. Dunn, L.A. Apparent Molar Volumes of Electrolytes.3: Some 1-1 and 2-1 Electrolytes in Aqueous Solution at 0, 5. 15. 35, 45, 55, and 65 Degrees C. *Trans. Faraday Soc.* **1968**, *64*, 2951–2961. [CrossRef]
144. Abdulagatov, I.M.; Azizov, N.D. Viscosity of aqueous calcium chloride solutions at high temperatures and high pressures. *Fluid Phase Equilibria* **2006**, *240*, 204–219. [CrossRef]
145. Herrington, T.M.; Roffey, M.G.; Smith, D.P. Densities of Aqueous Electrolytes $MnCl_2$, $CoCl_2$, $NiCl_2$, $ZnCl_2$, and $CdCl_2$ from 25 degrees to 72 degrees at 1 atm. *J. Chem. Eng. Data* **1986**, *31*, 221–225. [CrossRef]
146. Sonika; Thakur, R.C.; Chauhan, S.; Singh, K. Molecular Interactions of Transition Metal Chlorides in Water and Water–Ethanol Mixtures at 298–318 K on Viscometric Data. *Russ. J. Phys. Chem. A* **2018**, *92*, 2701–2709. [CrossRef]
147. Apelblat, A. A new two-parameter equation for correlation and prediction of densities as a function of concentration and temperature in binary aqueous solutions. *J. Mol. Liq.* **2016**, *219*, 313–331. [CrossRef]
148. Clegg, S.L.; Wexler, A.S. Densities and Apparent Molar Volumes of Atmospherically Important Electrolyte Solutions. 1. The Solutes H_2SO_4, HNO_3, HCl, Na_2SO_4, $NaNO_3$, NaCl, $(NH_4)_{(2)}SO_4$, NH_4NO_3, and NH_4Cl from 0 to 50 degrees C, Including Extrapolations to Very Low Temperature and to the Pure Liquid State, and $NaHSO_4$, NaOH, and NH_3 at 25 degrees C. *J. Phys. Chem. A* **2011**, *115*, 3393–3460.
149. Campbell, A.N.; Friesen, R.J. Conductance in the Range of Medium Concentration. *Can. J. Chem.-Rev. Can. Chim.* **1959**, *37*, 1288–1293. [CrossRef]
150. Laliberte, M.; Cooper, W.E. Model for calculating the density of aqueous electrolyte solutions. *J. Chem. Eng. Data* **2004**, *49*, 1141–1151. [CrossRef]

151. Sarma, T.S.; Ahluwalia, J.C. Experimental Studies on Structure of Aqueous Solutions of Hydrophobic Solutes. *Chem. Soc. Rev.* **1973**, *2*, 203–232. [CrossRef]
152. Sanyal, S.K.; Mandal, S.K. Viscosity B-Coefficients of Alkyl Carboxylates. *Electrochim. Acta* **1983**, *28*, 1875–1876.
153. Tamaki, K.; Ohara, Y.; Watanabe, S. Solution Properties of Sodium Perfluoralkanoates. Heats of Solution, Viscosity B-coefficients, and Surface Tensions. *Bull. Chem. Soc. Jpn.* **1989**, *62*, 2497–2501. [CrossRef]
154. Zhao, H. Viscosity B-coefficients and standard partial molar volumes of amino acids, and their roles in interpreting the protein (enzyme) stabilization. *Biophys. Chem.* **2006**, *122*, 157–183. [CrossRef]
155. Rajagopal, K.; Jayabalakrishnan, S.S. Volumetric, ultrasonic speed, and viscometric studies of salbutamol sulphate in aqueous methanol solution at different temperatures. *J. Chem. Thermodyn.* **2010**, *42*, 984–993. [CrossRef]
156. Kabiraz, D.C.; Biswas, T.K.; Huque, M.E. Physico-chemical properties of some electrolytes in water and aqueous sodiumdodecyl sulfate solutions at different temperatures. *J. Chem. Thermodyn.* **2011**, *43*, 1917–1923. [CrossRef]
157. Banipal, T.S.; Kaur, N.; Kaur, A.; Gupta, M.; Banipal, P.K. Effect of food preservatives on the hydration properties and taste behavior of amino acids: A volumetric and viscometric approach. *Food Chem.* **2015**, *181*, 339–346. [CrossRef]
158. Kumar, K.; Patial, B.S.; Chauhan, S. Interactions of Saccharides in Aqueous Glycine and Leucine Solutions at Different Temperatures of (293.15 to 313.15) K: A Viscometric Study. *J. Chem. Eng. Data* **2015**, *60*, 47–56. [CrossRef]
159. Sharma, S.K.; Singh, G.; Kumar, H.; Kataria, R. Densities, Sound Speed, and Viscosities of Some Amino Acids with Aqueous Tetra-Butyl Ammonium Iodide Solutions at Different Temperatures. *J. Chem. Eng. Data* **2015**, *60*, 2600–2611. [CrossRef]
160. Sarmad, S.; Zafarani-Moattar, M.T.; Nikjoo, D.; Mikkola, J.P. How Different Electrolytes Can Influence the Aqueous Solution Behavior of 1-Ethyl-3-Methylimidazolium Chloride: A Volumetric, Viscometric, and Infrared Spectroscopy Approach. *Front. Chem.* **2020**, *8*, 593786. [CrossRef]
161. Nain, A.K. Insight into solute-solute and solute-solvent interactions of l-proline in aqueous-D-xylose/L-arabinose solutions by using physicochemical methods at temperatures from 293.15 to 318.15 K. *J. Mol. Liq.* **2020**, *318*, 114190. [CrossRef]
162. Rajput, P.; Richu; Sharma, T.; Kumar, A. Temperature dependent physicochemical investigations of some nucleic acid bases (uracil, thymine and adenine) in aqueous inositol solutions. *J. Mol. Liq.* **2021**, *326*, 115210. [CrossRef]
163. Brinzei, M.; Ciocirlan, O. Volumetric and transport properties of ternary solutions of L-serine and L-valine in aqueous NaCl solutions in the temperature range (293.15–323.15) K. *J. Chem. Thermodyn.* **2021**, *154*, 106335. [CrossRef]
164. Bronshtein, I.N.; Semendyayev, A.K. *Handbook of Mathematics*, 3rd ed.; Springer: Berlin/Heidelberg, Germany, 1997.
165. Feakins, D.; Bates, F.M.; Waghorne, W.E.; Lawrence, K.G. Relative Viscosities and Quasi-Thermodynamics of Solutions of Ter-Butyl Alcohol in the Methanol-Water System—A Different View of the Alkyl Water Interaction. *J. Chem. Soc. Faraday Trans.* **1993**, *89*, 3381–3388. [CrossRef]
166. Haldar, P.; Das, B. Viscosities of some tetraalkylammonium bromides in 2-ethoxyethanol at 308.15, 313.15, 318.15, and 323.15 K. *Can. J. Chem.* **2005**, *83*, 499–504. [CrossRef]
167. Xu, W.G.; Qi, W.S.; Zhang, Z.H.; Tong, J. Volumetric and viscometric investigation of aqueous solution of Bmim BF4 from 288.15 to 318.15 K. *J. Mol. Liq.* **2019**, *278*, 600–606. [CrossRef]
168. Gupta, J.; Chand, D.; Nain, A.K. Study to reconnoiter solvation consequences of l-arginine/l-histidine and sodium salicylate in aqueous environment probed by physicochemical approach in the temperature range (293.15–318.15) K. *J. Mol. Liq.* **2020**, *305*, 112848. [CrossRef]
169. Ankita; Nain, A.K. Solute-solute and solute-solvent interactions of drug sodium salicylate in aqueous-glucose/sucrose solutions at temperatures from 293.15 to 318.15 K: A physicochemical study. *J. Mol. Liq.* **2020**, *298*, 112006. [CrossRef]
170. Vraneš, M.; Rackov, S.; Papović, S.; Pilić, B. The study of interactions in aqueous solutions of 1-alkyl-3-(3-butenyl)imidazolium bromide ionic liquids. *J. Chem. Thermodyn.* **2021**, *159*, 106479. [CrossRef]
171. Xu, J.; Zhao, L.Q.; Xiong, D.G.; Zhang, Z.; Yu, J.; Cai, J.X.; Yang, Z.Y. Density and viscosity behaviour of EBPBF4, TBPBF4 and HBPBF4 in acetonitrile solutions at different temperatures at T = (293.15–323.15) K. *J. Chem. Thermodyn.* **2021**, *158*, 106467. [CrossRef]
172. Sharma, T.; Bandral, A.; Bamezai, R.K.; Kumar, A. Interaction behavior of sucrose in aqueous tributylmethylammonium chloride solutions at various temperatures: A volumetric, ultrasonic and viscometric study. *Chem. Thermodyn. Therm. Anal.* **2022**, *6*, 100043. [CrossRef]
173. Yao, D.D.; Tang, P.C.; Song, P.; Sun, Z. Viscosities and Densities of Mixed Aqueous Alkali Metal Nitrate Solutions at T = (293.15–333.15) K. *J. Chem. Eng. Data* **2022**, *67*, 2088–2097. [CrossRef]
174. Tong, J.; Zheng, X.; Li, H.; Tong, J.; Liu, Q.S. Densities and Viscosities of Aqueous Amino Acid Ionic Liquids C(n)mim Ala (n = 3, 4, 5). *J. Chem. Eng. Data* **2017**, *62*, 2501–2508. [CrossRef]
175. Li, X.R.; Qin, Y.; Xu, W.G.; Liu, J.G.; Yang, J.Z.; Xu, Q.; Yan, C.W. Investigation of electrolytes of the vanadium redox flow battery (IV): Measurement and prediction of viscosity of aqueous $VOSO_4$ solution at 283.15 to 323.15 K. *J. Mol. Liq.* **2016**, *224*, 893–899. [CrossRef]
176. Banipal, T.S.; Singh, K.; Banipal, P.K.; Sood, A.K.; Singh, P.; Singh, G.; Patyar, P. Volumetric and Viscometric Studies of Some Metal Acetates in Aqueous Solutions at T = (288.15 to 318.15) K. *J. Chem. Eng. Data* **2008**, *53*, 2758–2765. [CrossRef]
177. Roy, M.N.; Sah, R.S.; Pradhan, P.P.; Roy, P.K. Ion-Solvent and Ion-Ion Interactions of Phosphomolybdic Acid in Aqueous Solution of Catechol at 298.15, 308.15, and 318.15 K. *Russ. J. Phys. Chem. A* **2009**, *83*, 1887–1895. [CrossRef]

178. Ameta, R.K.; Singh, M. Surface tension, viscosity, apparent molal volume, activation viscous flow energy and entropic changes of water plus alkali metal phosphates at T = (298.15, 303.15, 308.15) K. *J. Mol. Liq.* **2015**, *203*, 29–38. [CrossRef]
179. Gupta, J.; Nain, A.K. Molecular interactions of gentamicin sulphate in aqueous-l-asparagine/l-glutamine solutions at different temperatures: Volumetric, acoustic and viscometric properties. *J. Mol. Liq.* **2019**, *293*, 111547. [CrossRef]
180. Lin, L.N.; Brandts, J.F.; Brandts, J.M.; Plotnikov, V. Determination of the volumetric properties of proteins and other solutes using pressure perturbation calorimetry. *Anal. Biochem.* **2002**, *302*, 144–160. [CrossRef]
181. Mitra, L.; Smolin, N.; Ravindra, R.; Royer, C.; Winter, R. Pressure perturbation calorimetric studies of the solvation properties and the thermal unfolding of proteins in solution—Experiments and theoretical interpretation. *Phys. Chem. Chem. Phys.* **2006**, *8*, 1249–1265. [CrossRef]
182. Zhai, Y.; Okoro, L.; Cooper, A.; Winter, R. Applications of pressure perturbation calorimetry in biophysical studies. *Biophys. Chem.* **2011**, *156*, 13–23. [CrossRef] [PubMed]
183. Suladze, S.; Kahse, M.; Erwin, N.; Tomazic, D.; Winter, R. Probing volumetric properties of biomolecular systems by pressure perturbation calorimetry (PPC)—The effects of hydration, cosolvents and crowding. *Methods* **2015**, *76*, 67–77. [CrossRef]
184. Chialvo, A.A.; Cummings, P.T.; Kalyuzhnyi, Y.V. Solvation effect on kinetic rate constant of reactions in supercritical solvents. *Aiche J.* **1998**, *44*, 667–680. [CrossRef]
185. Popper, K.R. *Conjectures and Refutations: The Growth of Scientific Knowledge*; Routledge: Oxfordshire, UK, 2002.
186. Feynman, R.P. *What Do You Care What Other People Think?: Further Adventures of a Curious Character*; Norton & Company, Incorporated: New York, NY, USA, 1988.

liquids

MDPI

Article

Evaluation of Thermodynamic and Kinetic Contributions to Over-Extraction of Extractables by Nonpolar Organic Solvents in Comparison to Lipids in Exhaustive and Exaggerated Extractions of Medical Devices Based on Abraham Solvation Model and Solvent–Material Interactions Using Low-Density Polyethylene as a Representative Material

Jianwei Li

Chemical Characterization Solutions, LLC, P.O. Box 113, Newport, MN 55055, USA; jianwei.li@chemicalcharacterizationsolutions.com; Tel.: +1-651-302-3360

Abstract: The thermodynamic and kinetic contributions to the over-extraction of extractables by nonpolar organic solvents relative to biological lipids in exhaustive and exaggerated extractions of medical devices are studied based on the Abraham solvation model and solvent–material interactions, using low-density polyethylene (LDPE) as an exemplary material. The thermodynamic effect is evaluated by the partition constant of extractables between LDPE and extraction solvents, hexane and lipids, defined as the concentration in the polymer phase divided by the concentration in the solvent phase. The Abraham solvation model is used to correlate the measured LDPE-lipid partition constant ($log_{10}P_{ldpe/lipid}$) to construct the predictive model. Similar models are also derived from the thermodynamic cycle conversion, using the system constants of LDPE-water and Lipid-water partition systems. These constructed models, together with the predictive LDPE-hexane ($log_{10}P_{ldpe/hexane}$) model established from a previous study, are used to predict and compare the ranges and values of $P_{ldpe/s}$ (s = lipids and hexane) for the observed LDPE extractables over a wide hydrophobicity range in $log_{10}P_{o/w}$ from zero to 30. The solvent-LDPE interactions are examined by the degree of swelling of LDPE by hexane (or other nonpolar solvents) and lipids, including the solvent diffusion rates into the material. These parameters allow the evaluation of kinetic effect on the over-extraction. The extent of over-extraction is compiled directly by experimental "overall" or "specific" migration data or indirectly calculated by the diffusion coefficient of extractables when extracted by hexane or lipids. It is concluded from this study that the extractables distribution between LDPE and lipids highly favors the lipid phase thermodynamically ($P_{ldpe/lipid} < 1$), and the values of $P_{ldpe/lipid}$ are always lower than those of $P_{ldpe/hexane}$, thereby indicating that the thermodynamic effect is not the cause of over-extraction. It is the kinetic effect that dominantly contributes to the over-extraction, as supported by the material swelling and solvent diffusion rates. Finally, the extent of over-extraction has been established from a few folds to over a hundred-fold, and the median value is 7. Furthermore, the methods adopted and developed in this study can be invaluable tools in other disciplines such as the reliable prediction of extractables from other device materials and environmental sampling.

Keywords: extractables; exhaustive extraction; exaggerated extraction; lipids; low density polyethylene (LDPE); Abraham solvation model

1. Introduction

Chemical characterization of medical devices (or assessment of extractables and leachables, E&L) is currently a subject of intense interest and an essential aspect of the regulatory review and approval of medical devices in the U.S., the European Union (EU), and most major markets around the world [1,2]. Chemical characterization is a process that identifies and quantifies the constituent chemicals of a specific device or its component and materials, which may have a direct or indirect human exposure from its clinical use [3]. This process consists of the liquid–solid extraction of medical devices, components, or materials using appropriate extraction solvents and/or conditions, followed by instrumental analysis of extraction samples (also called extracts) to determine the concentration and identity of E&L chemical entities, both of which are then used for a toxicological risk (or safety) assessment.

Extractables are "substances that can be released from medical devices, components, or their materials using extraction solvents and/or extraction conditions that are expected to be at least as aggressive as the clinical use [3]". Leachables, which are typically a subset of extractables, are "substances that are released from a medical device during its clinical use [3]". As these medical and biomedical products use diverse categories of materials, including polymers, metals, and ceramics [4]; and many different compounds are used as polymer additives, including antioxidants, plasticizers, heat and UV light stabilizers, extrusion chemicals, and colorants [5], E&L are a result of polymer monomers/oligomers/additives, and residues from sterilization, cleaning, manufacturing processes, or chemical degradation of the material. Obviously, the classes of E&L compounds and their physicochemical properties can be rather broad [6], and the number of extractables has been reported to be over 10K [7]. For example, the range of hydrophobicity of likely extractables from different materials can be from −2 to 19 in $log_{10}P_{o/w}$ [6]. The term "extractables" is solely used in the remaining article, as nonpolar organic solvents are mainly used in extractables studies of medical devices by exaggerated and exhaustive extractions [3].

Nonpolar organic solvents, such as hexane, heptane, cyclohexane, etc., are used or recommended as the extraction vehicles in the exaggerated and exhaustive extraction of medical devices due to at least two reasons: (1) they are analytically expedient for instrumental analysis of extracts [3]; and (2) they are intended to simulate the polarity of the human lipid tissues, because the lipids are considered as nonpolar in general and are soluble in nonpolar organic solvents [8,9]. A similar practice has been adopted in the food industries over several decades. For example, vegetable oils such as olive oil are used as the food simulants for "overall" migration testing for fatty foods, according to Commission Regulation (EU) No 10/2011 on plastic food contact materials [10–12]. Because the "specific" migration testing in olive oil is often impossible, alternative simulants are being proposed: iso-octane and 95% ethanol [13–16]. The US FDA similarly recommends the use of some aqueous-based solvents (absolute or 95% ethanol) as alternatives for fatty-food simulants for polyolefins [11,17]. It is worth noting that lipids exist in biological tissues in many different physical forms, and the simple lipids are often part of large aggregates in storage tissues, such as oil bodies or adipose tissue (called as storage lipids). In contrast, other complex lipids are usually constituents of membranes, where they occur in a close association with such compounds as proteins and polysaccharides [18]. Because the recommended nonpolar extraction vehicles or simulants by ISO 10993-12 (2021) are vegetable oils such as olive oil [19], and vegetable oils and storage lipids are made of same mixtures of triglycerides, the lipids referred in this study belong to the storage lipids (or simply fats). Lipids, fats, and vegetable oils are considered the same in this study, and used interchangeably in the remaining article.

As extensively studied and documented already, the extraction solvent is perhaps the single most important variable affecting both the ultimate amount of extractables and their diffusion properties, acting through its solubility of extractables and its swelling action upon the polymer [20–23]. The equilibrium amount of extractables into a solvent system with a finite solvent-to-polymer volume ratio is governed by the partition constant or the respective solubilities of extractables in the solvent and in the polymer. When the system is far from equilibrium, the amount migrated is determined by the diffusion coefficients and associated parameters, such as the time of extraction and the thickness of the sample [21,24–26]. The combination of the effects of solubility, partitioning, and diffusion would determine ultimately the extractables profiles (the number and amounts of extractables, and the molecular weight distribution of all extractables from materials).

The partition constant, $P_{p/l}$ (typically defined as the equilibrium concentration in the polymer divided by the corresponding concentration in the solvent phase), is a fundamental physicochemical parameter describing the distribution of extractables between two contacting phases at equilibrium. The values for the partition constant range over several orders of magnitude depending on the polarities of polymeric materials involved, the extraction solvent, and the nature of extractables [27]. For instance, the partition constant of limonene, a nonpolar substance or hydrocarbon, in the LDPE/water system at 23 °C was found to be higher than 5000 [28]. This makes it retained in the polymer, whereas a much more polar compound like cis-hexenol shows a much lower value of 0.33, causing a considerable transfer into the water. The reverse is true for nonpolar extraction solvents such as hexane [29]. Numerous experimentally determined $P_{p/l}$ values between polyolefins and alcohols or alcohol/water mixtures have been reported [30]. In general, the partition constants between device materials and extraction solvents can be difficult to measure, due mostly to the time required to reach equilibrium and matrix interference for analysis. Several predictive approaches for calculating $P_{p/l}$ values have been proposed and published [29], but their accuracies are often questionable for a predictive purpose. It is currently assumed in the food industries that $P_{p/l} < 1$ for non-polar fat simulants or nonpolar organic solvents; and $P_{p/l} >> 1$ for polar aqueous simulants for polyolefin materials [11,24,31,32]. Additionally, if a significant fraction of extractables in the polymer can migrate into a solvent thermodynamically, the amount of extractables observed at specific time points is dominated by diffusion, and can often be described as approximately proportional to the square root of time [20,24–26]. It is noted that $P_{p/l}$ is used in this study, consistent with the notation typically used by the solvation model [33]; however, $K_{p/l}$ is typically used in food industries [24,32].

Regarding the extraction of nonpolar (device) materials by alkanes (e.g., hexane) and lipids, the release kinetics of extractions of polyolefins by alkanes is usually faster than that by lipids (or vegetable oils), because alkane solvents cause a significant swelling of nonpolar materials, and enhance the diffusion rate of material constituents [20,22]. The interactions between lipids and polyolefins are much weaker due to the high MW of lipids/oils [34]. Because of the difference in release kinetics, a much shorter extraction time and lower temperature of alkanes have been evaluated and used to establish the equivalence between alkane and lipid extraction, so that the same amount of extractables by oils can be predicted by alkanes in food industries [13,16,35–38]. For example, 10 days and 40 °C olive oil extraction on polyolefins can be replaced by 2 days and 20 °C extraction with isooctane [28]. In fact, the FDA guidelines to correct for the accelerating action of n-heptane over food oils is by 5-fold [20].

The amount of extractables migrated and measured into an extraction solvent at defined time points in an exaggerated or exhaustive extractions of medical devices may not be equilibrium values, because the extraction end point is determined by the 10% rule [3]. If a parallel extraction experiment is performed using lipids and alkanes to extract polyolefins, the amount by alkane is higher than that by oils at the same sampling points, thereby resulting in a so called "over-extraction" by alkane solvents relative to lipids. In fact, this over-extraction occurrence by alkane solvents has been demonstrated in several studies already [20,39–41]. For example, in the study of migration of a UV absorber additive (UVITEX or (2,5-bis(5-ter-butyl-benzoxazol-2-yl)thiophen)) from a polypropylene material into a pure extraction solvent glyceryl tripelargonate, it was demonstrated that this solvent could swell the materials and enhance the mobility of extractables to cause an over-extraction by 100% [39]. Furthermore, this effect was more pronounced for high molecular weight compounds and thick materials. In another study of extraction of polyolefins, including LDPE, HDPE, and PP, by n-heptane, ethanol, and vegetable oils, it was clearly shown that the heptane solvent could accelerate the diffusion process during extractions, and the total weight fraction of oligomers extracted by n-heptane is about

6–8 times greater than that extracted by corn oils [20]. Additionally, the difference in the total amount extracted was caused by different amounts of the higher molecular weight species being extracted by n-heptane. A similar accelerating effect of organic solvents was also demonstrated by a migration study of LDPE using Irganox 1076 as the model extractables, and organic solvents caused a significant solvent absorption, not observed for the triglycerides and olive oil [40]. Accordingly, an over-extraction should be expected relative to biological lipids, when nonpolar organic solvents are used to substitute lipids to perform the chemical characterization studies. This phenomenon could have a profound consequence on the results of the chemical characterization studies and subsequent toxicological risk assessments.

The general conclusion so far on the cause of over-extraction seems to be the kinetic effect [20]; however, this is based on a limited amount of partition data. In fact, the partition constants of LDPE-hexane or LDPE-lipids are scarce in the literature. A few studies can be found on measuring LDPE-oil partition constants of limited compounds [20,42–47], but the data accuracy is questionable due to the durations and techniques of these studies. Based on rather limited data, it was proposed that the logarithm of LDPE-lipid partition constant ($log_{10}P_{ldpe/lipid}$) is positively proportional to $log_{10}P_{o/w}$ with 16 compounds over a $log_{10}P_{o/w}$ range from 0.5 to 18 by olive oil [44,45]. The best and comprehensive measurement of LDPE-lipid partition constants were recently conducted by a polymer-lipid partitioning setup [47]. It was concluded that $log_{10}P_{ldpe/lipid}$ is independent of or (weakly) negatively proportional to $log_{10}P_{o/w}$. Overall, there is not a specific comprehensive study on the range of LDPE-lipid or LDPE-alkane partition constants of extractables from LDPE or other materials; and their dependence on $log_{10}P_{o/w}$. As the material extractables can be rather broad in physicochemical properties in chemical characterization studies [6], the scope of these partition constants remain an assumption and unknown, in particular on the difference between alkanes (hexane as an example in this study) and lipids/oils. The judgement on the thermodynamic and kinetic contributions to the over-extraction cannot be specifically delineated under the current situation. The partition constant between polyolefins, including LDPE, and lipids/oils/alkanes is currently assumed to be one in the diffusion-based predictive modeling of extractables release [48,49], but it remains unclear whether this assumption would be suitable limits for all extractables of different physicochemical properties. Although it is well recognized that methods for the evaluation of specific and accurate partition constants are required to predict patient exposure to polymeric medical device extractables using physics-based models [48], no such methods have been reported so far. Moreover, a reliable model for the prediction of LDPE-lipid distribution is also important in other fields such as environmental sampling [47,50–52].

The purpose of this study is four-fold: (1) to explore the correlation of $P_{ldpe/s}$ (s: hexane or lipids) by the Abraham solvation model for a predictive purpose; (2) to quantitatively explore the range of $P_{ldpe/s}$ of the LDPE extractables over a broad hydrophobicity range; (3) to evaluate the dependence of $P_{ldpe/s}$ on the hydrophobicity of extractables from a practical perspective, and the effect of $P_{ldpe/s}$ on over-extraction of medical devices by alkanes; and (4) to conduct a systematic study into the extent of over-extraction, using LDPE as an example. The first three goals are achieved by the Abraham solvation model, which has been used successfully to correlate the partition constant between polymeric materials and solvents with and without a material swelling [33]. The use of the solvation model to directly correlate the partition constants of LDPE-lipids has not been reported so far to the best of my knowledge. Additionally, LDPE is a nonpolar material, commonly used in food and pharmaceutical packaging, and medical devices [53,54], and can also be a suitable representation for other nonpolar materials such as linear low-density polyethylene (LLDPE), polydimethylsiloxane (PDMS or silicone), high-density polyethylene (HDPE), polystyrene (PS), etc.

The specific goals of this study are outlined as follows:

(1) Correlation of the partition constant between LDPE and lipids by the Abraham solvation model. The constructed (predictive) models are then used to predict the partition constant of extractables observed for LDPE over a wide hydrophobicity range.
(2) Calculation of the partition constant between LDPE and hexane for the same LDPE extractables.
(3) Assessment of the dependence of LDPE-lipid and LDPE-hexane partition constants on $log_{10}P_{o/w}$ using extractables observed for LDPE.
(4) Evaluation of the adequacy of the LDPE-lipid and LDPE-hexane predictive models.
(5) Comparison of the range and difference of two sets of partition constants (LDPE-lipid and LDPE-hexane) for the observed LDPE extractables to understand the thermodynamic effect on over-extraction.
(6) Study of LDPE-lipid interactions by the adsorption of lipids into LDPE in comparison to alkanes and other organic solvents.
(7) Assessment of the range of solvent diffusion coefficients by both types of solvents.
(8) Estimation of the extent of over-extractions by alkanes.

Based on the goals and conclusions of this study, two critical questions are answered: (1) What is the dominant factor in controlling over-extraction by alkanes (e.g., hexane)? (2) what is the extent of over-extraction? Furthermore, the conclusion of the study on the use of Abraham solvation model to correlate and predict the polymer-lipid partition constant can be applied not only to the prediction of extractables chemicals from medical device polymers, but also to the environmental passive sampling [47,50–52].

Elemental impurities and inorganic anions are not included in this study, as they are handled differently from organic extractables.

2. Methods

The sequence of the study can be briefly defined below:

- Development of the Abraham solvation working models to correlate the partition constant ($log_{10}P_{ldpe/lipid}$) between LDPE and lipids using: (1) multiple linear regression (MLR) analysis of the measured partition constant ($P_{ldpe/lipid}$), and (2) a thermodynamic cycle method [55,56]. The solutes used in the model construction are directly taken from the ref [47]. These predictive models are used to predict the $log_{10}P_{ldpe/lipid}$ of experimentally observed LDPE extractables over a wide range of hydrophobicity.
- Establishment of empirical relationships between the partition constant $log_{10}P_{ldpe/lipid}$ and $log_{10}P_{o/w}$ using experimentally observed LDPE extractables.
- Establishment of empirical relationships between the partition constant $log_{10}P_{ldpe/hexane}$ and $log_{10}P_{o/w}$ using the same experimentally observed LDPE extractables for a comparison purpose.
- Comparison of the dependence of $log_{10}P_{ldpe/s}$ on $log_{10}P_{o/w}$ (s: lipids or hexane) between hexane and lipids to assess the extractability of LDPE by hexane and lipids from a thermodynamic perspective.
- Justification of the suitability of the constructed solvation models in the prediction.
- Assessment of the practical implications of the difference between $log_{10}P_{ldpe/lipid}$ and $log_{10}P_{ldpe/hexane}$ values on over-extraction by hexane (or alkanes).
- Assessment of the material-solvent interactions between LDPE and alkanes/lipids and their influence on over-extraction.
- Compile and estimate the degree of over-extraction of LDPE material by alkane solvents relative to lipids/oils.

2.1. Abraham Solvation Model

The Abraham solvation parameter model describes the solute transfer between two condensed phases, or between a condensed phase and a gas phase [57–59]. Extensive specific chemical and biological processes have been successfully described by the basic model [60]. This model uses a common set of solute descriptors for all types of solute transfer processes. The basic model that describes solute transfer of neutral molecules between two condensed phases is as follows:

$$log_{10}P = c + eE + sS + aA + bB + vV \quad (1)$$

The dependent variable in Equation (1) is $log_{10}P$, where P is the solvent-to-solvent partition constant in molar concentration for the transfer of solutes between two condensed phases, including polymers. The intercept, c term, contains information for system properties only. (c, e, s, a, b, v) are the solvent system constants, and (E, S, A, B, V) are the solute descriptors: E is the solute excess molar refractivity in units of $(cm^3/mol)/10$, S is the solute dipolarity/polarizability, A and B are the overall or summation hydrogen bond acidity and basicity, and V is the McGowan characteristic volume in units of $(cm^3/mol)/100$. To date, solvent system constants have been determined for over 90 commonly used solvents (wet and dry) [59]. Equation (1) is used to correlate the LDPE-lipids (or hexane) partition constants in this study.

It is noted that the general solvation model has been used to describe the transfer between polymeric phases and liquid solvent phases with and without a material swelling [33]. One of the major goals of this study is to apply the solvation model to LDPE-lipid system to understand the thermodynamic effect on over-extraction of medical devices by nonpolar solvents, including alkanes.

2.2. Construction of Abraham Solvation Model for LDPE-Lipid Partition Constant

The LDPE-lipid partition constants ($log_{10}P_{ldpe/lipid}$) are difficult to measure and scarce in the literatures. However, a number of "nonpolar" organic compounds (total 78, polychlorinated biphenyls, polycyclic aromatic hydrocarbons, and organochlorines) have been measured and reported recently [47], shown in Supplemental Information Section (Table S1). Because the measured compounds are "nonpolar" or moderately polar, the descriptor spaces of these compounds are limited in the S, A, and B descriptors, and the range of the $log_{10}P_{ldpe/lipid}$ values is narrow as well. Although this situation is not ideal [57], a multiple linear regression (MLR) method is still used to correlate the partition constants with the descriptors by Equation (1), as described in Sections 3.4 and 3.5.

Additionally, a "thermodynamic cycle" method is used to derive the solvation model [55,56], using the system constants of LDPE-water partition systems ($log_{10}P_{ldpe/water}$) [56,61,62] and the Lipid-water partition system ($log_{10}P_{Lipid/water}$) from the literature [51]. The relevant system constants of partition systems are tabulated in Table 1. For example, the $log_{10}P_{ldpe/lipid}$ is derived through the thermodynamic cycle using the constants from Systems D and G in Table 1 as follows [51,56]:

$$\begin{aligned} log_{10}P_{\frac{ldpe}{lipid}} &= log_{10}P_{\frac{ldpe}{water}} - log_{10}P_{\frac{lipid}{water}} \\ &= \left[c + eE + sS + aA + bB + vV_{ldpe/water}\right] - \left[c + eE + sS + aA + bB + v\right]V_{lipd/water} \\ &= (\delta c) + (\delta e)E + (\delta s)S + (\delta a)A + (\delta b)B + (\delta v)V \\ &= -0.459 + 0.398{\cdot}E - 0.477{\cdot}S - 1.271{\cdot}A - 0.477{\cdot}B - 0.224{\cdot}V \end{aligned} \quad (2)$$

Equation (2) is shown and labelled as System I in Table 1. The results of a similar derivation using the LDPE-water Systems E and F are also shown in Table 1 (as J and K). The three systems are called as "Cycle-A", "Cycle-B", and "Cycle-C" models in Table 1.

Table 1. Tabulation of MLR Results and Descriptive Statistics plus Solvation System Constants of Various Partition Systems.

Method	Partition System		Solvation Model System Constants [a]						Descriptive Statistics [b]					
	System Label [c]	System Description [d]	*c*	*e*	*s*	*a*	*b*	*v*	*R*	*SE*	*F*	*LogP* Range	*SE*/Range (%)	*AR* (*AAR*)
MLR by Equation (1) (Sections 2.2 and 3.5)	A [e]	LDPE-Lipid (4–20 °C)	−0.529 (0.089)	0.526 (0.043)	−0.368 (0.095)	1.475 [h] (1.089)	−0.640 (0.082)	−0.455 (0.073)	0.893	0.118	57	1.265 (−1.55 to −0.287)	9.4	0.00 (0.08)
	B [f] (MLR)	LDPE-Lipid (4–20 °C)	−0.524 (0.089)	0.506 (0.041)	−0.329 (0.091)	NA	−0.619 (0.081)	−0.465 (0.073)	0.890	0.119	69		9.4	0.00 (0.08)
	C [g]	LDPE-Lipid (4–20 °C)	−0.997 (0.151)	0.005 [h] (0.026)					0.021	0.256	0.04		20.2	0.00 (0.19)
Compilation from Literatures and Thermodynamic Cycle Conversion (Sections 2.2 and 3.5)	D [i]	LDPE-Water (25 °C)	−0.529 (0.077)	1.098 (0.047)	−1.557 (0.081)	−2.991 (0.116)	−4.617 (0.110)	3.886 (0.058)						
	E [j]	LDPE-Water (20–25 °C)		1.002 (0.075)	−1.296 (0.128)	−1.820 (0.472)	−4.037 (0.153)	3.399 (0.072)						
	F [k]	LDPE-Water (20–25 °C)	−0.54 (0.05)	1.13 (0.03)	−1.58 (0.06)	−3.47 (0.08)	−4.53 (0.07)	3.92 (0.04)						
	G [l]	Lipid-Water (37 °C)	−0.07 (0.07)	0.7 (0.06)	−1.08 (0.08)	−1.72 (0.13)	−4.14 (0.09)	4.11 (0.06)						
	H [m]	Hexane-Water (w/d, 25 °C)	0.333	0.56	−1.71	−3.578	−4.939	4.463						
	I [n] (Cycle-A)	LDPE-lipid (25–37 °C)	−0.459 (0.104)	0.398 (0.076)	−0.477 (0.114)	−1.271 (0.174)	−0.477 (0.142)	−0.224 (0.084)						
	J [o] (Cycle-B)	LDPE-lipid (20–37 °C)	0.070 (0.07)	0.302 (0.096)	−0.216 (0.151)	−0.100 (0.490)	0.103 (0.178)	−0.711 (0.094)						
	K [p] (Cycle-C)	LDPE-lipid (20–37 °C)	−0.470 (0.086)	0.430 (0.067)	−0.500 (0.100)	−1.750 (0.153)	−0.390 (0.114)	−0.190 (0.072)						
	L [q] (Equation (3))	LDPE-Hexane (23 °C)	−1.104 [s] (0.153)	0.329 (0.109)	−0.033 (0.136)	−1.315 (0.242)	−0.977 (0.176)	0.229 (0.087)						
	M [r]	LDPE-Hexane (25–37 °C)	−0.862 [s]	0.538	0.153	0.587	0.322	−0.577						

[a] System constants obtained by MLR (95% confidence) and the thermodynamic cycle conversion, or taken directly from literatures, as described in Sections 2.2 and 3.5. The value in the parenthesis is the standard error of the constant, if available. [b] Notation of descriptive statistics: *R*-correlation coefficient; *SE*-standard error of the fit; *F*-Fisher statistic; *logP* Range-*logP* values from minimum to maximum; *SE*/Range-the standard error of the fit divided by the *logP* data range of the system, in percent; AR-average residual between predicted and measured values (AAR: absolute average residual between predicted and measured values in the parenthesis). [c] Partition system label. [d] System description. The value in the parenthesis is the temperature, range of temperatures in the initial study or the applicable range of temperature in the final system. [e] Regression results by Equation (1). The partition constant unit is kg/kg [47]. The value in the parenthesis of a system constant is the standard error from MLR (same for superscript f and g). [f] Regression results after the *A* parameter is removed in Equation (1). This correlation is called "MLR". [g] Regression by $log_{10}P_{o/w}$ model. The number in the *c* column is the intercept, and the number in the *e* column is the slope. [h] The *p*-value is greater than 0.05, indicating statistically insignificant. [i] LDPE-water partition system [56]. The concentration unit is the molar concentration. The partition constant unit is L/L. The value in the parenthesis is the standard error of the constant taken directly from the reference (same for superscript j, k, and l). [j] LDPE-water partition system [61]. The partition constant unit is L/kg. [k] LDPE-water partition system [62]. The partition constant unit is L/kg. [l] Lipid-water partition system [51]. The concentration unit is molar/liter. The partition constant unit is L/L. [m] Hexane-water partition system [59]. The concentration unit is mol/liter. The partition constant unit is L/L. [n] LDPE-lipid partition system by a thermodynamic cycle of LDPE-water and lipid-water systems [51,56]. The value in the parenthesis is the standard error of the constant derived from error propagation (same for superscript o, p, and q). LDPE (~0.92 g/mL) and lipid (0.9 g/mL) have nearly the same density, and the partition unit of L/L is considered the same as Kg/Kg in this study (also same for superscript o, and p). This correlation is called "Cycle-A". [o] Same as superscript n, using LDPE-water system [51,61]. This correlation is called "Cycle-B". [p] Same as superscript n, using LDPE-water system [51,62]. This correlation is called "Cycle-C". [q] LDPE-hexane partition system with a LDPE swelling [33]. The concentration unit is the molar concentration. LDPE and hexane have different density, and the partition unit of L/L is converted to Kg/Kg in this study to be consistent with the data in Ref. [47] (same for superscript r). [r] LDPE-hexane partition system without a swelling by a thermodynamic cycle of lipid-water [56] and hexane-water system [59]. [s] The correction on the *c* constant in Systems L and M is by subtracting $log_{10}(0.66/0.92) = -0.144$ to convert from L/L unit to kg/kg unit. The corrected *c* constant for System L is −0.960, and −0.718 for System M. [e,f,g] The number of solutes is 78.

2.3. Construction of Abraham Solvation Model for LDPE-Hexane Partition Constant

The model for the LDPE-hexane partition constants ($log_{10}P_{ldpe/hexane}$) has been constructed previously [33], as indicated by Equation (3) (and System L in Table 1).

$$logP_{ldpe/Hexane} = -1.104 + 0.329E - 0.033S - 1.315A - 0.977B + 0.229V \quad (3)$$

It is noted that the effect of solvent swelling in LDPE is included in the model. The suitability of Equation (3) is discussed further in Sections 2.6 and 4.4.

Furthermore, the $log_{10}P_{ldpe/hexane}$ solvation model (System M) is also derived through the thermodynamic cycle using the data (Systems D [51] and H [59]) in Table 1 without a consideration of solvent swelling. It is noted that the constants of Systems L and M are nearly opposite, indicating the effect of swelling on the extractables distribution.

2.4. Calculation of LDPE-Lipid and LDPE-Hexane Partition Constants of Observed LDPE Extractables

The solvation models (Systems B, I, J, and K in Table 1) are then used as the predictive models to calculate the LDPE-lipid and LDPE-hexane partition constants for the observed LDPE extractables (shown in Table S2). System B is called "MLR" model in Section 3.5.

2.5. Evaluation of Dependence of $log_{10}P_{ldpe/lipid}$ and $log_{10}P_{ldpe/hexane}$ on $log_{10}P_{o/w}$ of Observed LDPE Extractables

The dependence of $log_{10}P_{ldpe/lipid}$ and $log_{10}P_{ldpe/hexane}$ on $log_{10}P_{o/w}$ is empirically evaluated for the observed LDPE Extractables, because they do not have the same reference phase [33]. This is simply done by a linear correlation between of $log_{10}P_{ldpe/lipid}$ or $log_{10}P_{ldpe/hexane}$ and $log_{10}P_{o/w}$. Note that a more complex equation was used in ref [33] to empirically describe the relationship.

2.6. Suitability of Use of Constructed Models to Predict $log_{10}P_{ldpe/lipid}$ and $log_{10}P_{ldpe/hexane}$ for the Observed LDPE Extractables

To assess if the constructed solvation models (MLR, and Cycle-A to C, plus Equation (3)) are suitable in predicting $log_{10}P_{ldpe/lipid}$ and $log_{10}P_{ldpe/hexane}$ for the observed LDPE extractables, the partition constant ratio ($R_{h/l}$) between $P_{ldpe/hexane}$ and $P_{ldpe/lipid}$ is derived as follows:

$$log_{10}R_{h/l} = log_{10}\frac{P_{ldpe/hexane}}{P_{ldpe/lipid}} = log_{10}P_{ldpe/hexane} - log_{10}P_{ldpe/lipid} \quad (4)$$

Equation (4) can be thermodynamically corresponding to:

$$log_{10}R_{h/l} = log_{10}P_{lipid/water} - log_{10}P_{hexane/water} = log_{10}\frac{C_{lipid}}{C_{hexane}} \quad (5)$$

where C_{lipid} and C_{hexane} in Equation (5) are the concentration in lipid and hexane phases at equilibrium.

Equation (5) indicates that $log_{10}R_{h/l}$ can have the same reference phase, water, with $log_{10}P_{o/w}$ in principle and a linear relationship between $log_{10}R_{h/l}$ and $log_{10}P_{o/w}$ should be expected fundamentally, although they are derived differently. The correlation of $log_{10}P_{ldpe/hexane}$ was derived by the partition systems of LDPE-methanol and heptane-methanol [33]. The $log_{10}P_{ldpe/lipid}$ are derived from MLR and thermodynamic cycle conversion in this study. If a linear relationship could be established between $log_{10}R_{h/l}$ and $log_{10}P_{o/w}$, the constructed models for $log_{10}P_{ldpe/lipid}$ and $log_{10}P_{ldpe/hexane}$ are confirmed to be correct and suitable to predict LDPE-lipid and LDPE-hexane distributions. See Section 4.4 for more details.

2.7. Calculation of $log_{10}P_{o/w}$ of Measured and LDPE Extractables Compounds

The Abraham solvation parameter model for the octanol–water partition constant ($log_{10}P_{o/w}$), as the hydrophobicity parameter, has been constructed as follows [63]:

$$log_{10}P_{o/w} = 0.088 + 0.562E - 1.054S + 0.034A - 3.460B + 3.814V \tag{6}$$

The descriptive statistics in deriving Equation (6) are: $R = 0.995$, $SE = 0.12$, $F = 23{,}255$, and $n = 613$. This equation is used to calculate $log_{10}P_{o/w}$ of the measured compounds (Table S1) and representative LDPE extractables compounds (Table S2, see Section 3.6 for details).

2.8. Differentiation between Thermodynamic and Kinetic Contributions

The amount of extractables released into a solvent at a specific time is related to thermodynamic (time-independent) and kinetic (time-dependent) influences [14,25,40,48]. The thermodynamic contribution defines the limit of extraction by the partition constant ($P_{ldpe/s}$). If M_S^∞ denotes the amount of extractables released from LDPE at equilibrium and M_T is the total amount initially present in LDPE, the amount extracted at equilibrium can be written as follws [48]:

$$\frac{M_S^\infty}{M_T} = \frac{1}{V_{ldpe}P_{ldpe/s}/V_S + 1} \tag{7}$$

where V_{ldpe} and V_S are the volume of LDPE and solvent phase, respectively. Equation (7) indicates that, if $P_{ldpe/s}$ is much smaller than unity, $\frac{M_S^\infty}{M_T} \sim 1$ (or ~100%).

Under the limit set by the thermodynamic parameter, the kinetic contribution to the amount extracted at a specific time has been studied extensively [14,25,40,48]. Both exact and approximate equations have been developed. The most relevant conclusion to this study is that the amount extracted is exactly or approximately dependent on the square root of the time with the slope controlled in part by the diffusion coefficient of the extractables.

2.9. Estimation of Over-Extraction of LDPE by Alkanes Compared to Lipids

Over-extraction (OE) is defined in this study as the amount extracted by alkanes to that by lipids. No such data can be directly compiled from the literature. There were a few studies to compare the extraction equivalency between vegetable oils and alkanes, particularly isooctane, either by overall migration and specific migration (of an extractables compound). These data are directly used to estimate the over-extraction. Additionally, the extent of over-extraction is derived from the compiled diffusion coefficients of extractables from LDPE to alkanes and lipids. They are briefly described as follows:

(1) Direct calculation by:

$$OE = \frac{A_{Alkane}}{A_{Lipid}} \tag{8}$$

where A_{Alkane} is the amount migrated to alkanes; and A_{lipid} is the amount migrated to lipids and vegetable oils. The amount can be obtained by overall migration testing or specific migration testing over a specific time duration.

(2) Estimation by kinetic diffusion coefficients as follows:

$$OE = \sqrt{\frac{D_{Alkanes}}{D_{oils/ethanol}}} \tag{9}$$

where D_{Alkane} is the diffusion coefficient of extractables migrated to alkanes; and D_{lipid} is the diffusion coefficient migrated to lipids and vegetable oils. Equation (8) is derived by the square root dependence of extracted amount on the diffusion coefficient by each solvent [26]:

$$\frac{M_t}{M_\infty} = \left(\frac{4}{l}\right)\left(\frac{Dt}{\pi}\right)^{\frac{1}{2}} \tag{10}$$

where l is the thickness of the material, D is the diffusion coefficient, t is time, and $\frac{M_t}{M_\infty}$ is the fraction extracted at time t. Note that Equation (10) applies to devices of any shape [25,26]. For polyolefins, including LDPE, the diffusion rate by vegetable oils and lipids can be represented by that of ethanol solvent [13–15,35,64].

3. Experimental

3.1. Collection of LDPE-Lipid Partition Constants

The LDPE-lipid partition constants are collected from the literature [42–47], including olive oils and sunflower oils as the lipids [42–46]. The data from the ref [47] were measured by lipids for neutral organic compounds, mostly "nonpolar" environmental contaminants, and are used in the construction of the solvation model. They were measured at 4 and 20 °C by partitioning experiments between polymers (silicones and LDPE) and lipids (fish oil and triolein), and a total of 78 of them were measured. The unit of these partition constants is Kg/Kg [47]. These compounds and their LDPE-lipid partition constants, together with their descriptors and hydrophobicity, are Table S1. The values in Table S1 are the final generic (mean) $P_{p/l}$ (or K_{pl}) values, averaged from four mean values of two lipid types (fish oil and triolein) and temperatures (4 and 20 °C), since they show no dependence on temperature or storage lipid type. The range of $log_{10}P_{o/w}$ of these compounds is 2.75–7.92 (Equation (6)). It should be emphasized that the partition constant is insensitive to temperature [43,47,65–69] and lipid/oil types (fish oil, triolein fish, olive, rapeseed and sunflower oil) [47,50,51,70–73]. It is also assumed that LDPE material heterogeneity has no influence on the partition constant [52,54].

The LDPE-oil (olive oil and sunflower oil) partition constants are also included in Table S1 to understand the range of LDPE-lipid partition constant, considering lipids and vegetable oils are the same. However, they are not used in the model construction.

3.2. Collection of Observed LDPE Extractables

To evaluate the over-extraction from LDPE material, the most relevant LDPE extractables should be used in this study. Experimentally observed extractables from LDPE are collected from literature publications [74–80]. A total of 76 LDPE extractables of different classes of compounds by organic solvent extraction are compiled and tabulated in Table S2, together with their descriptors, hydrophobicity, etc. They are used to evaluate the range of LDPE-lipid and LDPE-hexane partition constants. The hydrophobicity range of the collected extractables compounds in $log_{10}P_{o/w}$ by Equation (6) is 0.3 to 29.5 (mostly below 17) and MW range from 72.1 to 1177.6.

3.3. Compilation of Abraham Solvation Model Constants of LDPE-Water and Lipid-Water Partition Systems

The Abraham model system constants for LDPE-water and LDPE-lipid partition systems are compiled to derive the system constants of LDPE-lipid partition system by a thermodynamic cycle method [55,56]. Table 1 shows the results of the collection, together with the system constants of other partition systems. It is noted in Table 1 that the range of temperature in Table 1 is from 20 to 37 °C in all studies. The correlation of measured partition constants by the solvation model is considered applicable within this range (and beyond), because the partition constant is not sensitive to temperature (Section 3.1). Additionally, the material sources for the studies in Table 1 could be different, but this is the reality that needs to be accepted.

It should be mentioned that the unit of partition constants in Table 1 can be kg/kg, L/kg or L/L. The density of LDPE is about 0.92 g/mL [52,54], and the density of lipids/fat is also about 0.92 g/mL. The values of LDPE-lipid partition constants are essentially the same by all three units and will not be distinguished further (see Table 1 for more details). However, the density of hexane is 0.66 g/mL, and significantly different from LDPE. This difference in density will be corrected in Equation (3) (System L and M in Table 1) to convert L/L unit to kg/kg unit, consistent with the data from Ref. [47]. The correction is on the c constant in Systems L and M by subtracting $log_{10}(0.66/0.92) = -0.144$.

3.4. Determination of Descriptors of Collected Compounds

To correlate $log_{10}P_{ldpe/lipid}$ by the solvation model, the descriptors (E, S, A, B, and V) of these compounds are required. For the compounds (Tables S1 and S2) collected in this study, their descriptors are calculated by the "RMG: Solvation Tools" (SoluteML module), an online computational program based on group contributions and machine learning approaches [81]. The inputs for the algorithm are the canonical SMILES notations. These notations are mostly obtained from "PubChem" database using CAS # search. The descriptors from Wayne State University experimental descriptor database are also adopted, as discussed in Section 4.1.1 [82].

The correlation matrix of descriptors for the solutes in Tables S1 and S2 are shown in Tables S3 and S4.

3.5. Construction and Correlation of LDPE-Lipid Partition Constant by Abraham Solvation Model

The Abraham solvation model, Equation (1), is used to construct the LDPE-lipid partition constants ($log_{10}P_{ldpe/lipid}$) in this study. This is performed in two ways, as described in Section 2.2. The first is by MLR with and without the aA term. The regression results and statistics are shown in Table 1. The model without the aA term is called MLR model in Table 1. The second is by the thermodynamic cycle conversion using the system constants of $log_{10}P_{lipid/water}$ and $log_{10}P_{ldpe/water}$. The results of the conversion are also indicated in Table 1. The models (Systems I, J and K) by thermodynamic cycle conversion are called "Cycle".

3.6. Calculation of $log_{10}P_{o/w}$ of Representative Extractables Compounds

Calculation of $log_{10}P_{o/w}$ of representative extractables compounds is performed using Equation (6) with descriptors in Tables S1 and S2, as described in Section 2.7. They are shown in Tables S1 and S2.

3.7. Calculation of $log_{10}P_{ldpe/lipid}$ and $log_{10}P_{ldpe/hexane}$ for Observed Extractables Compounds

The $log_{10}P_{ldpe/s}$ (s = hexane or lipids) for the observed LDPE extractables in Table S2 are computed using the system constants of MLR, Cycle-A to Cycle-C models, plus Equation (3) (System L) in Table 1.

The empirical correlation between $log_{10}P_{ldpe/s}$ (s = hexane or lipids) and $log_{10}P_{o/w}$ is described in Section 2.5. The results of the correlation for lipid and hexane are shown in Table 2. The established relationships can be used to determine and compare the range of $log_{10}P_{ldpe/s}$ for observed LDPE extractables.

The range of $log_{10}P_{ldpe/s}$ (s = hexane or lipids) for the observed LDPE extractables are calculated by the correlation results in Table 2, and are shown in Table 3.

Table 2. Tabulation of Linear Regression Results and Descriptive Statistics between $log_{10}P_{ldpe/s}$ and $log_{10}P_{o/w}$ by Collected LDPE Extractables.

Partition System [a]	Regression Results [b]		Descriptive Statistics [c]				
	Intercept	Slope	R	$Adj.\ R$	SE	F	n
LDPE-hexane (L)	−1.685 (0.152)	0.062 (0.014)	0.452	0.193	0.682	19	76
LDPE-hexane [d] (M)	−0.576 (0.092)	−0.136 (0.009)	0.878	0.768	0.411	249	76

Table 2. *Cont.*

Partition System [a]	Regression Results [b]		Descriptive Statistics [c]				
	Intercept	Slope	*R*	*Adj. R*	*SE*	*F*	*n*
LDPE-lipid (B)	−1.453 (0.174)	−0.117 (0.016)	0.640	0.402	0.779	51	76
LDPE-lipid (I)	−1.521 (0.211)	−0.058 (0.020)	0.319	0.090	0.949	8	76
LDPE-lipid (J)	−0.537 (0.113)	−0.183 (0.011)	0.895	0.798	0.506	297	76
LDPE-lipid (K)	−1.561 (0.222)	−0.048 (0.021)	0.258	0.054	0.995	5	76
$R_{h/l}$ (Equation (4), B)	−0.231 (0.080)	0.179 (0.008)	0.941	0.884	0.357	573	76
$R_{h/l}$ (Equation (4), I)	−0.164 (0.072)	0.120 (0.007)	0.899	0.806	0.323	313	76
$R_{h/l}$ (Equation (4), J)	−1.148 (0.071)	0.245 (0.007)	0.974	0.947	0.319	1345	76
$R_{h/l}$ (Equation (4), K)	−0.124 (0.077)	0.110 (0.007)	0.870	0.754	0.346	230	76

[a] The letter in the parentheses is the system label from Table 1. [b] The values in the parentheses are the errors of the regression coefficients (95% confidence). [c] All notations the same as those in Table 1 (95% confidence). [d] No swelling of LDPE by hexane.

Table 3. Tabulation of LDPE-Lipid and LDPE-Hexane Partition constants of LDPE Extractables.

$log_{10}P_{o/w}$	LDPE-Hexane [a]			LDPE-Lipid [b] (B)		LDPE-Lipid [c] (I)		LDPE-Lipid [d] (K)	
	Fit	Fit + SE	Fit − SE	Fit	Fit + SE	Fit	Fit + SE	Fit	Fit + SE
0	0.02	0.10	0.004	0.0354	0.2122	0.0302	0.268	0.0276	0.272
2.5	0.03	0.14	0.006	0.0180	0.1083	0.0217	0.192	0.0209	0.206
5	0.04	0.20	0.009	0.0092	0.0553	0.0155	0.138	0.0159	0.157
7.5	0.06	0.29	0.013	0.0047	0.0282	0.0111	0.099	0.0121	0.119
10	0.09	0.41	0.018	0.0024	0.0144	0.0080	0.071	0.0091	0.090
12.5	0.12	0.59	0.026	0.0012	0.0073	0.0057	0.051	0.0069	0.068
15	0.18	0.85	0.037	0.0006	0.0037	0.0041	0.036	0.0053	0.052
17.5	0.25	1.21	0.052	0.0003	0.0019	0.0029	0.026	0.0040	0.039
20	0.36	1.72	0.075	0.0002	0.0010	0.0021	0.019	0.0030	0.030
22.5	0.51	2.46	0.107	0.0001	0.0005	0.0015	0.013	0.0023	0.023
25	0.73	3.52	0.153	0.00004	0.0003	0.0011	0.010	0.0017	0.017
27.5	1.05	5.03	0.218	0.00002	0.0001	0.0008	0.007	0.0013	0.013
30	1.50	7.18	0.312	0.00001	0.0001	0.0006	0.005	0.0010	0.010

[a] Calculated by the partition system (L) in Table 1 and the coefficients in Table 2. [b] Calculated by the partition system (B, MLR) in Table 1 and the coefficients in Table 2. The values in the Table are mostly used for a comparison purpose with the two Cycle models and best applicable for solutes without hydrogen-bond acidity. [c] Calculated by the partition system (I, Cycle-A) in Table 1 and the coefficients in Table 2. [d] Calculated by the partition system (J, Cycle-C) in Table 1 and the coefficients in Table 2. [a,b,c,d] The upper and lower bounds (Fit + SE, and Fit − SE) are calculated by the *SE* of each fit in Table 2. The "Fit" column is the predictive values by the model. The number of significant digits is at least 1. The *SE* is based on 95% confidence in the regression in Table 2.

3.8. Correlation of $log_{10}R_{h/l}$ and $log_{10}P_{o/w}$ for Observed Extractables Compounds for Model Verification

The $log_{10}R_{h/l}$ values are plotted against $log_{10}P_{o/w}$ for the observed extractables compounds. The linear correlation results and descriptive statistics between them are shown in Table 2 to evaluate the model suitability for the predictive purpose.

3.9. Collection of Solvent Absorption and Swelling Data for LDPE

To fully elucidate the extractables migration processes from LDPE to hexane and lipid, it is necessary to consider the impact of LDPE-lipid and LDPE-hexane interactions, as indicated by the solvent absorption or swelling data [33,83]. Table 4 shows the results of collection of LDPE by lipids, oils, oils and organic solvents [40,47,84–94].

Table 4. Tabulation of LDPE Solvent Absorption and Diffusion Coefficients.

Material	Solvent	Temperature (°C)	Solvent Absorption (w/w%) [a]	Diffusion Coefficient ($D \times 10^{-12}$ cm^2/s)
LDPE	Fish Oil	20	1.55	2.53
		47		3.26
	Triolein	20	2.58	5.79
		47		29.6
	Olive oil	40	1.6–2	250–300
	Olive oil [b]	40	<2 (~1.5)	
	Ethanol [b]		~0.2	
	Isopropanol [b]		<0.4	
	Ethyl acetate [b]		3	
	Isooctane [b]		15	
	Cyclohexane [b]		25	
	Tributyrin [b]		<1	
	Tricaprylin [b]		<2	
	Cyclohexane	40	29	34,000
	Isooctane		12	5200
	n-Heptane		14	24,000
	Isooctane	25		199
	Ethanol [d]	25		8.66
	Olive oil	121	10	
	Isooctane	40	4	
		60	10	
		80	19	
	Cyclohexane	10	10.88	20,300
		15	12.67	25,100
		25	18.14	33,400
	Benzene	10	9.09	40,600
		15	10.74	45,900
		25	13.07	63,600

Table 4. *Cont.*

Material	Solvent	Temperature (°C)	Solvent Absorption (w/w%) [a]	Diffusion Coefficient (D × 10^{-12} cm^2/s)
LDPE	n-Hexane	25	11	2500
	n-Heptane	NA	5–15	
	Ethanol [d]		<1	
	Corn oil or Triglycerides		2–5	
LLDPE [c]	Olive oil	40	1.3	690
	Isooctane	40	10.65	
	95% Ethanol [d]	40	0.45	
PP [c]	Olive oil	40	~2%	

The data are collected from references [40,47,84–94]. LLDPE: linear low-density polyethylene; PP: polypropylene. [a] These values are presented as weight % (w/w, %), after a conversion if necessary. [b] These values are presented as volume % (v/v, %) in the original reference. The value in the parenthesis is the weight percent (w/w%). [c] Additional (similar) materials are collected and included in the table for information and comparison purposes. [d] Ethanol (or 95% ethanol) solvent is collected and included in the table for information and comparison purposes.

3.10. Estimation of Over-Extraction Data of LDPE by Alkanes

The extent of over-extraction (*OE*) of hexane (or alkanes in general) to lipid are estimated by direct overall migration or specific migration data or extractables diffusion coefficients, as described in Section 2.9. The *OE* values are tabulated in Table S5 [14,15,20,22,34,40,41,87,95–98]. It is also noted that, although the nonpolar organic solvent focused in this study is hexane, any reported data with alkane solvents are collected to determine *OE* because the extraction properties of all alkanes are similar toward nonpolar polymers such as LDPE (details are not shown). An example of this conclusion is shown in Figures S1–S3 between LDPE and HDPE. The kinetic extraction profiles of HDPE by solvents are similar to those of LDPE (see Section 4.8.2).

4. Results and Discussion

4.1. Correlation of LDPE-Lipid Partition Constant by Solvation Model

The construction of predictive models for LDPE-lipid partition constants is proceeded by two ways: (1) correlation of experimental data by the solvation model and (2) thermodynamic cycle conversion. They are discussed in the following sections.

4.1.1. The Space of Abraham Solute Descriptors

To construct the solvation models (Equation (1)) for the partition constant of $log_{10}P_{ldpe/lipid}$, the descriptors of these solutes (Table S1) are required and computed, as described in Section 3.4. As the range or space of each descriptor is important in the construction of a reliable prediction model [57], it is necessary to first examine them judiciously. Figure 1 shows the descriptor spaces for three sets of compounds: (a) the solutes in Wayne State University Database (WSUD, 293 compounds) [82], (b) solutes with measured $log_{10}P_{ldpe/lipid}$ (Table S1, 78 solutes), and (3) the experimentally observed LDPE extractables by solvent extraction (Table S2, 76 compounds). The descriptors of each set of compounds are plotted as box-overlap graph in Figure 1, including an inserted figure on the distribution of $log_{10}P_{ldpe/lipid}$ of the measured compounds (Table S1). The inclusion of WSUD is used as the reference for the adequate range of each descriptor, as the descriptors in WSUD were experimentally measured using the same protocol and calibration scheme, and they are considered as adequate in the construction of the solvation models due to their range and accuracy. The range of each descriptor in WSUD is as follows: *E* from −0.45 to 4.47, *S* from −0.11 to 2.51, *A* from 0 to 1.38. *B* from 0 to 2.06, and *V* from 0.51 to 3.79. Moreover, the hydrophobicity range of compounds in WSUD in $log_{10}P_{o/w}$ is from −1 to 13.6, calculated by Equation (6).

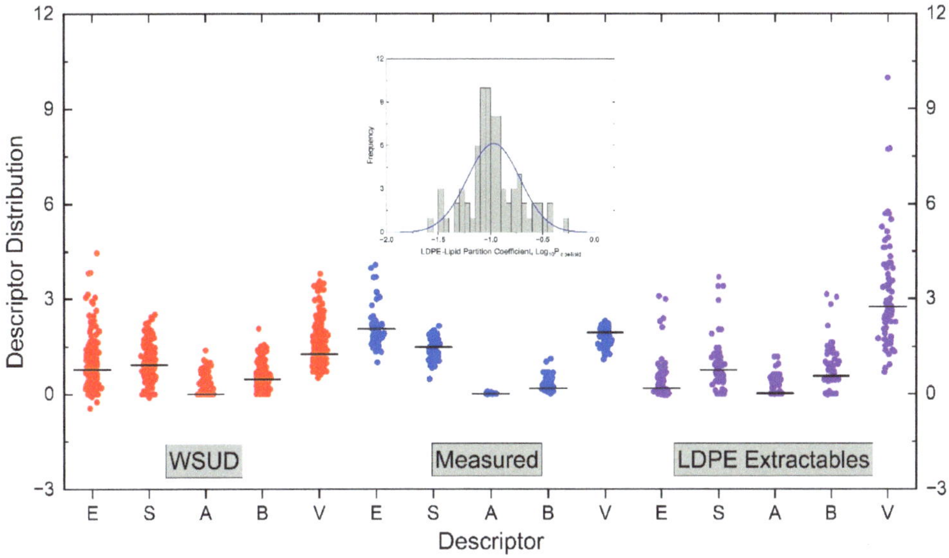

Figure 1. Illustration of Abraham solute descriptors for three sets of compounds by box-overlap graph: WSUD: Wayne state university database [82]; compounds with measured $log_{10}P_{ldpe/lipid}$ [47]; observed LDPE extractables (Section 3.2). The descriptors of WSUD are measured values; while those of measured solutes and observed LDPE extractables are computed by an online program: RMG [81]. The solid black horizontal line in each distribution is the median value. The inserted graph on top is the distribution of the measured $log_{10}P_{ldpe/lipid}$ values for 78 compounds.

It is seen in Figure 1 that the span of each descriptor in WSUD is wide, only the *A* descriptor has a small median value. The span of each descriptor is at least 1.4. They are used as the references for the other two sets of solutes. The descriptors for the measured solutes in Table S1 are in general narrower, particularly for the *A* descriptor (its span is less than 0.1, from zero to 0.08), because no hydroxy or carboxylic compounds are included in the measurement [47]. The consequence of this narrow *A* descriptor is the statistical uncertainty in the *a* constant in the multiple linear regression (MLR) by Equation (1) to construct the solvation model. The *B* descriptor is reasonably wide and acceptable, but the median value is small. It is also seen in Figure 1 that the space of each descriptor of the observed LDPE extractables are in general wide and acceptable compared to the WSUD, although the *A* descriptor is also in a narrow space with a small median value. The *V* descriptor of the observed LDPE extractables is wide, due to the observation of high MW polymer additives, such as Irganox 1010, in extraction solvents.

It should be mentioned that the range of descriptors of the measured solutes are more important than those for the observed LDPE extractables, because they are used to construct the predictive models. The correlation matrix for the measured solutes and observed LDPE extractables are shown in Tables S3 and S4. It can be seen in Table S3 that, although the *A* descriptor space is narrow, the correlation among the descriptors is in general excellent [57].

4.1.2. Construction by the Multiple Linear Regression (MLR)

The experimental $log_{10}P_{ldpe/lipid}$ data collected for 78 organic compounds are used to construct the predictive model (Table S1). The distribution of the $log_{10}P_{ldpe/lipid}$ data are shown as an inserted figure in Figure 1. It can be seen in Figure 1 that the distribution of $log_{10}P_{ldpe/lipid}$ is narrow from −1.55 to −0.29, about 1.3 log units. The range of $log_{10}P_{ldpe/lipid}$ values and the descriptor spaces for the measured values are not considered ideal for constructing the correlation model [57]; however, this is the only available dataset with adequate accuracy and number of solutes. The range of $log_{10}P_{ldpe/lipid}$, of course, should be narrow, because the two partitioning phases are chemically similar.

The experimental data (Table S1) are correlated with model Equation (1) by MLR. Two correlation options are adopted, with and without the aA term (Sections 2.2 and 3.5). The correlation results and descriptive statistics are shown in Table 1 (Systems A and B). It can be seen in Table 1 that the a constant is 1.475 if included in the regression, and it is not statistically significant, as indicated by the p-value (>0.05), consisting with the lack of A descriptor values in the dataset. When the aA term is excluded in the linear regression, the correlation results and statistics are barely changed. The SE is about 0.12 (SE/range < 10%) in both cases, and the F-values are 57 (with aA term) and 69 (without aA term), respectively. The correlation coefficient is low because the range of the $log_{10}P_{ldpe/lipid}$ is narrow. Overall, the correlation statistics by Equation (1) is considered acceptable with the limited range of the data.

The predicted $log_{10}P_{ldpe/lipid}$ values by MLR for the compounds in Table S1 are plotted against the measured values, and they are shown in Figure 2 (plot A). Also included in Figure 2 are the standardized residual distribution as plot B. It is seen in Figure 2 that the data points in plot A are symmetrically distributed along the diagonal line, and the absolute standardized residuals are within 3 (plot B). Thus, the constructed MLR model is considered as adequate in predicting the $log_{10}P_{ldpe/lipid}$ of compounds without a significant HB acidity contribution. The model is used to explore the range of $log_{10}P_{ldpe/lipid}$ for the observed LDPE extractables in Section 4.2 (Table S2).

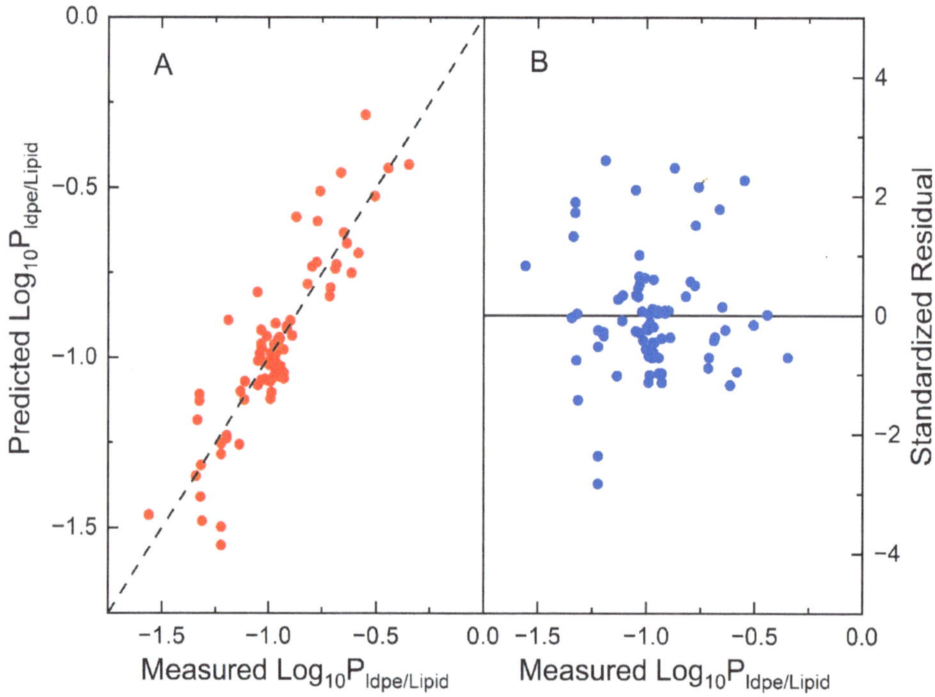

Figure 2. Illustration of the correlation results by the solvation model for the measured compounds [47]. Plot (**A**): plot of the predicted LDPE-lipid partition constants ($log_{10}P_{ldpe/lipid}$) against the measured values. Plot (**B**): the standardized residual distribution.

4.1.3. Construction by the Thermodynamic Cycle Conversion

The distribution of extractables between LDPE and lipids can also be derived using LDPE-water and lipid-water partition system by a thermodynamic cycle method

(Sections 2.2 and 3.5), because the LDPE and lipids are considered as immiscible. Although there is a slight lipid penetration to the LDPE phase, it is negligibly low [47], as discussed in Section 4.8.1. Due to the difficulty in measuring the partition constants, the thermodynamic cycle method can be a useful and acceptable approach to evaluate this distribution. Table 1 also tabulates the system constants for LDPE-water, lipid-water, hexane-water partition systems. Three sets of LDPE-water systems are listed in Table 1 (Systems D, E, and F), and they are all used to derive the system constants for the LDPE-lipid system (Cycle-A to C).

Comparing the MLR model and Cycle-A to C models (total four models) in Table 1, it is apparent that the a constant in the MLR model should be negative, as it is consistently negative in the Cycle models. Furthermore, the sign of the a constant should be consistent with that of the b constant in general. The e (or δe) constant in all equations is positive and the v (or δv) constant in all models is negative. It is noted in Cycle-B that the b constant (0.103) is positive, but the error bars for both a and b are larger than their values, indicating that they are statistically insignificant. But it is used further for a comparison purpose, as discussed in the following sections.

It is worth noting regarding the four models that the positive (e or δe) constant suggests that LDPE is more prone to generate dispersion interactions than lipids to interact with the solute through n and π-electron pairs [57]. The negative (s or δs) constant suggests LDPE having a lower ability to interact with dipolar/polarizable solutes. The negative (a or δa) and (b or δb) constants reveal LDPE having a lower ability to interact with solutes through the hydrogen bonding interactions, and the negative v (or δv) reflects LDPE have a lower ability to interact with solutes from the dispersion interactions and cavity effects. These results are sensible considering the fact that the lipid phase has water and polar components (the ester function group) [51,70], and it makes sense that all S, A, and B constants are negative. The ester groups in lipids are hydrogen bond acceptors and can interact with more polar solutes. The negative v constant in all models indicates that the $log_{10}P_{ldpe/lipid}$ is becoming smaller as the hydrophobicity increase. This is a contrast to the of $log_{10}P_{ldpe/hexane}$ from the previous study [33].

To evaluate the similarity and differences among the four models, the partition constants of measured compounds are computed by each model (Table 1) and descriptors in Table S1. The calculated values are plotted against the predicted values by MLR model (without the aA term), as it best represents the measured data. It is noted that these compounds do not possess significant hydrogen-bonding acidity. The results of the calculation are shown in Figure 3, together with the calculation results using the MLR model, adding the aA term with a being -1.271 (taken from Cycle-A). It is obvious in Figure 3 that the predicted values by MLR with aA term are all aligned along the diagonal line (plot A), because these compounds have essentially no hydrogen bonding acidity. The data points by the Cycle-A model are in general distributed above the diagonal line, except for several compounds at higher $log_{10}P_{ldpe/lipid}$ values. The data points by Cycle-B model are in general distributed above the diagonal line as well, but no trending relationship with the MLR model can be observed. The data points by Cycle-C model are distributed essentially the same as that of Cycle-A, except that all predicted $log_{10}P_{ldpe/lipid}$ values are higher than those predicted by the MLR model. If the highest predictive limit of $log_{10}P_{ldpe/lipid}$ is the criterion of model acceptance, Cycle-C model is considered as the best. This point is discussed further in Section 4.5 using the measured LDPE extractables (Table S2).

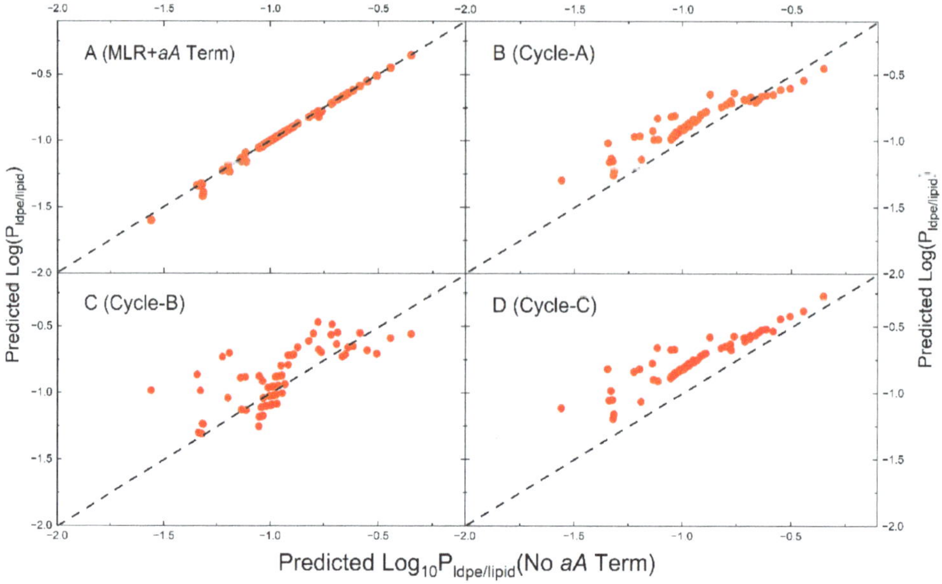

Figure 3. Comparison of predicted $log_{10}P_{ldpe/lipid}$ values for the measured solutes by four models. The X-axis values in all plots are calculated by MLR model (System B in Table 1). Plot (**A**): the predicted values by the MLR model with an addition of the *aA* term (A = −1.271 from System I); Plot (**B**): the predicted values by the Cycle-A model (System I); Plot (**C**): the predicted values by the Cycle-B model (System J); Plot (**D**): the predicted values by the Cycle-C model (System K). A diagonal line is shown in each plot.

4.2. Calculation of LDPE-Lipid and LDPE-Hexane Partition Constant Range for Collected LDPE Extractables

Based on the predictive models established in the section above (Table 1), the range of $P_{ldpe/lipid}$ can be determined for the collected LDPE extractables (Table S2). Figure 4 shows the results of the calculations and plots against $log_{10}P_{o/w}$ values. Also included in the figure are the predicted $P_{ldpe/hexane}$ values for the same LDPE extractables using two equations, one established in a previous publication [33], and the other one by a thermodynamic cycle between $log_{10}P_{ldpe/water}$ and $log_{10}P_{hexane/water}$ (System L and M in Table 1), as discussed in Section 2.3. It is noted that the effect of LDPE swelling by hexane is absent in System M. It is seen in Figure 3 that the logarithm of $P_{ldpe/lipid}$ is trending towards a decrease with an increase in the $log_{10}P_{o/w}$ by MLR and Cycle models (plot A, B, C and D). The extent of the decrease is strongly correlated with v constant. The more negative v constant in Cycle-B model (plot B), the sharper decrease with $log_{10}P_{o/w}$. However, the logarithm of $P_{ldpe/hexane}$ shows an increasing trend with the $log_{10}P_{o/w}$ with a LDPE swelling (plot E). If the effect of swelling by hexane on LDPE is not considered, $P_{ldpe/hexane}$ decreases obviously with the $log_{10}P_{o/w}$ (plot F).

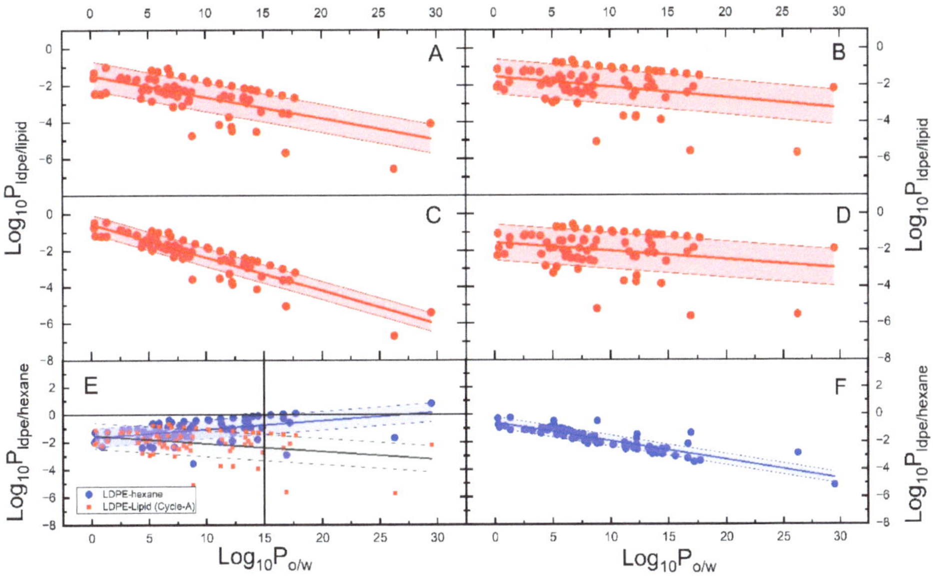

Figure 4. Illustration of the relationship between $log_{10}P_{ldpe/s}$ (s = lipids and hexane) against $log_{10}P_{o/w}$ as the independent variable for the observed LDPE extractables. Plot (**A**): $log_{10}P_{ldpe/lipid}$ by MLR model; Plot (**B**): $log_{10}P_{ldpe/lipid}$ by Cycle-A model; Plot (**C**): $log_{10}P_{ldpe/lipid}$ by Cycle-B model; Plot (**D**): $log_{10}P_{ldpe/lipid}$ by Cycle-C model; Plot (**E**): $log_{10}P_{ldpe/hexane}$ by System L in Table 1 (with swelling); Plot (**F**): $log_{10}P_{ldpe/hexane}$ by System M in Table 1 (no swelling). The solid line in each plot is the linear regression line between $log_{10}P_{ldpe/s}$ (s = lipids and hexane) vs. $log_{10}P_{o/w}$. The lines above and below the regression lines are the limits of the data by *SE* of the regression with 95% confidence. Also included in Plot (**E**) is the $log_{10}P_{ldpe/lipid}$ by Cycle-A model for the evaluation of the data overlap between $log_{10}P_{ldpe/lipid}$ and $log_{10}P_{ldpe/hexane}$. The vertical line at $log_{10}P_{o/w}$ = 15 in plot (**E**) indicates the full divergence between $log_{10}P_{ldpe/lipid}$ and $log_{10}P_{ldpe/hexane}$. The horizontal line at $log_{10}P_{ldpe/s}$ = 0 in plot (**E**) indicates the $log_{10}P_{ldpe/s}$ = 1 to show the distribution of data points.

As the representation of extractables physicochemical properties in this study needs to be $log_{10}P_{o/w}$, and there is not a direct thermodynamic relationship between $log_{10}P_{o/w}$ and $log_{10}P_{ldpe/s}$ [33], a simple empirical equation is used instead to quantitatively describe the relationship to enable the prediction of LDPE-solvent partition constant for LDPE extractables of different hydrophobicity. The (empirical) linear regression results and statistics between $log_{10}P_{ldpe/s}$ (s = hexane or lipid) and $log_{10}P_{o/w}$ are shown in Table 2. It is seen in the table that all intercepts and slopes are statistically significant, as all *p*-values are much smaller than 0.05. As expected, the *SE* is large and *F*-values are small (<100), because this is an empirical relationship. It is noted that the *F*-value by hexane without swelling is the largest. It is also worth noting in Figure 4 that the shaded area above and below the linear regression line bounded by the *SE* covers the range of $P_{ldpe/lipid}$ and $P_{ldpe/hexane}$ values across the hydrophobicity range in each plot.

4.3. Dependence of $log_{10}P_{ldpe/lipid}$ on $log_{10}P_{o/w}$ for LDPE Extractables

It has been reported that $log_{10}P_{ldpe/lipid}$ could be linearly related to $log_{10}P_{o/w}$ as follows [44,45]:

$$log_{10}P_{ldpe/lipid} = 1.17 + 0.056 \cdot log_{10}P_{o/w} \quad (11)$$

Equation (11) was established with 16 compounds and the range of $log_{10}P_{o/w}$ was from 0.5 to 18. The lipid was represented by olive oil. However, a different conclusion has been demonstrated in a recent study [47], and $log_{10}P_{ldpe/lipid}$ has no relationship, or (weakly) negative relationship with $log_{10}P_{o/w}$.

The $log_{10}P_{ldpe/lipid}$ data in Table S1 are linearly correlated with $log_{10}P_{o/w}$ as well, and the results are tabulated in Table 1, as System C. It is seen in Table 1 that the slope of the regression is 0.005 with a *p*-value of >0.05, thereby indicating that a statistically significant relationship between $log_{10}P_{ldpe/lipid}$ and $log_{10}P_{o/w}$ cannot be established.

The regression results between $log_{10}P_{ldpe/lipid}$ and $log_{10}P_{o/w}$ for the observed LDPE extractables by the four models in Table 2 all indicate a small negative slope, with large *SE* and small *F*-values, which indicates that $log_{10}P_{ldpe/lipid}$ is weakly and negatively proportional to $log_{10}P_{o/w}$. However, $log_{10}P_{ldpe/hexane}$ shows a weakly and positively correlation with $log_{10}P_{o/w}$ (System L). The slopes and intercepts of the four models are used to determine the upper limits of $P_{ldpe/lipid}$ in Section 4.5.

Overall, $log_{10}P_{ldpe/lipid}$ is weakly and negatively dependent on $log_{10}P_{o/w}$ by the compounds in Tables S1 and S2 with a large *SE*. The opposite is true for $log_{10}P_{ldpe/hexane}$. The conclusions of this section have been reported for other solvents [22,23], as shown in Figure S4.

4.4. Model Validation

To confirm if the four models (MLR and Cycle-A to -C) are acceptable to predict $log_{10}P_{ldpe/lipid}$ for the measured LDPE extractables, the ratio ($R_{h/l}$) of $P_{ldpe/hexane}$ and $P_{ldpe/lipid}$ are determined by $log_{10}P_{ldpe/hexane}$ and $log_{10}P_{ldpe/lipid}$ using Equation (4), as discussed in Section 2.6. It is then plotted against $log_{10}P_{o/w}$, as suggested by Equation (5). If the $P_{ldpe/hexane}$ and $P_{ldpe/lipid}$ were correct, the $log_{10}R_{h/l}$ should be a linear relationship with $log_{10}P_{o/w}$, because it theoretically related to the same water phase by $log_{10}P_{lipid/water}$ and $log_{10}P_{hexane/water}$. The derivation of Equation (3) ($log_{10}P_{ldpe/hexane}$, System L in Table 1) was not based on water-related phase [33]. The MLR model is not based on a partition system involving water. Figure 5 plots the $log_{10}R_{h/l}$ against $log_{10}P_{o/w}$ by MLR and Cycle-A to -C (plots A, B, C and D) for the $P_{ldpe/lipid}$, and Equation (3) as the $log_{10}P_{ldpe/hexane}$ (with swelling). Also included in Figure 5 are the plot (E and F) using System M (Table 1) to calculate $log_{10}P_{ldpe/hexane}$ (no swelling).

It can be seen in Figure 5 that there is an excellent linear relationship between $log_{10}R_{h/l}$ and $log_{10}P_{o/w}$ using Equation (3) to calculate $log_{10}P_{ldpe/hexane}$; however, no relationship is expected when System M (Table 1) is used to represent $log_{10}P_{ldpe/hexane}$. The plots in Figure 5 clearly demonstrate the influence of LDPE swelling by hexane on the extractables distribution (System M is not discussed further). Furthermore, the data in plots A–D show no sign of curvatures. The data in plots A–D are linearly correlated and the results of correlation and descriptive statistics are shown in Table 2 as well. It can be seen in Table 2 that the intercepts and slopes are statistically significant. The *SE* is between 0.32–0.36, and *F*-values are all high. Overall, although the $log_{10}P_{ldpe/hexane}$ and $log_{10}P_{ldpe/lipid}$ are not derived based on the same (water) phase (either both or one system), the excellent linear relationship still exists. This clearly indicates that the MLR and Cycle models (and Equation (3)) are all suitable for the prediction of $log_{10}P_{ldpe/hexane}$ and $log_{10}P_{ldpe/lipid}$ of the observed LDPE extractables.

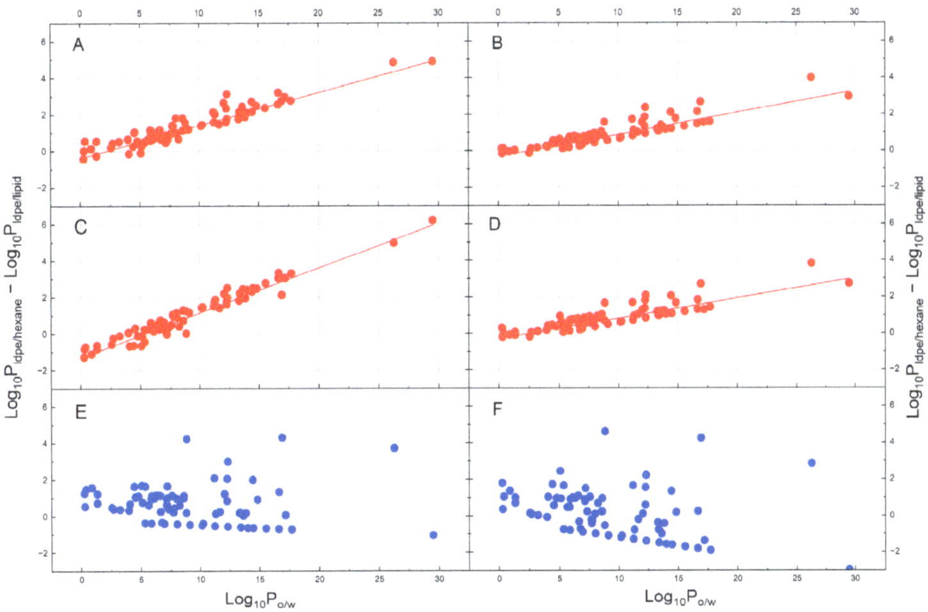

Figure 5. Illustration of the relationship between $log_{10}R_{h/l}$ (h = hexane and l = lipids, Equation (4)) against $log_{10}P_{o/w}$ for the observed LDPE extractables. The $log_{10}P_{ldpe/hexane}$ by System L (with swelling) is used in the calculation of plots (**A**–**D**). Plot (**A**): $log_{10}P_{ldpe/lipid}$ by MLR model; Plot (**B**): $log_{10}P_{ldpe/lipid}$ by Cycle-A model; Plot (**C**): $log_{10}P_{ldpe/lipid}$ by Cycle-B model; Plot (**D**): $log_{10}P_{ldpe/lipid}$ by Cycle-C model. The $log_{10}P_{ldpe/hexane}$ by System M (no swelling) is used in the calculation of plots (**E**,**F**). Plot (**E**): $log_{10}P_{ldpe/lipid}$ by MLR model; Plot (**F**): $log_{10}P_{ldpe/lipid}$ by Cycle-A model.

4.5. Comparison of MLR and Cycle Models in Predicting LDPE-Lipid Partition Constants for Observed LDPE Extractables

To further evaluate the similarities and differences among the four $log_{10}P_{ldpe/lipid}$ models, the $log_{10}P_{ldpe/lipid}$ values for the observed LDPE extractables (Table S2) are computed by each model. The predicted values by Cycle models are plotted against those by the MLR model in Figure 6. Also included in Figure 6 is a plot (A) by the calculated values with the same MLR model with an addition of the aA term (a = −1.271, same as in Figure 3). It is seen in Figure 6 (plot A) that some of the data points by the MLR model with the aA term are below the diagonal line, because this term is negative and decreases $log_{10}P_{ldpe/lipid}$ values for solutes with a significant hydrogen- bond acidity, as expected. The data points predicted by the Cycle-A and Cycle-C models are in general distributed above the diagonal line, except for a few compounds below the line, and they are essentially the same. The prediction results by Cycle-B show lower results for alkane extractables due likely to the small negative a and positive b constant in the model. Thus, Cycle-A and C models generally predict higher values than those by the MLR model.

Finally, model Cycle-A and Cycle-C are used to predict the upper limits of LDPE-lipid partition constant in the following section.

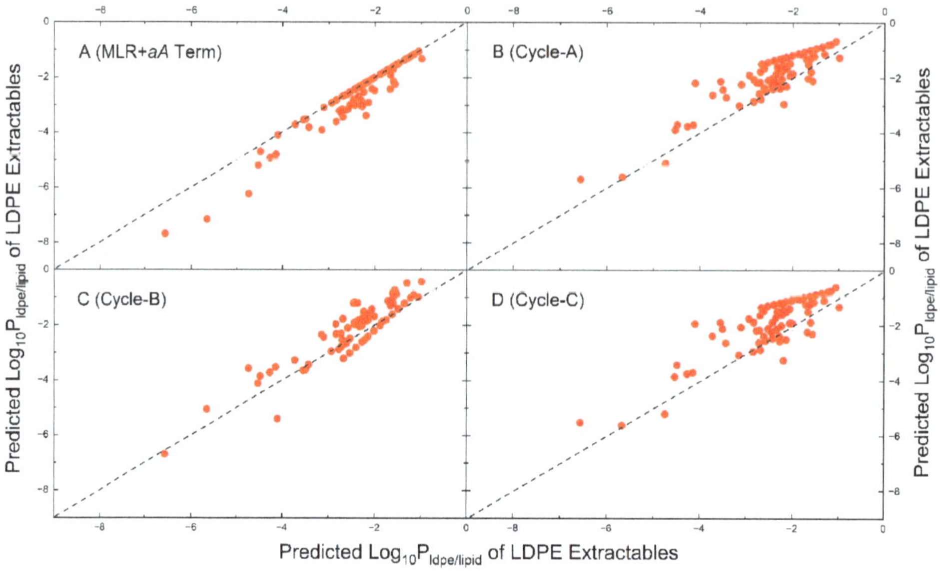

Figure 6. Comparison of calculated $log_{10}P_{ldpe/lipid}$ values for the observed LDPE extractables by four models. The notations and descriptions are the same as those in Figure 3.

4.6. Comparison of LDPE-Hexane and LDPE-Lipid Partition Constants for LDPE Extractables

The ranges and values of LDPE-hexane and LDPE-lipid partition constants are determined by the empirical regression coefficients in Table 2 (systems L, B, I, J, and K) using $log_{10}P_{o/w}$ from zero to 30. The results of the calculation are tabulated in Table 3. The lower and upper bounds are also included using the *SE* value of each regression (95% confidence).

It can be seen in Table 3 that the range of $P_{ldpe/hexane}$ predictive values is from 0.02 at $log_{10}P_{o/w} = 0$ to 1.50 at $log_{10}P_{o/w} = 30$. The upper bound at the same hydrophobicity range is from 0.10 to 7.18. The range of $P_{ldpe/lipid}$ predictive values by Cycle-A model, for example, is from 0.0302 at $log_{10}P_{o/w} = 0$ to 0.0006 at $log_{10}P_{o/w} = 30$. This range of $P_{ldpe/lipid}$ values is considered small, thereby indicating that extractables distribution between LDPE and lipids are highly favorable towards the lipid phase, due to the lipid physicochemical properties, as discussed in Section 4.1.3.

It should be emphasized that the $P_{ldpe/lipid}$ values are in general small, and the exact values may not be as important as the upper bounds from a practical extraction perspective (see Equation (7)). That is why the upper bound of the $P_{ldpe/lipid}$ values are shown in Table 2, and why the four predictive models are compared in Figures 4 and 6 to understand which one predicts the highest values and upper bounds. The range of the upper bound of $P_{ldpe/lipid}$ values by Cycle-A model is from 0.268 to 0.005 over the same hydrophobicity range. Additionally, the range of the upper bound of $P_{ldpe/lipid}$ values by Cycle-C model is from 0.272 to 0.01. Based on these values, Cycle-C model should be used to predict the $P_{ldpe/lipid}$ values, consistent with the conclusion in Section 4.1.

It is thus concluded that the assumption of $P_{ldpe/s}$ (s = lipids and hexane) is less than 1 or equal to 1 is not appropriate for hexane solvent when $log_{10}P_{o/w}$ is greater than ~15. It is also noted that the predictive values of $P_{ldpe/lipid}$ in this section and the developed methods in this study can be critically important in environmental sampling. Additionally, the $P_{ldpe/hexane}$ values derived in this study can be valuable in the prediction of extractables release [48].

4.7. Thermodynamic Contribution to Over-Extraction of LDPE by Hexane

Based on the ranges and values of $P_{ldpe/hexane}$ and $P_{ldpe/lipid}$ from Section 4.6, the distribution of extractables between LDPE and hexane/lipids highly favor the solvent phase in general, except for the very hydrophobic solutes by hexane solvent. As illustrated in plot E of Figure 4, the $P_{ldpe/hexane}$ values are overlapped nearly completely with the $P_{ldpe/lipid}$ values, when their $log_{10}P_{o/w}$ values are less than about 6. The $P_{ldpe/hexane}$ values are totally larger than $P_{ldpe/lipid}$ values, when $log_{10}P_{o/w}$ is higher than ~15 (see also Table 3). It is therefore concluded that the extractables distribution between LDPE and solvents, hexane and lipids, are approximately the same and favors the solvent phase ($P_{ldpe/s} < 1$), when their $log_{10}P_{o/w}$ values are less than about 6, after which the distribution to lipids are becoming more favorable thermodynamically than to hexane. Overall, the partitioning of LDPE extractables into lipids is the same as or more favorable than into hexane. This point can be further illustrated by calculating the amount extracted at equilibrium by both solvents using the data in Table 3 by Equation (7). For example, the upper limit of $P_{ldpe/s}$ value by hexane at $log_{10}P_{o/w}$ = 15 is 0.85, while that of lipids is 0.052. The fractions extracted by hexane and lipids are 54% and 95%, respectively, assuming that the LDPE and solvent phase volumes are the same. The fractions extracted by hexane and lipids at $log_{10}P_{o/w}$ = 30 are 12% and 99%, respectively, by the same calculation. Therefore, lipids are better extraction solvents than hexane from a thermodynamic consideration. The over-extraction by alkane solvents or any nonpolar organic solvents towards LDPE should not be the thermodynamic effect, but the kinetic effect, consistent with the previous assumption [20] and further supported by the discussion in Section 4.8.

4.8. Determination of Over-Extraction of LDPE by Hexane over Lipid

The thermodynamics of extractables distribution between LDPE and solvents is the focus up to this point. The contribution to over-extraction is concluded to be mainly the kinetic effect from LDPE to lipids relative to hexane. However, this effect is closely related to solvent-material interactions or swelling of material by alkanes and lipids [33,83].

4.8.1. Material-Solvent Interactions

During an extraction of materials with solvents, the polymer will swell to a certain extent, depending on the degree of polymer-solvent network interactions. The degree of swelling for LDPE by nonpolar solvents and lipids (or vegetable oils) are collected and tabulated in Table 4, together with the diffusion rates of solvents into materials. Also included in the table are the swelling data by alcohols and two more related materials (LLDPE and PP) for a comparison purpose. It can be seen in Table 4 that the degree of swelling of LDPE by alcohols (semipolar) is <2%, typically 10–30% by nonpolar solvents, and less than 3% by lipids/oils (at ambient temperature). The degree of material-solvent interactions and swelling strongly affect the effective diffusion coefficient of both extraction solvents and LDPE extractables, as interpretated by the free-volume theory [83]. Of course, the viscosity of the solvents also plays a role in the diffusion rate.

The solvent diffusion rates collected in Table 4 are plotted in Figure 7. It can be seen in the figure that the change in solvent diffusion rate by alkanes to lipids can be over a thousand-fold using the median values. The same range has been documented for larger MW extractables [20]. The increased diffusion rate of extractables by organic solvents relative to lipids can affect their release kinetics, as discussed in the next section.

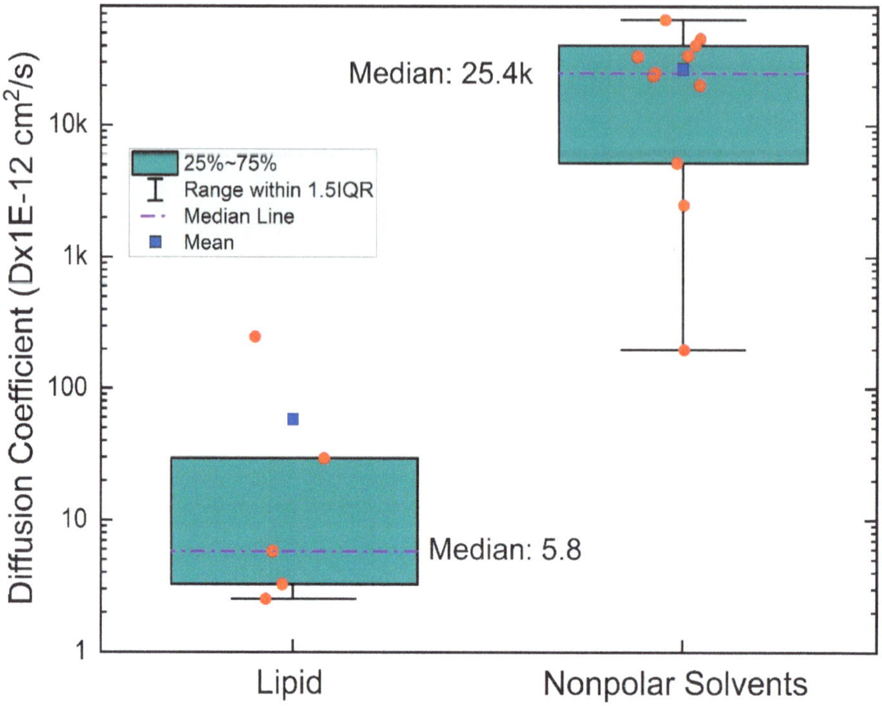

Figure 7. Illustration of box-whisker plots for the solvent diffusion coefficient. The solvent in the left plot is lipids/oils; and the solvent in the right plot is hexane (or nonpolar organic solvents). The median values are indicated in the graph.

4.8.2. Time-Dependent LDPE Extractables Release Profiles

To demonstrate the solvent effect in the release of LDPE extractables into lipids and alkane solvents, Figures 8–10 show the extraction kinetics of butylated hydroxytoluene (BHT), n-octadecane, and n-dotriacontane, respectively [46]. Similar graphs by HDPE material are shown in Figures S1–S3.

It is seen in Figure 8 that the release of BHT is a diffusion-controlled process by all solvents, because an approximately linear relationship can be observed by the log-log scale [25,26]. The full extraction can be achieved by all solvents; however, the time to reach a full extraction is different. For example, it takes 0.4 day by heptane solvent, 40 days by ethanol/corn oil/trioctanoin, and nearly 400 days by water. Additionally, the release rate by ethanol, corn oil, and trioctanoin are essentially the same. This is the reason why ethanol (or 95% ethanol) is used in the food industries to simulate fats for polyolefin materials [13,15]. It is also the basis that Equation (8) is used to estimate the over estimation in this study.

Similar conclusions can be drawn for the extraction of n-octadecane (Figure 9) and n-dotriacontane (Figure 10). The release kinetics are becoming slower due to the increase in the MW from BHT to n-dotriacontane.

It is also pointed out from the three figures that the over-extraction should be the vertical distance between the release curves of heptane and lipids/oils/ethanol within their linear ranges.

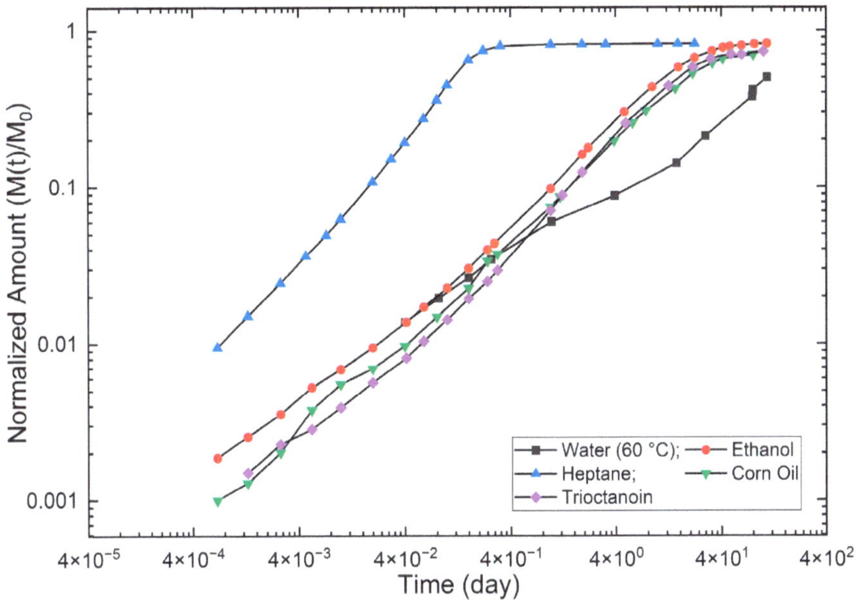

Figure 8. Illustration of BHT extraction release kinetic profiles from LDPE material by different solvents [46]. The X-axis and Y-axis are in 10-based log scale to show a much broad range of time duration (day). The Y-axis is the fraction (or amount at time t, $M_{(t)}$) extracted to the total amount in the material (M_t). The solvents are indicated in the figure. The temperature of the extraction is 30 °C, except for water (60 °C). The thickness of the material is 0.061 cm.

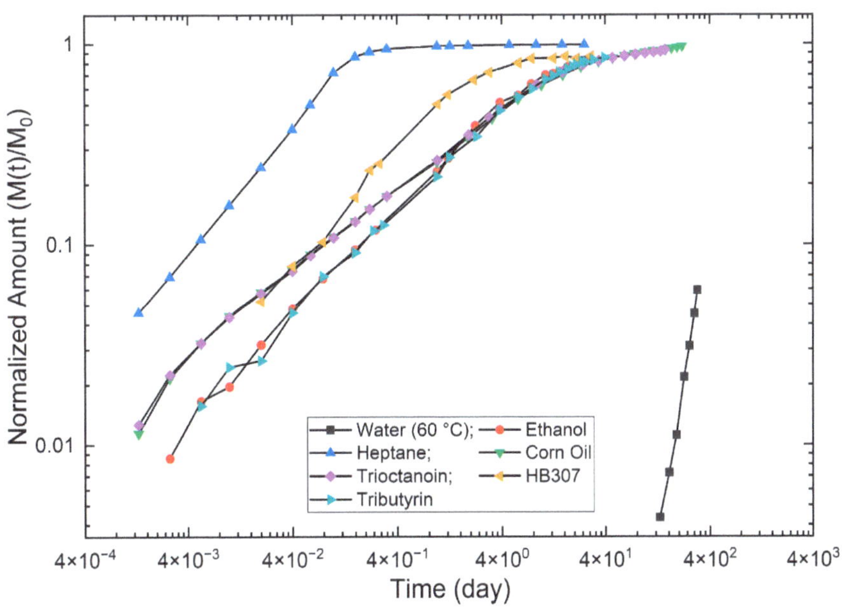

Figure 9. Illustration of C18 (n-octadecane) extraction release kinetic profiles from LDPE material by different solvents [46]. All notations the same as those in Figure 8. The thickness of the material is 0.072 cm.

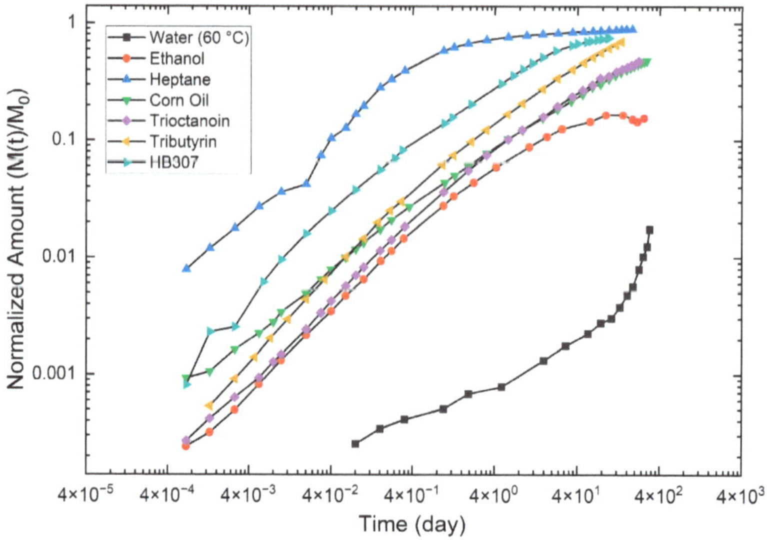

Figure 10. Illustration of C32 (n-dotriacontane) extraction release kinetic profiles from LDPE material by different solvents [46]. All notations the same as those in Figure 8. The thickness of the material is 0.028 cm.

4.8.3. Over-Extraction by Alkanes

The over-extraction data are compiled, as outlined in Section 3.10, and they are presented in Table S5. The data in Table S5 are plotted in Figure 11 as a histogram plot. It can be seen in Figure 11 that the median value of over-extraction is about 7-fold, and as high as 100-fold can be expected for high MW extractables. This is a result of a nearly 1000-fold increase in the corresponding diffusion coefficients of high MW extractables into alkane solvents relative to lipids [20].

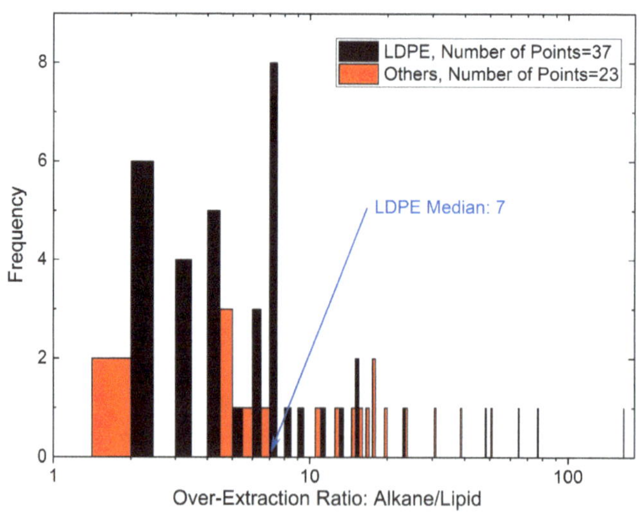

Figure 11. Illustration of histograms of over-extraction of LDPE and other nonpolar materials by hexane (or alkanes). The data are shown in Table S5. The median value by LDPE material is 7. The number of data points is indicated in the figure.

5. Conclusions

The over-extraction of extractables from medical devices by nonpolar organic solvents relative to biological lipids are evaluated from both thermodynamic and kinetic perspectives, based on the Abraham solvation model and solvent-material interactions, using LDPE as representative material. Abraham solvation models are established by either MLR or thermodynamic cycle conversion to correlate the LDPE-lipid partition constant using the measured LDPE-lipid partition constants. The constructed models for $log_{10}P_{ldpe/lipid}$ and $log_{10}P_{ldpe/hexane}$ are used to calculate and compare the ranges and values of $P_{ldpe/s}$ (s = lipids and hexane) for the observed LDPE extractables. The kinetic contributions to over-extraction are evaluated by the material swelling and solvent diffusion rate. It is concluded from the study that the kinetic effect, rather than the thermodynamic effect, is the main reason for the over-extraction. Finally, the extent of over-extraction can be up to over 100-fold with a median value of 7-fold.

Supplementary Materials: The following supporting information can be downloaded at: https://www.mdpi.com/article/10.3390/liquids4010006/s1. Figure S1: Time-Dependent BHT Release Profiles by Different Solvents from HDPE Material., Figure S2: Time-Dependent C18 Release Profiles by Different Solvents from HDPE Material., Figure S3: Time-Dependent C32 Release Profiles by Different Solvents from HDPE Material., Figure S4: Illustration of the Dependence of LDPE (or HDPE)-Solvent Partition Coefficient on $log_{10}P_{o/w}$ of Extractables.; Table S1: Tabulation of Analytes with Measured Partition coefficients (P_(ldpe/lipid)) and Their Physicochemical Properties., Table S2: Tabulates the collected LDPE extractables and their physicochemical properties by organic solvent extractions., Table S3: Correlation Matrix of Descriptors of Solutes with Measured $P_{ldpe/lipid}$, Table S4: Correlation Matrix of Observed LDPE Extractables Descriptors., Table S5: Tabulates the over extraction data, compiled from literatures.

Funding: This research received no external funding.

Data Availability Statement: Data are contained within the article and Supplementary Materials.

Conflicts of Interest: The Author Jianwei Li was employed by the company Chemical Characterization Solutions. The author declares that the research was conducted in the absence of any commercial or financial relationships that could be construed as a potential conflict of interest. The funders had no role in the design of the study; in the collection, analyses, or interpretation of data; in the writing of the manuscript; or in the decision to publish the results.

References

1. Reeve, L.; Baldrick, P. Biocompatibility assessments for medical devices—Evolving regulatory considerations. *Expert Rev. Med. Devices* **2017**, *14*, 161–167. [CrossRef] [PubMed]
2. Cuadros-Rodríguez, L.; Lazúen-Muros, M.; Ruiz-Samblás, C.; Navas-Iglesias, N. Leachables from plastic materials in contact with drugs. State of the art and review of current analytical approaches. *Int. J. Pharm.* **2020**, *583*, 119332. [CrossRef] [PubMed]
3. *ISO10993-18:2020*; Biological Evaluation of Medical Devices—Part 18: Chemical Characterization of Medical Device Materials within a Risk Management Process. Available online: https://www.iso.org/standard/64750.html (accessed on 27 November 2023).
4. Kiradzhiyska, D.D.; Mantcheva, R.D. Overview of Biocompatible Materials and Their Use in Medicine. *Folia Med.* **2019**, *61*, 34–40. [CrossRef] [PubMed]
5. Marturano, V.; Cerruti, P.; Ambrogi, V. Polymer additives. *Phys. Sci. Rev.* **2017**, *2*, 20160130. [CrossRef]
6. Li, B.J. Exploring three-dimensional space of extractables and leachables in volatility, hydrophobicity, and molecular weight and assessment of roles of gas and liquid chromatographic methods in their comprehensive analysis. *J. Pharm. Biomed. Anal.* **2023**, *223*, 115142. [CrossRef] [PubMed]
7. Geueke, B.; Groh, K.J.; Maffini, M.V.; Martin, O.V.; Boucher, J.M.; Chiang, Y.-T.; Gwosdz, F.; Jieh, P.; Kassotis, C.D.; Łańska, P.; et al. Systematic evidence on migrating and extractable food contact chemicals: Most chemicals detected in food contact materials are not listed for use. *Crit. Rev. Food Sci. Nutr.* **2022**, *63*, 9425–9435. [CrossRef] [PubMed]
8. Finley, J.W.; deMan, J.M. Lipids. In *Principles of Food Chemistry*; Food Science Text Series; Springer: Cham, Switzerland, 2018. [CrossRef]
9. Reis, A.; Rudnitskaya, A.; Blackburn, G.J.; Fauzi, N.M.; Pitt, A.R.; Spickett, C.M. A comparison of five lipid extraction solvent systems for lipidomic studies of human LDL. *J. Lipid Res.* **2013**, *54*, 1812–1824. [CrossRef] [PubMed]

10. Muncke, J. Hazards of Food Contact Material: Food Packaging Contaminants. In *Encyclopedia of Food Safety*; Motarjemi, Y., Ed.; Academic Press: Cambridge, MA, USA, 2014; pp. 430–437. [CrossRef]
11. Bhunia, K.; Sablani, S.S.; Tang, J.; Rasco, B. Migration of Chemical Compounds from Packaging Polymers during Microwave, Conventional Heat Treatment, and Storage. *Compr. Rev. Food Sci. Food Saf.* **2013**, *12*, 523–545. [CrossRef]
12. Gavriil, G.; Kanavouras, A.; Coutelieris, F.A. Food-packaging migration models: A critical discussion. *Crit. Rev. Food Sci. Nutr.* **2018**, *58*, 2262–2272. [CrossRef]
13. de Kruijf, N.; Rijk, R. The suitability of alternative fatty food simulants for overall migration testing under both low- and high-temperature test conditions. *Food Addit. Contam.* **1997**, *14*, 775–789. [CrossRef]
14. Baner, A.; Bieber, W.; Figge, K.; Franz, R.; Piringer, O. Alternative fatty food simulants for migration testing of polymeric food contact materials. *Food Addit. Contam.* **1992**, *9*, 137–148. [CrossRef] [PubMed]
15. Cooper, I.; Goodson, A.; O'Brien, A. Specific migration testing with alternative fatty food simulants. *Food Addit. Contam.* **1998**, *15*, 72–78. [CrossRef] [PubMed]
16. De Kruijf, N.; Rijk, M.A.H. Iso-octane as fatty food simulant: Possibilities and limitations. *Food Addit. Contam.* **1988**, *5*, 467–483. [CrossRef]
17. Guidance for Industry: Preparation of Premarket Submissions for Food Contact Substances (Chemistry Recommendations). Available online: https://www.fda.gov/regulatory-information/search-fda-guidance-documents/guidance-industry-preparation-premarket-submissions-food-contact-substances-chemistry (accessed on 27 November 2023).
18. Fahy, E.; Subramaniam, S.; Brown, H.A.; Glass, C.K.; Merrill, A.H., Jr.; Murphy, R.C.; Raetz, C.R.H.; Russell, D.W.; Seyama, Y.; Shaw, W.; et al. A comprehensive classification system for lipids. *J. Lipid Res.* **2005**, *46*, 839–861. [CrossRef] [PubMed]
19. *ISO 10993-12:2021*; Biological Evaluation of Medical Devices—Part 12: Sample Preparation and Reference Materials. Available online: https://www.iso.org/standard/75769.html (accessed on 7 January 2023).
20. Chang, S.-S.; Pummer, W.J.; Maurey, J.R. Fat-simulating and accelerating solvents for polyolefins and MWD of solvent extracts of polyethylenes. *Polymer* **1983**, *24*, 1267–1272. [CrossRef]
21. Limm, W.; Hollifield, H.C. Modelling of additive diffusion in polyolefins. *Food Addit. Contam.* **1996**, *13*, 949–967. [CrossRef] [PubMed]
22. Koszinowski, J. Diffusion and solubility of *n*-alkanes in polyolefines. *J. Appl. Polym. Sci.* **1986**, *32*, 4765–4786. [CrossRef]
23. Koszinowski, J. Diffusion and Solubility of Hydroxy Compounds in Polyolefines. *J. Appl. Polym. Sci.* **1986**, *31*, 2711–2720. [CrossRef]
24. Poças, M.F.; Oliveira, J.C.; Oliveira, F.A.R.; Hogg, T. A critical survey of predictive mathematical models for migration from packaging. *Crit. Rev. Food Sci. Nutr.* **2008**, *48*, 913–928. [CrossRef] [PubMed]
25. Wan, J.A.L.; Chatwin, P.C.; Katan, L.L. Migration from plastic packages into their contents. I. The role of mathematical models. *Philos. Trans. R. Soc. Lond. Ser. A Phys. Eng. Sci.* **1995**, *350*, 379–406. [CrossRef]
26. Chatwin, P.C.; Katan, L.L. The role of mathematics and physics in migration predictions. *Packag. Technol. Sci. Int. J.* **1989**, *2*, 75–84. [CrossRef]
27. Franz, R.; Stormer, A. Migration of Plastic Constituents. In *Plastic Packaging—Interactions with Food and Pharmaceuticals*, 2nd ed.; Piringer, O.G., Baner, A.L., Eds.; WILEY-VCH Verlag GmbH & Co. KGaA: Weinheim, Germany, 2008; Chapter 11; pp. 349–415. ISBN 978-3-527-31455-3.
28. Franz, R. Migration of plastic constituents. In *Plastic Packaging Materials for Food, Barrier Function, Mass Transport, Quality Assurance, and Legislation*; Piringer, O.G., Baner, A.L., Eds.; WILEY-VCH: Weinheim, Germany, 2000; Chapter 10; pp. 287–357.
29. Baner, A.L. Partition constant. In *Plastic Packaging Materials for Food, Barrier Function, Mass Transport, Quality Assurance, and Legislation*; Piringer, O.G., Baner, A.L., Eds.; WILEY-VCH: Weinheim, Germany, 2000; Chapter 4; pp. 78–123.
30. Piringer, O.G.; Baner, A.L. Appendix II. In *Plastic Packaging—Interactions with Food and Pharmaceuticals*, 2nd ed.; WILEY-VCH Verlag GmbH & Co. KGaA: Weinheim, Germany, 2008; pp. 558–590. ISBN 978-3-527-31455-3.
31. Piringer, O. Mathematical modeling of chemical migration from food contact materials. In *Chemical Migration and Food Contact Materials*; Barnes, K.A., Sinclair, C.R., Watson, D.H., Eds.; Woodhead: Cambridge, UK, 2007; pp. 180–202.
32. Helmroth, E.; Rijk, R.; Dekker, M.; Jongen, W. Predictive modelling of migration from packaging materials into food products for regulatory purposes. *Trends Food Sci. Technol.* **2002**, *13*, 102–109. [CrossRef]
33. Li, B.J. Correlation of Extractables Transfer from Low-Density Polyethylene into Extraction Solvents with Abraham Solvation Parameter Model and Assessment of Thermodynamic and Kinetic Effects in Achieving Exhaustive Extractions in Chemical Characterization of Medical Devices. *J. Solut. Chem.* **2023**, *52*, 967–1000. [CrossRef]
34. Figge, K. Migration of additives from plastics films into edible oils and fat simulants. *Food Cosmet. Toxicol.* **1972**, *10*, 815–828. [CrossRef] [PubMed]
35. Gramiccioni, L.; Di Prospero, P.; Milana, M.; Di Marzio, S.; Marcello, I. Global migration from plastic materials into olive oil and isooctane: An experimental comparison. *Food Chem. Toxicol.* **1986**, *24*, 23–26. [CrossRef] [PubMed]
36. Garde, J.A.; Catalá, R.; Gavara, R.; Garde, R.C.J.A. Global and specific migration of antioxidants from polypropylene films into food simulants. *J. Food Prot.* **1998**, *61*, 1000–1006. [CrossRef] [PubMed]
37. Hamdani, M.; Feigenbaum, A. Migration from plasticized poly(vinyl chloride) into fatty media: Importance of simulant selectivity for the choice of volatile fatty simulants. *Food Addit. Contam.* **1996**, *13*, 717–729. [CrossRef]

38. Monteiro, M.; Nerín, C.; Reyes, F. Migration of Tinuvin P, a UV stabilizer, from PET bottles into fatty-food simulants. *Packag. Technol. Sci. Int. J.* **1999**, *12*, 241–248. [CrossRef]
39. Reynier, A.; Dole, P.; Feigenbaum, A. Integrated approach of migration prediction using numerical modelling associated to experimental determination of key parameters. *Food Addit. Contam.* **2002**, *19*, 42–55. [CrossRef]
40. Helmroth, I.E.; Dekker, M.; Hankemeier, T. Influence of solvent absorption on the migration of Irganox 1076 from LDPE. *Food Addit. Contam.* **2002**, *19*, 176–183. [CrossRef]
41. Beldì, G.; Pastorelli, S.; Franchini, F.; Simoneau, C. Time- and temperature-dependent migration studies of Irganox 1076 from plastics into foods and food simulants. *Food Addit. Contam. Part A Chem. Anal. Control Expo. Risk Assess.* **2012**, *29*, 836–845. [CrossRef] [PubMed]
42. Hernandez-Muñoz, P.; Catalá, R.; Gavara, R. Food aroma partition between packaging materials and fatty food simulants. *Food Addit. Contam.* **2001**, *18*, 673–682. [CrossRef] [PubMed]
43. Paseiro-Cerrato, R.; Tongchat, C.; Franz, R. Study of the partition coefficients $K_{p/f}$ of seven model migrants from LDPE polymer in contact with food simulants. *Food Addit. Contam. Part A Chem. Anal. Control Expo. Risk Assess.* **2016**, *33*, 885–892. [CrossRef] [PubMed]
44. Seiler, A.; Bach, A.; Driffield, M.; Losada, P.P.; Mercea, P.; Tosa, V.; Franz, R. Correlation of foodstuffs with ethanol–water mixtures with regard to the solubility of migrants from food contact materials. *Food Addit. Contam. Part A* **2014**, *31*, 498–511. [CrossRef]
45. Correlation of Partition Constants KPolymer/Food and KOctanol/Water for Potential Migrants in Food Contact Polymers. Available online: https://www.ivv.fraunhofer.de/content/dam/ivv/en/documents/Forschungsfelder/Produktsicherheit-und-analytik/Correlation_of_partition_coefficients.pdf (accessed on 27 November 2023).
46. Migration of Low Molecular Weight Additives in Polyolefins and Copolymers, National Bureau of Standards (NBS), Final Project Report, NBSIR 82-2472. 1982. Available online: https://nvlpubs.nist.gov/nistpubs/Legacy/IR/nbsir82-2472.pdf (accessed on 27 November 2023).
47. Smedes, F.; Rusina, T.P.; Beeltje, H.; Mayer, P. Partitioning of hydrophobic organic contaminants between polymer and lipids for two silicones and low density polyethylene. *Chemosphere* **2017**, *186*, 948–957. [CrossRef] [PubMed]
48. Turner, P.; Elder, R.M.; Nahan, K.; Talley, A.; Shah, S.; Duncan, T.V.; Sussman, E.M.; Saylor, D.M. Leveraging Extraction Testing to Predict Patient Exposure to Polymeric Medical Device Leachables Using Physics-based Models. *Toxicol. Sci.* **2020**, *178*, 201–211. [CrossRef] [PubMed]
49. Begley, T.; Castle, L.; Feigenbaum, A.; Franz, R.; Hinrichs, K.; Lickly, T.; Mercea, P.; Milana, M.; O'brien, A.; Rebre, S.; et al. Evaluation of migration models that might be used in support of regulations for food-contact plastics. *Food Addit. Contam.* **2005**, *22*, 73–90. [CrossRef] [PubMed]
50. Jahnke, A.; McLachlan, M.S.; Mayer, P. Equilibrium sampling: Partitioning of organochlorine compounds from lipids into polydimethylsiloxane. *Chemosphere* **2008**, *73*, 1575–1581. [CrossRef]
51. Geisler, A.; Endo, S.; Goss, K.-U. Partitioning of organic chemicals to storage lipids: Elucidating the dependence on fatty acid composition and temperature. *Environ. Sci. Technol.* **2012**, *46*, 9519–9524. [CrossRef]
52. Rusina, T.P.; Smedes, F.; Klanova, J.; Booij, K.; Holoubek, I. Polymer selection for passive sampling: A comparison of critical properties. *Chemosphere* **2007**, *68*, 1344–1351. [CrossRef]
53. Paxton, N.C.; Allenby, M.C.; Lewis, P.M.; Woodruff, M.A. Biomedical applications of polyethylene. *Eur. Polym. J.* **2019**, *118*, 412–428. [CrossRef]
54. Lohmann, R. Critical Review of Low-Density Polyethylene's Partitioning and Diffusion Coefficients for Trace Organic Contaminants and Implications for Its Use As a Passive Sampler. *Environ. Sci. Technol.* **2012**, *46*, 606–618. [CrossRef] [PubMed]
55. Ulrich, N.; Böhme, A.; Strobel, A.B.; Egert, T. Predicting partitioning from low density polyethylene to blood and adipose tissue by linear solvation energy relationship models. *J. Biomed. Mater. Res. B Appl. Biomater.* **2023**, *111*, 2044–2054. [CrossRef] [PubMed]
56. Egert, T.; Langowski, H.C. Linear solvation energy relationships (LSERs) for robust prediction of partition coefficients between low density polyethylene and water. Part II: Model evaluation and benchmarking. *Eur. J. Pharm. Sci.* **2022**, *172*, 106138. [CrossRef] [PubMed]
57. Poole, C.F. Solvation parameter model: Tutorial on its application to separation systems for neutral compounds. *J. Chromatogr. A* **2021**, *1645*, 462108. [CrossRef] [PubMed]
58. Poole, C.F. Partition constant database for totally organic biphasic systems. *J. Chromatogr. A* **2017**, *1527*, 18–32. [CrossRef]
59. Abraham, M.H.; Acree, W.E., Jr. Descriptors for the Prediction of Partition Coefficients and Solubilities of Organophosphorus Compounds. *Sep. Sci. Technol.* **2013**, *48*, 884–897. [CrossRef]
60. Jouyban, A.; Acree WMichael, H. Abraham and His Developed Parameters: Various Applications in Medicine, Chemistry and Biology. *Pharm. Sci.* **2022**, *28*, 170–173. [CrossRef]
61. Khawar, M.I.; Nabi, D. Relook on the Linear Free Energy Relationships Describing the Partitioning Behavior of Diverse Chemicals for Polyethylene Water Passive Samplers. *ACS Omega* **2021**, *6*, 5221–5232. [CrossRef]
62. Liu, Z.; Sun, X.; Xu, Y. Recalibrating polyparameter linear free energy relationships and reanalyzing mechanisms for partition of nonionic organic compounds to low-density polyethylene passive sampler. *J. Chromatogr. A* **2023**, *1700*, 464039. [CrossRef]
63. Abraham, M.H.; Chadha, H.S.; Whiting, G.S.; Mitchell, R.C. Hydrogen bonding. 32. An analysis of water-octanol and water-alkane partitioning and the Δlog P parameter of seiler. *J. Pharm. Sci.* **1994**, *83*, 1085–1100. [CrossRef] [PubMed]

64. Lickly, T.D.; Bell, C.D.; Lehr, K.M. The migration of Irganox 1010 antioxidant from high-density polyethylene and polypropylene into a series of potential fatty-food simulants. *Food Addit. Contam.* **1990**, *7*, 805–814. [CrossRef] [PubMed]
65. Garde, J.A.; Catalá, R.; Gavara, R.; Hernandez, R.J. Characterizing the migration of antioxidants from polypropylene into fatty food simulants. *Food Addit. Contam.* **2001**, *18*, 750–762. [CrossRef] [PubMed]
66. Li, B. Determination of the mass transport properties of chemical additives in polypropylene material/simulated food system. *Food Addit. Contam. Part A* **2019**, *36*, 625–637. [CrossRef] [PubMed]
67. Zhu, T.; Jafvert, C.T.; Fu, D.; Hu, Y. A novel method for measuring polymer–water partition coefficients. *Chemosphere* **2015**, *138*, 973–979. [CrossRef] [PubMed]
68. Booij, K.; Hofmans, H.E.; Fischer, C.V.; Van Weerlee, E.M. Temperature-dependent uptake rates of nonpolar organic compounds by semipermeable membrane devices and low-density polyethylene membranes. *Environ. Sci. Technol.* **2003**, *37*, 361–366. [CrossRef] [PubMed]
69. Zhang, K.; Fahr, A.; Abraham, M.H.; Acree, W.E., Jr.; Tobin, D.J.; Liu, X. Comparison of lipid membrane–water partitioning with various organic solvent–water partitions of neutral species and ionic species: Uniqueness of cerasome as a model for the stratum corneum in partition processes. *Int. J. Pharm.* **2015**, *494*, 1–8. [CrossRef] [PubMed]
70. Cao, Y.; Marra, M.; Anderson, B.D. Predictive relationships for the effects of triglyceride ester concentration and water uptake on solubility and partitioning of small molecules into lipid vehicles. *J. Pharm. Sci.* **2004**, *93*, 2768–2779. [CrossRef]
71. Allan, I.J.; Vrana, B.; Ruus, A. Passive Sampling Helps the Appraisal of Contaminant Bioaccumulation in Norwegian Fish Used for Regulatory Chemical Monitoring. *Environ. Sci. Technol.* **2022**, *56*, 7945–7953. [CrossRef]
72. Gargas, M.L.; Burgess, R.J.; Voisard, D.E.; Cason, G.H.; Andersen, M.E. Partition coefficients of low-molecular-weight volatile chemicals in various liquids and tissues. *Toxicol. Appl. Pharmacol.* **1989**, *98*, 87–99. [CrossRef]
73. Ruus, A.; Allan, I.J.; Bæk, K.; Borgå, K. Partitioning of persistent hydrophobic contaminants to different storage lipid classes. *Chemosphere* **2021**, *263*, 127890. [CrossRef] [PubMed]
74. Alan, D. Hendricker, Controlled Extraction Study on Low Density Polyethylene (LDPE). 2015. Available online: https://pqri.org/wp-content/uploads/2015/11/LDPE_Poster.pdf (accessed on 27 November 2023).
75. Cozzi, A.C.; Briasco, B.; Salvarani, E.; Mannucci, B.; Fangarezzi, F.; Perugini, P. Evaluation of Mechanical Properties and Volatile Organic Extractable to Investigate LLDPE and LDPE Polymers on Final Packaging for Semisolid Formulation. *Pharmaceutics* **2018**, *10*, 113. [CrossRef] [PubMed]
76. Jenke, D.; Castner, J.; Egert, T.; Feinberg, T.; Hendricker, A.; Houston, C.; Hunt, D.G.; Lynch, M.; Shaw, A.; Nicholas, K.; et al. Extractables characterization for five materials of construction representative of packaging systems used for parenteral and ophthalmic drug products. *PDA J. Pharm. Sci. Technol.* **2013**, *67*, 448–511. [CrossRef] [PubMed]
77. Vera, P.; Canellas, E.; Barknowitz, G.; Goshawk, J.; Nerín, C. Ion-Mobility Quadrupole Time-of-Flight Mass Spectrometry: A Novel Technique Applied to Migration of Nonintentionally Added Substances from Polyethylene Films Intended for Use as Food Packaging. *Anal. Chem.* **2019**, *91*, 12741–12751. [CrossRef] [PubMed]
78. Schwope, A.; Till, D.; Ehntholt, D.; Sidman, K.; Whelan, R.; Schwartz, P.; Reid, R. Migration of BHT and Irganox 1010 from low-density polyethylene (LDPE) to foods and food-simulating liquids. *Food Chem. Toxicol.* **1987**, *25*, 317–326. [CrossRef]
79. Chapke, K.; Gandhi, K.; Lata, K.; Sharma, R.; Mann, B.; Singh, N. Migration study of chemical additives from low density polyethylene (LDPE) into dahi. *J. Food Sci. Technol.* **2022**, *59*, 3283–3295. [CrossRef] [PubMed]
80. Wessling, C.; Nielsen, T.; Leufvén, A.; Jägerstad, M. Mobility of α-tocopherol and BHT in LDPE in contact with fatty food simulants. *Food Addit. Contam.* **1998**, *15*, 709–715. [CrossRef] [PubMed]
81. RMG: Group Contribution Method for Abraham Solute Parameter Prediction. Available online: https://rmg.mit.edu/database/solvation/search/ (accessed on 27 November 2023).
82. Poole, C.F. Wayne State University experimental descriptor database for use with the solvation parameter model. *J. Chromatogr. A* **2020**, *1617*, 460841. [CrossRef]
83. Li, J.; Sobańtka, A. A Systematic Analysis of the Effect of Extraction Solvents on the Chemical Composition of Extraction Solutions and the Analytical Implications in Extractables and Leachables Studies. *J. Pharm. Biomed. Anal.* **2023**, *222*, 115081. [CrossRef]
84. Igwe, I.O. Uptake of aromatic solvents by polyethylene films. *J. Appl. Polym. Sci.* **2007**, *104*, 3849–3854. [CrossRef]
85. Riquet, A.-M.; Scholler, D.; Feigenbaum, A. Tailoring fatty food simulants made from solvent mixtures (2): Determining the equivalent migration behaviour of olive oil and of solvents in the case of polyolefins. *Food Addit. Contam.* **2002**, *19*, 582–593. [CrossRef] [PubMed]
86. Fava, L.P.P. Study of diffusion through LDPE film of Di-*n*-butyl phthalate. *Food Addit. Contam.* **1999**, *16*, 353–359. [CrossRef] [PubMed]
87. Hedenqvist, M.; Gedde, U. Parameters affecting the determination of transport kinetics data in highly swelling polymers above Tg. *Polymer* **1999**, *40*, 2381–2393. [CrossRef]
88. Harding, S.G.; Johns, M.L.; Pugh, S.R.; Fryer, P.J.; Gladden, L.F. Magnetic resonance imaging studies of diffusion in polymers. *Food Addit. Contam.* **1997**, *14*, 583–589. [CrossRef] [PubMed]
89. Riquet, A.M.; Wolff, N.; Laoubi, S.; Vergnaud, J.M.; Feigenbaum, A. Food and packaging interactions: Determination of the kinetic parameters of olive oil diffusion in polypropylene using concentration profiles. *Food Addit. Contam.* **1998**, *15*, 690–700. [CrossRef] [PubMed]

90. Nasiri, A.; Peyron, S.; Gastaldi, E.; Gontard, N. Effect of nanoclay on the transfer properties of immanent additives in food packages. *J. Mater. Sci.* **2016**, *51*, 9732–9748. [CrossRef]
91. Rusina, T.P.; Smedes, F.; Klanova, J. Diffusion Coefficients of Polychlorinated Biphenyls and Polycyclic Aromatic Hydrocarbons in Polydimethylsiloxane and Low-Density Polyethylene Polymers. *J. Appl. Polym. Sci.* **2010**, *116*, 1803–1810. [CrossRef]
92. Randová, A.; Bartovská, L.; Hovorka, Š.; Bartovský, T.; Izák, P.; Poloncarzová, M.; Fries, K. Diffusion coefficients in systems LDPE+cyclohexane and LDPE+benzene. *e-Polymers* **2010**, *10*, 742–747. [CrossRef]
93. Mauricio-Iglesias, M.; Guillard, V.; Gontard, N.; Peyron, S. Application of FTIR and Raman microspectroscopy to the study of food/packaging interactions. *Food Addit. Contam. Part A Chem. Anal. Control Expo. Risk Assess.* **2009**, *26*, 1515–1523. [CrossRef]
94. Helmroth, I.E.; Dekker, M.; Hankemeier, T. Additive Diffusion from LDPE Slabs into Contacting Solvents as a Function of Solvent Absorption. *J. Appl. Polym. Sci.* **2003**, *90*, 1609–1617. [CrossRef]
95. Ibarra, V.G.; Sendón, R.; García-Fonte, X.; Losada, P.P.; de Quirós, A.R.B. Migration studies of butylated hydroxytoluene, tributyl acetylcitrate and dibutyl phthalate into food simulants. *J. Sci. Food Agric.* **2019**, *99*, 1586–1595. [CrossRef] [PubMed]
96. de Kruijf, N.; Rijk, M.; Soetardhi, L.; Rossi, L. Selection and application of a new volatile solvent as a fatty food simulant for determining the global migration of constituents of plastics materials. *Food Chem. Toxicol.* **1983**, *21*, 187–191. [CrossRef] [PubMed]
97. Hernández-Muñoz, P.; Catalá, R.; Gavara, R. Simple method for the selection of the appropriate food simulant for the evaluation of a specific food/packaging interaction. *Food Addit. Contam.* **2002**, *19*, 192–200. [CrossRef] [PubMed]
98. Kirchkeszner, C.; Petrovics, N.; Tábi, T.; Magyar, N.; Kovács, J.; Szabó, B.S.; Nyiri, Z.; Eke, Z. Swelling as a promoter of migration of plastic additives in the interaction of fatty food simulants with polylactic acid- and polypropylene-based plastics. *Food Control* **2022**, *132*, 108354. [CrossRef]

liquids

MDPI

Article

Calculation of the Three Partition Coefficients $logP_{ow}$, $logK_{oa}$ and $logK_{aw}$ of Organic Molecules at Standard Conditions at Once by Means of a Generally Applicable Group-Additivity Method

Rudolf Naef [1,*] and William E. Acree, Jr. [2]

[1] Department of Chemistry, University of Basel, 4003 Basel, Switzerland
[2] Department of Chemistry, University of North Texas, Denton, TX 76203, USA; acree@unt.edu
* Correspondence: rudolf.naef@unibas.ch; Tel.: +41-619-119-273

Abstract: Assessment of the environmental impact of organic chemicals has become an important subject in chemical science. Efficient quantitative descriptors of their impact are their partition coefficients $logP_{ow}$, $logK_{oa}$ and $logK_{aw}$. We present a group-additivity method that has proven its versatility for the reliable prediction of many other molecular descriptors for the calculation of the first two partition coefficients and indirectly of the third with high dependability. Based on the experimental $logP_{ow}$ data of 3332 molecules and the experimental $logK_{oa}$ data of 1900 molecules at 298.15 K, the respective partition coefficients have been calculated with a cross-validated standard deviation S of only 0.42 and 0.48 log units and a goodness of fit Q^2 of 0.9599 and 0.9717, respectively, in a range of ca. 17 log units for both descriptors. The third partition coefficient $logK_{aw}$ has been derived from the calculated values of the former two descriptors and compared with the experimentally determined $logK_{aw}$ value of 1937 molecules, yielding a standard deviation σ of 0.67 log units and a correlation coefficient R^2 of 0.9467. This approach enabled the quick calculation of 29,462 $logP_{ow}$, 27,069 $logK_{oa}$ and 26,220 $logK_{aw}$ values for the more than 37,100 molecules of ChemBrain's database available to the public.

Keywords: group-additivity method; Gauss–Seidel diagonalisation; partition coefficient; $logP_{ow}$; $logK_{oa}$; $logK_{aw}$

1. Introduction

Environmental considerations of organic molecules as potential contaminants have become an important subject in recent years. Several descriptors have been applied to quantify their impact on the natural environment, among them the octanol/water partition coefficient $logP_{ow}$ (more recently named $logK_{ow}$), a standard model for the description of the lipophilicity of drugs in medicinal and agricultural chemistry, whereby octanol is the substitute for the natural organic matter, and the octanol/air partition coefficient K_{oa} and the air/water partition coefficient $logK_{aw}$ both indicate the role of the chemicals for air-breathing organisms [1–3]. In view of the time consumption and costs of their experimental determination, fast mathematical methods for the prediction of their value attributed to a molecule have been developed. An excellent comprehensive overview of the various methods for the prediction of the $logK_{ow}$—among many other descriptors—is given by Nieto-Draghi et al. [4]. Cappelli et al. [5] analysed a series of free programs based on atom/fragment contributions, hydrophobicity contributions of atoms, the number of carbon atoms and heteroatoms as well as Monte Carlo methods to calculate $logP_{ow}$ and found correlation coefficients R^2 of between 0.7 and 0.8 and root mean square errors (RMSE) from 0.8 to 1.5. A number of authors [6–13] have successfully carried out $logP_{ow}$ calculations for a large variability of compounds based on various group-additivity methods. Plante and Werner [14] presented a $logP_{ow}$ prediction method based on the combination of the calculated data of the four different open-source group-additivity calculation methods

AlogP, XlogP2, SlogP and XlogP3 into a single model, providing a best RMSE of 0.63. Ulrich et al. [15] used deep neural networks (DNNs) for the $\log P_{ow}$ calculations, based on ca. 14,000 different SMILES representations of molecules including potential tautomers, whereby, however, a substantial number of compounds might have been presented as duplicates and triplicates to the DNNs. Their best prediction performance yielded an RMSE of 0.47. Recently, an entirely different path was followed by Sun et al. [16]: since $\log P_{ow}$ is proportional to the Gibbs free energy of the transfer from one solvent to another, it can be calculated using the free solvation energy in these solvents. Sun used the molecular mechanics–Poisson Boltzmann surface area (MM-PBSA) method for the determination of the free energies of solvation. Their best RMSE for the 707 compounds test set was 0.91.

Many publications [17–28] dealing with the prediction of the coefficient K_{oa}, based on various QSPR methods, are limited to specific chemical families, thus lacking general applicability. Li et al. [29] used a group-additivity method based on five fragment constants and one structural correction factor for the evaluation of $\log K_{oa}$, limited to halogenated aromatic pollutants. Recently, Ebert et al. [30] suggested a general-purpose fragment model for the calculation of the air/water partition coefficient $\log K_{aw}$ resembling the atom group-additivity method presented in one of our earlier papers [13] for the calculation of—among several further descriptors—the octanol/water partition coefficient $\log P_{ow}$.

The goal of the present paper was to suggest the extension of a simple tool, which has already served well for the prediction of the octanol/water coefficient $\log P_{ow}$ described in [13], to enable it to calculate all three mentioned partition coefficients at once by means of a uniform computer algorithm based on the atom group-additivity method detailed in [13]. Since under common standard conditions, any third partition coefficient can be directly calculated from the other two if we neglect the effect of the contamination of water in octanol (and vice versa) influencing the determination of the $\log P_{ow}$ values, which will be addressed later on, it made sense to select the two coefficients for which any group parameters could be founded on the most reliable as well as the largest number of experimental data. It turned out that the experimental data for the partition coefficients $\log P_{ow}$ and $\log K_{oa}$ provided excellent basis sets for the evaluation of their respective tables of atom and special group parameters. Accordingly, from the subsequently calculated values of a molecule's $\log P_{ow}$ and $\log K_{oa}$, its air/water partition coefficient $\log K_{aw}$ should easily be evaluable following the equation $\log K_{aw} \approx \log P_{ow} - \log K_{oa}$.

2. Method

The calculation method is based on a regularly updated object-oriented database of more than 37,100 compounds stored in their geometry-optimised 3D structure, encompassing pharmaceuticals, herbicides, pesticides, fungicides, textile and other dyes, ionic liquids, liquid crystals, metal–organics, lab intermediates and many more, collecting—among further molecular experimental and calculated descriptors—a large set of experimental $\log P_{ow}$, $\log K_{oa}$ and $\log K_{aw}$ data, outlined in the respective sections below. It should be stressed that for the calculation of the partition coefficients, the 3D geometry-optimised form of the compounds is not required—except for the algorithm-based determination of intramolecular hydrogen bridges, the impact of which will be discussed further down. In order to avoid structural ambiguities in the presentation of the chemical structures to the computer algorithm defining the molecules´ atom groups, a special algorithm ensured at the time of the input of a new compound that any six-membered aromatic ring system is defined by six aromatic bonds instead of alternating single–double-bonds.

2.1. Definition of the Atom and Special Groups

The details of the atom group-additivity model applied in the present study have been outlined in [13]. Accordingly, the definition of the atom types and their immediate atomic neighbourhood and meaning are retained as described in Table 1 of [13] and are also valid for both the $\log P_{ow}$ and $\log K_{oa}$ descriptors. However, since these atom groups are not able to cover certain additional structural effects such as intramolecular hydrogen-bridge bonds and the influence of saturated cyclic compared to saturated noncyclic systems, a number of additional special groups had to be introduced. In a paper applying a different group-additivity method for the calculation of $logP_{ow}$, Klopman et al. [8] discovered that the inclusion of a correction value per carbon atom in pure saturated and unsaturated hydrocarbons improved compliance with the experiment. This has indeed been confirmed in the present study.

In order to take account of these and further potential structure-related peculiarities, the list of atom groups has been extended by "special groups" for which the column-title terms "atom type" and "neighbours" in the subsequent tables should not be taken literally, but which the computer algorithm treats in the same way as ordinary atom groups. In Table 1, the respective special groups, their nomenclature and meaning are detailed. In order to enable a future comparison of the contributions of the special group parameter sets within this study, the same special groups have been applied for the calculation of both descriptors $logP_{ow}$ and $logK_{oa}$.

Table 1. Special groups and their meaning.

Atom Type	Neighbours	Meaning
H	H Acceptor	Correction value for intramolecular H bridge between acidic H (on O, N or S) and basic acceptor (O, N or F)
(COH)n	n > 1	Correction value for each additional hydroxy group
(COOH)n	n > 1	Correction value for each additional carboxylic acid group
Alkane	No. of C atoms	Correction value for each C atom in a pure alkane
Unsaturated HC	No. of C atoms	Correction value for each C atom in an aromatic hydrocarbon
Endocyclic bonds	No. of single bonds	Correction value for each single endocyclic bond

At present, the list of elements is limited to H, B, C, N, O, P, S, Si and halogen, but an extension is always possible, provided that corresponding molecules with experimental descriptor data are available.

2.2. Calculation of the Atom and Special Group Contributions

Since the algorithm for the evaluation of the parameter values of the atom groups has been outlined in detail in [13], its four steps may just be summarised as follows: the first step encompasses the selection of all the compounds from a database of, at present, more than 37,100 compounds for which the experimental descriptor data in question are known and their storage is in a temporary compounds list. In the second step, the molecules in the temporary list are broken down into their constituting atom groups, whereby their central atoms, called "backbone atoms", are characterised in that they are bound to at least two covalently bound neighbour atoms. The atom groups' atom types and neighbour terms are generated according to the rules described in [13] and their occurrence is registered. Any molecule carrying an atom group that is not found in the pre-defined group parameters table is discarded from the temporary compound list. The third step generates an $M \times (N + 1)$ matrix, wherein M is the number of molecules, N + 1 is the number of pre-defined atom groups plus the container for the molecule´s descriptor value, and the matrix element (i,j) contains the number of registered occurrences of the jth atom group in the ith molecule. Atom groups and their related jth column, which are not present in any molecule of the temporary molecules list, are removed from the $M \times (N + 1)$ matrix. In the final step, this adjusted matrix is normalised into an $Ax = B$ matrix, followed by its balancing by means of fast Gauss–Seidel calculus [31] to receive the atom and special group parameters x. These parameters are then added to their related atom and special group in the corresponding parameter table assigned to the specific descriptor.

The group parameter calculation is then immediately followed by the computation of each molecule's descriptor value in question, on the basis of these group parameters according to Equation (1) outlined in the next section, and compared with its experimental value to receive the related statistics data, which are finally added at the bottom of the parameter table. Following the above-mentioned procedure resulted in the two parameter sets in Tables 2 and 3, designed for the calculation of the molecules' $\log P_{ow}$ and $\log K_{oa}$ values, respectively.

Table 2. Atom and special groups and their contribution in $\log P_{ow}$ calculations.

Entry	Atom Type	Neighbours	Contribution	Occurrences	Molecules
1	Const		0.73	3332	3332
2	B(−)	F4	2.71	10	10
3	C sp3	H3C	0.27	2614	1498
4	C sp3	H3N	0.14	457	320
5	C sp3	H3N(+)	−1.35	2	2
6	C sp3	H3O	−0.26	375	285
7	C sp3	H3P	−0.3	4	4
8	C sp3	H3S	−0.34	61	53
9	C sp3	H3Si	0.76	44	5
10	C sp3	H2C2	0.44	3262	1046
11	C sp3	H2CN	0.42	741	429
12	C sp3	H2CN(+)	−0.86	32	25
13	C sp3	H2CO	−0.1	799	604
14	C sp3	H2CS	−0.33	97	69
15	C sp3	H2CF	−0.29	5	5
16	C sp3	H2CCl	0.33	84	67
17	C sp3	H2CBr	0.41	54	48
18	C sp3	H2CJ	1.08	6	6
19	C sp3	H2CP	2.77	1	1
20	C sp3	H2N2	2.05	3	3
21	C sp3	H2NO	0.46	4	4
22	C sp3	H2NS	0.72	3	3
23	C sp3	H2O2	−0.17	6	6
24	C sp3	H2S2	−0.86	6	6
25	C sp3	HC3	0.45	417	269
26	C sp3	HC2N	0.58	200	157
27	C sp3	HC2N(+)	−0.73	25	24
28	C sp3	HC2O	0.1	383	241
29	C sp3	HC2S	−0.21	8	8
30	C sp3	HC2F	−0.36	2	2
31	C sp3	HC2Cl	0.69	64	22
32	C sp3	HC2Br	0.81	26	22
33	C sp3	HCN2	1.2	6	5
34	C sp3	HCNO	1.15	17	17
35	C sp3	HCNS	0.9	25	25
36	C sp3	HCO2	−0.02	31	22

Table 2. *Cont.*

Entry	Atom Type	Neighbours	Contribution	Occurrences	Molecules
37	C sp3	HCOS	0.6	3	3
38	C sp3	HCOCl	0.19	3	1
39	C sp3	HCOBr	1.03	1	1
40	C sp3	HCOP	0.31	1	1
41	C sp3	HCF2	−0.02	2	2
42	C sp3	HCCl2	0.93	13	12
43	C sp3	HOF2	−0.04	1	1
44	C sp3	C4	0.54	144	111
45	C sp3	C3N	0.71	37	36
46	C sp3	C3N(+)	−0.43	6	6
47	C sp3	C3O	0.04	54	52
48	C sp3	C3S	−0.1	17	17
49	C sp3	C3F	0.94	4	4
50	C sp3	C3Cl	0.8	21	8
51	C sp3	C3Br	0.59	5	4
52	C sp3	C2N2	−1.17	1	1
53	C sp3	C2NO	0.52	5	5
54	C sp3	C2O2	1.65	5	5
55	C sp3	C2F2	0.67	2	2
56	C sp3	C2Cl2	0.84	9	9
57	C sp3	CNO2	1.46	1	1
58	C sp3	CF3	0.86	80	76
59	C sp3	CF2Cl	1.1	3	2
60	C sp3	CFCl2	1.1	3	2
61	C sp3	CCl3	1.6	23	21
62	C sp3	CCl2Br	0	1	1
63	C sp3	CBr3	2.44	1	1
64	C sp3	OF3	0.8	2	2
65	C sp3	SF3	1.04	8	8
66	C sp3	SFCl2	1.9	1	1
67	C sp3	SCl3	0.76	3	3
68	C sp2	H2=C	0.25	97	87
69	C sp2	H2=N	−0.62	1	1
70	C sp2	HC=C	0.24	449	285
71	C sp2	HC=N	−1.98	18	18
72	C sp2	HC=N(+)	0.94	10	10
73	C sp2	H=CN	−0.08	146	109
74	C sp2	H=CN(+)	−0.6	18	18
75	C sp2	HC=O	−0.73	45	45
76	C sp2	H=CO	0.32	14	13
77	C sp2	H=CS	0.02	17	16

Table 2. *Cont.*

Entry	Atom Type	Neighbours	Contribution	Occurrences	Molecules
78	C sp2	H=CCl	0.51	8	6
79	C sp2	H=CBr	0.59	1	1
80	C sp2	HN=N	−0.06	65	52
81	C sp2	HN=O	−0.63	16	15
82	C sp2	HO=O	−0.4	10	10
83	C sp2	H=NS	−0.51	4	4
84	C sp2	C2=C	0.38	160	133
85	C sp2	C2=N	−0.25	105	102
86	C sp2	C2=N(+)	2.45	1	1
87	C sp2	C2=O	−0.86	242	194
88	C sp2	C=CN	0.76	76	64
89	C sp2	C=CN(+)	−0.56	3	3
90	C sp2	C=CO	0.64	41	36
91	C sp2	C=CS	−0.16	17	15
92	C sp2	C=CF	−0.01	3	3
93	C sp2	C=CCl	0.81	31	21
94	C sp2	C=CBr	0.94	4	4
95	C sp2	C=CJ	0.89	1	1
96	C sp2	C=CP	0	1	1
97	C sp2	=CN2	1.36	19	19
98	C sp2	=CN2(+)	0.74	11	11
99	C sp2	CN=N	0.24	67	63
100	C sp2	CN=N(+)	−0.67	1	1
101	C sp2	CN=O	−0.69	449	364
102	C sp2	C=NO	−0.76	1	1
103	C sp2	=CNO	−0.01	4	4
104	C sp2	=CNO(+)	−0.37	2	2
105	C sp2	CN=S	−0.36	8	8
106	C sp2	C=NS	0.07	5	4
107	C sp2	=CNS	0.37	4	4
108	C sp2	=CNCl	1.94	1	1
109	C sp2	=CNBr	0.7	5	3
110	C sp2	C=NCl	1.75	1	1
111	C sp2	CO=O	−0.13	700	613
112	C sp2	CO=O(-)	−2.16	35	35
113	C sp2	C=OS	−0.99	4	4
114	C sp2	C=OCl	0.28	4	4
115	C sp2	=COCl	1.27	1	1
116	C sp2	=CS2	0	3	3
117	C sp2	=CSBr	−2.41	1	1
118	C sp2	=CF2	0.26	1	1

Table 2. *Cont.*

Entry	Atom Type	Neighbours	Contribution	Occurrences	Molecules
119	C sp2	=CCl2	1.21	12	10
120	C sp2	=CBr2	1.36	1	1
121	C sp2	N2=N	0.79	26	25
122	C sp2	N2=N(+)	0.74	1	1
123	C sp2	N2=O	0.07	135	134
124	C sp2	N=NO	0.11	1	1
125	C sp2	N2=S	0.11	9	8
126	C sp2	N=NS	0.24	25	24
127	C sp2	N=NCl	1.13	3	3
128	C sp2	N=NBr	0.24	3	2
129	C sp2	NO=O	0.2	117	114
130	C sp2	=NOS	−0.19	1	1
131	C sp2	N=OS	0.05	7	7
132	C sp2	NO=S	0.97	1	1
133	C sp2	=NS2	−1.65	2	2
134	C sp2	NS=S	−1.02	5	3
135	C sp2	=NSCl	1.17	1	1
136	C sp2	O2=O	0	3	3
137	C sp2	O=OCl	−0.13	3	3
138	C aromatic	H:C2	0.25	9963	2133
139	C aromatic	H:C:N	−0.49	283	193
140	C aromatic	H:C:N(+)	0.22	33	27
141	C aromatic	H:N2	−0.91	9	9
142	C aromatic	:C3	0.25	389	170
143	C aromatic	C:C2	0.32	2023	1351
144	C aromatic	C:C:N	−0.38	74	62
145	C aromatic	C:C:N(+)	−3.29	4	3
146	C aromatic	:C2N	0.39	653	534
147	C aromatic	:C2N(+)	−0.15	194	161
148	C aromatic	:C2:N	−0.09	93	72
149	C aromatic	:C2:N(+)	−3.54	19	19
150	C aromatic	:C2O	0.57	1076	742
151	C aromatic	:C2S	0.08	208	170
152	C aromatic	:C2F	0.27	126	86
153	C aromatic	:C2Cl	0.78	1718	565
154	C aromatic	:C2Br	0.9	248	111
155	C aromatic	:C2J	1.26	50	34
156	C aromatic	:C2P	1.08	1	1
157	C aromatic	C:N2	−1.81	9	9
158	C aromatic	:C:N2	−0.13	1	1
159	C aromatic	:CN:N	0.49	38	34

Table 2. *Cont.*

Entry	Atom Type	Neighbours	Contribution	Occurrences	Molecules
160	C aromatic	:CN:N(+)	−0.83	1	1
161	C aromatic	:C:NO	0.97	21	15
162	C aromatic	:C:NS	−0.16	5	5
163	C aromatic	:C:NF	−0.23	4	3
164	C aromatic	:C:NCl	0.16	18	16
165	C aromatic	:C:NBr	0.06	1	1
166	C aromatic	N:N2	−0.05	51	41
167	C aromatic	N:N2(+)	0	1	1
168	C aromatic	:N2O	1.53	8	8
169	C aromatic	:N2S	0.8	3	3
170	C aromatic	:N2Cl	0.89	6	6
171	C(+) aromatic	H:N2	0.21	25	25
172	C sp	H#C	−0.27	28	28
173	C sp	C#C	0.2	86	57
174	C sp	C#N	−0.7	136	130
175	C sp	N#N	0.04	3	3
176	C sp	#NS	−0.59	5	5
177	C sp	=N=O	0.64	4	4
178	C sp	=N=S	1.53	27	26
179	N sp3	H2C	−1.57	86	84
180	N sp3	H2C(pi)	−1.05	326	292
181	N sp3	H2N	−0.85	20	20
182	N sp3	H2S	−1.55	34	34
183	N sp3	HC2	−1.3	74	73
184	N sp3	HC2(pi)	−0.93	225	203
185	N sp3	HC2(2pi)	−0.47	311	272
186	N sp3	HCN	−1.1	4	3
187	N sp3	HCN(pi)	−0.49	14	13
188	N sp3	HCN(2pi)	1.65	42	42
189	N sp3	HCO(pi)	−1.32	9	9
190	N sp3	HCS	−1.69	4	4
191	N sp3	HCS(pi)	−0.98	47	47
192	N sp3	HCP	−1.78	3	3
193	N sp3	HCP(pi)	−0.41	1	1
194	N sp3	C3	−1.03	122	108
195	N sp3	C3(pi)	−0.73	153	138
196	N sp3	C3(2pi)	−0.72	149	136
197	N sp3	C3(3pi)	−0.75	23	23
198	N sp3	C2N	−1.57	1	1
199	N sp3	C2N(pi)	−1.41	31	28
200	N sp3	C2N(2pi)	−0.67	51	47

Table 2. *Cont.*

Entry	Atom Type	Neighbours	Contribution	Occurrences	Molecules
201	N sp3	C2N(3pi)	−0.44	10	10
202	N sp3	C2O(pi)	−0.31	5	5
203	N sp3	C2S	−1.42	5	5
204	N sp3	C2S(pi)	0.03	7	6
205	N sp3	C2S(2pi)	0.76	2	2
206	N sp3	C2P	−0.33	5	3
207	N sp3	CN2(2pi)	1.36	1	1
208	N sp3	CS2	0.27	1	1
209	N sp3	CS2(pi)	−0.29	1	1
210	N sp2	H=C	−0.67	12	11
211	N sp2	C=C	−0.72	200	180
212	N sp2	C=N	0.01	13	12
213	N sp2	=CN	0.49	96	78
214	N sp2	C=N(+)	−6.61	1	1
215	N sp2	=CN(+)	−1.02	2	2
216	N sp2	=CO	−0.64	47	41
217	N sp2	C=O	−1.05	2	2
218	N sp2	=CS	−1.44	5	4
219	N sp2	N=N	−0.78	25	18
220	N sp2	N=O	0.16	40	37
221	N aromatic	C2:C(+)	0	50	25
222	N aromatic	:C2	0.38	354	258
223	N aromatic	:C:N	−0.35	4	2
224	N(+) sp3	H3C	−1.03	26	26
225	N(+) sp3	H2C2	1.2	5	5
226	N(+) sp3	HC3	2.68	1	1
227	N(+) sp3	C4	3.03	1	1
228	N(+) sp2	C=CO(−)	−2.3	10	10
229	N(+) sp2	CO=O(−)	0.27	235	198
230	N(+) sp2	NO=O(−)	−0.19	2	2
231	N(+) sp2	O2=O(−)	0.44	55	29
232	N(+) aromatic	H:C2	2.5	3	3
233	N(+) aromatic	C:C2	−0.48	7	6
234	N(+) aromatic	:C2O(−)	1.73	19	19
235	N(+) sp	=C=N(−)	1.8	1	1
236	N(+) sp	=N2(−)	0	1	1
237	O	HC	−0.96	481	344
238	O	HC(pi)	−0.72	627	557
239	O	HN	−0.15	11	11
240	O	HN(pi)	−0.24	6	6
241	O	C2	0.06	156	115

Table 2. *Cont.*

Entry	Atom Type	Neighbours	Contribution	Occurrences	Molecules
242	O	C2(pi)	−0.13	726	588
243	O	C2(2pi)	−0.51	301	280
244	O	CN	0.4	3	3
245	O	CN(pi)	0.82	4	4
246	O	CN(+)(pi)	0.01	55	29
247	O	CN(2pi)	0.53	13	12
248	O	CS	−0.13	13	8
249	O	CS(pi)	−0.1	3	3
250	O	CP	0.23	132	68
251	O	CP(pi)	−0.49	36	26
252	O	CSi	−0.15	8	2
253	O	N2(2pi)	1.91	5	5
254	O	NP(pi)	−1.95	14	14
255	O	Si2	0.09	18	4
256	S2	HC	0.65	14	12
257	S2	HC(pi)	0.14	31	31
258	S2	C2	1.39	48	45
259	S2	C2(pi)	0.98	68	63
260	S2	C2(2pi)	0.98	55	54
261	S2	CN	0	3	3
262	S2	CN(2pi)	2.3	1	1
263	S2	CS	0.87	2	1
264	S2	CS(pi)	1.97	4	2
265	S2	CP	1.12	17	15
266	S2	CP(pi)	0.48	3	2
267	S2	N2	−2.2	2	2
268	S2	N2(2pi)	5.96	1	1
269	S4	C2=O	−1.13	11	11
270	S4	C2=O2	−0.5	16	16
271	S4	CO=O2	−0.48	2	1
272	S4	CN=O2	−0.05	85	80
273	S4	C=O2F	0.24	2	2
274	S4	NO=O2	0	3	3
275	S4	N2=O2	0.77	5	5
276	S4	O2=O	0.83	2	2
277	S4	O2=O2	0.5	2	2
278	S4	O2=O2(−)	−1.14	3	3
279	P4	CO2=O	−1.11	2	2
280	P4	CO2=S	0.26	1	1
281	P4	CO=OS	−2.58	1	1
282	P4	CO=OF	−0.88	3	3

Table 2. *Cont.*

Entry	Atom Type	Neighbours	Contribution	Occurrences	Molecules
283	P4	COS=S	−2.04	1	1
284	P4	O3=O	−0.56	29	29
285	P4	O3=S	1.12	18	18
286	P4	O2S=S	0.7	12	11
287	P4	O=OS2	−0.54	2	2
288	P4	N3=O	−0.31	1	1
289	P4	N2O=O	0.24	2	2
290	P4	NO=OS	−1.5	2	2
291	Si	C4	−0.51	1	1
292	Si	C3O	−1.7	2	1
293	Si	C2O2	0.13	17	4
294	Si	O4	0	2	2
295	Halide		1.1	20	19
296	H	H Acceptor	0.51	164	154
297	(COH)n	n > 1	0.26	137	74
298	(COOH)n	n > 1	−0.15	26	25
299	Alkane	No. of C atoms	0.09	290	32
300	Unsaturated HC	No. of C atoms	0.02	1584	135
301	Endocyclic bonds	No. of single bds	−0.14	2338	384
A	Based on	Valid groups	214		3332
B	Goodness of fit	R^2	0.9648		3246
C	Deviation	Average	0.31		3246
D	Deviation	Standard	0.39		3246
E	K-fold cv	K	10		3164
F	Goodness of fit	Q^2	0.9599		3164
G	Deviation	Average (cv)	0.33		3164
H	Deviation	Standard (cv)	0.42		3164

Table 3. Atom and special groups and their contribution in logK_{oa} calculations.

Entry	Atom Type	Neighbours	Contribution	Occurrences	Molecules
1	Const		1.46	1900	1900
2	C sp3	H3C	−0.07	1800	875
3	C sp3	H3N	3.42	131	87
4	C sp3	H3N(+)	1.42	1	1
5	C sp3	H3O	2.24	292	219
6	C sp3	H3S	1.51	30	26
7	C sp3	H3P	−0.42	3	3
8	C sp3	H3Si	0.42	68	11
9	C sp3	H2C2	0.43	1732	538
10	C sp3	H2CN	3.91	191	129

Table 3. *Cont.*

Entry	Atom Type	Neighbours	Contribution	Occurrences	Molecules
11	C sp3	H2CN(+)	1.64	6	5
12	C sp3	H2CO	2.61	535	342
13	C sp3	H2CS	1.76	57	44
14	C sp3	H2CP	2.58	3	3
15	C sp3	H2CF	−0.77	3	3
16	C sp3	H2CCl	0.71	75	56
17	C sp3	H2CBr	1.05	23	18
18	C sp3	H2CJ	1.13	5	5
19	C sp3	H2CSi	2.91	4	4
20	C sp3	H2N2	4.89	8	3
21	C sp3	H2NO	5.65	9	8
22	C sp3	H2NS	4.67	5	5
23	C sp3	H2O2	4.78	6	4
24	C sp3	H2S2	3.74	4	4
25	C sp3	HC3	0.64	268	180
26	C sp3	HC2N	4.08	64	53
27	C sp3	HC2N(+)	2.05	1	1
28	C sp3	HC2O	2.86	169	135
29	C sp3	HC2S	1.76	9	7
30	C sp3	HC2F	−1.66	1	1
31	C sp3	HC2Cl	1.21	43	17
32	C sp3	HC2Br	1.31	14	9
33	C sp3	HC2J	1.95	1	1
34	C sp3	HCNO	8.18	3	3
35	C sp3	HCNS	2.08	1	1
36	C sp3	HCO2	5.73	6	6
37	C sp3	HCF2	−0.18	7	7
38	C sp3	HCFCl	0.02	2	2
39	C sp3	HCCl2	1.18	15	14
40	C sp3	HCClBr	0.77	1	1
41	C sp3	HOF2	1.79	3	3
42	C sp3	C4	0.73	98	84
43	C sp3	C3N	4.1	13	13
44	C sp3	C3O	3.11	40	37
45	C sp3	C3S	2.6	3	3
46	C sp3	C3Cl	0.87	37	15
47	C sp3	C2NO	5.94	1	1
48	C sp3	C2O2	5.94	6	6
49	C sp3	C2F2	0.23	58	10
50	C sp3	C2Cl2	1.24	18	17
51	C sp3	CNO2	9.56	1	1

Table 3. *Cont.*

Entry	Atom Type	Neighbours	Contribution	Occurrences	Molecules
52	C sp3	COF2	3.06	3	3
53	C sp3	CF3	−0.06	55	51
54	C sp3	CF2Cl	−0.02	4	3
55	C sp3	CFCl2	0.37	3	2
56	C sp3	CCl3	1.62	17	16
57	C sp3	CBr3	0.57	1	1
58	C sp3	O2F2	6.85	1	1
59	C sp3	OF3	1.86	3	3
60	C sp2	H2=C	−0.19	88	76
61	C sp2	HC=C	0.34	233	141
62	C sp2	HC=N	0.85	8	8
63	C sp2	HC=O	1	27	27
64	C sp2	H=CN	1.1	19	13
65	C sp2	H=CO	0.48	15	14
66	C sp2	H=CS	−1.08	9	7
67	C sp2	H=CCl	0.44	12	10
68	C sp2	H=CBr	0.6	3	2
69	C sp2	H=CSi	2.17	1	1
70	C sp2	HN=N	1.73	53	30
71	C sp2	HN=O	2.27	3	3
72	C sp2	HO=O	0.92	4	4
73	C sp2	H=NS	2.88	1	1
74	C sp2	C2=C	0.8	103	79
75	C sp2	C2=N	1.62	34	30
76	C sp2	C=CN	1.6	19	16
77	C sp2	C2=O	1.08	87	75
78	C sp2	C=CO	1.22	27	26
79	C sp2	C=CP	−0.09	1	1
80	C sp2	C=CS	−0.41	14	10
81	C sp2	C=CCl	0.62	39	24
82	C sp2	C=CBr	1.01	12	5
83	C sp2	=CN2	2.98	2	2
84	C sp2	CN=N	2.75	7	7
85	C sp2	CN=O	2.64	93	88
86	C sp2	C=NO	1.26	5	5
87	C sp2	=CNO	−1.24	3	3
88	C sp2	C=NS	0.46	6	6
89	C sp2	=CNCl	3.35	6	3
90	C sp2	CO=O	1.73	244	210
91	C sp2	C=OS	−0.61	3	2
92	C sp2	=CS2	−0.66	1	1

Table 3. *Cont.*

Entry	Atom Type	Neighbours	Contribution	Occurrences	Molecules
93	C sp2	=CF2	−1.14	1	1
94	C sp2	=CCl2	1.14	16	14
95	C sp2	N2=N	3.36	9	9
96	C sp2	N2=O	3.65	43	40
97	C sp2	N=NO	2.64	4	4
98	C sp2	N=NS	0.71	7	7
99	C sp2	NO=O	2.81	38	36
100	C sp2	N=OS	0.93	17	17
101	C sp2	NO=S	4.26	1	1
102	C sp2	=NOS	0.68	3	3
103	C sp2	NS=S	6.03	3	2
104	C sp2	=NSCl	−5.44	2	2
105	C sp2	O2=O	2.56	3	3
106	C aromatic	H:C2	0.31	5436	1136
107	C aromatic	H:C:N	0.53	81	49
108	C aromatic	H:N2	0.17	6	6
109	C aromatic	:C3	0.89	441	148
110	C aromatic	C:C2	0.79	1163	657
111	C aromatic	C:C:N	0.68	42	30
112	C aromatic	:C2N	1.35	164	146
113	C aromatic	:C2N(+)	2.09	96	69
114	C aromatic	:C2:N	1.01	13	10
115	C aromatic	:C2O	1.27	769	453
116	C aromatic	:C2P	3.53	5	3
117	C aromatic	:C2S	−0.19	38	33
118	C aromatic	:C2Si	−0.25	1	1
119	C aromatic	:C2F	0.13	99	41
120	C aromatic	:C2Cl	0.91	1844	550
121	C aromatic	:C2Br	1.24	391	143
122	C aromatic	:C2J	2.14	10	9
123	C aromatic	C:N2	0.77	11	10
124	C aromatic	:CN:N	0.8	4	4
125	C aromatic	:C:NO	1.2	28	24
126	C aromatic	:C:NCl	0.9	14	12
127	C aromatic	N:N2	1.18	60	36
128	C aromatic	:N2O	1.15	11	11
129	C aromatic	:N2S	−0.6	8	8
130	C aromatic	:N2Cl	0.43	9	8
131	C sp	H#C	−0.45	18	17
132	C sp	C#C	0.67	18	17
133	C sp	C#N	0.73	46	43

Table 3. *Cont.*

Entry	Atom Type	Neighbours	Contribution	Occurrences	Molecules
134	C sp	N#N	5.32	1	1
135	C sp	#NP	−5.58	1	1
136	C sp	=N=S	−0.13	2	2
137	N sp3	H2C	−2.18	17	16
138	N sp3	H2C(pi)	1.02	57	53
139	N sp3	H2N	3.57	5	5
140	N sp3	H2S	1.81	1	1
141	N sp3	HC2	−5.94	12	11
142	N sp3	HC2(pi)	−2.38	93	70
143	N sp3	HC2(2pi)	0.08	65	56
144	N sp3	HCN(pi)	0.02	5	4
145	N sp3	HCN(2pi)	1.2	4	4
146	N sp3	HCO(pi)	1.13	1	1
147	N sp3	HCP	−4.1	3	3
148	N sp3	HCP(pi)	1.51	1	1
149	N sp3	HCS(pi)	−1.54	8	8
150	N sp3	C3	−9.44	17	17
151	N sp3	C3(pi)	−6.39	58	55
152	N sp3	C3(2pi)	−4.82	49	45
153	N sp3	C3(3pi)	−3.61	9	9
154	N sp3	C2N	−5.12	1	1
155	N sp3	C2N(pi)	−2.54	15	14
156	N sp3	C2N(+)(pi)	−1.93	7	2
157	N sp3	C2N(2pi)	−3.84	36	36
158	N sp3	C2N(3pi)	−0.65	13	12
159	N sp3	C2P	0	1	1
160	N sp3	C2P(pi)	−2.97	1	1
161	N sp3	C2P(2pi)	−4.07	1	1
162	N sp2	H=C	0.51	1	1
163	N sp2	C=C	−0.97	54	48
164	N sp2	C=N	0.61	6	4
165	N sp2	=CN	0.03	54	49
166	N sp2	=CN(+)	9.74	2	2
167	N sp2	=CO	−3.65	30	26
168	N sp2	N=N	−1.3	4	3
169	N sp2	N=O	−2.02	13	13
170	N aromatic	:C2	0.54	194	109
171	N aromatic	:C:N	0.47	4	1
172	N(+) sp2	CO=O(−)	−0.36	104	76
173	N(+) sp2	NO=O(−)	0	9	4
174	N(+) sp2	O2=O(−)	−1.09	63	35

Table 3. *Cont.*

Entry	Atom Type	Neighbours	Contribution	Occurrences	Molecules
175	O	HC	−0.66	143	121
176	O	HC(pi)	1.39	175	159
177	O	HN(pi)	4.18	2	2
178	O	HP	2.11	4	2
179	O	HSi	1.91	3	2
180	O	C2	−4.17	139	105
181	O	C2(pi)	−2.68	392	317
182	O	C2(2pi)	−0.92	255	228
183	O	CN(pi)	0.51	20	16
184	O	CN(+)(pi)	0.1	63	35
185	O	CN(2pi)	3.07	8	8
186	O	CO(pi)	−1.03	2	1
187	O	CS	−0.88	11	6
188	O	CP	−1.2	183	93
189	O	CP(pi)	−0.01	70	54
190	O	CSi	−2.38	9	3
191	O	NP(pi)	4.65	1	1
192	O	P2	1.7	1	1
193	O	Si2	0	21	6
194	P4	C3=O	−5.7	1	1
195	P4	CNO=O	1.2	1	1
196	P4	CO2=O	1.47	3	3
197	P4	CO2=S	−1.5	3	3
198	P4	CO=OS	1.99	1	1
199	P4	CO=OF	1.94	1	1
200	P4	COS=S	−0.86	1	1
201	P4	NO2=O	3.42	1	1
202	P4	NO2=S	1.88	3	3
203	P4	NO=OS	1.2	2	2
204	P4	O3=O	0.09	29	29
205	P4	O3=S	−0.4	32	30
206	P4	O2=OS	0.43	5	5
207	P4	O2=OF	−0.17	1	1
208	P4	O=OS2	1.58	3	3
209	P4	O2S=S	−0.27	18	17
210	P4	=OS3	1.46	1	1
211	S2	HC	−1.08	2	2
212	S2	HC(pi)	1.54	1	1
213	S2	C2	−1.5	14	14
214	S2	C2(pi)	0.4	41	39
215	S2	C2(2pi)	2.82	24	23

Table 3. *Cont.*

Entry	Atom Type	Neighbours	Contribution	Occurrences	Molecules
216	S2	CS	−0.58	4	2
217	S2	CS(pi)	−2.98	2	1
218	S2	CP	−0.12	33	28
219	S2	CP(pi)	1.78	3	2
220	S4	C2=O	0.6	2	2
221	S4	C2=O2	2.13	3	3
222	S4	CN=O2	3.26	9	9
223	S4	CO=O2	−0.05	1	1
224	S4	O2=O	−0.35	2	2
225	S4	O2=O2	0.24	3	3
226	S6	CF5	1.92	3	3
227	Si	C4	−1.33	3	3
228	Si	C3O	−0.65	7	4
229	Si	C2O2	0.1	19	6
230	Si	CO3	0	3	3
231	H	H Acceptor	−1.51	47	45
232	(COH)n	n > 1	0.06	22	15
233	(COOH)n	n > 1	1.2	6	6
234	Alkane	No. of C atoms	−0.05	268	34
235	Unsaturated HC	No. of C atoms	−0.03	1512	140
236	Endocyclic bonds	No. of single bds	−0.11	1109	210
A	Based on	Valid groups	167		1900
B	Goodness of fit	R^2	0.9765		1829
C	Deviation	Average	0.34		1829
D	Deviation	Standard	0.44		1829
E	K-fold cv	K	10		1765
F	Goodness of fit	Q^2	0.9717		1765
G	Deviation	Average (cv)	0.37		1765
H	Deviation	Standard (cv)	0.48		1765

2.3. Calculation of Descriptors $logP_{ow}$ and $logK_{oa}$

Based on the aforementioned respective atom-group-parameter tables, the descriptors $logP_{ow}$ and $logK_{oa}$ of a molecule can now easily be calculated by simply summing up the contribution of each atom and special group occurring in the molecule, following Equation (1), wherein i and j are the number of atom groups A_i and special groups B_j, respectively, a_i is the contribution of atom group A_i, b_j is the contribution of special group B_j and c is the constant listed at the top of the respective parameter tables.

$$\text{Descriptor calc} = \sum a_i A_i + \sum b_j B_j + c \quad (1)$$

In Table 4, a typical example is presented with endosulfan sulphate (Figure 1), demonstrating the ease of the calculation of $logK_{oa}$ for which the experimental value was 9.68 [30]. Note that the term "endocyclic bonds" only concerns C-C single bonds.

Table 4. Example calculation of the $logK_{oa}$ of endosulfan sulphate.

Atom Type	C sp3	C sp3	C sp3	C sp3	C sp2	O	S4	Endocycl. Bonds	Const	Sum
Neighbors	H2CO	HC3	C3Cl	C2Cl2	C=CCl	CS	O2=O2	n C-C		
Contribution	2.61	0.64	0.87	1.24	0.62	−0.88	0.24	−0.11	1.46	
n Groups	2	2	2	1	2	2	1	9		
n × Contribution	5.22	1.28	1.74	1.24	1.24	−1.76	0.24	−0.99	1.46	9.67

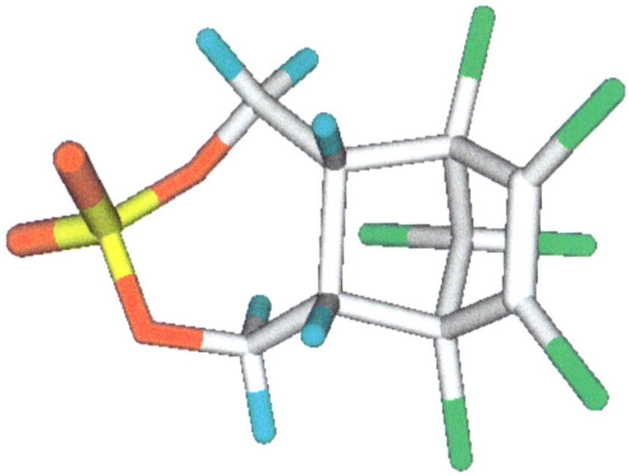

Figure 1. Endosulfan sulphate (graphics by ChemBrain IXL).

Evidently, the group-additivity method is limited to the calculation of a molecule's $logP_{ow}$ or $logK_{oa}$ for which a parameter value in the respective Table 2 or Table 3 is known for each atom group that is found in the molecule. Beyond this, since the reliability of these parameter values increases with the number of independent molecules upon which their calculation is based, the lowest reliability limit for these parameters was set to three molecules, which, as a consequence, excluded any atom group based on less than three molecules from further calculations. Accordingly, only atom groups for which the number of molecules is three or more, shown in the rightmost columns of Tables 2 and 3, have been accepted as "valid" for descriptor calculations. This explains the lower number of molecules for which the $logP_{ow}$ and $logK_{oa}$ have been calculated (lines B, C and D in Tables 2 and 3) than the number upon which the evaluation of the complete set of parameters is based (line A).

2.4. Cross-Validation Calculations

The plausibility of the group-parameter calculations was immediately checked by applying a 10-fold cross-validation algorithm, which comprises 10 recalculations of the complete set of group parameters, whereby, before each recalculation, every other 10th compound of the total compounds' list is temporarily removed from the calculation and separated into a test list, thus ensuring that each molecule has played the role of a test sample once. The combined test data were then statistically worked up and their results added to Tables 2 and 3 at the bottom in lines E, F, G and H. It may be noticed that the total number of test compounds shown in the right-most column of the statistics lines is lower than that in the training set in lines B, C and D; this is a consequence of the requirement that only "valid" atom groups are to be used for descriptor calculations, and due to the 10% lower number of training samples in each recalculation, the number of "valid" atom groups (as defined in the prior section) tends to decrease to an unpredictable degree. Atom groups, which are represented by less than three molecules, as shown in the right-most column, and are thus not "valid" for descriptor calculations, are therefore remnants of the parameter calculation based on the complete compound set (line A in Tables 2 and 3). Nevertheless, they have deliberately been left in Tables 2 and 3 for use in future calculations with additional molecules potentially carrying under-represented atom groups in this ongoing project.

3. Sources

3.1. Sources of $logP_{ow}$ Values

The majority of the experimental $logP_{OW}$ data originates from the comprehensive collection of Klopman et al. [8], supplemented by works of Sangster [32] and Lipinski et al. [33], already cited in [13]. Additional data have been provided for unsubstituted and substituted, saturated and unsaturated hydrocarbons, alcohols and esters in the works of Tewari et al. [34]; for heterocycles, hetarenes and carboxylic acids by Ghose et al. [6,7]; complemented for amines, amides and nitro derivatives by Leo [10]. Further data for the aforementioned compound classes have been found in papers by Abraham et al. [35], for certain drugs by Hou and Xu [12] and Wang et al. [11], for organophosphorus derivatives by Czerwinski et al. [36], for the specific energetic compound 2,4-dinitroanisole by Boddu et al. [37], for a number of fluorobenzenes, -anilines and -phenols by Li et al. [38] and finally for a number of pesticides and oil constituents in a paper by Saranjampour et al. [39].

3.2. Sources of $logK_{oa}$ Values

Recently, Ebert et al. [40] published a comprehensive collection of more than 2000 experimental $logK_{oa}$ values upon which the present study is essentially based. This set of data has been complemented with data for 75 chloronaphthalene derivatives by Puzyn et al. [41], for 14 PAHs by Odabasi et al. [42], for some methylsiloxanes and dimethylsilanol by Xu and Kropscott [43] and for ethyl nitrate by Easterbrook et al. [44].

3.3. Sources of $logK_{aw}$ Values

Ebert's paper [30], cited in the introductory section, presented in their supplementary information a large collection of experimental $logK_{aw}$ data, which served as reference values for the calculated data. Sander [45] provided an extensive library of Henry's law constants for more than 2600 compounds, which, after translation into $logK_{aw}$ values at 298.15 K, complemented Ebert's data set.

4. Results

4.1. Partition Coefficient $logP_{ow}$

As shown at the bottom of Table 2, the number of molecules upon which the present group parameter set is based is 3332, substantially larger than the 2780 samples in our earlier paper [13]. Beyond this, the significantly better statistical results in Table 2 (lines B to H) with, e.g., a cross-validated (cv) standard deviation S of 0.42 (line H) vs. the earlier value of 0.51 is the result of the removal of molecules from the parameter computation for which the experimental value deviates by more than three times the value of S. The 122 molecules thus removed (3.5% of the total set) have been collected in an outlier list, available in the Supplementary Materials. The larger number of compounds for the group parameter computation not only significantly improved the statistical results but also enlarged the list of "valid" atom groups from 195 to 214, enabling the calculation of the $logP_{OW}$ value of at present 29,462 molecules (79.4% of the total dataset). The correlation coefficients R^2 of 0.9648, the (cross-validated) Q^2 of 0.9599 and the cv standard error S of 0.42, based on 3246 and 3164 molecules, respectively, are significantly better than in our earlier paper [13] and compare very well not only with Klopman's [8] results, which are based on a group-additivity method comparable to ours and have R^2 and Q^2 values of 0.93 and 0.926, respectively, but also with the statistical results of more elaborate calculation methods published recently [4,5,14–16]. As shown in the correlation diagram of Figure 2 and the histogram for Figure 3, the experimental $logP_{OW}$ values range from −4.6 to +12.53 with a fairly even Gaussian error distribution.

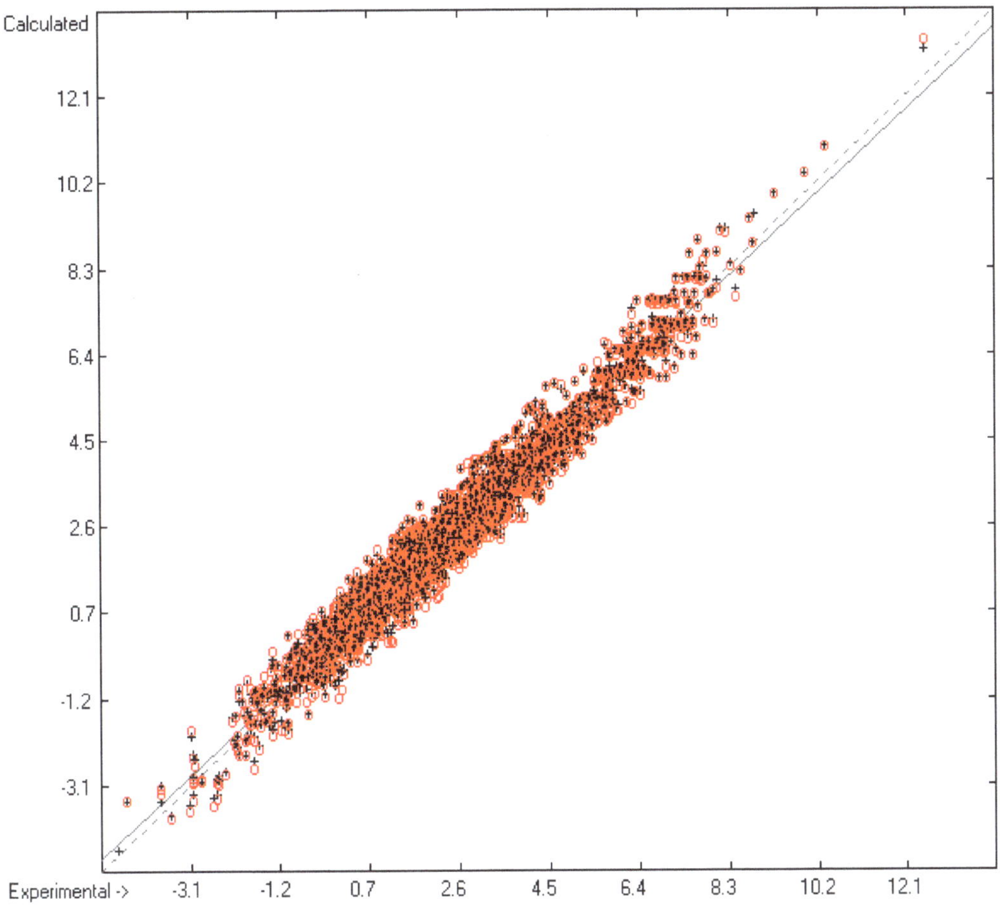

Figure 2. Correlation diagram of the $\log P_{ow}$ data. Cross-validation data are superpositioned as red circles (10-fold cross-valid.: N = 3246; Q^2 = 0.9599; regression line: intercept = 0.1052; slope = 0.9636).

It is worth mentioning that the observation discussed in our earlier paper (see Table 9 in [13]) concerning the two forms of amino acids (nonionic or zwitterionic) is not only confirmed by the new and extended group parameter set of Table 2 but also that the $\log P_{ow}$ differences in nearly all cases even more clearly distinguish the two forms. On the other hand, the ambiguous results concerning the keto/enol forms of the compounds listed in Table 10 in [13] could not be lifted by the new parameter set, which is not surprising in view of the sometimes strong solvent dependence of the equilibrium, as exemplified with acetylacetone [46]. In view of the discussion of certain particularities concerning the subsequent calculation of the third partition coefficient $\log K_{aw}$ in Section 4.3, it should be stressed at this point that the calculated $\log P_{ow}$ values for the hydrocarbons do not show any abnormal or systematic deviations from experimental values.

4.2. Partition Coefficient $\log K_{oa}$

The calculation of the group parameter set of Table 3 used for the prediction of the $\log K_{oa}$ values is essentially based on the curated data set provided in Ebert's paper [40], whereby compounds with just one "backbone atom" such as halomethanes or hydrocyanide had to be omitted as they are obviously not calculable using the present method. After the removal of another 129 compounds as outliers (6.36% of the total), following the same exclusion criterion as in the previous section, 1900 samples with their experimental data (line A in Table 3) remained for the computation of the group parameter values. Again, the outliers have been collected in a separate list available in the Supplementary Materials for readers who might want to re-evaluate their logK_{oa} values.

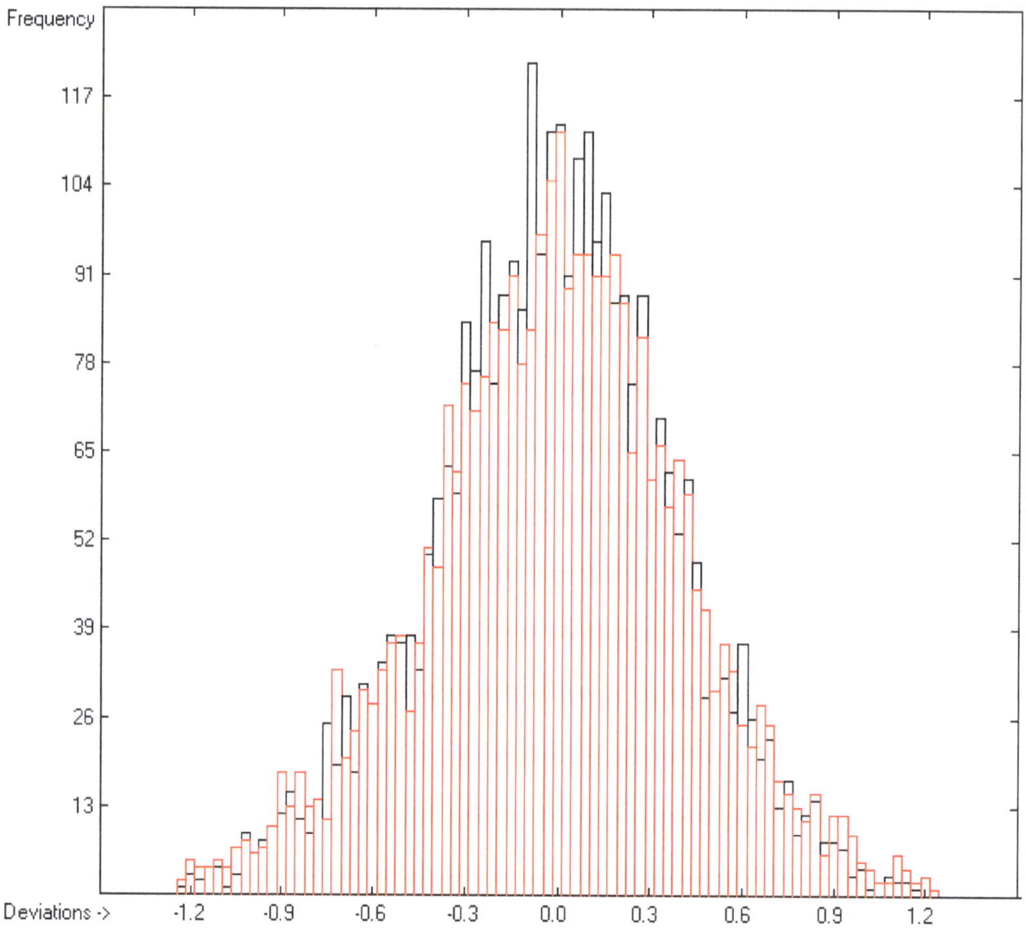

Figure 3. Histogram of the logP_{ow} data. Cross-validation data are superpositioned as red bars. (σ = 0.39; S = 0.42; experimental values range from −4.6 to +12.53).

The subsequent calculation of the logK_{oa} values of 1829 training and 1765 test molecules based on 167 "valid" atom and special groups (line A) revealed excellent statistical results with a correlation coefficient R^2 of 0.9765, a standard deviation s of 0.44 (lines B and D) and a cross-validated Q^2 of 0.9717 with a corresponding S of 0.48 (lines F and H), visualised in the correlation diagram on Figure 4 and the histogram on Figure 5. These statistical data even outperform those given in Ebert's paper and thus also the competing methods mentioned therein such as COSMOtherm [47] and EPI-Suite KOAWIN [48], not only confirming the versatility but also the reliability of the present group-additivity approach, which allowed the calculation of the logK_{oa} value for 27,044 molecules (72.9% of the entire database). Again, it should be kept in mind that just like in Section 4.1, any particularly large or systematic deviations between the experimental and calculated logK_{oa} values for the hydrocarbons could not be observed.

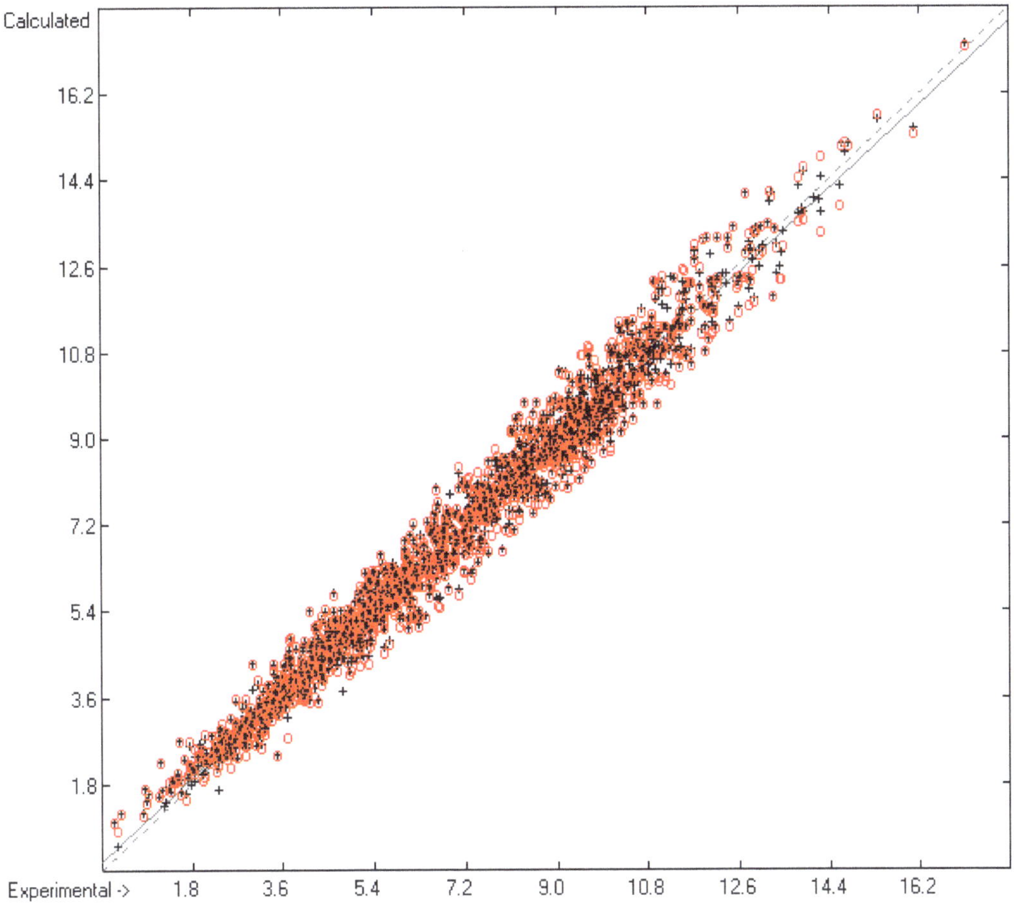

Figure 4. Correlation diagram of the $\log K_{oa}$ data. Cross-validation data are superpositioned as red circles (10-fold cross-valid.: N = 1829; Q^2 = 0.9717; regression line: intercept = 0.1997; slope = 0.9729; MAPD = 6.39%).

4.3. Partition Coefficient logK$_{aw}$

Once the partition coefficients $\log P_{ow}$ and $\log K_{oa}$ were calculated by means of the group-additivity method based on Tables 2 and 3, respectively, it was easy to determine the $\log K_{aw}$ values, applying Equation (2) on each molecule in the database for which both descriptors had been calculated, adding up to 26,220 molecules. In order to assess the quality of the $\log K_{aw}$ values, it is important to recognise the flaws of this approach: while the $\log P_{ow}$ values were experimentally measured in a mixture of water-saturated octanol and octanol-saturated water, the $\log K_{oa}$ measurements occurred in dry octanol, an aspect that has been discussed in detail by Ebert et al. [40]. Hence, Equation (2) serves only as an approximation. In addition, since both descriptors on the right side of the equation appear with their own standard error, the error-propagation rule stipulates a standard error of $\log K_{aw}$ that is clearly larger than either of the two constituting descriptors. Entering the standard errors S for the test molecules of 0.42 (for $\log P_{ow}$) and 0.48 (for $\log K_{oa}$) into an error-propagation calculation, the expected standard error S for $\log K_{aw}$ is 0.638.

$$\log K_{aw}\ (\text{calc}) = \log P_{ow}\ (\text{calc}) - \log K_{oa}\ (\text{calc}) \quad (2)$$

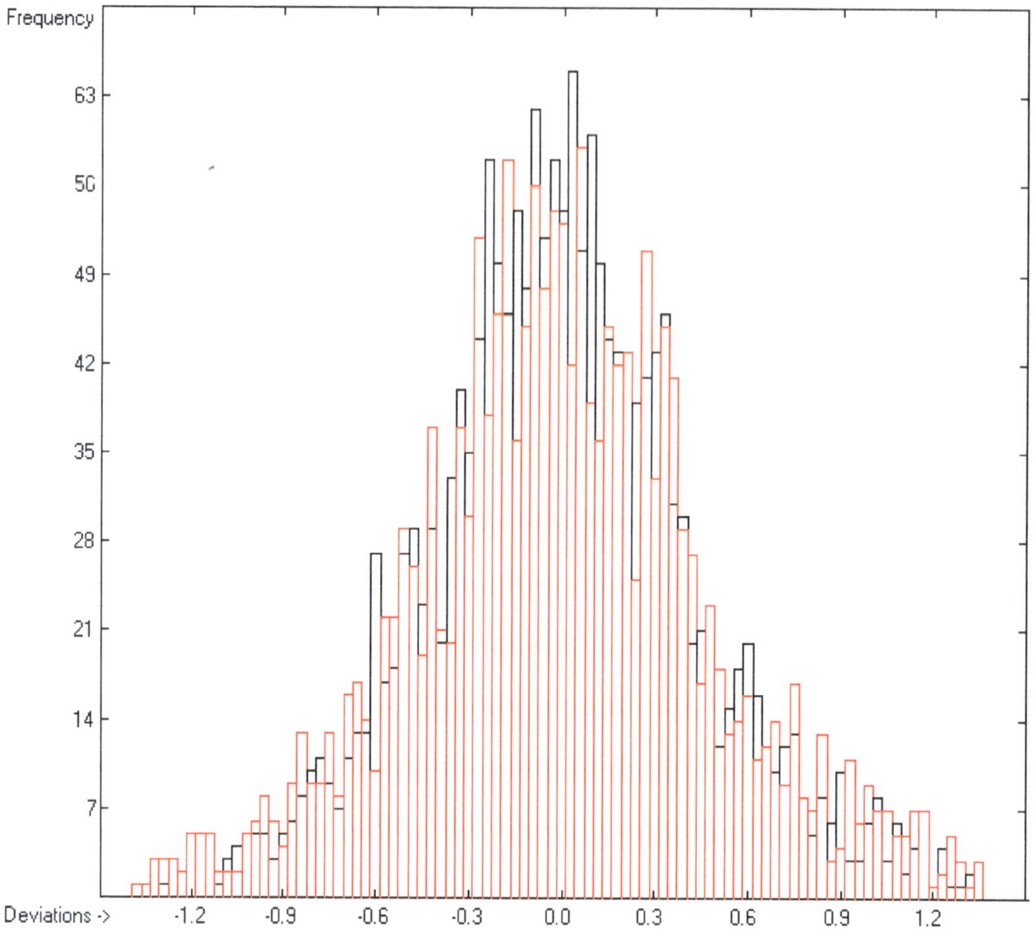

Figure 5. Histogram of the $logK_{oa}$ data. Cross-validation data are superpositioned as red bars (σ = 0.44; S = 0.48; experimental values range from 0.28 to 17.15).

In order to test the reliability of the thus-calculated $logK_{aw}$ values, a representative number of experimentally determined $logK_{aw}$ data, extracted from the comprehensive databases of Ebert et al. [30] and Sander [45], were added to the database. In the latter case, the Henry's law solubility constants H_s^{cp} were translated into the corresponding $logK_{aw}$ values at 298.15 K. The comparison of the calculated with the experimental $logK_{aw}$ values is visualised in the correlation diagram of Figure 6 and the histogram in Figure 7.

The complete set of experimental data was separated from the outliers, applying the same exclusion conditions as for the $logP_{ow}$ and $logK_{oa}$ values, and the outliers were collected in a corresponding list, available in the Supplementary Materials. Comparison of the remaining dataset with the calculated values yielded a standard error of 0.67, slightly higher than that predicted by the error-propagation calculation. A detailed analysis of the experimental data revealed two potential explanations for the inordinate scatter: (1) Within a series of substitution isomers, e.g., the tetra- or hexachlorobiphenyls, the tri- or pentachlorodiphenyl ethers or the dichloroanisoles, the experimental $logK_{aw}$ values varied in a range of up to and over 1 unit, which was hard to assign to the specific positioning of the substituents. At any rate, the group-additivity-based calculation of the $logP_{ow}$ and $logK_{oa}$ values was not able to distinguish between these substitution isomers. (2) Sander's comprehensive database of Henry's law constants [45], listing the experimental H_s^{cp} values for a compound originating from various authors, showed for many compounds large differences between their H_s^{cp} values, in some cases exceeding one unit after translation into $\log K_{aw}$, e.g., for undecane, acetylacetone or anthraquinone.

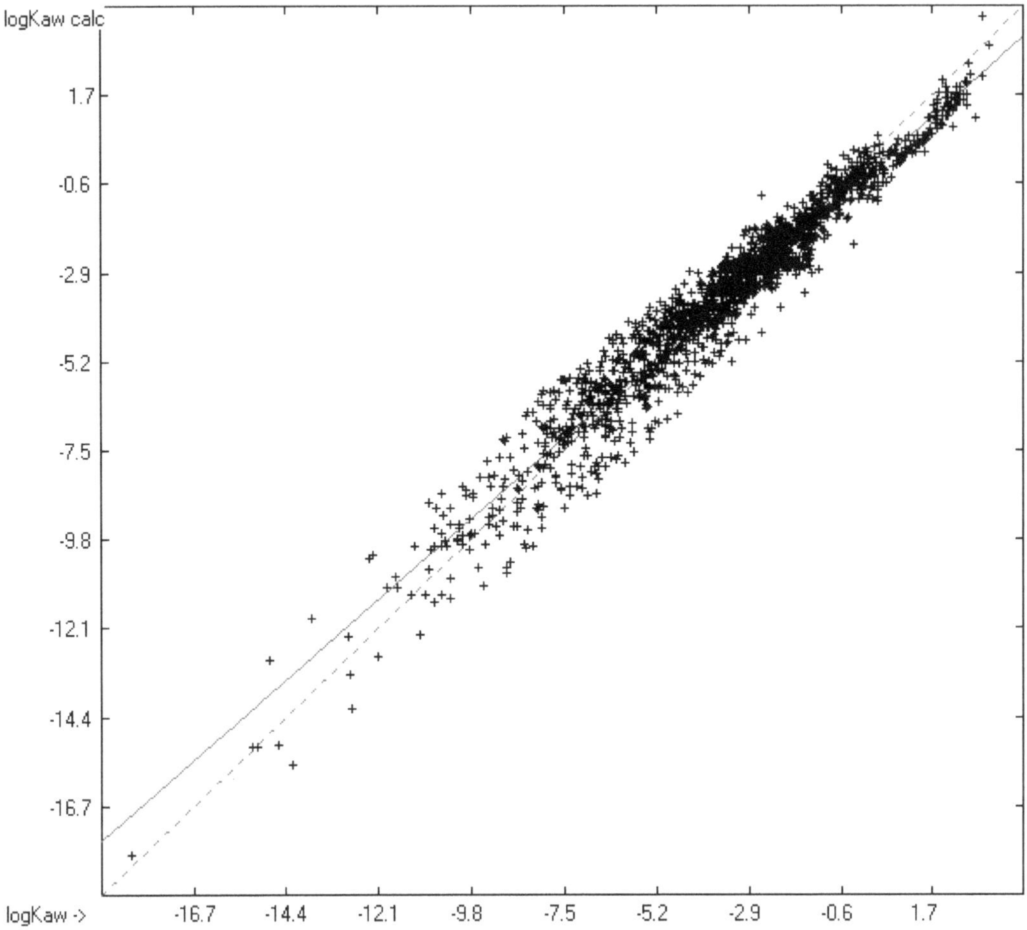

Figure 6. Correlation diagram of the $\log K_{aw}$ data (N = 1937; Q^2 = 0.9467; regression line: intercept = −0.4196; slope = 0.9044).

A thorough analysis of the correlation diagram in Figure 6 and the histogram in Figure 7 revealed an interesting peculiarity, visible as an indentation at the upper end of the correlation diagram and as a weak hump on the right side of the histogram: except for some siloxanes with experimental $\log K_{aw}$ values above 1.6 and normal scatter about calculated values, the predicted $\log K_{aw}$ for the remaining compounds with experimental $\log K_{aw}$ values above −1.0 were nearly systematically too low by ca. 0.5–1 units. It turned out that they were all pure hydrocarbons, in particular alkanes, alkenes and alkynes. The correlation diagram of the $\log K_{aw}$ data in Figure 8, focussing on these hydrocarbons, confirms this observation.

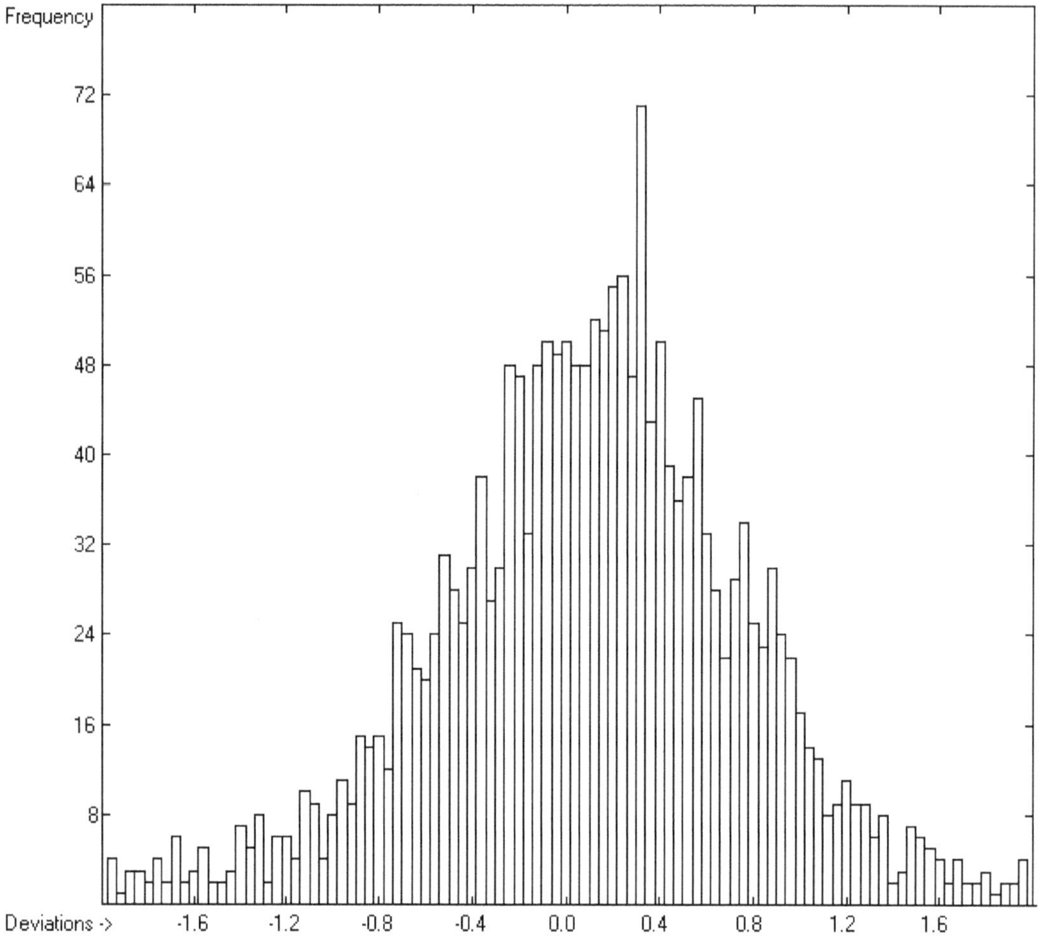

Figure 7. Histogram of the $\log K_{aw}$ data (S = 0.67; experimental values range from −17.99 to +3.71).

Since, as mentioned in Sections 4.1 and 4.2, no particularly large or systematic deviations between the experimental and calculated $\log P_{ow}$ and $\log K_{oa}$ data for the hydrocarbons could be detected, a potential explanation for this peculiarity might be based on the experimental conditions for the determination of the $\log P_{ow}$ values as mentioned by Ebert et al. [40]: since water-saturated octanol is a more polar solvent than pure octanol, while octanol-saturated water is less polar than pure water, the experimental $\log P_{ow}$ values, measured in an octanol/water mixture, tend to be shifted to smaller absolute values than theory would predict. While this is true for all measured solutes, it is possibly most effective for the least polar solutes such as the mentioned hydrocarbons, thus leading to experimental $\log P_{ow}$ values that are particularly low for the hydrocarbons. As a consequence, their calculation based on the group-additivity method predicts equally low $\log P_{ow}$ values, which again lead to low $\log K_{aw}$ data when Equation (2) is applied and when compared with experimental $\log K_{aw}$ values that are determined under pure air/water conditions.

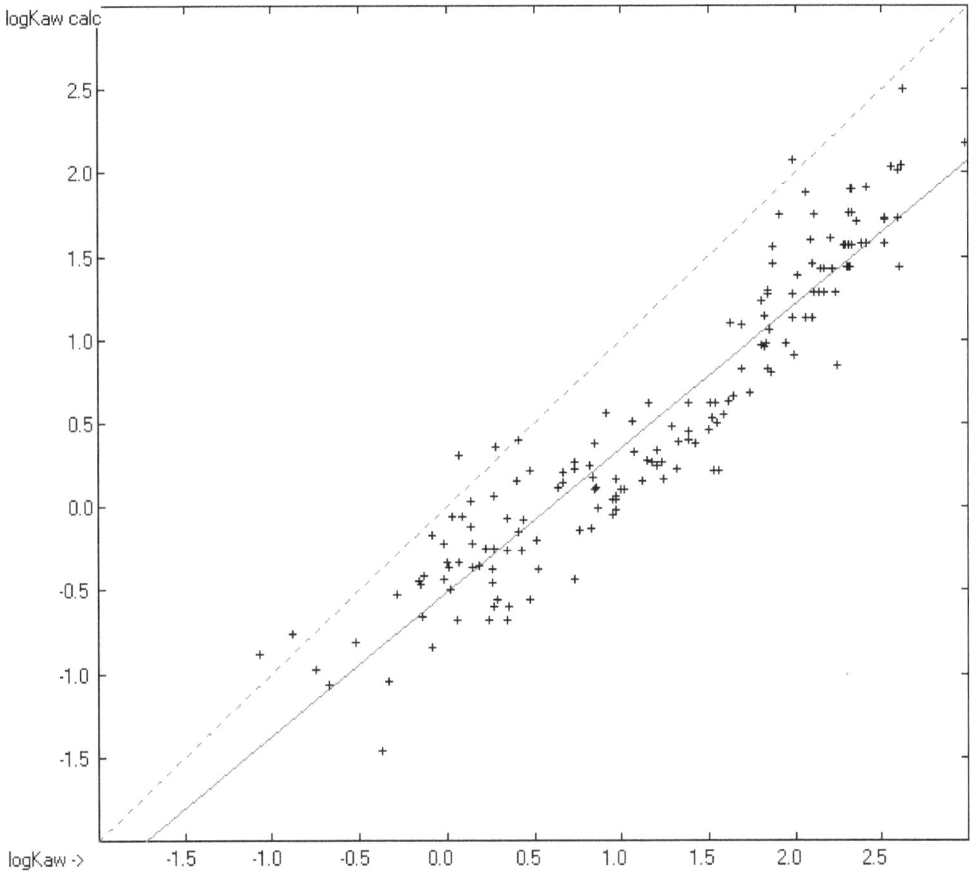

Figure 8. Correlation diagram of the $logK_{aw}$ data for alkanes, alkenes and alkynes (N = 170).

4.4. Interpretation of the Special Groups' Contributions to $logP_{ow}$ and $logK_{oa}$, and Ultimately for $logK_{aw}$

While the atom group parameters are descriptor-specific and their comparison between descriptors does not make sense, special groups serve as differentiators of molecules that carry these groups from those that do not. Therefore, their meanings have overlapping descriptors; their values, however, must be viewed in the context of the value range of the descriptors. In the present case, the value ranges of $logP_{ow}$ and $logK_{oa}$ are similar (ca. 17 log units) and in the same area, and thus, a direct comparison of the special group contributions in Tables 2 and 3 is permissible and leads to a few interesting observations: While the groups "(COH)n", "Alkane", "Unsaturated HC" and "Endocyclic bonds" in both tables only contribute to a minor degree (but nevertheless improve the statistical results) and consequently show only minor differences between the two tables, a significant difference was found for the groups "H/H Acceptor" and "(COOH)n". The former special group, taking account of intramolecular hydrogen bridges, indicates a small but clearly higher chance of being a compound carrying an intramolecular H-bridge towards the octanol side in an octanol/water mixture, thus raising the $logP_{ow}$ value. In contrast, the same H-bridge-carrying molecule has its inclination significantly shifted to the air side in an octanol/air environment compared to that without H-bridges, expressed in its lower $logK_{oa}$ value. The reason may be found in the lower solvent–solute interaction caused by the H-bridge being bound intramolecularly, leading in both cases to a preference for the less polar of the respective two media. A typical example is the compound couple 2- and 3-nitroaniline, sampled in Table 5, where the former molecule carries a H-bridge between an amino-H and an oxygen of the nitro group.

Table 5. Experimental (calculated) $logP_{ow}$ and $logK_{oa}$ values of 2- and 3-nitroaniline.

Descriptor	2-Nitroaniline	3-Nitroaniline
$logP_{ow}$	1.85 (1.70)	1.37 (1.19)
$logK_{oa}$	6.46 (5.29)	7.62 (6.80)

An inverse effect can be found with molecules carrying two or more carboxylic acid functions: While the additional contribution of a second or third COOH function shows little effect in an octanol/water environment with a slightly increased shift towards water, leading to a lower $logP_{ow}$ value, in an octanol/air environment, each additional COOH group drastically tilts the equilibrium towards the octanol side, thus strongly raising the $logK_{oa}$ value. This may be demonstrated by the couple of hexanoic/1,6-hexanedioic acid, where both have the same carbon-chain length but where the second molecule carries two carboxylic acid functions, which tilts the octanol/air equilibrium by a factor of more than 10,000 towards the octanol side as shown in Table 6. Now, it is well known that monocarboxylic acids usually exist as dimeric associates in all three aggregate states. This association effect on the solubility is inherently taken into account in the atom group parameter evaluation of the COOH function. On the other hand, dicarboxylic acids do not only form dimers but also cyclical and linear oligomeric associates, with drastic consequences on their solubility in the various solvents. It is these additional associations that are considered by the special group "(COOH)n".

Table 6. Experimental (calculated) $logP_{ow}$ and $logK_{oa}$ of hexanoic and 1,6-hexanedioic acid.

Descriptor	Hexanoic Acid	1,6-Hexanedioic Acid
$logP_{ow}$	1.92 (1.91)	0.08 (0.64)
$logK_{oa}$	6.31 (6.23)	10.74 (10.62)

As a consequence, solutes with a low tendency to interact with solvents, either inherent or induced by intramolecular hydrogen bridges, show a trend to higher $logK_{aw}$ values; the additional intermolecular association of di- and tricarboxylic acids, on the other hand, results in a significantly lower $logK_{aw}$ value, as exemplified in Table 7, where the respective calculated data in Tables 5 and 6 have been applied in Equation (2). The experimental $logK_{aw}$ values have been extracted from Ebert et al. [30].

Table 7. Experimental and calculated $logK_{aw}$ of some examples.

Compound	$logK_{aw}$ Exp	$logK_{aw}$ Calc
2-Nitroaniline	−4.77	−3.59
3-Nitroaniline	−6.49	−5.61
Hexanoic Acid	−4.531	−4.32
1,6-Hexanedioic Acid	−11.15	−9.98

5. Conclusions

The present study, which is part of an ongoing project, put to use a tool for the simple and reliable calculation of the two partition coefficients $logP_{ow}$ and $logK_{oa}$ that has proven its unmatched versatility in the equally reliable prediction of now up to 19 physical, thermodynamic, solubility-, optics-, charge- and environment-related molecular descriptors [13,49–55], based on a common group-additivity method. The large database of more than 3300 and 1900 experimental data, respectively, upon which the group parameters for the $logP_{ow}$ and $logK_{oa}$ calculations are founded enabled their prediction for nearly 29,500 and more than 27,000 molecules, respectively, of the presently more than 37,100 compounds in ChemBrain's database. In addition, these results also allowed the trustworthy calculation of the third partition coefficient $logK_{aw}$ for more than 26,000 compounds. The big advantage of the present approach is its ease of use by simply adding, by means of paper and pencil, the parameters of the atoms and groups found in a particular molecule, which are listed in the respective Tables 2 and 3.

The mentioned project's software is called ChemBrain IXL, available from Neuronix Software, version ChemBrain IXL 5.9.70.1 (www.neuronix.ch (accessed on 27 November 2023), Rudolf Naef, Lupsingen, Switzerland).

Supplementary Materials: The following supporting information can be downloaded at https://www.mdpi.com/article/10.3390/liquids4010011/s1: The lists of compounds used in the present work, collected in their 3D structure together with their experimental data, are available as standard SDF files for use in external chemical software under the names of "S01. Compounds List for logPow-Parameters Calculations.sdf", "S02. Compounds List for logKoa-Parameters Calculations.sdf" and "S03. Compounds List with exp logKaw Data". The compounds used in the correlation diagrams and histograms are listed with their names and experimental and calculated data under the respective names of "S04. Compounds with Experimental vs. Calculated logPow Values.doc", "S05. Compounds with Experimental vs. Calculated logKoa Values.doc", "S06. Compounds with Experimental vs. Calculated logKaw Values.doc" and "S07. Alkanes, Alkenes and Alkynes with Exp. vs. Calc. logKaw Values.doc". In addition, for each of the three partition coefficients, a list of their outliers has been added under the names of "S08. Outliers of logPow.doc", "S09. Outliers of logKoa.doc" and "S10. Outliers of logKaw.doc". Beyond this, the Supplementary Materials encompass all the figures and tables cited in the text as .tif and .doc files, respectively.

Author Contributions: R.N. developed the project ChemBrain and its software upon which this paper is based, and also fed the database, calculated and analysed the results and wrote the paper. W.E.A.J. suggested the extension of ChemBrain's tool and contributed experimental data and the great majority of the literature references. Beyond this, R.N. is indebted to W.E.A.J. for the many valuable discussions. All authors have read and agreed to the published version of the manuscript.

Funding: This research received no external funding.

Data Availability Statement: The data presented in this study are available in the Supplementary Materials.

Acknowledgments: R.N. is indebted to the library of the University of Basel for allowing him full and free access to the electronic literature database.

Conflicts of Interest: The author declares no conflicts of interest.

References

1. Simonich, S.L.; Hites, R.A. Organic Pollutant Accumulation in Vegetation. *Environ. Sci. Technol.* **1995**, *29*, 2905–2914. [CrossRef]
2. McLachlan, M.S. Bioaccumulation of Hydrophobic Chemicals in Agricultural Food Chains. *Environ. Sci. Technol.* **1996**, *30*, 252–259. [CrossRef]
3. Doucette, W.J.; Shunthirsasingham, C.; Dettenmaier, E.M.; Zaleski, R.T.; Fantke, P.; Arnot, J.A. A Review of Measured Bioaccumulation Data on Terrestrial Plants for Organic Chemicals:Metrics, Variability, and the Need for Standardized Measurement Protocols. *Environ. Toxicol. Chem.* **2018**, *37*, 21–33. [CrossRef] [PubMed]
4. Nieto-Draghi, C.; Fayet, G.; Creton, B.; Rozanska, X.; Rotureau, P.; de Hemptinne, J.-C.; Ungerer, P.; Rousseau, B.; Adamo, C. A General Guidebook for the Theoretical Prediction of Physicochemical Properties of Chemicals for Regulatory Purposes. *Chem. Rev.* **2015**, *115*, 13093–13164. [CrossRef] [PubMed]
5. Cappelli, C.I.; Benfenati, E.; Cester, J. Evaluation of QSAR models for predicting the partition coefficient (log P) of chemicals under the REACH regulation. *Environ. Res.* **2015**, *43*, 26–32. [CrossRef]
6. Ghose, A.K.; Crippen, G.M. Atomic physicochemical parameters for three-dimensional structure-directed quantitative structure-activity relationships I. Partition coefficients as a measure of hydrophobicity. *J. Computer. Chem.* **1986**, *7*, 565–577. [CrossRef]
7. Ghose, A.K.; Pritchett, A.; Crippen, G.M. Atomic physicochemical parameters for three dimensional structure directed quantitative structure-activity relationships III: Modeling hydrophobic interactions. *J. Comput. Chem.* **1988**, *9*, 80–90. [CrossRef]

8. Klopman, G.; Li, J.-Y.; Wang, S.; Dimayuga, M. Computer automated log P calculations based on an extended group contribution approach. *J. Chem. Inf. Comput. Sci.* **1994**, *34*, 752–781. [CrossRef]
9. Visvanadhan, V.N.; Ghose, A.K.; Revankar, G.R.; Robins, R.K. Atomic physicochemical parameters for three dimensional structure directed quantitative structure-activity relationships. 4. Additional parameters for hydrophobic and dispersive interactions and their application for an automated superposition of certain naturally occurring nucleoside antibiotics. *J. Chem. Inf. Comput. Sci.* **1989**, *29*, 163–172. [CrossRef]
10. Leo, A.J. Calculating log P_{oct} from structures. *Chem. Rev.* **1993**, *93*, 1281–1306. [CrossRef]
11. Wang, R.; Fu, Y.; Lai, L. A new atom-additive method for calculating partition coefficients. *J. Chem. Inf. Comput. Sci.* **1997**, *37*, 615–621. [CrossRef]
12. Hou, T.J.; Xu, X.J. ADME evaluation in drug discovery. 2. Prediction of partition coefficient by atom-additive approach based on atom-weighted solvent accessible surface areas. *J. Chem. Inf. Comput. Sci.* **2003**, *43*, 1058–1067. [CrossRef]
13. Naef, R. A Generally Applicable Computer Algorithm Based on the Group Additivity Method for the Calculation of Seven Molecular Descriptors: Heat of Combustion, Log$P_{O/W}$, LogS, Refractivity, Polarizability, Toxicity and LogBB of Organic Compounds; Scope and Limits of Applicability. *Molecules* **2015**, *20*, 18279–18351. [CrossRef] [PubMed]
14. Plante, J.; Werner, S. JPlogP: An improved logP predictor trained using predicted data. *J. Cheminform.* **2018**, *10*, 61. [CrossRef] [PubMed]
15. Ulrich, N.; Goss, K.-U.; Ebert, A. Exploring the octanol–water partition coefficient dataset using deep learning techniques and data augmentation. *Com. Chem.* **2021**, *4*, 90. [CrossRef] [PubMed]
16. Sun, Y.; Hou, T.; He, X.; Man, V.H.; Wang, J. Development and test of highly accurate endpoint free energy methods. 2: Prediction of logarithm of n-octanol–water partition coefficient (logP) for druglike molecules using MM-PBSA method. *J. Comput. Chem.* **2023**, *44*, 1300–1311. [CrossRef] [PubMed]
17. Chen, J.; Harner, T.; Schramm, K.W.; Quan, X.; Xue, X.; Wu, W.; Kettrup, A. Quantitative relationships between molecular structures, environmental temperatures and octanol/air partition coefficients of PCDD/Fs. *Sci. Total Environ.* **2002**, *300*, 155–166. [CrossRef] [PubMed]
18. Chen, J.; Harner, T.; Yang, P.; Quan, X.; Chen, S.; Schramm, K.W.; Kettrup, A. Quantitative predictive models for octanol/air partition coefficients of polybrominated diphenyl ethers at different temperatures. *Chemosphere* **2003**, *51*, 577–584. [CrossRef] [PubMed]
19. Chen, J.; Harner, T.; Schramm, K.W.; Quan, X.; Xue, X.; Kettrup, A. Quantitative relationships between molecular structures, environmental temperatures and octanol/air partition coefficients of polychlorinated biphenyls. *Comput. Biol. Chem.* **2003**, *27*, 405–421. [CrossRef]
20. Zhao, H.; Chen, J.; Quan, X.; Qu, B.; Liang, X. Octanol/air partition coefficients of polybrominated biphenyls. *Chemosphere* **2009**, *74*, 1490–1494. [CrossRef]
21. Staikova, M.; Wania, F.; Donaldson, D. Molecular polarizability as a single parameter predictor of vapour pressures and octanoleair partitioning coefficients of non-polar compounds: A priori approach and results. *Atmos. Environ.* **2004**, *38*, 213–225. [CrossRef]
22. Zhao, H.; Zhang, Q.; Chen, J.; Xue, X.; Liang, X. Prediction of octanol/air partition coefficients of semivolatile organic compounds based on molecular connectivity index. *Chemosphere* **2005**, *59*, 1421–1426. [CrossRef]
23. Zeng, X.L.; Zhang, X.L.; Wang, Y. Qspr modeling of n-octanol/air partition coefficients and liquid vapor pressures of polychlorinated dibenzo-p-dioxins. *Chemosphere* **2013**, *91*, 229–232. [CrossRef] [PubMed]
24. Liu, H.; Shi, J.; Liu, H.; Wang, Z. Improved 3D-QSPR analysis of the predictive octanol/air partition coefficients of hydroxylated and methoxylated polybrominated diphenyl ethers. *Atmos. Environ.* **2013**, *77*, 840–845. [CrossRef]
25. Jiao, L.; Gao, M.; Wang, X.; Li, H. QSPR study on the octanol/air partition coefficient of polybrominated diphenyl ethers by using molecular distance-edge vector index. *Chem. Cent. J.* **2014**, *8*, 36. [CrossRef]
26. Chen, Y.; Cai, X.; Jiang, L.; Li, Y. Prediction of octanol-air partition coefficients for polychlorinated biphenyls (PCBs) using 3D-SQAR models. *Ecotoxicol. Environ. Saf.* **2016**, *124*, 202–212. [CrossRef] [PubMed]
27. Fu, Z.; Chen, J.; Li, X.; Wang, Y.; Yu, H. Comparison of prediction methods for octanol-air partition coefficients of diverse organic compounds. *Chemosphere* **2016**, *148*, 118–125. [CrossRef] [PubMed]
28. Jin, X.; Fu, Z.; Li, X.; Chen, J. Development of polyparameter linear free energy relationship models for octanol/air partition coefficients of diverse chemicals. *Environ. Sci. Process. Impact.* **2017**, *19*, 300–306. [CrossRef] [PubMed]
29. Li, X.; Chen, J.; Zhang, L.; Qiao, X.; Huang, L. The fragment constant method for predicting octanol/air partition coefficients of persistent organic pollutants at different temperatures. *J. Phys. Chem. Ref. Data* **2006**, *35*, 1365–1384. [CrossRef]
30. Ebert, R.-U.; Kühne, R.; Schüürmann, G. Henry's Law Constant—A General-Purpose Fragment Model to Predict Log Kaw from Molecular Structure. *Environ. Sci. Technol.* **2023**, *57*, 160–167. [CrossRef]
31. Hardtwig, E. *Fehler—Und Ausgleichsrechnung*; Bibliographisches Institut AG: Mannheim, Germany, 1968.
32. Sangster, J. Octanol-water partition coefficients of simple organic compounds. *J. Phys. Chem. Ref. Data* **1989**, *18*, 1111–1229. [CrossRef]
33. Lipinski, C.A.; Lombardo, F.; Dominy, B.W.; Feeney, P.J. Experimental and computational approaches to estimate solubility and permeability in drug discovery and development settings. *Adv. Drug Deliv. Rev.* **1997**, *23*, 3–25. [CrossRef]
34. Tewari, Y.B.; Miller, M.M.; St. Wasik, P.; Martire, D.E. Aqueous Solubility and Octanol/Water Partition Coefficient of Organic Compounds at 25.0 °C. *J. Chem. Eng. Data* **1982**, *27*, 451–454. [CrossRef]

35. Abraham, M.H.; Chadha, H.S.; Whiting, G.S.; Mitchell, R.C. Hydrogen Bonding. 32. An Analysis of Water-Octanol and Water-Alkane Partitioning and the Δlog P Parameter of Seiler. *J. Pharm. Sci.* **1994**, *83*, 1085–1100. [CrossRef] [PubMed]
36. St. Czerwinski, E.; Skvorak, J.P.; Maxwell, D.M.; Lenz, D.E.; St. Baskin, I. Organophosphorus Compounds on Biodistribution and Percutaneous Toxicity. *J. Biochem. Mol. Tox.* **2006**, *20*, 241–246. [CrossRef] [PubMed]
37. Boddu, V.M.; Abburi, K.; St. Maloney, W.; Damavarapu, R. Thermophysical Properties of an Insensitive Munitions Compound, 2,4-Dinitroanisole. *J. Chem. Eng. Data* **2008**, *53*, 1120–1125. [CrossRef]
38. Li, X.-J.; Shan, G.; Liu, H.; Wang, Z.-Y. Determination of lgK_{OW} and QSPR Study on Some Fluorobenzene Derivatives. *Chin. J. Struct. Chem.* **2009**, *28*, 1236–1241.
39. Saranjampour, P.; Vebrosky, E.N.; Armbrust, K.L. Salinity Impacts on Water Solubility and n-Octanol/Water Partition Coefficients of Selected Pesticides and Oil Constituents. *Environ. Toxicol. Chem.* **2017**, *36*, 2274–2280. [CrossRef] [PubMed]
40. Ebert, R.-U.; Kühne, R.; Schüürmann, G. Octanol/Air Partition Coefficient. A General-Purpose Fragment Model to Predict Log Koa from Molecular Structure. *Environ. Sci. Technol.* **2023**, *57*, 976–984. [CrossRef] [PubMed]
41. Puzyn, T.; Falandysz, J.; Rostkowski, P.; Piliszek, S.; Wilczynska, A. Computational estimation of logarithm of octanol/air partition coefficients and subcooled vapour pressures for each of 75 chloronaphtalene congeners. *Phys.-Chem. Prop. Distr. Model. Organohal. Compds.* **2004**, *66*, 2354–2360. [CrossRef]
42. Odabasi, M.; Cetin, E.; Sofuoglu, A. Determination of octanol–air partition coefficients and supercooled liquid vapor pressures of PAHs as a function of temperature: Application to gas–particle partitioning in an urban atmosphere. *Atm. Environ.* **2006**, *40*, 6615–6625. [CrossRef]
43. Xu, S.; Kropscott, B. Method for Simultaneous Determination of Partition Coefficients for Cyclic Volatile Methylsiloxanes and Dimethylsilanediol. *Anal. Chem.* **2012**, *84*, 1948–1955. [CrossRef] [PubMed]
44. Easterbrook, K.D.; Vona, M.A.; Osthoff, H.D. Measurement of Henry's law constants of ethyl nitrate in deionized water, synthetic sea salt solutions, and n-octanol. *Chemosphere* **2024**, *346*, 140482. [CrossRef] [PubMed]
45. Sander, R. Compilation of Henry's law constants (version 5.0.0) for water as solvent. *Atmos. Chem. Phys.* **2023**, *23*, 10901–12440. [CrossRef]
46. Allen, G.; Dwek, R.A. An n.m.r. study of keto-enol tautomerism in β-diketones. *J. Chem. Soc. B* **1966**, *1966*, 161–163. [CrossRef]
47. COSMOlogic GmbH Co. KG. A Dassault Systemes Company, Version 19.0.4, *COSMOthermX*. 2019. Available online: www.cosmologic.de (accessed on 4 December 2023).
48. US EPA. *Estimation Programs Interface Suite for Microsoft Windows*; Version 4.11, Module KOAWIN v. 1.11; United States Environmental Protection Agency: Washington, DC, USA, 2015.
49. Naef, R.; Acree, W.E., Jr. Calculation of Five Thermodynamic Molecular Descriptors by Means of a General Computer Algorithm Based on the Group-Additivity Method: Standard Enthalpies of Vaporization, Sublimation and Solvation, and Entropy of Fusion of Ordinary Organic Molecules and Total Phase-Change Entropy of Liquid Crystals. *Molecules* **2017**, *22*, 1059. [CrossRef]
50. Naef, R.; Acree, W.E. Application of a General Computer Algorithm Based on the Group-Additivity Method for the Calculation of Two Molecular Descriptors at Both Ends of Dilution: Liquid Viscosity and Activity Coefficient inWater at Infinite Dilution. *Molecules* **2018**, *23*, 5. [CrossRef]
51. Naef, R.; Acree, W.E., Jr. Calculation of the Surface Tension of Ordinary Organic and Ionic Liquids by Means of a Generally Applicable Computer Algorithm Based on the Group-Additivity Method. *Molecules* **2018**, *23*, 1224. [CrossRef] [PubMed]
52. Naef, R. Calculation of the Isobaric Heat Capacities of the Liquid and Solid Phase of Organic Compounds at 298.15K by Means of the Group-Additivity Method. *Molecules* **2020**, *25*, 1147. [CrossRef]
53. Naef, R.; Acree, W.E., Jr. Calculation of the Vapour Pressure of Organic Molecules by Means of a Group-Additivity Method and Their Resultant Gibbs Free Energy and Entropy of Vaporization at 298.15 K. *Molecules* **2021**, *26*, 1045. [CrossRef] [PubMed]
54. Naef, R.; Acree, W.E., Jr. Revision and Extension of a Generally Applicable Group-Additivity Method for the Calculation of the Standard Heat of Combustion and Formation of Organic Molecules. *Molecules* **2021**, *26*, 6101. [CrossRef] [PubMed]
55. Naef, R.; Acree, W.E., Jr. Revision and Extension of a Generally Applicable Group Additivity Method for the Calculation of the Refractivity and Polarizability of Organic Molecules at 298.15 K. *Liquids* **2022**, *2*, 327–377. [CrossRef]

liquids

MDPI

Article

Dissolution Thermodynamics and Preferential Solvation of Phenothiazine in Some Aqueous Cosolvent Systems

Fleming Martínez [1,*], María Ángeles Peña [2] and Abolghasem Jouyban [3,4]

[1] Grupo de Investigaciones Farmacéutico-Fisicoquímicas, Departamento de Farmacia, Facultad de Ciencias, Universidad Nacional de Colombia, Sede Bogotá, Cra. 30 No. 45-03, Bogotá 11001, Colombia
[2] Departamento de Ciencias Biomédicas, Facultad de Farmacia, Universidad de Alcalá, Alcalá de Henares, 28805 Madrid, Spain; angeles.pena@uah.es
[3] Pharmaceutical Analysis Research Center, Tabriz University of Medical Sciences, Tabriz 5166/15731, Iran; ajouyban@hotmail.com
[4] Faculty of Pharmacy, Tabriz University of Medical Sciences, Tabriz 5166/15731, Iran
* Correspondence: fmartinezr@unal.edu.co; Tel.: +571-3165000 (ext. 14608)

Abstract: Published equilibrium mole fraction solubilities of phenothiazine in ethanol, propylene glycol and water as mono-solvents at several temperatures were investigated to find standard apparent thermodynamic quantities of dissolution mixing and solvation based on the van't Hoff and Gibbs equations. Further, by processing the reported mole fraction solubility values of phenothiazine in some aqueous cosolvent mixtures at $T/\mathrm{K} = 298.2$, the inverse Kirkwood–Buff integrals treatment demonstrated preferential hydration of phenothiazine in water-rich mixtures and preferential solvation of this agent by cosolvents in mixtures of $0.24 < x_1 < 1.00$ in the {ethanol (1) + water (2)} mixed system and mixtures of $0.18 < x_1 < 1.00$ in the {propylene glycol (1) + water (2)} mixed system.

Keywords: phenothiazine; {cosolvent (1) + water (2)} mixtures; cosolvency; dissolution thermodynamics; preferential solvation; IKBI

1. Introduction

Phenothiazine (molecular structure shown in Figure 1; IUPAC name: 10*H*-phenothiazine, molar mass: 199.27 g·mol^{-1}, CAS number: 92-84-2, PubChem CID: 7108) is a heterocyclic compound related to thiazine [1]. Some phenothiazine derivatives exhibit highly bioactive behavior. In particular, its derivatives chlorpromazine and promethazine are commonly employed in psychiatry and allergy treatment, respectively [2]. It is noteworthy that the widely used agent methylene blue also has the basic heterocyclic structure of phenothiazine [3]. From a non-medical point of view, it is interesting to note that phenothiazine is commonly used during acrylic acid polymerization as an anaerobic inhibitor. Moreover, it is also used as an in-process inhibitor in some purification processes of acrylic acid [4]. Although this agent was prepared a long time ago, its physicochemical properties have not been completely studied, including the equilibrium solubility in mono- or mixed solvents. Thus, the solubility of phenothiazine in 38 organic solvents at $T/\mathrm{K} = 298.2$ was reported by Hoover et al. [5], whereas its solubility in neat water, ethanol and propylene glycol at several temperatures as well as in some aqueous mixtures of ethanol and propylene glycol at $T/\mathrm{K} = 298.2$ was reported by Ahmadian et al. [6]. These studies were performed in order to understand the effect of different kinds of solvents on its equilibrium solubility. However, a complete thermodynamic study including the effect of mixture composition on preferential solvation has not been reported yet. As has been described in the specialized literature, systematic data about the equilibrium solubility of drugs or drug-alike compounds in different aqueous cosolvent mixtures, as well as a deep understanding of the respective solute dissolution mechanisms, are very important from theoretical and practical points of view [7–10].

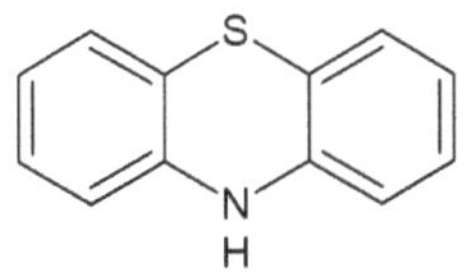

Figure 1. Molecular structure of phenothiazine.

Owing to the lack of this information for phenothiazine, based on published solubility and phase-equilibrium values [6,11], the main aims of this research were as follows: (i) analyze the effect of temperature on the solubility of phenothiazine in ethanol, propylene glycol and water as mono-solvents; (ii) calculate the apparent standard dissolution, mixing and solvation thermodynamic parameters of these neat solvents; (iii) correlate equilibrium solubility data in aqueous cosolvent mixtures with several well-known thermodynamic models; and (iv) evaluate the preferential solvation parameters of phenothiazine in aqueous binary mixtures of ethanol or propylene glycol at $T/\text{K} = 298.2$.

2. Results and Discussion

2.1. Ideal Mole Fraction Solubility and Asymmetrical Activity Coefficients of Phenothiazine in Neat Solvents

The ideal solubility (x_s^{id}) of crystalline drugs is calculated as follows:

$$\ln x_s^{id} = -\frac{\Delta_f H (T_f - T)}{R T_f T} + \left(\frac{\Delta C_p}{R}\right)\left[\frac{(T_f - T)}{T} + \ln\left(\frac{T}{T_f}\right)\right] \quad (1)$$

where $\Delta_f H$ is the molar enthalpy of fusion of the solid drug at the melting point, T_f is the absolute melting point, T is the absolute solution temperature, R is the universal gas constant (8.314 $\text{J·mol}^{-1}\text{·K}^{-1}$), and ΔC_p denotes the difference between the molar heat capacity of the hypothetical super-cooled liquid form of the solute and the molar heat capacity of its crystalline form at the solution temperature [11]. Since ΔC_p is not frequently reported for drugs in the literature, it may be approximated to the entropy of fusion, $\Delta_f S$, which, in turn, is calculated as $\Delta_f H / T_f$. Ideal solubilities of phenothiazine at temperatures from T/K = (298.2 to 338.2) summarized in Table 1 were calculated based on the calorimetric values reported by Sabbah and El Watik [12], namely, $\Delta_f H/\text{kJ·mol}^{-1} = 22.66$ and $T_f/\text{K} = 458.4$.

Table 1. Ideal solubility and activity coefficients of phenothiazine in some pure solvent mixtures at several temperatures and $p/\text{kPa} = 101.3$. [a,b].

	T/K [b]				
	298.2	**308.2**	**318.2**	**328.2**	**338.2**
			x_s^{id}		
	5.52×10^{-2}	6.90×10^{-2}	8.55×10^{-2}	0.1053	0.1289
Solvent [a,b]			γ_s		
EtOH	6.31	6.90	7.01	7.27	
PG	13.28	12.78	13.04	13.81	15.13
Water	3.54×10^5	3.48×10^5	3.39×10^5	2.94×10^5	2.63×10^5

[a] p is the atmospheric pressure in Tabriz, Iran. [b] Standard uncertainty in p is $u(p)/\text{kPa} = 3.0$. Average relative uncertainty in γ_s is $u_r(\gamma_3) = 0.027$. Standard uncertainty in T is $u(T)/\text{K} = 0.2$.

Table 1 also summarizes the asymmetrical activity coefficients (γ_s) of phenothiazine in ethanol, propylene glycol and water as mono-solvents at several temperatures. These γ_s values were calculated as the quotient x_s^{id}/x_s from the ideal solubilities (x_s^{id}) summarized in Table 1 and the experimental mole fraction solubilities reported by Ahmadian et al. for these neat solvents [6]. Reported solubility values were determined by means of the shake-flask method [6,13]. As observed, phenothiazine γ_s values vary from a maximum of 3.54×10^5 in neat water at $T/K = 298.2$ to a minimum of 6.31 in neat ethanol at $T/K = 298.2$. In all solvents at all temperatures, phenothiazine exhibits γ_s values higher than the unity because the ideal solubilities are higher than the experimental ones. Moreover, γ_3 values decrease with the temperature increase in water but increase in ethanol and propylene glycol. This implies some distancing of real behavior from the ideal dissolution hypothetical process when the temperature increase for both cosolvents.

On the other hand, Equation (2) allows a rough estimation of the magnitudes of solute–solvent intermolecular interactions from γ_s values [14].

$$\ln \gamma_s = (e_{11} + e_{ss} - 2e_{1s})\frac{V_s\varphi_1^2}{RT} \tag{2}$$

Subscript 1 stands for the solvent system and e_{11}, e_{ss} and e_{1s} represent the magnitudes of solvent–solvent, solute–solute and solvent–solute interaction energies, respectively. V_s denotes the molar volume of the hypothetical super-cooled liquid phenothiazine, whereas φ_1 denotes the volume fraction of every solvent. For low x_s solubility values, $V_s\varphi_1^2/RT$ may be considered as constant despite the solvent. Hence, γ_s values should depend mainly on $e_{11,}$ e_{ss} and e_{1s} [14]. As is well known, e_{11} and e_{ss} are unfavorable for drug solubility and dissolution, whereas e_{1s} favors the respective drug solubility and dissolution rate increasing. The contribution of e_{33} was planned as constant in the different solvents studied. Thus, from a qualitative point of view, based on the energetic quantities described in Equation (2), the next analysis could be established by considering that e_{11} was highest in neat water (because of its high Hildebrand solubility parameter, $\delta_1/\text{MPa}^{1/2} = 47.8$), followed by propylene glycol ($\delta_1/\text{MPa}^{1/2} = 30.2$) and, finally, ethanol ($\delta_1/\text{MPa}^{1/2} = 26.5$) [15,16]. It is noteworthy that the Hildebrand solubility parameter is one of the most used polarity indices in pharmaceutical sciences [7–9]. Thus, because pure water reveals γ_s values higher than 2.5×10^5, it should have high e_{11} and low e_{1s} values, whereas, in pure propylene glycol exhibiting γ_s values near 12.0 and ethanol exhibiting γ_s values near 7.0, the e_{11} values should be relatively low, while the e_{1s} values should be high. In this way, a higher solvation of phenothiazine in ethanol and propylene glycol was expected.

2.2. *Apparent Thermodynamic Properties of the Dissolution, Mixing and Solvation of Phenothiazine in Mono-Solvents*

2.2.1. Apparent Thermodynamic Functions of Phenothiazine Dissolution in Mono-Solvents

All the apparent standard thermodynamic quantities relative to phenothiazine dissolution processes were calculated at $T/K = 298.2$. Hence, all the apparent standard enthalpy changes ($\Delta_{soln}H^\circ$) were obtained by means of the modified van't Hoff equation, as shown in Equation (3) [17]:

$$\left(\frac{\partial \ln x_s}{\partial(1/T - 1/298.2)}\right)_P = -\frac{\Delta_{soln}H^\circ}{R} \tag{3}$$

The apparent standard Gibbs energy changes relative to all the phenothiazine dissolution processes ($\Delta_{soln}G^\circ$) were calculated by means of the following equation [17]:

$$\Delta_{soln}G^\circ = -R \cdot 298.2 \cdot \text{intercept} \tag{4}$$

The intercepts used in Equation (4) were those obtained as a result of the linear regressions of $\ln x_s$ as a function of $(1/T - 1/298.2)$. Therefore, Figure 2 depicts the phenothiazine solubility linear van't Hoff behavior of all three neat solvents. It is noteworthy that linear regressions with $r^2 > 0.98$ were observed in all cases, as shown in Table 2 [18–20].

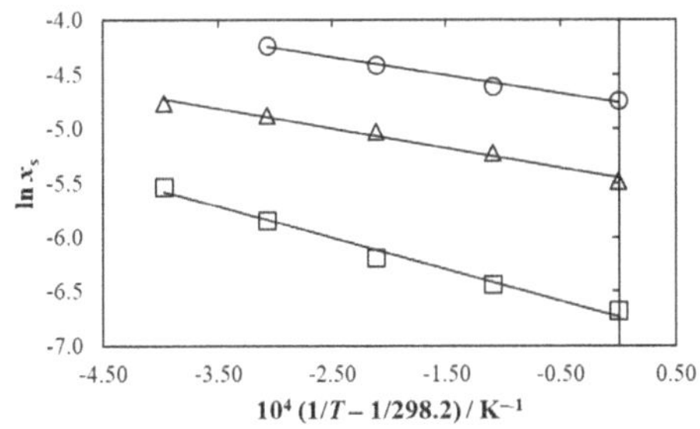

Figure 2. van't Hoff plot of the solubility of phenothiazine (s) in some pure solvents. ○: ethanol, Δ: x_1 = propylene glycol, □: water (logarithmic value + 9.0).

Table 2. Coefficients and statistical parameters of the van't Hoff linear regressions of mole fraction solubility of phenothiazine (3) in some pure solvents at different temperatures and p/kPa = 101.3. [a,b].

Solvent [a,b]	Intercept	Slope	Adjusted r^2	Typical Error	F
EtOH	−4.758 ± 0.023	−1671 ± 116	0.986	0.026	207.6
PG	−5.444 ± 0.033	−1804 ± 132	0.979	0.041	187.7
Water	−15.729 ± 0.050	−2899 ± 202	0.981	0.063	205.9

[a] p is the atmospheric pressure in Tabriz, Iran. [b] Standard uncertainty in T is $u(T)$ = 0.2 K. Standard uncertainty in p is $u(p)$/kPa = 3.0. Average relative uncertainty in phenothiazine solubility is $u_r(x_s)$ = 0.027.

Finally, the apparent standard changes in entropy for all the studied phenothiazine dissolution processes ($\Delta_{soln}S^\circ$) were calculated based on the respective $\Delta_{soln}H^\circ$ and $\Delta_{soln}G^\circ$ values by using Equation (5) [17]:

$$\Delta_{soln}S^o = \frac{(\Delta_{soln}H^\circ - \Delta_{soln}G^\circ)}{298.2} \quad (5)$$

Table 3 summarizes all the apparent standard thermodynamic quantities for the dissolution processes of phenothiazine in all the neat solvents at T/K = 298.2, including those corresponding to the ideal dissolution process. As expected, all the apparent standard Gibbs energies and apparent enthalpies of dissolution of phenothiazine were positive in every solvent. The apparent standard entropy of dissolution was negative in neat water but positive in ethanol and propylene glycol. Thus, the global dissolution processes of phenothiazine are always endothermic in nature and entropy-driven for those occurring in ethanol and propylene glycol (with $\Delta_{soln}H^\circ > 0$ and $\Delta_{soln}S^\circ > 0$), whereas, in neat water, neither enthalpy nor entropy-driving are observed because $\Delta_{soln}H^\circ > 0$ and $\Delta_{soln}S^\circ < 0$. As observed, the highest $\Delta_{soln}H^\circ$ and lowest $\Delta_{soln}S^\circ$ values were observed in neat water. The negative apparent dissolution entropy obtained in neat water could be a consequence of the possible hydrophobic hydration around the phenylene groups of phenothiazine (Figure 1).

Table 3. Apparent thermodynamic functions relative to dissolution processes of phenothiazine (3) in some pure solvents at T/K = 298.2 and p/kPa = 101.3. [a,b].

Solvent [a,b]	$\Delta_{soln}G^\circ$/ kJ·mol^{-1} [b]	$\Delta_{soln}H^\circ$/ kJ·mol^{-1} [b]	$\Delta_{soln}S^\circ$/ J·mol^{-1}·K^{-1} [b]	$T\Delta_{soln}S^\circ$/ kJ·mol^{-1} [b]	ζ_H [c]	ζ_{TS} [c]
EtOH	11.80	13.89	7.03	2.10	0.869	0.131
PG	13.50	15.00	5.05	1.50	0.909	0.091
Water	38.99	24.11	−49.93	−14.89	0.618	0.382
Ideal	7.20	17.76	35.42	10.56	0.627	0.373

[a] p is the atmospheric pressure in Tabriz, Iran. [b] Standard uncertainty in T is $u(T)$ = 0.2 K. Standard uncertainty in p is $u(p)$/kPa = 3.0. Average relative standard uncertainty in apparent thermodynamic quantities of real dissolution processes are $u_r(\Delta_{soln}G^\circ)$ = 0.032, $u_r(\Delta_{soln}H^\circ)$ = 0.070, $u_r(\Delta_{soln}S^\circ)$ = 0.072, and $u_r(T\Delta_{soln}S^\circ)$ = 0.072. [c] ζ_H and ζ_{TS} are the relative contributions by enthalpy and entropy to the apparent Gibbs energy of dissolution.

Moreover, to calculate the magnitude contributions by enthalpy (ζ_H) and entropy (ζ_{TS}) toward the dissolution processes, the subsequent equations were used [21]:

$$\zeta_H = \frac{|\Delta_{\text{soln}} H^\circ|}{|\Delta_{\text{soln}} H^\circ| + |T\Delta_{\text{soln}} S^\circ|} \tag{6}$$

$$\zeta_{TS} = \frac{|T\Delta_{\text{soln}} S^\circ|}{|\Delta_{\text{soln}} H^\circ| + |T\Delta_{\text{soln}} S^\circ|} \tag{7}$$

As observed in Table 3, the higher contributions to the positive apparent standard molar Gibbs energies of phenothiazine dissolution processes were given by the positive enthalpies.

2.2.2. Apparent Thermodynamic Quantities of Phenothiazine Mixing in Mono-Solvents

The overall dissolution processes of phenothiazine in all mono-solvent systems may be represented by the following hypothetical stages:

$\text{Solute}_{\text{(Solid state)}}$ at $T \rightarrow \text{Solute}_{\text{(Solid state)}}$ at $T_{\text{fus}} \rightarrow \text{Solute}_{\text{(Liquid state)}}$ at $T_{\text{fus}} \rightarrow \text{Solute}_{\text{(Liquid state)}}$ at $T \rightarrow \text{Solute}_{\text{(Solution state)}}$ at T

Here, the hypothetical stages were considered as follows: (i) the heating and fusion of phenothiazine at $T_f/\text{K} = 458.4$, (ii) the cooling of the liquid fused phenothiazine to the considered temperature (i.e., $T/\text{K} = 298.2$), and (iii) the subsequent mixing of both the hypothetical super-cooled liquid phenothiazine and the pure solvents at $T/\text{K} = 298.2$ [22]. This treatment allowed us to calculate every individual thermodynamic contribution toward the overall phenothiazine dissolution processes using the following equations:

$$\Delta_{\text{soln}} H^\circ = \Delta_{\text{fus}} H^T + \Delta_{\text{mix}} H^\circ \tag{8}$$

$$\Delta_{\text{soln}} S^\circ = \Delta_{\text{fus}} S^T + \Delta_{\text{mix}} S^\circ \tag{9}$$

where $\Delta_{\text{fus}} H^T$ and $\Delta_{\text{fus}} S^T$ represent the thermodynamic quantities relative to phenothiazine melting and its cooling at $T/\text{K} = 298.2$, which, in turn, were calculated as follows [23]:

$$\Delta_f H^T = \Delta_f H^T - \Delta C_p (T_f - T) \tag{10}$$

$$\Delta_f S^T = \Delta_f S^T - \Delta C_p \ln\left(\frac{T_f}{T}\right) \tag{11}$$

where ΔC_p denotes the difference in the heat capacities of hypothetical liquid and solid states at the temperature of melting. Owing to the difficulties of ΔC_p experimental determinations, the entropy of fusion (ΔS_f) was used instead [23]. Table 4 summarizes all the apparent standard thermodynamic quantities of mixing of the hypothetical super-cooled liquid phenothiazine with all the pure solvents, as calculated at $T/\text{K} = 298.2$. As observed, the Gibbs energies of mixing were positive in all three solvents. The contributions by the thermodynamic quantities of mixing subprocesses to the overall dissolution processes of phenothiazine were variable depending on the solvent. Thus, $\Delta_{\text{mix}} H^\circ$ values were negative for ethanol and propylene glycol but positive for pure water. Moreover, $\Delta_{\text{mix}} S^\circ$ values were negative in all three solvents. In this way, the mixing processes of phenothiazine in ethanol and propylene glycol were enthalpy-driven because of the exothermic character exhibited. However, in pure water, neither enthalpy nor entropy-driving was observed for mixing. Furthermore, in order to evaluate the relative contributions of enthalpy (ζ_H) and entropy (ζ_{TS}) to the mixing processes in these mono-solvent systems, two equations analogous to Equations (6) and (7) were also employed. As observed in Table 4, in all cases, the main contributor to the Gibbs energies of mixing was entropy.

Table 4. Apparent thermodynamic functions relative to mixing processes of phenothiazine (3) in some pure solvents at $T/\text{K} = 298.2$ and $p/\text{kPa} = 101.3$. [a,b].

Solvent [a,b]	$\Delta_{mix}G^\circ$/ kJ·mol^{-1} [b]	$\Delta_{mix}H^\circ$/ kJ·mol^{-1} [b]	$\Delta_{mix}S^\circ$/ J·mol^{-1}·K^{-1} [b]	$T\Delta_{mix}S^\circ$/ kJ·mol^{-1} [b]	ζ_H [c]	ζ_{TS} [c]
EtOH	4.60	−3.87	−28.39	−8.47	0.313	0.687
PG	6.30	−2.76	−30.38	−9.06	0.233	0.767
Water	31.80	6.35	−85.35	−25.45	0.200	0.800

[a] p is the atmospheric pressure in Tabriz, Iran. [b] Standard uncertainty in T is $u(T) = 0.2$ K. Standard uncertainty in p is $u(p)/\text{kPa} = 3.0$. Average relative standard uncertainty in apparent thermodynamic quantities of mixing processes are $u_r(\Delta_{mix}G^\circ) = 0.035$, $u_r(\Delta_{mix}H^\circ) = 0.080$, $u_r(\Delta_{mix}S^\circ) = 0.082$, and $u_r(T\Delta_{mix}S^\circ) = 0.082$. [c] ζ_H and ζ_{TS} are the relative contributions by enthalpy and entropy to the apparent Gibbs energy of mixing.

As described earlier in the literature, the net value of $\Delta_{mix}H^\circ$ values depends on the contribution of various kinds of intermolecular interactions. Hence, the cavity formation in the solvent system, required for the solute accommodation, is endothermic because some quantity of energy must be supplied to overcome the respective cohesive forces of the solvent. This contribution diminishes the phenothiazine equilibrium solubility. Conversely, the solvent–solute interactions, which result mainly from van der Waals and Lewis acid–base interactions, like hydrogen bonding, are clearly exothermic in nature. This contribution increases the phenothiazine solubility and dissolution rate. Furthermore, the structuring of water molecules around the phenylene groups of phenothiazine (Figure 1) diminishes the net $\Delta_{mix}H^\circ$ quantity to small or even negative values in water-rich mixtures, as indicated above [24]. However, this event is not observed with phenothiazine in pure water.

2.2.3. Apparent Thermodynamic Quantities of Phenothiazine Solvation in Mono-Solvents

In addition to the previously exposed fusion-mixing process, the phenothiazine dissolution process may also be represented by the following hypothetic stages: $\text{Solute}_{(\text{Solid})} \rightarrow \text{Solute}_{(\text{Vapor})} \rightarrow \text{Solute}_{(\text{Solution})}$, where the respective partial processes of the global dissolution processes are phenothiazine sublimation and solvation [25]. This treatment permits the calculation of the partial thermodynamic contributions to dissolution processes by means of Equations (12) and (13), respectively, while the standard Gibbs energy of solvation is calculated by means of Equation (14):

$$\Delta_{soln}H^\circ = \Delta_{subl}H^\circ + \Delta_{solv}H^\circ \tag{12}$$

$$\Delta_{soln}S^\circ = \Delta_{subl}S^\circ + \Delta_{solv}S^\circ \tag{13}$$

$$\Delta_{soln}G^\circ = \Delta_{subl}G^\circ + \Delta_{solv}G^\circ \tag{14}$$

where $\Delta_{subl}H^\circ/\text{kJ mol}^{-1} = 111.45$ at $T/\text{K} = 298.2$, which was reported by Sabbah and El Watik [12]; moreover, from phenothiazine vapor pressure ($P_v/\text{mmHg} = 5.0 \times 10^{-3}$ at $T/\text{K} = 580$ [12]), the other thermodynamic quantities of sublimation at $T/\text{K} = 298.2$ were calculated as summarized in Table 5. Thermodynamic functions of phenothiazine solvation in all three pure solvents are also summarized in Table 5. Owing to the negative values observed in all thermodynamic quantities, it follows that drug solvation in all three solvents is exothermic and, thus, enthalpy-driven. On the other hand, with the aim of comparing the relative contributions by enthalpy (ζ_H) and entropy (ζ_{TS}) to the solvation processes, two equations analogous to Equations (6) and (7) were employed again. The ζ_H and ζ_{TS} values shown in Table 5 show that enthalpy was the main contributing force to the standard Gibbs energy of solvation in all mono-solvents.

Table 5. Apparent thermodynamic functions relative to sublimation and solvation processes of phenothiazine (3) in some pure solvents at T/K = 298.2 and p/kPa = 101.3. [a,b].

			Sublimation			
	$\Delta_{subl}G°$/ kJ·mol^{-1}	$\Delta_{subl}H°$/ kJ·mol^{-1}	$\Delta_{subl}S°$/ J·mol^{-1}·K^{-1}	$T\Delta_{subl}S°$/ kJ·mol^{-1}	ζ_H	ζ_{TS}
	53.57	111.45	194.15	57.88	0.658	0.342
			Solvation			
Solvent [a,b]	$\Delta_{solv}G°$/ kJ·mol^{-1} [b]	$\Delta_{solv}H°$/ kJ·mol^{-1} [b]	$\Delta_{solv}S°$/ J·mol^{-1}·K^{-1} [b]	$T\Delta_{solv}S°$/ kJ·mol^{-1} [b]	ζ_H [c]	ζ_{TS} [c]
EtOH	−41.77	−97.56	−187.12	−55.79	0.636	0.364
PG	−40.07	−96.45	−189.10	−56.38	0.631	0.369
Water	−14.57	−87.34	−244.07	−72.77	0.546	0.454

[a] p is the atmospheric pressure in Tabriz, Iran. [b] Standard uncertainty in T is $u(T)$ = 0.2 K. Standard uncertainty in p is $u(p)$/kPa = 3.0. Average relative standard uncertainty in apparent thermodynamic quantities of solvation processes are $u_r(\Delta_{mix}G°)$ = 0.040, $u_r(\Delta_{mix}H°)$ = 0.090, $u_r(\Delta_{mix}S°)$ = 0.093, and $u_r(T\Delta_{mix}S°)$ = 0.093. [c] ζ_H and ζ_{TS} are the relative contributions by enthalpy and entropy to the apparent Gibbs energy of solvation.

2.3. Effect of Mixed Solvents Polarity on Phenothiazine Solubility in Aqueous Cosolvent Mixtures

Figure 3 depicts the phenothiazine solubility profiles as a function of the Hildebrand solubility parameters (δ_{1+2}) of both {ethanol (1) + water (2)} and {propylene glycol (1) + water (2)} mixtures at T/K = 298.2. As widely described, δ_{1+2} is a very important polarity descriptor of aqueous cosolvent mixtures [7–10]. Numerical values of this descriptor were calculated considering the Hildebrand solubility parameter of all pure solvents mentioned above and the volume fraction (f_i) of each solvent [7,26]:

$$\delta_{1+2} = \sum_{i=1}^{2} f_i\delta_i \tag{15}$$

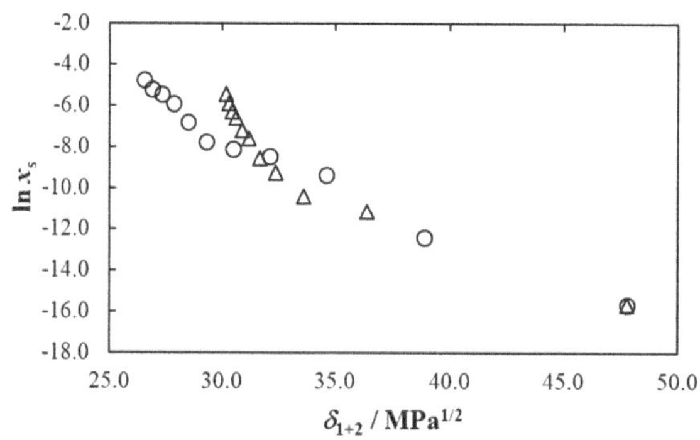

Figure 3. Logarithmic mole fraction solubility of phenothiazine (ln x_s) as a function of the Hildebrand solubility parameter in some {cosolvent (1) + water (2)} mixtures at T/K = 298.2. ○: {ethanol (1) + water (2), Δ: {propylene glycol (1) + water (2).

As observed, both solubility curves exhibited maximum phenothiazine solubilities in neat cosolvents. Because organic solutes normally reach their maximum solubilities in solvent systems exhibiting similar polarity [7,8], it was expected that the phenothiazine δ_3 value would be lower than 26.5 MPa$^{1/2}$ at T/K = 298.2. Table 6 shows a phenothiazine δ_3 value of 25.2 MPa$^{1/2}$, as calculated by means of the Fedors' method [27], which was lower than the ethanol δ_1 value of 26.5 MPa$^{1/2}$ [15,16].

Table 6. Application of the Fedors' method to estimate the molar volume and Hildebrand solubility parameter of phenothiazine.

Group	Group Number	$\Delta U°$/kJ·mol^{-1}	V/cm^3·mol^{-1}
Phenylene	2	2 × 31.9 = 63.8	2 × 52.4 = 104.8
Ring closure of 6 atoms	1	1.05	16
(-NH-)	1	8.4	4.5
(-S-)	1	14.15	12
		$\sum \Delta U°$ = 87.4	$\sum V$ = 137.3
		δ_3 = (87,400/137.3)$^{1/2}$ = 25.2 MPa$^{1/2}$	

2.4. *Preferential Solvation Analysis of Phenothiazine in Aqueous Cosolvent Mixtures*

The preferential solvation parameter of phenothiazine (component 3) by ethanol or propylene glycol (component 1) in the {cosolvent (1) + water (2)} mixtures at saturation was defined as follows:

$$\delta x_{1,3} = x_{1,3}^{L} - x_1 = -\delta x_{2,3} \tag{16}$$

where $x_{1,3}^{L}$ is the local mole fraction of cosolvent (1) in the molecular environment around phenothiazine (3) and x_1 is the bulk mole fraction of cosolvent in the initial aqueous cosolvent mixture in the absence of phenothiazine. If $\delta x_{1,3}$ values were positive, phenothiazine (3) was preferentially solvated by cosolvent (1), but if they were negative, phenothiazine was preferentially solvated by water (2). Thus, the respective $\delta x_{1,3}$ values were obtained by means of the inverse Kirkwood–Buff integrals (IKBI) for both solvent components, based on the following [28–30]:

$$\delta x_{1,3} = \frac{x_1 x_2 (G_{1,3} - G_{2,3})}{x_1 G_{1,3} + x_2 G_{2,3} + V_{\text{cor}}} \tag{17}$$

with

$$G_{1,3} = RT\kappa_T - \overline{V}_3 + x_2 \overline{V}_2 \left(\frac{D}{Q} \right) \tag{18}$$

$$G_{2,3} = RT\kappa_T - \overline{V}_3 + x_1 \overline{V}_1 \left(\frac{D}{Q} \right) \tag{19}$$

$$V_{\text{cor}} = 2522.5 \cdot \left\{ r_3 + 0.1363 \cdot \left(x_{1,3}^{L} \overline{V}_1 + x_{2,3}^{L} \overline{V}_2 \right)^{1/3} - 0.085 \right\}^3 \tag{20}$$

Here, κ_T represents the isothermal compressibility of every {cosolvent (1) + water (2)} mixture. $\overline{V}_1$ and $\overline{V}_2$ denote the partial molar volumes of cosolvent and water in the aqueous cosolvent mixtures. $\overline{V}_3$ denotes the partial molar volume of phenothiazine. The function D corresponds to the first derivative of the variation of standard molar Gibbs energies of transfer of phenothiazine from neat water to {cosolvent (1) + water (2)} mixtures regarding the cosolvent proportion in the mixtures free of solute, as shown in Equation (21). The function Q involves the second derivative of the variation of excess molar Gibbs energy of mixing of cosolvent and water (G_{1+2}^{Exc}/kJ·mol^{-1}) regarding the water proportion in the aqueous cosolvent mixtures, as shown in Equation (22). V_{cor} is the correlation volume and r_3 is the hydrodynamic molecular radius of phenothiazine, which is commonly calculated by means of Equation (23), where N_{Av} is the number of Avogadro.

$$D = \left(\frac{\partial \Delta_{\text{tr}} G_{3,2\to 1+2}^{o}}{\partial x_1} \right)_{T,p} \tag{21}$$

$$Q = RT + x_1 x_2 \left(\frac{\partial^2 G_{1+2}^{Exc}}{\partial x_2^2} \right)_{T,p} \tag{22}$$

$$r_3 = \left(\frac{3 \cdot 10^{21} V_3}{4 \pi N_{Av}} \right)^{1/3} \quad (23)$$

As reported in the literature, the definitive V_{cor} values require iteration because they depend on the local mole fractions of cosolvent and water around the phenothiazine molecules in the equilibrated solutions. Hence, these iterations were performed by substituting $\delta x_{1,3}$ and V_{cor} values in Equations (16), (17) and (20) in order to recalculate the $x^{L}_{1,3}$ value until almost invariant values of V_{cor} were obtained.

Figure 4 depicts the apparent Gibbs energies of transfer of phenothiazine from neat water to both {cosolvent (1) + water (2)} systems ($\Delta_{tr}G^{o}_{3,2\to1+2}$) at $T/K = 298.2$. These $\Delta_{tr}G^{o}_{3,2\to1+2}$ values were calculated from the mole fraction solubilities reported in [6], as follows:

$$\Delta_{tr}G^{o}_{3,2\to1+2} = RT \ln \left(\frac{x_{3,2}}{x_{3,1+2}} \right) \quad (24)$$

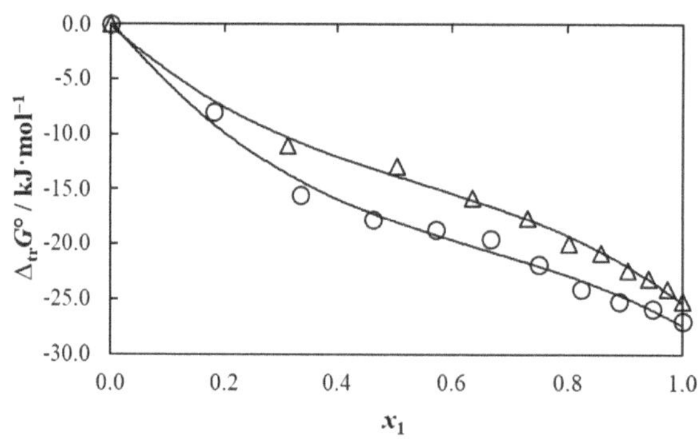

Figure 4. Gibbs energy of transfer of phenothiazine (3) from neat water (2) to some {cosolvent (1) + water (2)} mixtures at $T/K = 298.2$. ○: {ethanol (1) + water (2)}, Δ: {propylene glycol (1) + water (2)}.

$\Delta_{tr}G^{o}_{3,2\to1+2}$ values were correlated according to the regular third-degree polynomial presented as Equation (25). Coefficients and statistical parameters of Equation (25) for both aqueous cosolvent systems are summarized in Table 7.

$$\Delta_{tr}G^{o}_{3,2\to1+2} = a + bx_1 + cx_1^2 + dx_1^3 \quad (25)$$

Table 7. Coefficients and statistical parameters of Equation (23) applied to Gibbs energy of transfer from neat water (2) to {cosolvent (1) + water (2)} mixtures at $T/K = 298.2$.

Coefficient or Parameter	Ethanol (1) + Water (2)	Propylene Glycol (1) + Water (2)
a	0.24 ± 0.91	-0.13 ± 0.54
b	-63.63 ± 7.89	-47.22 ± 4.97
c	73.10 ± 18.65	57.85 ± 11.95
d	-37.09 ± 12.07	-35.85 ± 7.64
Adjusted r^2	0.985	0.994
Statistical error	0.938	0.545
F-statistical	254.6	613.3

In this way, the D values shown in Tables 8 and 9 were calculated from the first derivative of Equation (23) by considering the variation of aqueous cosolvent mixture composition in incremental $x_1 = 0.05$ steps for all the mixture composition intervals. The required Q, $RT \cdot \kappa_T$, $\overline{V}_1$ and $\overline{V}_2$ values corresponding to {ethanol (1) + water (2)} and {propylene glycol (1) + water (2)} mixtures were taken from the literature [31]. Because $\overline{V}_3$ values are not available for phenothiazine in these {cosolvent (1) + water (2)} mixtures, they were considered as those calculated based on the Fedors' method, i.e., 137.3 $cm^3 \cdot mol^{-1}$, as shown in Table 6 [27].

Table 8. Some properties associated with the preferential solvation of phenothiazine (3) in {ethanol (1) + water (2)} mixtures at T/K = 298.2.

x_1 [a]	D/ $kJ \cdot mol^{-1}$	$G_{1,3}$/ $cm^3 \cdot mol^{-1}$	$G_{2,3}$/ $cm^3 \cdot mol^{-1}$	V_{cor}/ $cm^3 \cdot mol^{-1}$	100 $\delta x_{1,3}$
0.00	−63.63	−600.1	−136.2	698	0.00
0.05	−56.60	−573.1	−204.4	708	−3.61
0.10	−50.12	−534.1	−269.7	731	−5.47
0.15	−44.20	−488.0	−327.1	772	−4.87
0.20	−38.84	−440.3	−374.3	827	−2.41
0.25	−34.03	−395.1	−411.6	885	0.64
0.30	−29.78	−354.9	−441.0	942	3.43
0.35	−26.09	−320.9	−465.7	994	5.70
0.40	−22.95	−293.2	−489.7	1042	7.48
0.45	−20.37	−271.5	−517.2	1088	8.93
0.50	−18.34	−255.0	−552.9	1132	10.23
0.55	−16.88	−243.0	−602.6	1176	11.53
0.60	−15.96	−234.4	−672.0	1221	12.94
0.65	−15.61	−227.7	−764.5	1266	14.37
0.70	−15.81	−220.5	−875.2	1307	15.44
0.75	−16.57	−210.1	−981.9	1342	15.42
0.80	−17.88	−195.0	−1048.3	1365	13.66
0.85	−19.75	−176.9	−1048.5	1377	10.39
0.90	−22.17	−159.3	−990.6	1386	6.54
0.95	−25.16	−145.0	−903.7	1395	2.97
1.00	−28.69	−134.4	−813.4	1409	0.00

[a] x_1 is the mole fraction of ethanol (1) in the {ethanol (1) + water (2)} mixtures free of phenothiazine (3).

$G_{1,3}$ and $G_{2,3}$, shown in Tables 8 and 9, were negative in all cases, indicating the affinity of phenothiazine to both cosolvents and water. An approximated hydrodynamic radius of phenothiazine (r_3) was calculated as 0.379 nm by means of Equation (23). In turn, the preferential solvation parameters of phenothiazine by ethanol or propylene glycol molecules are also summarized in Tables 8 and 9. According to Figure 5, the initial addition of a cosolvent to water gave negative $\delta x_{1,3}$ values of phenothiazine in the interval from neat water to the mixture of x_1 = 0.24 for aqueous ethanol mixtures and the mixture of x_1 = 0.18 for aqueous propylene glycol mixtures. The maximum negative values of this parameter were obtained for the mixture of x_1 = 0.10, with $\delta x_{1,3} = -5.47 \times 10^{-2}$ for ethanolic mixtures and $\delta x_{1,3} = -2.34 \times 10^{-2}$ for propylene glycol mixtures. These values were higher than the considered absolute values of 1.00×10^{-2}. Therefore, they were a consequence of real preferential solvation effects of water on phenothiazine, rather than a consequence of the uncertainties propagation in the respective IKBI calculations [32,33]. The cosolvent action of both cosolvents for increasing the phenothiazine solubility in these water-rich mixtures was associated with the breaking of the "iceberg"-like ordered structure exhibited by water molecules around the phenylene groups of phenothiazine.

Table 9. Some properties associated with the preferential solvation of phenothiazine (3) in {propylene glycol (1) + water (2)} mixtures at T/K = 298.2.

x_1 [a]	D/ kJ·mol^{-1}	$G_{1,3}$/ cm^3·mol^{-1}	$G_{2,3}$/ cm^3·mol^{-1}	V_{cor}/ cm^3·mol^{-1}	100 $\delta x_{1,3}$
0.00	−47.22	−480.7	−136.2	698	0.00
0.05	−41.71	−426.2	−195.4	730	2.10
0.10	−36.73	−377.9	−241.7	780	−2.34
0.15	−32.29	−335.9	−276.3	842	−1.36
0.20	−28.39	−300.1	−301.1	907	0.03
0.25	−25.02	−270.1	−318.1	970	1.36
0.30	−22.20	−245.5	−329.5	1029	2.43
0.35	−19.91	−225.8	−337.7	1084	3.24
0.40	−18.16	−210.4	−345.0	1137	3.82
0.45	−16.94	−198.5	−353.9	1188	4.25
0.50	−16.27	−189.7	−366.8	1238	4.62
0.55	−16.13	−183.1	−385.8	1286	4.96
0.60	−16.53	−178.3	−413.3	1335	5.31
0.65	−17.47	−174.7	−451.7	1383	5.67
0.70	−18.94	−171.7	−503.3	1430	6.01
0.75	−20.96	−168.8	−570.9	1476	6.25
0.80	−23.51	−165.5	−657.8	1519	6.28
0.85	−26.60	−161.3	−767.8	1558	5.92
0.90	−30.22	−155.6	−906.1	1593	4.96
0.95	−34.39	−147.5	−1079.6	1619	3.11
1.00	−39.09	−136.1	−1297.9	1635	0.00

[a] x_1 is the mole fraction of propylene glycol (1) in the {propylene glycol (1) + water (2)} mixtures free of phenothiazine (3).

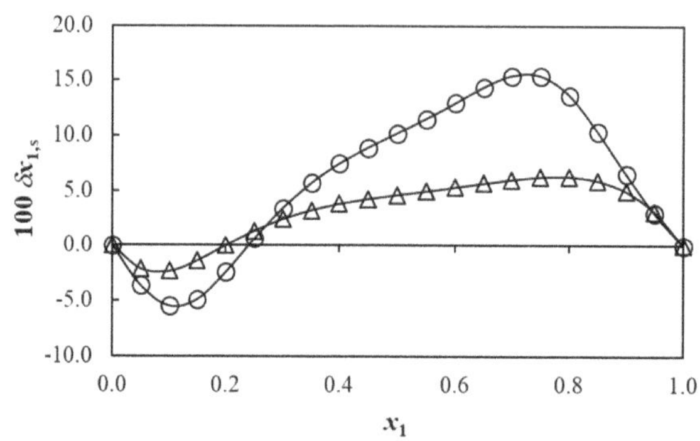

Figure 5. Preferential solvation parameters of phenothiazine (s) in some {cosolvent (1) + water (2)} mixtures at T/K = 298.2. ○: {ethanol (1) + water (2)}, Δ: {propylene glycol (1) + water (2)}.

In mixtures of 0.24 < x_1 < 1.00 for {ethanol (1) + water (2)} and 0.18 < x_1 < 1.00 for {propylene glycol (1) + water (2)} mixtures, the $\delta x_{1,3}$ values were positive, indicating the preferential solvation of phenothiazine by ethanol or propylene glycol. Maximum $\delta x_{1,3}$ values were obtained for the mixture of x_1 = 0.70 ($\delta x_{1,3}$ = 0.1544) for aqueous ethanol mixtures and the mixture of x_1 = 0.80 ($\delta x_{1,3}$ = 6.28 × 10^{-2}) for aqueous propylene glycol mixtures. The maximum positive $\delta x_{1,3}$ values were also higher than $|1.00 \times 10^{-2}|$, as a consequence of real preferential solvation effects of both cosolvents [32,33]. From a mechanistic viewpoint, the preferential solvation by both cosolvents could have been due to phenothiazine acting as a Lewis acid in front of the ethanol or propylene glycol molecules owing to the unshared electrons of the hydroxyl atoms of these cosolvents. Notably, these cosolvents are more basic than water, noted in the magnitude of their Kamlet–Taft hydrogen bond acceptor parameters, namely, $\beta = 0.75$ for ethanol, 0.78 for propylene glycol and 0.47 for water [16]. By comparing the phenothiazine behavior in aqueous mixtures of ethanol and propylene glycol, it follows that the region of preferential hydration is wider with ethanol and preferential hydration magnitude is also higher with this cosolvent. Nonetheless, the region of preferential solvation with a cosolvent is wider with propylene glycol, although the preferential solvation magnitude is higher with ethanol.

3. Conclusions

Reported equilibrium mole fraction solubilities of phenothiazine in ethanol, propylene glycol and water were processed to find apparent thermodynamic quantities of dissolution mixing and solvation based on van't Hoff and Gibbs equations. IKBI treatment demonstrated the preferential hydration of phenothiazine in water-rich mixtures and preferential solvation by cosolvents in mixtures of $0.24 < x_1 < 1.00$ for the {ethanol (1) + water (2)} system and mixtures of $0.18 < x_1 < 1.00$ for the {propylene glycol (1) + water (2)} system.

Author Contributions: Conceptualization, F.M., M.Á.P. and A.J.; methodology, F.M. and A.J.; software, F.M. and A.J.; validation, F.M., M.Á.P. and A.J.; formal analysis, F.M., M.Á.P. and A.J.; investigation, F.M., M.Á.P. and A.J.; resources, F.M. and A.J.; data curation, F.M. and A.J.; writing—original draft preparation, F.M. and A.J.; writing—review and editing, F.M., M.Á.P. and A.J.; visualization, F.M., M.Á.P. and A.J.; supervision, F.M., M.Á.P. and A.J.; project administration, F.M., M.Á.P. and A.J.; funding acquisition, F.M. and A.J. All authors have read and agreed to the published version of the manuscript.

Funding: This research was funded by the Research Directorate of the National University of Colombia with grant number HERMES 56224.

Data Availability Statement: Data are contained within the article.

Acknowledgments: We thank the Department of Pharmacy of the Universidad Nacional de Colombia for facilitating computational resources.

Conflicts of Interest: The authors declare no conflicts of interest.

References

1. O'Neil, M.J. *The Merck Index: An Encyclopedia of Chemicals, Drugs, and Biologicals*, 13th ed.; Merck & Co., Inc.: Whitehouse Station, NJ, USA, 2001.
2. Sweetman, S.C. *Martindale: The Complete Drug Reference*, 36th ed.; Pharmaceutical Press: London, UK, 2009.
3. Ohlow, M.J.; Moosmann, B. Phenothiazine: The seven lives of pharmacology's first lead structure. *Drug Discov. Today* **2011**, *16*, 119–131. [CrossRef] [PubMed]
4. Levy, L.B. Inhibition of acrylic acid polymerization by phenothiazine and p-methoxyphenol. II. Catalytic inhibition by phenothiazine. *J. Polymer Sci. A Polymer Chem.* **1992**, *30*, 569–576. [CrossRef]
5. Hoover, K.R.; Acree, W.E., Jr.; Abraham, M.H. Mathematical correlation of phenothiazine solubilities in organic solvents with the Abraham solvation parameter model. *Phys. Chem. Liq.* **2006**, *44*, 367–376. [CrossRef]
6. Ahmadian, S.; Panahi-Azar, V.; Fakhree, M.A.A.; Acree, W.E., Jr.; Jouyban, A. Solubility of phenothiazine in water, ethanol, and propylene glycol at (298.2 to 338.2) K and their binary and ternary mixtures at 298.2 K. *J. Chem. Eng. Data* **2011**, *56*, 4352–4355. [CrossRef]
7. Rubino, J.T. Cosolvents and cosolvency. In *Encyclopedia of Pharmaceutical Technology*; Swarbrick, J., Boylan, J.C., Eds.; Marcel Dekker, Inc.: New York, NY, USA, 1988; Volume 3, pp. 375–398.
8. Martin, A.; Bustamante, P.; Chun, A.H.C. *Physical Pharmacy: Physical Chemical Principles in the Pharmaceutical Sciences*, 4th ed.; Lea & Febiger: Philadelphia, PA, USA, 1993.
9. Yalkowsky, S.H. *Solubility and Solubilization in Aqueous Media*; American Chemical Society and Oxford University Press: New York, NY, USA, 1999.
10. Jouyban, A. *Handbook of Solubility Data for Pharmaceutical*; CRC Press: Boca Raton, FL, USA, 2010.

11. Delgado, D.R.; Martínez, F. Solution thermodynamics and preferential solvation of sulfamerazine in methanol + water mixtures. *J. Solution Chem.* **2015**, *44*, 360–377. [CrossRef]
12. Sabbah, R.; El Watik, L. Etude thermodynamique de la phénoxazine et de la phenothiazine. *Thermochim. Acta* **1992**, *197*, 381–390. [CrossRef]
13. Zarghampour, A.; Jouyban, K.; Jouyban-Gharamaelki, V.; Jouyban, A.; Rahimpour, E. A description on the shake-flask and laser monitoring-based techniques for determination of the drug's solubility. *Pharm. Sci.* **2024**, *30*, 274–278. [CrossRef]
14. Kristl, A.; Vesnaver, G. Thermodynamic investigation of the effect of octanol–water mutual miscibility on the partitioning and solubility of some guanine derivatives. *J. Chem. Soc. Faraday Trans.* **1995**, *91*, 995–998. [CrossRef]
15. Barton, A.F.M. *Handbook of Solubility Parameters and Other Cohesion Parameters*, 2nd ed.; CRC Press: Boca Raton, FL, USA, 1991.
16. Marcus, Y. *The Properties of Solvents*; John Wiley & Sons: Chichester, UK, 1998.
17. Jouyban, A.; Acree, W.E., Jr.; Martínez, F. Solution Thermodynamics and preferential solvation of atenolol in {ethanol (1) + water (2)} cosolvent mixtures. *J. Appl. Solution Chem. Model.* **2018**, *7*, 1–8. [CrossRef]
18. Bevington, P.R. *Data Reduction and Error Analysis for the Physical Sciences*; McGraw-Hill Book, Co.: New York, NY, USA, 1969; pp. 56–65.
19. Carstensen, J.T. *Modeling and Data Treatment in the Pharmaceutical Sciences*; Technomic Publishing Co., Inc.: Lancaster, PA, USA, 1996; pp. 127–159.
20. Barrante, J.R. *Applied Mathematics for Physical Chemistry*, 2nd ed.; Prentice Hall, Inc.: Upper Saddle River, NJ, USA, 1998; 227p.
21. Perlovich, G.L.; Kurkov, S.V.; Kinchin, A.N.; Bauer-Brandl, A. Thermodynamics of solutions III: Comparison of the solvation of (+)-naproxen with other NSAIDs. *Eur. J. Pharm. Biopharm.* **2004**, *57*, 411–420. [CrossRef] [PubMed]
22. Delgado, D.R.; Almanza, O.A.; Martínez, F.; Peña, M.A.; Jouyban, A.; Acree, W.E., Jr. Solution thermodynamics and preferential solvation of sulfamethazine in (methanol + water) mixtures. *J. Chem. Thermodyn.* **2016**, *97*, 264–276. [CrossRef]
23. Jouyban, K.; Agha, E.M.H.; Hemmati, S.; Martinez, F.; Kuentz, M.; Jouyban, A. Solubility of 5-aminosalicylic acid in *N*-methyl-2-pyrrolidone + water mixtures at various temperatures. *J. Mol. Liq.* **2020**, *310*, 113143. [CrossRef]
24. Romero, S.; Reillo, A.; Escalera, B.; Bustamante, P. The behaviour of paracetamol in mixtures of aprotic and amphiprotic-aprotic solvents. Relationship of solubility curves to specific and nonspecific interactions. *Chem. Pharm. Bull.* **1996**, *44*, 1061–1066. [CrossRef]
25. Jiménez, J.A.; Martínez, F. Temperature dependence of solubility of acetaminophen in propylene glycol + ethanol mixtures. *J. Solution Chem.* **2006**, *35*, 335–352. [CrossRef]
26. Connors, K.A. *Thermodynamics of Pharmaceutical Systems: An Introduction for Students of Pharmacy*; Wiley–Interscience: Hoboken, NJ, USA, 2002.
27. Fedors, R.F. A method for estimating both the solubility parameters and molar volumes of liquids. *Polym. Eng. Sci.* **1974**, *14*, 147–154. [CrossRef]
28. Marcus, Y. *Solvent Mixtures: Properties and Selective Solvation*; Marcel Dekker, Inc.: New York, NY, USA, 2002.
29. Marcus, Y. On the preferential solvation of drugs and PAHs in binary solvent mixtures. *J. Mol. Liq.* **2008**, *140*, 61–67. [CrossRef]
30. Marcus, Y. Preferential solvation of drugs in binary solvent mixtures. *Pharm. Anal. Acta* **2017**, *8*, 1000537. [CrossRef]
31. Jiménez, D.M.; Cárdenas, Z.J.; Delgado, D.R.; Martínez, F.; Jouyban, A. Preferential solvation of methocarbamol in aqueous binary co-solvent mixtures at 298.15 K. *Phys. Chem. Liq.* **2014**, *52*, 726–737. [CrossRef]
32. Ben–Naim, A. Preferential solvation in two- and in three-component systems. *Pure Appl. Chem.* **1990**, *62*, 25–34. [CrossRef]
33. Marcus, Y. Solubility and solvation in mixed solvent systems. *Pure Appl. Chem.* **1990**, *62*, 2069–2076. [CrossRef]

liquids

MDPI

Article

Vaporization Enthalpies and Vapor Pressures of 5α-Androstane and 5α-Cholestane by Correlation Gas Chromatography

Christian Fischer-Lodike, Mohammad Albinsaad and James S. Chickos *

Department of Chemistry and Biochemistry, 1 University Blvd, University of Missouri-St. Louis, St. Louis, MO 63121, USA; cfischerlodike@gmail.com (C.F.-L.)
* Correspondence: jsc@umsl.edu; Tel.: +1-314-516-5377

Abstract: Vaporization enthalpies and vapor pressures of 5α-androstane and 5α-cholestane are reported using correlation gas chromatography (CGC). The results for 5α-cholestane are compared to both estimated and experimental values reported previously for 5α-cholestane. The results are generally in agreement with the literature within the reported uncertainties. A simple method for reducing the amount of curvature in logarithm plots of vapor pressures as a function of K/T when using n-alkanes as standards in CGC experiments is also reported. This may prove useful in evaluating vapor pressures of rigid hydrocarbons at high temperatures.

Keywords: 5α-androstane; 5α-cholestane; vaporization enthalpy; vapor pressure; correlation gas chromatography

1. Introduction

5α-Cholestane, (5R,8R,9S,10S,13R,14S,17R)-10,13-dimethyl-17-((R)-6-methylheptan-2-yl)-hexadecahydro-1H-cyclopenta[a]phenanthrene, is a 27-carbon saturated tetracyclic hydrocarbon, produced by the diagenesis of cholesterol. It is one of the most abundant biomarkers in the rock record and is generally interpreted by paleontologists as an indicator of animal life [1]. While the stereochemistry of cholesterol during diagenesis to cholestane is typically maintained, cholestane with many different configurations is found in the fossil record. In 5α-cholestane, the stereochemical configuration of cholesterol is maintained.

5α-Androstane, (8S,9S,10S,13S,14S)-hexadecahydro-10,13-dimethyl-1H-cyclopenta[a] phenanthrene, is a structurally related steroid in which the alkyl side chain in 5α-cholestane has been replaced by hydrogen. In spite of not containing any oxygen, it appears to exhibit androgenic properties [2]. 5α-Androstane is used as a standard in various applications in chromatography including GC, GC/MS, HPLC, LC/MS, and other analytical instrumentation [3]. The structures of 5α-androstane and 5α-cholestane are illustrated in Figure 1.

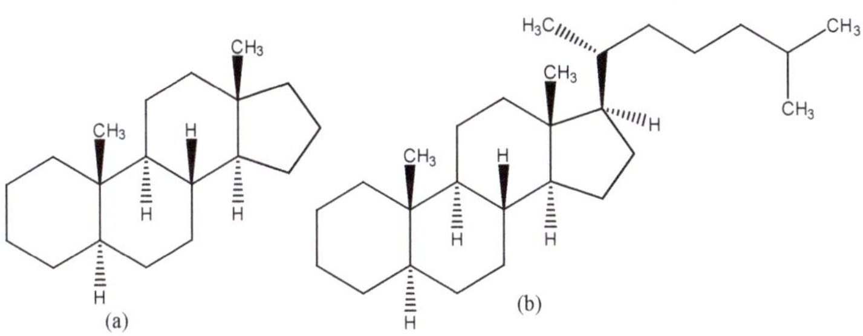

Figure 1. The structures of 5α-androstane (**a**) and 5α-cholestane (**b**).

Correlation gas-chromatography has proven quite useful in providing reasonable pure component vaporization enthalpies and vapor pressure measurements as a function of temperature on systems that are not easily measured by other more conventional techniques. This has included the study of relatively large molecules, individual component analysis of multicomponent systems and of substances available in very limited quantities [4,5]. A major limitation of the method is the availability of appropriate compounds with reliable properties to serve as standards. For hydrocarbons, the n-alkanes have proven to be reliable so far, for evaluating vaporization enthalpies regardless of structure but are not without their limitations. Accurate evaluation of vapor pressures are dependent on the structure of the targets. The use of n-alkanes as standards for rigid targets generally produce reasonable vaporization enthalpies but somewhat smaller vapor pressures at ambient temperatures. Vapor pressures for spherical hydrocarbons are often underestimated at higher temperatures as a result of more significant curvature observed for alkane standards than for the targets in plots of $\ln(p/p^o)$ versus K/T [6,7]. The observed curvature is such as to increase predicted normal boiling temperatures.

This study examines the behavior of both 5α-cholestane and 5α-androstane, two interesting examples of rigid tetracyclic molecules that are relatively planar; one of these, cholestane, has been studied previously [8,9]. This provides a rare opportunity to evaluate how well n-alkanes can reproduce some thermodynamic properties of an important class of large polycyclic hydrocarbons evaluated previously by an alternate method and to examine factors that might attenuate the curvature mentioned above. In addition to being polycyclic, both substances also contain the presence of two quaternary carbon centers, known to both reduce vaporization enthalpy and increase volatility [10]. It was therefore surprising at first, to observe that the retention times of 5α-cholestane, a C_{27} hydrocarbon, exhibited gas chromatographic retention times greater than the corresponding C_{28} alkane, n-octacosane. Results for 5α-androstane, a smaller $C_{19}H_{32}$ hydrocarbon and close relative to cholestane, exhibited similar retention time behavior. The properties of 5α-androstane were chosen for study since this substance has not been studied previously and could serve as a useful standard for evaluating similar properties of polycyclic diterpenes, a class of substances whose thermodynamic properties that have not yet been investigated.

2. Experimental

2.1. Compounds: Identity and Purity Controls

Purities and origin of the standards and targets used in this study are summarized in Table 1. Cholestane was purchased from Aldrich and an analytical sample of 5α-androstane, dissolved in methylene chloride (2000 μg/mL), was obtained from Supelco. Both were used as supplied. Cholestane, a single peak in the GC, was identified as 5α-cholestane by it melting temperature, [T_{fus} = (79.6–80.3) °C; (lit. (79.5–80) °C [11]; (mp reference: vanillin, T_{fus} = (81.2–82.4); lit.: T_{fus} = (82) [12]]; its infrared spectrum exhibited four sharp peaks at 1170, 956, 927 and 729 cm^{-1} absent in the 5β isomer and the absence of a peak at 735 cm^{-1} characteristic of 5β-cholestane. Infrared spectra of 5α- and 5β-cholestane available at the National Institute of Advanced Industrial Science and Technology (Tokyo, Japan) were used for comparisons [13]. An infrared spectrum of the material studied is available as Figure S1 (Supplementary Materials). Since the size of 5α-cholestane and 5α-androstane differ significantly, a different series of n-alkanes were used as standards.

Table 1. Origin of the standards and their analysis.

Compound	CAS Registry No.	Supplier [a]	Mass Fraction [b] Supplier
n-Hexadecane	544-76-3	Sigma Aldrich	0.99+
n-Heptadecane	62978-7	Aldrich	0.99
n-Octadecane	593-45-3	Fischer	0.99
n-Nonadecane	629-78-7	Sigma Aldrich	0.99

Table 1. *Cont.*

Compound	CAS Registry No.	Supplier [a]	Mass Fraction [b] Supplier
n-Eicosane	112-95-8	Aldrich	0.99
n-Docosane	629-97-0	Aldrich	0.99
n-Tetracosane	646-31-1	Aldrich	0.98
n-Pentacosane	629-99-2	Aldrich	0.99
n-Octacosane	630-02-4	Aldrich	0.99
n-Nonacosane	630-03-5	Aldrich	0.99
n-Triacontane	638-68-5	Sigma	0.99
n-Hentriacontane	630-04-6	Fluka	0.995
5α-Androstane/CH_2Cl_2	438-22-2	Sigma Aldrich	0.99+ [c]
5α-Cholestane	481-21-0	Sigma Aldrich	0.98

[a] Aldrich: Milwaukee WI., USA; Sigma: St. Louis MO. USA; Fischer: St. Louis MO, USA. Fluka: Buchs, Switzerland; Sigma Aldrich: Milwaukee WI., USA. [b] Values provided by the suppliers. [c] Analytical sample, analysis by gas chromatography (this work).

2.2. Methods

All chromatography was performed on an HP5890 gas chromatograph using a Supelco SPB-5 capillary column (0.32 mm, 1.0 µm film thickness; (St. Louis, MO, USA)) connected to a computer running Chemstation. Temperature was controlled to ±0.1 K by the instrument and monitored independently by a high temperature probe connected to a Go Link interface. The carrier gas was helium run at a pressure of 50 kPa above ambient pressure. All experiments were performed in duplicate. The solvent used for each experiment was methylene chloride. At the column temperatures employed, methylene chloride was not retained by the column. It was therefore used as a measure of the time necessary to traverse the column. Column residence times for each analyte, t_r, were calculated as the difference between the analyte's retention time and the retention time of the solvent. Data was collected for each run consecutively. Duplicate runs were performed under similar conditions of temperature and pressure but generally on different days. All retention times and correlations are reported in the Supplementary Materials, Tables S1A–S6A and S1B–S6B, respectively. Compounds are arranged according to their elution off the column.

2.3. Thermochemical Method: Vaporization Enthalpies

The basic premise in correlation gas chromatography is that the residence time t_r of an analyte, i, is inversely proportional to its vapor pressure off the column, $t_o/t_r(i)$. Granting this assertion, a plot of $\ln(t_o/t_{r(i)})$ versus K/T over the temperature range, T = 30 K where t_o = 60 s, a reference time, should result in a linear relationship. The slope of this line results in the negative enthalpy of transfer of the analyte from the stationary phase of the column to the gas phase, divided by the gas constant, $-\Delta_{\text{trn}}^{\text{g}}H_{\text{m}}(T_{\text{m}})_{(i)}/R$ (R = 8.314 J·mol^{-1}·K^{-1}). The enthalpy of transfer is related to the vaporization enthalpy, $\Delta_{\text{l}}^{\text{g}}H_m(T_m)_{(i)}$, by Equation (1), where $\Delta_{\text{l}}^{\text{g}}H_{\text{m}}(T_{\text{m}})_{(i)}$ is the enthalpy of interaction of each analyte with the column [14]. In the few cases where it has been evaluated, generally at elevated temperatures, $\Delta H_{\text{intr}}\left(T_{\text{m}}\right)_{(i)}$ has been endothermic and small in comparison to the vaporization enthalpy [4,15].

$$\Delta_{\text{trn}}^{\text{g}}H_{\text{m}}(T_{\text{m}}) = \Delta_{\text{l}}^{\text{g}}H_{\text{m}}(T_{\text{m}}) + \Delta H_{\text{intr}}(T_{\text{m}}) \tag{1}$$

Provided the standards are chosen appropriately, a second plot of their $\Delta_{\text{l}}^{\text{g}}H_{\text{m}}(298.15\text{ K})$ versus $\Delta_{\text{trn}}^{\text{g}}H_{\text{m}}(T_{\text{m}})(i)$ has also been observed to be linear. The equation of this line together with $\Delta_{\text{trn}}^{\text{g}}H_{\text{m}}(T_{\text{m}})(i)$ of the target(s) result(s) in evaluation of the vaporization enthalpy of the target(s). For hydrocarbons, provided the standards bracket both the vaporization enthalpies and retention times of the targets, vaporization enthalpy results do not appear to be very sensitive to the structure of the standards.

2.4. Thermochemical Method: Vapor Pressures

It has been observed empirically that a plot of $\ln(p/p^o)$ (where p^o is the reference pressure, 101,325 Pa) of the pure standards against their corresponding values of $\ln(t_o/t_r)$, also results in a linear relationship at T = 298.15 K. While (p_i/p^o) and (t_o/t_r) of the hydrocarbon standards differ, their logarithms appear to correlate over a wide range of temperatures in a linear fashion. Using their relationship and values of $\ln(t_o/t_r)$ of the targets, values of $\ln(p/p^o)$ of the targets can be evaluated. Obviously, the closer the standards resemble the targets, the more accurate the resulting vapor pressures. For hydrocarbons, results do not appear to be terribly sensitive to structure, although it does play a role. Hence the significance of comparing results evaluated by different methods. In this study, all correlations for both steroids studied were performed from T = (298.15 to 550) K; all correlation coefficients, r^2, exceeded 0.999 over this entire temperature range. Vapor pressures evaluated as a function of temperature in this work are expressed in the form of a second order polynomial, Equation (2). Results were fit to Equation (2) from T = (298.15 to 400) K and also up to 550 K.

$$\ln(p/p^o) = A_S + B_S(K/T) + C_S(K/T)^2; p^o = 101,325\,\text{Pa} \tag{2}$$

2.5. Uncertainties

Uncertainties reported in this work all refer to one standard deviation unless expressed otherwise [16]. All calculations were performed using Excel (2019) or Sigma Plot (V. 14). Linear plots were evaluated by linear regression. Uncertainties evaluated by correlation are a reflection of the quality of the correlation; actual uncertainties may differ. Uncertainties reported from logarithmic terms are reported as an average value.

2.6. Estimations of Vaporization Enthalpy

A simple relationship for estimating vaporization enthalpies of hydrocarbons, generally within ±5% of the experimental value, is given by Equation (3) [17]. The terms n_C and n_Q refer to the number of carbon atoms and number of quaternary sp^3 hybridized carbon atoms in the molecule. For 5α-cholestane and 5α-androstane, n_C = 27 and 19, respectively, and n_Q = 2 for both. Predictions are useful for providing a reasonable value or otherwise aiding in identifying unusual interactions or erroneous results.

$$\Delta_l^g H_m(298.15\,\text{K})/\text{kJmol}^{-1} = 4.69(n_C - n_Q) + 1.3n_Q + 3.0 \tag{3}$$

2.7. Vaporization Enthalpies and Vapor Pressures of the Standards and of 5α-Cholestane

The vaporization enthalpies of the n-alkanes used as standards are listed in the second column of Table 2. Vaporization enthalpies of the n-alkanes reported in Table 2A are from the critically evaluated values reported by Ruzicka and Majer [18]; those in Table 2C are from this laboratory [19,20]. Vapor pressures as a function of temperature for liquid n-alkanes in Table 2A are expressed in the form of the Cox Eq., Equation (4) [17], while those in Table 2C are expressed in the form of a third order polynomial, Equation (5) [19,20]. Previous studies of 5α-cholestane have included both the solid and liquid phase. Table 2, section B lists the constants for the Cox Eq., Equation (6), used to predict the vapor pressures of both phases [8] while Section 2D lists the vaporization enthalpy evaluated by the Antoine equation, Equation (7). The Antoine constants reported in the compendium by Stephenson and Malanowski appear to have been derived from vapor pressures reported by the American Petroleum Institute Project. 42 [9,21] (API). Also included in Table 2, section E, are the constants of a second order polynomial, Equation (8), evaluated by Mokbel et al. [8], from fits of experimental heat capacities of both the solid and liquid phases of cholestane as a function of temperature.

$$\ln(p/p^o) = [1 - T_b/T]\exp[A_0 + A_1\cdot\left(\frac{T}{K}\right) + A_2\cdot\left(\frac{T}{K}\right)^2; p^o = 101,325\,\text{Pa} \tag{4}$$

$$\ln(p/p^{o}) = A_T + B_T\cdot(K/T) + C_T.(K/T)^2 + D_T\cdot\left(K/T\right)^3;\ p^{\circ} = 101{,}325\ \text{Pa} \quad (5)$$

$$\ln\left(p/p^{\text{ref}}\right) = [1 - T_{\text{fus}}/T]\exp[A\prime_0 + A\prime_1\cdot(T/K) + A\prime_2.(T/K)^2;\ p^{\text{ref}} = 0.008\ \text{Pa} \quad (6)$$

$$\log(p/p^{o}) = A_A - B_A/(C_A + T/K);\ p^{o} = 1.0\ \text{kPa} \quad (7)$$

$$Cp(T/K) = [A_{cp} + B_{cp}\cdot T/K + C_{cp}\cdot(T/100\ K)^2]\cdot R;\ R = 8.314\ J\cdot K^{-1}\cdot mol^{-1} \quad (8)$$

Table 2. Vaporization enthalpies of the compounds and constants for equations used to evaluate vapor pressures; sections (A) and (B): vapor pressure constants for the Cox equation, Equation (4); section (C): vapor pressure constants for a third order polynomial, Equation (6); section (D): vapor pressure constants for the Antoine equation, Equation (7); section (E): constants for evaluating both liquid and solid heat capacities of 5α-cholestane, Equation (8).

(A) Standards [a]	$\Delta_l^g H_m$ **(298.15 K)** [b] **kJ·mol^{-1}**	A_0	A_1	$A_2 \times 10^6$	T_o/K/p^o/kPa	
n-Heptadecane	86.47 ± 1.7	3.21826	−0.002036553	1.383899	575.375/101.325	
n-Octadecane	91.44 ± 1.8	3.24741	−0.002048039	1.362445	590.023/101.325	
n-Nonadecane	96.44 ± 1.9	3.27626	−0.002062714	1.346737	603.989/101.325	
n-Eicosane	101.80 ± 0.5	3.31181	−0.002102218	1.348780	617.415/101.325	
(B) x = l; cr	$\Delta_x^g H_m$**(351.8 K)**	$A\prime_0$	$A\prime_1\cdot(10^4)$	$A\prime_2\cdot(10^7)$	T_{fus}/K	$\Delta T/K_{range}$
5α-Cholestane (l) [c]	108.4	3.948254	−11.945709	6.8443691		352–478
5α-Cholestane (cr) [c]	133.8	3.936924	−3.2336427	0	351.75	308–352
(C) Standards [d]	$\Delta_l^g H_m$**(298.15 K)** [e] **kJ·mol^{-1}**	A_t	B_t	C_t	D_t	
n-Heneicosane	106.8 ± 2.5	6.6591	−98.135	−2,907,500	199,890,000	
n-Docosane	111.9 ± 2.7	6.5353	110.72	−3,117,600	217,130,000	
n-Tetracosane	121.9 ± 2.8	6.2817	530.15	−3,528,600	250,720,000	
n-Pentacosane	126.8 ± 2.9	6.1496	741.19	−3,730,700	267,380,000	
n-Hexacosane	131.7 ± 3.2	6.0704	910.53	−3,919,300	282,440,000	
n-Octacosane	141.9 ± 4.9	5.8835	1279.4	−4,312,000	313,890,000	
n-Nonacosane	147.1 ± 5.1	5.8413	1431.2	−4,504,300	328,710,000	
n-Tricontane	152.3 ± 5.3	5.7696	1601.6	−4,699,800	344,040,000	
n-Henitricontane [f]	157.3 ± 0.7	5.679	1791.2	−4,900,200	360,370,000	
(D) Antoine Eq. [g]		A_A	B_A	C_A	p^o/kPa	T_{range}/K
5α-Cholestane (l)		11.35684	6038.938	0	1	481–538
(E) Heat capacity [c]	T_{range}/K	A_{cp}	B_{cp}	C_{cp}	Cp(298 K)/J·K^{-1}·mol^{-1}	
5α-Cholestane (l)	368–387	9.22607	0.256109	−0.881339	646.4 [h]	
5α-Cholestane (cr) [c]	308–333	−9.44908	0.257208	0	559 [h]	

[a] Reference [18]. [b] Uncertainties represent probable error. [c] Reference [8]. [d] Reference [19]. [e] Uncertainties represent one standard deviation. [f] Reference [20]. [g] Reference [9]; data fit to the Antoine equation by R. M. Stephenson and S. Malanowski, ref. [21]. [h] Reference [8]; extrapolated to T = 298.15 K.

3. Results

3.1. Vaporization Enthalpies

Enthalpies of transfer, $\Delta_{trn}^g H_m(T_m)$, were evaluated as the product of the gas constant, R, and the slopes of the lines obtained from linear plots of $\ln(t_o/t_a)$ vs. K/T. All correlation coefficients, r^2, exceeded 0.997. Table 3 (A, B and C) summarizes the results of one of two duplicate sets of correlations for 5α-cholestane and one of a duplicate set for 5α-androstane. The first duplicate set for 5α-cholestane was based on the usual practice of bracketing the retention time of the target. In view of the long retention time exhibited by 5α-cholestane and its relatively small vaporization enthalpy relative to those of the standards, this prompted the use of a second set of correlations to bracket its vaporization enthalpy better. Equations (9) and (10) and the corresponding correlation coefficients listed below each respective correlation describe the quality of the fits for one of the two duplicate sets of correlations performed. The second correlation for both is available in the Supplementary Materials.

Table 3. Correlations of $\Delta_{trn}^{g}H_m(T/K)$ of the standards with their respective $\Delta_{l}^{g}H_m(298\ K)$ values.

Run 1	-slope [a]/K	Intercept [a]	$\Delta H_{trn}(538K)^a$/ kJ·mol^{-1}	$\Delta_l^g H_m(298K)$/ kJ·mol^{-1}	
				Lit. [a]	Calc. [a]
Pentacosane	9754 ± 97	16.95 ± 0.18	81.09 ± 0.81	126.8 ± 2.9	126.6 ± 3.9
Hexacosane	10,123 ± 96	17.38 ± 0.18	84.15 ± 0.80	131.7 ± 3.2	131.8 ± 4.0
Octacosane	10,846.0 ± 97	18.21 ± 0.18	90.17 ± 0.82	141.9 ± 4.9	142.0 ± 4.1
5-α-Cholestane	9720 ± 81	15.89 ± 0.15	80.81 ± 0.68		126.2 ± 3.9
Nonacosane	11,197 ± 88	18.61 ± 0.16	93.08 ± 0.73	147.1 ± 5.1	146.9 ± 4.2
Tricosane	11,632 ± 95	19.16 ± 0.18	96.70 ± 0.79	152.3 ± 5.3	153.0 ± 4.3
Hentriacontane	11,892 ± 95	19.40 ± 0.17	98.86 ± 0.78	157.3 ± 1.2	156.7 ± 4.3

$$\Delta_l^g H_m(298\ K)/kJ{\cdot}mol^{-1} = (1.692 \pm 0.032)\Delta_{trn}^{g}H_m(538\ K) - (10.59 \pm 2.9) \quad r^2 = 0.9986 \quad (9)$$

Run 3	-slope [a]/K	Intercept [a]	$\Delta H_{trn}(518K)^a$/ kJ·mol^{-1}	$\Delta_l^g H_m(298K)$/ kJ·mol^{-1}	
				Lit. [a]	Calc. [a]
n-Tetracosane	8596.9 ± 62	15.21 ± 0.12	71.47 ± 0.52	121.9 ± 2.8	121.7 ± 1.6
n-Pentacosane	8961.0 ± 61	15.62 ± 0.12	74.50 ± 0.51	126.8 ± 2.9	126.8 ± 1.6
n-Hexacosane	9327.2 ± 55	16.05 ± 0.11	77.54 ± 0.46	131.7 ± 3.2	131.9 ± 1.6
n-Octacosane	10,047.5 ± 50	16.87 ± 0.10	83.53 ± 0.41	141.9 ± 4.9	142.0 ± 1.7
5-α-Cholestane	8870.8 ± 57	14.45 ± 0.09	73.75 ± 0.47		125.6 ± 1.6
n-Nonacosane	10,405.7 ± 45	17.29 ± 0.09	86.51 ± 0.38	147.1 ± 5.1	147.0 ± 1.7

$$\Delta_l^g H_m(298\ K)/kJ{\cdot}mol^{-1} = (1.677 \pm 0.015)\Delta H_{trn}(518\ K) + (1.90 \pm 1.2) \quad r^2 = 0.9998 \quad (10)$$

Run 5	-slope [a]/K	Intercept [a]	$\Delta H_{trn}(473K)^a$/ kJ·mol^{-1}	$\Delta_l^g H_m(298K)$/ kJ·mol^{-1}	
				Lit.	Calc. [a]
n-Heptadecane	67,817 ± 16	13.61 ± 0.03	56.37 ± 0.13	86.47 ± 1.7 [b]	86.3 ± 1.9
n-Octadecane	7183 ± 19	14.09 ± 0.04	59.72 ± 0.16	91.44 ± 1.8 [b]	91.5 ± 2.0
n-Nonadecane	7582 ± 29	14.56 ± 0.06	63.03 ± 0.24	96.44 ± 2.0 [b]	96.7 ± 2.1
n-Eicosane	7987 ± 31	15.05 ± 0.07	66.40 ± 0.25	101.8 ± 0.5 [b]	101.9 ± 2.2
5-α-Androstane	6891 ± 21	12.58 ± 0.04	57.29 ± 0.17		87.7 ± 1.9
n-Heneicosane	8344 ± 20	15.44 ± 0.04	69.37 ± 0.16	106.8 ± 2.5 [a]	106.6 ± 2.1
n-Docosane	87,956 ± 31	16.04 ± 0.07	73.12 ± 0.26	111.9 ± 2.7 [a]	112.4 ± 2.2

$$\Delta_l^g H_m(298\ K)/kJ{\cdot}mol^{-1} = (1.560 \pm 0.023)\Delta_{trn}^{g}H_m(473\ K) - (1.681 \pm 1.44) \quad r^2 = 0.9994 \quad (11)$$

[a] Uncertainties represent one standard deviation, references [19,20]. [b] Uncertainties represent probable error, ref. [18].

Correlations were performed in a similar manner for 5α-androstane; $r^2 > 0.999+$. Since this molecule lacks the alkyl side chain, a smaller set of n-alkane standards were required. 5α-Androstane also exhibited longer retention times than might be expected for a C_{19} hydrocarbon. The results obtained for 5α-cholestane in runs 1 and 3 of Table 3 suggested that a second set of correlations, bracketing the vaporization enthalpy for 5α-androstane better, need not be necessary. Additional details for both substances are available in the Supplementary Materials. The results of the correlations are discussed below. Table 4 summarizes the results of all correlations involving the evaluation of vaporization enthalpies at $T = 298.15$ K.

Table 4. Summary of the vaporization enthalpies of 5α-cholestane and androstane at T = 298.15 K (kJ·mol^{-1}) [a].

	Run 1	Run 2	Run 3	Run 4	Avg.	Estimate [b]	Cox Eq. [c]
5α-Cholestane	126.2 ± 3.9	124.1 ± 4.3	125.6 ± 1.6	126.6 ± 5.9	125.6 ± 3.9	122.9 ± 6.1	116.7 ± 3.3
	Run 5	Run 6					
5α-Androstane	87.7 ± 1.9	87.8 ± 2.0			87.8 ± 2.0	85.3 ± 4.3	

[a] Uncertainties represent one standard deviation. [b] Reference [17]. [c] Evaluated by extrapolation of the Cox Eq. [8].

3.2. Vapor Pressures

Values of t_o/t_r of the standards obtained from duplicate runs evaluated under similar experimental conditions were averaged and correlated against their corresponding values of $\ln(p/p^o)$ in the form, $\ln(t_o/t_r)_{avg}$, over the temperature range T = (298.15 to 550) K. In separate correlations for each steroid, plots of $\ln(p/p^o)$ vs. $\ln(t_o/t_r)_{avg}$ of the standards resulted in linear relationships that persisted over the entire temperature ranges evaluated, $r^2 > 0.999+$. Results obtained at T = 298.15 K are illustrated in Table 5 (A, B and C). Results are characterized by Equations (12) to (14) and by the correlation coefficients listed beside them. Additional vapor pressures were evaluated at 10 K increments starting at T = 310 K. Polynomials for both targets were evaluated for each set of correlations from T = (298.15 to 400) K and from T = (298.15 to 550) K for both 5α-cholestane and 5α-androstane. The constants evaluated for both are reported in Table 6. The average absolute fractional deviation (AAFD) in vapor pressures evaluated using results over the temperature range T= (298.15 to 400) K between combined runs (1&2) and (3&4) for 5α-cholestane was 0.026. When evaluated over the temperature range, T = (298.15 to 550) K, AAFD = 0.01; consequently runs (1&2) and (3&4) for each temperature range was averaged. The recommended constants for both 5α-cholestane (runs 1–4) and 5α-androstane (runs 5&6) (298 to 400, and 298.15 to 550) K are reported in Table 7.

Table 5. Correlations of $\ln(p/p^o)$ versus $\ln(t_o/t_r)_{avg}$ at T = 298.15 K of the standards [a].

A (Runs 1 and 2)	$\ln(t_o/t_r)_{avg}$	$\ln(p/p^o)$	$\ln(p/p^o)$ Calc	$10^6 \cdot p_{298.15\ K}$/Pa Calc	$10^6 \cdot p_{298.15\ K}$/Pa Lit. [b]
n-Pentacosane	−16.034	−23.244	−23.24 ± 0.45	8.3 ± 3.9	8.1
n-Hexacosane	−16.807	−24.309	−24.31 ± 0.46	2.8 ± 1.3	2.8
n-Octacosane	−18.370	−26.490	−26.51 ± 0.48	0.31 ± 0.16	0.32
5-α-Cholestane	−16.889		−24.43 ± 0.47	2.5 ± 1.2	7.8 [c]
n-Nonacosane	−19.149	−27.627	−27.61 ± 0.49	0.10 ± 0.05	0.10
n-Tricontane	−20.035	−28.748	−28.86 ± 0.51	0.03 ± 0.02	0.033
n-Hentriacontane	−20.665	−29.841	−29.75 ± 0.51	0.012 ± 0.007	0.011 [d]
$\ln(p/p^o) = (1.314 \pm 0.010) \cdot \ln(t_o/t_r)_{avg} - (2.22 \pm 0.11)$				(12)	r^2 = 0.9997
B (Runs 3 and 4)	$\ln(t_o/t_r)_{avg}$	$\ln(p/p^o)$	$\ln(p/p^o)$ Calc	$10^6 \cdot p_{298.15\ K}$/Pa Calc	$10^6 \cdot p_{298.15\ K}$/Pa Lit. [b]
n-Tetracosane	−14.746	−22.175	−22.22 ± 0.44	24 ± 6	24
n-Pentacosane	−15.447	−23.244	−23.17 ± 0.45	8.1 ± 2.1	8.1
n-Hexacosane	−16.306	−24.309	−24.32 ± 0.46	2.7 ± 0.7	2.8
n-Octacosane	−17.957	−26.49	−26.55 ± 0.48	0.31 ± 0.09	0.32
5-α-Cholestane	−16.408		−24.46 ± 0.46	2.5 ± 0.7	7.8 [c]
n-Nonacosane	−18.732	−27.627	−27.59 ± 0.49	0.11 ± 0.03	0.10
$\ln(p/p^o) = (1.37 \pm 0.012) \cdot \ln(t_o/t_r)_{avg} - (2.84 \pm 0.19)$				(13)	r^2 = 0.9998

Table 5. *Cont.*

C Runs (5 and 6)	$\ln(t_o/t_r)_{avg}$	$\ln(p/p^o)$	$\ln(p/p^o)$ Calc	$10^2 \cdot p_{298.15\ K}$/Pa Calc	$10^2 \cdot p_{298.15\ K}$/Pa Lit. [e]
n-Heptadecane	−9.187	−14.314	−14.308	6.3 ± 1.3	6.2
n-Octadecane	−10.059	−15.435	−15.439	2.0 ± 0.4	2.0
n-Nonadecane	−10.924	−16.551	−16.562	0.64 ± 0.1	0.66
n-Eicosane	−11.793	−17.696	−17.688	0.20 ± 0.05	0.21
5α-Androstane	−10.594		−16.133	1.0 ± 0.2	0.23 [f]
n-Heneicosane	−12.600	−18.836	−18.736	0.070 ± 0.02	0.067 [d]
n-Docosane	−13.520	−19.972	−19.928	0.021 ± 0.005	0.021 [b]
$\ln(p/p^o) = (1.320 \pm 0.014) \cdot \ln(t_o/t_r)_{avg} - (2.16 \pm 0.15)$				(14)	$r^2 = 0.9997$

[a] Uncertainties represent on standard deviation unless indicated otherwise. [b] Reference [19] unless indicated otherwise. [c] Extrapolation of the Cox Eq. [8]. [d] Reference [20] [e] Reference [18] unless noted otherwise. [f] Estimate of the subcooled liquid [22].

Table 6. Constants of Equation (2) evaluated by correlations evaluated from T = (298.15 to 400, 550) K; p^o = 101325 Pa [a].

Runs	A_s	$-B_s$	$-C_s$	T_{range}/K	T_B/K	T_B/K (Lit)
			5α-Cholestane			
1&2	11.45 ± 0.20	6016.6 ± 140	1,395,542 ± 14,123	298.15 to 400	700	740.4 [b]
1&2	8.42 ± 0.20	3863.5 ± 161	1,773,500 ± 31,755	298.15 to 550	743	665 [c]
3&4	11.59 ± 0.21	6097.3 ± 144	1,387,138 ± 24,713	298.15 to 400	698	
3&4	8.42 ± 0.21	3843.0 ± 171	1,783,043 ± 33,662	298.15 to 550	743	
			5α-Androstane			
5&6	9.045 ± 0.140	4261.0 ± 95	969,095 ± 16,268	298.15 to 400	638.8	587.6 [c]
5&6	6.578 ± 0.180	2500.4 ± 145	1,279,139 ± 28,441	298.15 to 550	670.1	609 ± 9 [d]

[a] Uncertainties represent one standard deviation; T_B/K the boiling temperature at p^o = 101325 Pa. [b] Estimated, subcooled liquid, ref. [22]. [c] Extrapolation of the Cox equation, ref. [8]. [d] Estimated, reference [23].

Table 7. Combined constants of Equation (2) for vapor pressures evaluated in Runs 1–4 for 5α-cholestane and Runs 5 and 6 for 5α-androstane [a].

Runs	A_s	$-B_s$	$-C_s$	T_{range}/K	T_B/K	T_B/K (Lit)
			5α-Cholestane			
1–4	**11.518 ± 0.210**	**6057.3 ± 142**	**1,391,260 ± 24,410**	298.15 to 400	699	665 [b], 740.4 [c]
1–4	8.419 ± 0.210	3853.7 ± 166	1,778,173 ± 32,710	298.15 to 550	743	714 [d]
			5α-Androstane			
5&6	**9.045 ± 0.140**	**4261.0 ± 95**	**969,095 ± 16,268**	298.15 to 400	638.8	588 [b], 609 [c]

[a] Recommended values in bold. [b] Reference [22]. [c] Extrapolation of the Cox Eq, [8]. [d] Estimate, ref. [23].

4. Discussion

4.1. Vaporization Enthalpy of 5α-Cholestane Using Literature Data

Both 5α-cholestane and 5α-androstane are solids at room temperature. The vaporization enthalpies at T = 298.15 K are for the subcooled liquid and have not been reported previously. A value of 108.4 kJ·mol^{-1} has been reported for 5α-cholestane at the fusion temperature, also used as the triple point temperature, T_{tp} = 351.75 K [8]. Using the experimental vapor pressures reported over the temperature range, T = (391.7 to 461.8) K for 5α-cholestane from reference [8], a vaporization enthalpy of (98.1 ± 0.25) kJ·mol^{-1} is calculated at the mean temperature of measurement, T_m = 426.8 K. Adjustment to the fusion temperature, T_{tp} = 351.75 K, using Equation (15) [24], ΔT/K = (426.8 − 351.75) and an extrapolated experimental $Cp_{(l)}$(298 K) value of 646.4 J·K^{-1}·mol^{-1} [8], results in an estimated temperature adjustment of (13.4 ± 2.7) kJ·mol^{-1}. Applied to a vaporization enthalpy of (98.1 ± 0.25) kJ·mol^{-1} results in a value of (111.5 ± 2.7) kJ·mol^{-1} at T_{fus}; this compares to a vaporization enthalpy of 108.4 kJ·mol^{-1} reported previously [8], adjusted in a different manner. The uncertainty in the temperature adjustment assumes a 20% overall uncertainty using Equation (15). The two vaporization enthalpies at T_{fus} differ by less than 3% and differences are mainly due to the manner in which the vaporization enthalpies are adjusted to T_{fus}. Applying Equation (15) from T_m = 426.8 to T = 298.15 K, results in a temperature adjustment of (23 ± 4.6) kJ·mol^{-1} and a vaporization enthalpy of (121.1 ± 4.6) kJ·mol^{-1}. This compares to the value (125.4 ± 3.1) kJ·mol^{-1} (Table 4) evaluated directly by correlation in this work. Details regarding these evaluations are provided in the material accompanying Table S7 (Supplementary Materials, pp. S7–S8). A vaporization enthalpy of 117 kJ·mol^{-1} at T = 298.15 K is evaluated by extrapolation of the Cox Eq. to evaluate the vaporization enthalpy at T = 298.15 K [8].

$$\Delta_l^g H_m(298.15\ K) = \Delta_l^g H_m(T_{fus}) + \left[10.58 + 0.26Cp_{(1)}(298\ \text{K})\right] \cdot [\Delta T/\text{K}] \quad (15)$$

The vaporization enthalpy evaluated using the vapor pressures reported by the API study at the mean temperature of measurement, T = 509.7 K, is (115.2 ± 0.2) kJ·mol^{-1}. Evaluations of $\Delta_l^g H_m(509.7\ \text{K})$ using vapor pressures from both the Cox Eq. and Equation (2) and the constants reported in Tables 2 and 7 (using values evaluated from (298.15 to 400) K) for 5α-cholestane to this mean temperature result in vaporization enthalpies of [(87.5 ± 0.5) and (95.8 ± 0.4)] kJ·mol^{-1}, respectively. The uncertainties reported here are the result of the slight curvature associated with the 30 K temperature intervals used in evaluations of the slopes of the line generated by the Clausius Clapeyron Eq. Vapor pressures and corresponding temperatures reported by the API study may be located in Table S8 as a footnote [9].

4.2. Liquid Vapor Pressures of 5α-Cholestane

A comparison of the vapor pressures evaluated by correlation with literature values for 5α-cholestane over the two temperature ranges studied using Equation (2) and the two sets of constants reported in Table 7 are provided in Figures 2 and 3. Figure 2 illustrates the results using correlations from T = (298.15 to 400) K followed by extrapolation to the temperature where $\ln(p/p^o) = 0$. Figure 3 illustrates the same comparison when the range of the correlation is extended to 550 K. The agreement between this work and the values calculated by the Cox Eq. at the higher temperatures are clearly better in Figure 3. This is also reflected in one of the predictions of the normal boiling temperatures reported in Table 7. Despite this agreement, we believe the results using the smaller temperature range in Figure 2 are more reliable for the following two reasons. A vaporization enthalpy at T = 298.15 K of 126.9 kJ·mol^{-1} is evaluated using the constants of Equation (2) and the vapor pressures generated by correlation from T = (298.15 to 400) K while a value of 131.4 kJ·mol^{-1} is calculated using constants evaluated from T = (298.15 to 550) K. These two values compare to a mean value of (125.9 ± 3.9) kJ·mol^{-1} (Table 4) evaluated directly by correlation.

Additionally, comparisons of the overall absolute average fractional deviation (AAFD) using the combined experimental vapor pressures from both the API database and Mokbel et al. [8] (T = (391.65 to 537.65)) was 0.2 using the larger temperature range and 0.15 using constants of Equation (2) evaluated over the shorter temperature range. AAFDs varied from a maximum of 0.44 to a minimum of 0.047 for vapor pressures evaluated using constants generated from the larger temperature range and 0.15 to 0.022 for those evaluated using the shorter temperature range. Table S8 provides a detailed comparison of the vapor pressures and Table S9 their differences (Supplementary Materials).

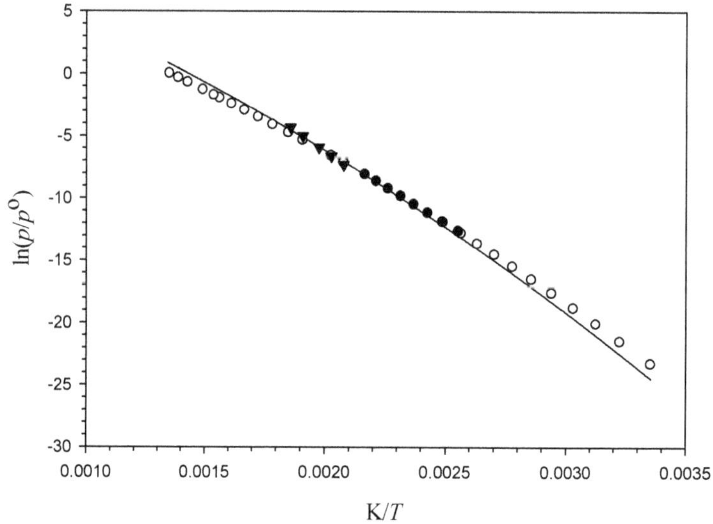

Figure 2. A plot of $\ln(p/p^{o})$ vs. K/T for cholestane from T = (298.15 to 740.4) K, the normal boiling temperature calculated by the Cox Eq. The line represents results calculated by Equation (2) using data evaluated from T = (298.15 to 400) K by correlation. The solid circles represent experimental data from Mokbel et al. [8] while the other circles are extrapolated data calculated by the Cox equation, Equation (6). The triangles represent experimental boiling temperatures evaluated at reduced pressures from API Project 42 [9].

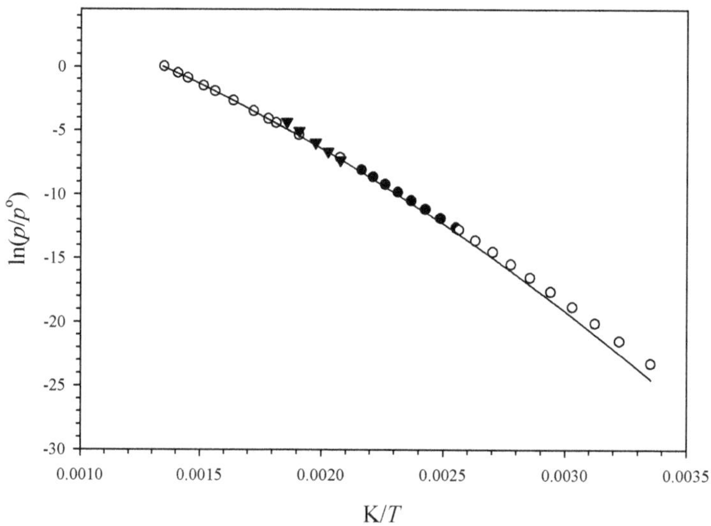

Figure 3. A plot of $\ln(p/p^{o})$ vs. K/T for 5α-cholestane from T = (298.15 to 740.4) K, the normal boiling temperature calculated by the Cox Eq. The line represents results calculated by Equation (2) using data evaluated from T = (298.15 to 550) K by correlation. The solid circles represent experimental data from Mokbel et al. [8] while the other circles are extrapolated data calculated by the Cox equation, Equation (6). The triangles represent experimental boiling temperatures at reduced pressures from API Project 42 [9].

4.3. Sublimation Enthalpy and Vapor Pressure of Solid 5α-Cholestane

Combining the vaporization enthalpy of (111.5 ± 2.6) kJ·mol^{-1} derived from the data from [8] at T_{fus} described in Section 4.1 with the fusion enthalpy, (25.4 ± 0.8) kJ·mol^{-1}, results in a sublimation enthalpy of (136.9 ± 2.8) kJ·mol^{-1}. Adjusting the sublimation enthalpy to T = 298.15 using Equation (16) [24] results in a sublimation of (141.4 ± 2.9) kJ·mol^{-1} at T = 298.15 (138.1 kJ·mol^{-1} [8]) reported previously). An approximate vapor pressure of the solid can also be evaluated using the Clausius Clapeyron Eq., Equation (17). Using a sublimation enthalpy of (141.4 ± 2.9) kJ·mol^{-1} and a vapor pressure of 0.008 Pa evaluated at T_{tp} = 351.75 K (Table S7, p. 8) [8], an approximate vapor pressure of $p_{cr} \approx 1.3 \times 10^{-6}$ Pa at T = 298.15 K is calculated that compares to $p_{cr} \approx 1.9 \times 10^{-6}$ Pa reported previously [8]).

Repeating the same calculations using the vaporization enthalpy at T = 298.15 K and vapor pressure evaluated at T_{tp} in this work, (125.6 ± 3.9; p_{298K} = 0.0044 Pa), results in a sublimation enthalpy of (146.0 ± 4.4) kJ·mol^{-1} and a vapor pressure of the solid of 5.6×10^{-7} Pa. Details are provided in the Supplementary Materials, p. 8.

$$\Delta_{cr}^{g}H_m(298.15\text{ K}) = \Delta_{cr}^{g}H_m(351.75\text{ K}) + (0.75 + 0.15\cdot(Cp_{(cr)}298.15\text{ K})\cdot(T_{tp} - 298.15)/1000 \quad (16)$$

$$\ln(p_{298}/p_{tp}) = \Delta_{cr}^{g}H_m(351.75\text{ K})\cdot[1/T_{298} - 1/T_{tp}]/R \quad (17)$$

4.4. Vaporization Enthalpy of 5α-Androstane

5α-Androstane has not been studied previously. The results of two correlations resulted in a vaporization enthalpy of (87.8 ± 0.8) kJ·mol^{-1}. Like 5α-cholestane, 5α-androstane is also a solid at T = 298.15 K, melting at: T_{fus} = 352.2 K. The vaporization enthalpy at T = 298.15 K refers to the subcooled liquid. The vaporization enthalpy at T_{fus} was also calculated using Equation (15) and an estimated liquid heat capacity of 435.3 J·K^{-1}·mol^{-1} evaluated previously by synthetic analysis [25]. A vaporization enthalpy of (81.1 ± 1.6) kJ·mol^{-1} is calculated at the melting point by rearranging Equation (15) and solving for $\Delta_l^g H_m(T_{fus})$, The uncertainty of ±1.6 kJ·mol^{-1} is based on a similar assumption of a total uncertainty of 20% associated with the temperature adjustment using Equation (15) [24]; this also includes the reported uncertainty associated with the vaporization enthalpy of 5α-androstane.

4.5. Liquid Vapor Pressures of 5α-Androstane

A similar comparison of the vapor pressures evaluated for 5α-androstane over the two temperature ranges studied using Equation (2) and the two sets of constants reported in Table 6 are provided in Figure 4. The recommended constants of Equation (2) reported in Table 7 are based on the results observed for 5α-cholestane. As also observed for 5α-cholestane, the vaporization enthalpy evaluated at T = 298.15 K using the shorter temperature range is closer to the value evaluated directly by correlation. Values of (89.6 ± 0.4 and 92.3 ± 0.5) kJ·mol^{-1} are calculated by the constants evaluated at T = (298.15 to 400 and to 550) K, respectively. These compare to the value of (87.8 ± 0.8) kJ·mol^{-1} obtained by the direct enthalpy correlations reported above. The uncertainties reported are a reflection of the curvature in the plot evaluated over the temperature range, T = (283.15 to 315.15) K.

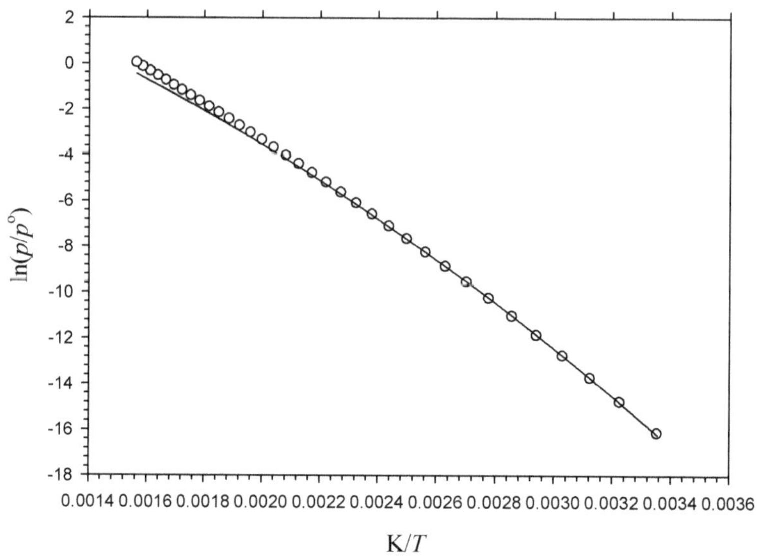

Figure 4. A plot of $\ln(p/p^o)$ vs. K/T for 5α-androstane evaluated by correlation. The circles represent values calculated by Equation (2) using correlated values from T = (298.15 to 400) K while the line was generated by data evaluated by correlation up to 550 K. The additional curvature associated with the extended range correlated is evident at the higher temperatures.

4.6. Retention Times of 5α-Cholestane and 5α-Androstane

The long retention times observed for both 5α-cholestane and 5α-androstane relative to similar sizes of n-alkanes needed to bracket them was mentioned in the introduction. An explanation for this becomes apparent once the vaporization enthalpies of both are known. 5α-Androstane, a $C_{19}H_{32}$ hydrocarbon, with a vaporization enthalpy of (87.8 ± 2.0) kJ·mol^{-1}, exhibits a retention time that is greater than n-eicosane ($C_{20}H_{42}$), an alkane with a vaporization enthalpy value of (101.8 ± 0.5) kJ·mol^{-1}. Similarly for 5α-cholestane, a $C_{27}H_{48}$ hydrocarbon with a vaporization enthalpy of (125.9 ± 3.9) kJ·mol^{-1} exhibits a retention time greater than n-octacosane, $C_{28}H_{58}$, with a vaporization enthalpy of (141.9 ± 4.9) kJ·mol^{-1} [20]. Despite the fact that both 5α-androstane and 5α-cholestane are more volatile than n-eicosane and n-octacosane at T = 298.15 K, ((1.0 ± 0.2) × 10^{-2} Pa versus (6.6 × 10^{-3}) Pa [18] for the former, and (2.4; 7.8 [8]) × 10^{-6} Pa versus (3.2 [19]; 5.8 [26]) × 10^{-7} Pa for the latter), the larger vaporization enthalpies of the standards (a measure of the rate of change in vapor pressure with temperature) at the high temperatures required for both to elute off the column, result in vapor pressures that quickly overcome these differences.

5. Conclusions

Vaporization enthalpies and vapor pressures for 5α-cholestane and 5α-androstane are reported and compared to literature values. While agreement is not perfect, given the size of the molecules studied, it is quite good. In addition, this work identifies a possible solution for attenuating the curvature observed at higher temperatures when forced to use n-alkanes as standards in plots of $\ln(p/p^o)$ versus K/T when dealing with large polycyclic hydrocarbons. By reducing the temperature range of the correlations, the resulting curvature and the corresponding normal boiling temperature are reduced. This should prove useful when studying hydrocarbon targets that are structurally rigid.

Supplementary Materials: The following are available online at https://www.mdpi.com/article/10.3390/liquids4030025/s1. Tables S1A–S4A: Retention times of 5α-cholestane and standards;

Tables S5A and S6A: Retention times of 5α-androstane and standards. Tables S1B–S6B: Correlations of $\Delta^{g}_{trn}H_{m}(T_{m}\text{ K})$ with $\Delta^{g}_{l}H_{m}(298\text{ K})$ of the Standards. Table S7: Experimental vapor pressures and vaporization enthalpy of 5α-cholestane at the mean temperature of measurement, 426.8 K [8]. Table S8: Comparisons of vapor pressures evaluated in runs 1–4 with literature data. Table S9: A comparison of experimental data on the effect of temperature range ($T_{r)}$ on the constants of the correlation equation of runs 1–4 used to evaluate vapor pressures of 5α-cholestane. Figure S1: ATR IR spectrum of 5α-cholestane.

Author Contributions: C.F.-L., M.A.: Investigation, J.S.C.: Supervision, analysis, writing, and validation. All authors have read and agreed to the published version of the manuscript.

Funding: This research received no external funding.

Data Availability Statement: Data is contained within the article and in the Supplementary Materials.

Conflicts of Interest: The authors declare no conflicts of interest.

References

1. Bobrovskiy, I.; Hope, J.M.; Ivantsov, A.; Andrey, I.; Nettersheim, B.J.; Nettersheim, B.J.; Hallmann, C.; Brocks, J.J. Ancient steroids establish the Ediacaran fossil Dickinsonia as one of the earliest animals. *Science* **2018**, *361*, 1246–1249. [CrossRef] [PubMed]
2. Segaloff, A.; Gabbard, R.B.A. 5α-Androstane-An Androgenic Hydrocarbon. *Endocrinology* **1960**, *67*, 887–889. [PubMed]
3. Analytical Standards and Certified Reference Materials, 2012–2015, Fluka Analytical; Sigma Aldrich Schweiz, Industriestrasse 25 CH-9470 Buchs SG Schweiz.
4. Zafar, A.; Chickos, J.S. The vapor pressure and vaporization enthalpy of squalene and squalene by correlation gas chromatography. *J. Chem. Thermodyn.* **2019**, *135*, 192–197. [CrossRef]
5. Thorton, M.; Gobble, C.; Chickos, J. The vaporization enthalpy and vapor pressure of S (+) methamphetamine at T = 298.15 K by correlation gas chromatography. *J. Chem. Thermodyn.* **2014**, *73*, 51–56.
6. Umnahanant, P.; Zafar, A.; Kankala, V.; Chickos, J. Vapor pressure and vaporization enthalpy studies of (+) longifolene, (-) isolongifolene and β-myrcene by correlation gas chromatography. *J. Chem. Thermodyn.* **2019**, *131*, 583–591. [CrossRef]
7. Orf, M.; Kurian, M.; Nelson, C.; Simmons, D.; Espinosa, L.; Chickos, J. Correlation gas chromatographic study of the vaporization enthalpies and vapor pressures of the major sesquiterpene hydrocarbons in patchouli oil. *J. Chem. Eng. Data* **2021**, *66*, 2538–2549. [CrossRef]
8. Mokbel, I.; Ruzicka, K.; Majer, V.; Ruzicka, V.; Ribeiro, M.; Jose, J.; Zabransky, M. Vapor pressures and thermal data for three high-boiling compounds of petroleum interest: 1-phenyldodecane, 5α-cholestane, adamantane. *Fluid Phase Equil.* **2000**, *169*, 191–207. [CrossRef]
9. American Petroleum Institute. *Properties of Hydrocarbons of High Molar Mass*; Research Project 42; Pennsylvania State University: State College, PA, USA, 1967.
10. Barton, D.; Chickos, J.S. The vapor pressure and vaporization enthalpy of (-) β-Elemene and (-) β-Bisabolene by correlation gas chromatography. *J. Chem. Thermodyn.* **2020**, *148*, 106139. [CrossRef]
11. 5alpha-Cholestane. Available online: https://pubchem.ncbi.nlm.nih.gov/compound/5alpha-Cholestane (accessed on 4 May 2024).
12. Melting Point Standards. Available online: https://www.sigmaaldrich.com/deepweb/assets/sigmaaldrich/product/documents/258/468/melting-point-standards-fl8678en-ms.pdf (accessed on 4 May 2024).
13. Spectral Database for Organic Compounds. Available online: https://sdbs.db.aist.go.jp (accessed on 19 December 2020).
14. Peacock, L.A.; Fuchs, R. Enthalpy of vaporization measurements by gas chromatography. *J. Am. Chem. Soc.* **1977**, *99*, 5524–5525. [CrossRef]
15. Lipkind, D.; Chickos, J.S. An Examination of Factors Influencing the Thermodynamics of Correlation-Gas Chromatography as Applied to Large Molecules and Chiral Separations. *J. Chem. Eng. Data* **2010**, *55*, 698–707. [CrossRef]
16. Shoemaker, D.P.; Garland, C.W.; Steinfield, J.I.; Niebler, J.W. *Experiments in Physical Chemistry*, 4th ed.; McGraw Hill Co.: New York, NY, USA, 1981.
17. Chickos, J.S. *Computational Thermochemistry, Prediction and Estimation of Molecular Thermodynamics*; ACS Symposium Series 677; Irikura, K.K., Frurip, D.J., Eds.; ACS: Washington, DC, USA, 1996; Chapter 4; pp. 63–91.
18. Ruzicka, K.; Majer, V. Simultaneous Treatment of Vapor Pressures and Related Thermal Data Between the Triple and Normal Boiling Temperatures for n-Alkanes C_5–C_{20}. *J. Phys. Chem. Ref. Data* **1994**, *23*, 1–39. [CrossRef]
19. Chickos, J.S.; Hanshaw, W. Vapor pressures and vaporization enthalpies of the n-alkanes from C_{21}-C_{30} at T = 298.15 by correlation-gas chromatography. *J. Chem. Eng. Data* **2004**, *49*, 77–85. [CrossRef]
20. Chickos, J.S.; Hanshaw, W. Vapor pressures and vaporization enthalpies of the n-alkanes from C_{31}-C_{38} at T = 298.15 by correlation-gas chromatography. *J. Chem. Eng. Data* **2004**, *49*, 620–630. [CrossRef]
21. Stephenson, R.M.; Malanowski, S. *Handbook of the Thermodynamics of Organic Compounds*; Elsevier: New York, NY, USA, 1987.

22. EPI Suite version 4.11 (Estimation Programs Interface). Available online: http://www.epa.gov/oppt/exposure/pubs/episuitedl.htm (accessed on 18 March 2020).
23. Advanced Chemistry Development (ACD/Labs). *ACD/Labs Software*, V11.02; (© 1994–2023 ACD/Labs) through CAS SciFinder; ACD/Labs: Toronto, ON, Canada, 1994.
24. Acree, W., Jr.; Chickos, J.S. Phase Transition Enthalpy Measurements of Organic and Organometallic Compounds. Sublimation, Vaporization and Fusion Enthalpies From 1880 to 2015. Part 1. C_1–C_{10}. *J. Phys. Chem. Ref. Data* **2016**, *45*, 033101.
25. Chickos, J. An update on liquid heat capacity estimations of cyclic organic compounds by group additivity and their application in estimations of complex molecules by synthetic and retrosynthetic analysis. *J. Chem. Thermodyn.* **2023**, *182*, 107039. [CrossRef]
26. Chirico, R.D.; Nguyen, A.; Steele, W.V.; Strube, M.M. Vapor pressure data of n- alkanes revisited. New high precision vapor pressure data on *n*-decane, n-eicosane and *n*-octacosane. *J. Chem. Eng. Data* **1989**, *34*, 149–156. [CrossRef]

liquids

MDPI

Article

Abraham General Solvation Parameter Model: Predictive Expressions for Solute Transfer into Isobutyl Acetate

Ramya Motati, Trisha Kandi, Jilawan Francis, Jocelyn Chen, Emily Yao, Saikiran Motati, Audrey Chen, Dhishithaa Kumarandurai, Nikita Shanmugam and William E. Acree, Jr. *

Department of Chemistry, University of North Texas, Denton, TX 76203, USA; ramyamotati@my.unt.edu (R.M.)
* Correspondence: bill.acree@unt.edu

Abstract: Mole fraction of solubilities are reported for the: *o*-acetoacetanisidide, anthracene, benzoin, 4-*tert*-butylbenzoic acid, 3-chlorobenzoic acid, 3-chlorobenzoic acid, 2-chloro-5-nitrobenzoic acid, 4-chloro-3-nitrobenzoic acid, 3,4-dichlorobenzoic acid, 2,3-dimethoxybenzoic acid, 3,4-dimethoxybenzoic acid, 3,5-dimethoxybenzoic acid, 3,5-dinitrobenzoic acid, diphenyl sulfone, 2-ethylanthraquinone, 2-methoxybenzoic acid, 4-methoxybenzoic acid, 2-methylbenzoic acid, 3-methylbenzoic acid, 2-methyl-3-nitrobenzoic acid, 3-methyl-4-nitrobenzoic acid, 4-methyl-3-nitrobenzoic acid, 2-naphthoxyacetic acid, 3-nitrobenzoic acid, 4-nitrobenzoic acid, salicylamide, thioxanthene-9-one, 3,4,5-trimethoxybenzoic acid, and xanthene dissolved in isobutyl acetate at 298.15 K. The results of our experimental measurements, combined with the published literature data, were used to obtain Abraham model equations for isobutyl acetate. The mathematical correlations presented in the current study describe the observed molar solubility ratios of the solutes dissolved in isobutyl acetate to within an overall standard deviation of 0.12 log units or less.

Keywords: isobutyl acetate solvent; solubility; crystalline organic solutes; Abraham model correlations

1. Introduction

Several million tons of organic solvents are consumed annually by the chemical manufacturing sector. Organic solvents serve as the reaction media in the synthesis of new chemical products, as the mobile phase in high-performance liquid chromatographic separations, and as extractants in biphasic liquid–liquid partitioning systems. Considerable effort has been devoted to developing and revising guidelines to aid industrial workers in selecting the most appropriate organic solvent for a particular application. The guidelines are revised continually as compounds are added to or removed from the list of "acceptable" solvents for use in manufacturing processes. Workers are instructed to focus not only the safety and economic aspects, but to also consider the solvents' polarity and hydrogen-bonding characteristics, as these are important items as well. Solute–solvent interactions that result from hydrogen-bond formation and from differences in dipole moments can not only increase a given solute's solubility, but can also alter the thermodynamic states of chemical reactants, synthesized products, and reaction intermediates. Product yields and selectivity, as well as chemical reaction rates, can be controlled to some extent through the solvent selection process. Slight increases in product yield and reaction rates can make a manufacturing process significantly more profitable.

It is only natural that industrial researchers utilize all of the available resources at their disposal in designing synthetic processes. Design engineers ideally prefer to utilize actual experimental data whenever available, as this provides less uncertainty in the required input values/parameters. Experimental data are often unavailable for the many possible solute–solvent combinations currently encountered in the various manufacturing processes. This is particularly true in the case of more environmentally friendly organic mono-solvents and solvent mixtures that are now being suggested as possible solvent alternatives to replace the more toxic, hazardous compounds that have been traditionally used in industrial manufacturing processes. Increased environmental awareness and worker safety, combined with much larger disposal costs, have encouraged the manufacturing sector to reduce (and in certain cases to completely eliminate) the use of specific organic solvents. The disposal of toxic organic compounds in accordance with prevailing governmental regulations can represent a significant expenditure, which is then passed on to the consumer in the form of a higher product purchase price.

Over the last 50 years, a multitude of predictive methods have been developed to provide the scientific and manufacturing communities with the needed input values if measured data are unavailable. The expressions derived from such predictive methods can be used to prioritize potential organic solvent candidates for possible thermodynamic and solubility determinations. Predictive expressions must yield reasonably accurate estimates of the desired property to be of any real value. The more reliable of the proposed models have incorporated mathematical terms to account for the various molecular interactions believed to be present.

The objective of the current study is to develop Abraham model expressions [1–3], as follows:

$$\log P \text{ or } \log (C_{S,organic}/C_{S,water}) = c_p + e_p \times \mathbf{E} + s_p \times \mathbf{S} + a_p \times \mathbf{A} + b_p \times \mathbf{B} + v_p \times \mathbf{V} \quad (1)$$

$$\log K \text{ or } \log (C_{S,organic}/C_{S,gas}) = c_k + e_k \times \mathbf{E} + s_k \times \mathbf{S} + a_k \times \mathbf{A} + b_k \times \mathbf{B} + l_k \times \mathbf{L} \quad (2)$$

for estimating the logarithm water-to-isobutyl acetate partition coefficients, log P, the logarithm of gas-to-isobutyl acetate partition coefficients, log K, and the logarithm of molar solubility ratios, log $(C_{S,organic}/C_{S,water})$ and log $(C_{S,organic}/C_{S,gas})$, for solid nonelectrolyte organic compounds and inorganic gases dissolved in the isobutyl acetate mono-solvent. The subscripts "organic," "water," and "gas" are used to define the two solubility ratios that denote the phase to which each molar solute concentration pertains.

Each multiplicative term on the right-hand side of Equations (1) and (2) describes a different type of solute–solvent molecular interaction. Each interaction is quantified as the product of a solvent property (c_p, e_p, s_p, a_p, b_p, v_p, c_k, e_k, s_k, a_k, b_k, and l_k) times the complementary solute property (**E**, **S**, **A**, **B**, **V**, and **L**). The solvent properties are defined as follows: (e_p and e_k) give the tendency of the solvent to interact with solute molecules through polarizability-type interactions involving electron pairs; (s_p and s_k) are measures of the solvent phase's dipolarity/polarity; (a_p and a_k) and (b_p and b_k) quantify the solvent's ability to function as a hydrogen-bond acceptor and hydrogen-bond donor in its interactions with the surrounding solute molecules, respectively; and (v_p and l_k) represent the ease of breaking the solvent–solvent interaction in regards to the formation of a solvent cavity in which the dissolved solute will reside. The numerical values of the aforementioned properties are deduced by curve-fitting measured partition coefficient data and molar solubility ratios for a series of solutes with known solute descriptors (**E**, **S**, **A**, **B**, **V**, and **L**) in accordance with Equations (1) and (2). In the present study, we focus our attention on the lowercase solvent properties. A detailed discussion of the solute descriptors and their determination based on measured chromatographic retention times, partition coefficients, and molar solubilities is available in several published review articles [4–8] and research papers [2,3,9–12].

There are published experimental data for the following: hydrogen [13], nitrogen [13], carbon dioxide [13], carbon monoxide [13], naphthalene [14], fluorene [15], benzoic acid [16], acetylsalicylic acid [17], isophthalic acid [16], forchlorfenuron [18], metronidiazole benzoate [19], iminostilbene [20], isovanillin [21], lansoprazole [22], 2,2′-bis(2-hydroxyethoxy)-1,1′-binaphthalene [23], lovastatin [24], etodolac [25], simvastatin [26,27], 18β-glycyrrhetinic acid [28], and exo-5,6-dehydronorcantharidin [29] dissolved in isobutyl acetate. An experimental gas-to-liquid partition coefficient for isobutyl acetate dissolved in itself is also available from the vapor pressure measurements of Monton and coworkers [30]. Unfortunately, the number of experimental values is not sufficient for us to develop meaningful Abraham model expressions. To supplement the published literature values, we have determined the mole fraction of the solubilities of the following: *o*-acetoacetanisidide, anthracene, benzoin, 4-*tert*-butylbenzoic acid, 3-chlorobenzoic acid, 3-chlorobenzoic acid, 2-chloro-5-nitrobenzoic acid, 4-chloro-3-nitrobenzoic acid, 3,4-dichlorobenzoic acid, 2,3-dimethoxybenzoic acid, 3,4-dimethoxybenzoic acid, 3,5-dimethoxybenzoic acid, 3,5-dinitrobenzoic acid, diphenyl sulfone, 2-ethylanthraquinone, 2-methoxybenzoic acid, 4-methoxybenzoic acid, 2-methylbenzoic acid, 3-methylbenzoic acid, 2-methyl-3-nitrobenzoic acid, 3-methyl-4-nitrobenzoic acid, 4-methyl-3-nitrobenzoic acid, 2-naphthoxyacetic acid, 3-nitrobenzoic acid, 4-nitrobenzoic acid, salicylamide, thioxanthene-9-one, 3,4,5-trimethoxybenzoic acid, and xanthene dissolved in isobutyl acetate at 298.15 K based on UV/visible spectrophotometric measurements.

Our solute selection was based, in part, on the compounds' availability within the laboratory from prior solubility studies, and the fact that their solute descriptor values were determined from large numbers of experimental measurements. Furthermore, our past studies have shown that these compounds are not prone to form solid solvates with alkyl acetates, and their solubilities can be readily determined by either UV/visible absorbance measurements or volumetric acid–base titrations. The senior author is also currently preparing a volume for the IUPAC-NIST Solubility Data Series (project number #2021-023-1-500) that will update his earlier volume on the solubility of benzoic acid and substituted benzoic acids dissolved in both neat organic solvents and organic solvent mixtures [31]. The solubility data from the current study will likely be part of the new volume. We further note that the carboxylic acid functional group is present in many medicinal compounds, such as nonsteroidal anti-inflammatory drugs (e.g., acetylsalicylic acid, salicylic acid, naproxen, ketoprofen, ibuprofen, etodolac, flurbiprofen, ketorolac, and tolfenamic acid), statins (e.g., atorvastatin, Lipitor®, fluvastatin, pitavastatin, pravastatin, and cerivastatin) and β-lactam antibiotics (e.g., amoxicillin, oxacillin, flucloxacillin, and benzylpenicillin). The results of our experimental measurements, combined with the published literature data, were used to obtain Abraham model equations for isobutyl acetate. The derived Abraham model correlations reported in the current study are based on 49 experimental data points.

2. Chemical Materials and Experimental Methodology

The crystalline organic solutes selected for the solubility measurements include 21 carboxylic acids, as well as 8 noncarboxylic acid solutes of varying polarity and molecular sizes. All chemicals used in the current study were purchased from commercial sources in the highest purity available. Several of the noncarboxylic acid solutes were further purified by recrystallization from either acetone or anhydrous methanol prior to performing the solubility measurements. All solid compounds were dried for at least two days at 333 K to remove trace moisture. The purification details and chemical suppliers are given in Table 1, along with the final purities, as determined by either gas–liquid chromatographic analysis (noncarboxylic acid solutes and flame ionization detector) or non-aqueous acid–base volumetric titration based on a modification of the published method used by Fritz and Lisicki [32]. Our modified titration procedure replaced benzene with toluene as a component in the titration solvent for health reasons.

Table 1. Chemical sources and final mass fraction purities of chemicals used in the solubility studies.

Chemical	Supplier	Purification Method	Purity (Mass Fraction)
Isobutyl acetate	TCI America Chemical Company, Portland, OR, USA	Stored over activated molecular sieves and distilled	0.998
o-Acetoacetanisidide	Acros Organics, Morris Plains, NJ, USA	Dried for two days at 333 K	0.996
Anthracene	Aldrich Chemical Company, Milwaukee, WI, USA	Recrystallized from anhydrous acetone	0.997

Table 1. *Cont.*

Chemical	Supplier	Purification Method	Purity (Mass Fraction)
Benzoin	Aldrich Chemical Company	Recrystallized from anhydrous methanol	0.997
Diphenyl sulfone	Aldrich Chemical Company	Recrystallized from anhydrous methanol	0.996
2-Ethylanthraquinone	Aldrich Chemical Company	Recrystallized from anhydrous methanol	0.996
Salicylamide	Aldrich Chemical Company	Recrystallized from anhydrous methanol	0.997
Thioxanthen-9-one	Aldrich Chemical Company	Recrystallized from anhydrous methanol	0.997
Xanthene	Aldrich Chemical Company	Recrystallized from anhydrous methanol	0.996
4-*tert*-Butylbenzoic acid	TCI America Chemical Company	Dried for two days at 333 K	0.998
3-Chlorobenzoic acid	Aldrich Chemical Company	Dried for two days at 333 K	0.997
4-Chlorobenzoic acid	Acros Organics	Dried for two days at 333 K	0.996
2-Chloro-5-nitrobenzoic acid	Acros Organics	Dried for two days at 333 K	0.998
4-Chloro-3-nitrobenzoic acid	Acros Organics	Dried for two days at 333 K	0.998
3,4-Dichlorobenzoic acid	Aldrich Chemical Company	Dried for two days at 333 K	0.998
2,3-Dimethoxybenzoic acid	Thermo Scientific, Ward Hill, MA, USA	Dried for two days at 333 K	0.997
3,4-Dimethoxybenzoic acid	Across Organics	Dried for two days at 333 K	0.998
3,5-Dimethoxybenzoic acid	Across Organics	Dried for two days at 333 K	0.995
3,5-Dinitrobenzoic acid	Aldrich Chemical Company	Dried for two days at 333 K	0.997
2-Methoxybenzoic acid	Aldrich Chemical Company	Dried for two days at 333 K	0.998
4-Methoxybenzoic acid	Aldrich Chemical Company	Dried for two days at 333 K	0.998
2-Methylbenzoic acid	Aldrich Chemical Company	Dried for two days at 333 K	0.998
3-Methylbenzoic acid	Aldrich Chemical Company	Dried for two days at 333 K	0.998
2-Methyl-3-nitrobenzoic acid	Aldrich Chemical Company	Dried for two days at 333 K	0.997
3-Methyl-4-nitrobenzoic acid	Aldrich Chemical Company	Dried for two days at 333 K	0.997
4-Methyl-3-nitrobenzoic acid	Aldrich Chemical Company	Dried for two days at 333 K	0.998
2-Naphthoxyacetic acid	Sigma-Aldrich Chemical Company, Milwaukee, WI, USA	Dried for two days at 333 K	0.995
3-Nitrobenzoic acid	Aldrich Chemical Company	Dried for two days at 333 K	0.996
4-Nitrobenzoic acid	Acros Organics	Dried for two days at 333 K	0.998
3,4,5-Trimethoxybenzoic acid	Aldrich Chemical Company	Dried for two days at 333 K	0.998
Toluene	Aldrich Chemical Company	None	0.998, anhydrous
Sodium methoxide, 25 mass % solution in methanol	Aldrich Chemical Company	None	
2-Propanol	Aldrich Chemical Company	None	0.99

The experimental mole fraction solubilities were determined using a spectrophotometric method of chemical analysis based on the Beer–Lambert law that establishes a mathematical relationship between the measured solution absorbance and the molar concentration of the dissolved solid solute. The saturated solutions were prepared by placing excess solid solute and approximately 20 mLs of isobutyl acetate into amber glass bottles.

The resulting solutions were then equilibrated at 298.15 ± 0.05 in a constant-temperature water bath. The sealed solutions were periodically shaken to facilitate the mixing and dissolution of the solid solute. After a three-day equilibration period, weighed aliquots of the saturated solutions were transferred into tared volumetric flasks, and the transferred sample was diluted quantitatively with 2-propanol. The absorbances of the diluted solutions were recorded with a Milton Roy Spectronic 1000 Plus single-beam spectrophotometer (Milton Roy, Rochester, NY, USA). An additional dilution was sometimes required to ensure that the sample's measured absorbance fell within the range of values determined for the standard solutions. Our earlier publications contain the analysis wavelength and concentration range for each solute considered in the current study. For the convenience of the reader, this information is summarized in Table 2.

Table 2. Analysis wavelengths and concentration ranges of standard solutions used in the spectrophotometric determination of solubility.

Chemical	Analysis Wavelength	Molar Concentration Range
o-Acetoacetanisidide	282 (nm)	5.81×10^{-5} to 1.94×10^{-4}
Anthracene	356 (nm)	6.86×10^{-5} to 2.28×10^{-4}
Benzoin	313 (nm)	1.21×10^{-3} to 4.03×10^{-3}
Diphenyl sulfone	267 (nm)	2.61×10^{-4} to 8.70×10^{-4}
2-Ethylanthraquinone	325 (nm)	1.25×10^{-4} to 4.17×10^{-4}
Salicylamide	300 (nm)	1.06×10^{-4} to 3.55×10^{-4}
Thioxanthen-9-one	378 (nm)	6.05×10^{-5} to 2.02×10^{-4}
Xanthene	280 (nm)	1.79×10^{-4} to 5.95×10^{-4}
4-*tert*-Butylbenzoic acid	275 (nm)	2.81×10^{-4} to 9.37×10^{-4}
3-Chlorobenzoic acid	280 (nm)	5.05×10^{-4} to 1.68×10^{-3}
4-Chlorobenzoic acid	272 (nm)	4.63×10^{-4} to 1.54×10^{-3}
2-Chloro-5-nitrobenzoic acid	280 (nm)	8.69×10^{-5} to 2.90×10^{-4}
4-Chloro-3-nitrobenzoic acid	292 (nm)	3.55×10^{-4} to 1.18×10^{-3}
3,4-Dichlorobenzoic acid	280 (nm)	4.66×10^{-4} to 1.55×10^{-3}
2,3-Dimethoxybenzoic acid	293 (nm)	2.27×10^{-4} to 7.56×10^{-4}
3,4-Dimethoxybenzoic acid	286 (nm)	9.25×10^{-5} to 3.08×10^{-4}
3,5-Dimethoxybenzoic acid	305 (nm)	2.17×10^{-4} to 7.23×10^{-4}
3,5-Dinitrobenzoic acid	267 (nm)	6.35×10^{-5} to 2.12×10^{-4}
2-Methoxybenzoic acid	295 (nm)	1.69×10^{-4} to 5.63×10^{-4}
4-Methoxybenzoic acid	273 (nm)	9.77×10^{-5} to 3.26×10^{-4}
2-Methylbenzoic acid	279 (nm)	4.29×10^{-4} to 1.43×10^{-3}
3-Methylbenzoic acid	280 (nm)	4.17×10^{-4} to 1.39×10^{-3}
2-Methyl-3-nitrobenzoic acid	290 (nm)	3.36×10^{-4} to 1.12×10^{-3}
3-Methyl-4-nitrobenzoic acid	295 (nm)	1.70×10^{-4} to 5.67×10^{-4}
4-Methyl-3-nitrobenzoic acid	295 (nm)	3.29×10^{-4} to 1.10×10^{-3}
2-Naphthoxyacetic acid	284 (nm)	1.35×10^{-4} to 6.74×10^{-4}
3-Nitrobenzoic acid	280 (nm)	1.67×10^{-4} to 5.57×10^{-4}
4-Nitrobenzoic acid	272 (nm)	4.75×10^{-5} to 1.58×10^{-4}
3,4,5-Trimethoxybenzoic acid	289 (nm)	1.38×10^{-4} to 4.60×10^{-4}

We noted that isobutyl acetate is optically transparent at the analysis wavelengths listed in Table 2. We checked for possible interferences from the small amount of isobutyl acetate in the dilute samples that were subjected to absorbance measurements. We found that, to within experimental uncertainty, identical absorbances were obtained for standard solutions that contained no isobutyl acetate and for standard solutions that contained up to 5 percent isobutyl acetate by volume. We also checked for both solvate formation and solid-to-solid phase transitions by determining the melting point temperatures of the solid residue that remained in the amber glass bottles after the solubility measurements were complete. While there is no evidence for either phenomenon in the published literature for the organic solutes considered in the current study, we wanted to confirm that the solid phase remained the same during the initial equilibration period and during the additional equilibration time before the replicate follow-up set of solubility measurements. The attainment of equilibrium was verified by performing a second set of solubility measurements two (or three) days after the initial set of measurements. In all cases, the measured point temperature of the recovered solid material was within the experimental error of the melting point temperature of the purchased commercial sample or the recrystallized compound prior to being placed in contact with the isobutyl acetate mono-solvent. The measured melting point temperatures, given in Table 3, show no indication of solid–solvate formation or polymorphism.

Table 3. Comparison of the melting point temperatures of the crystalline solutes prior to contact with isobutyl acetate, $T_{mp,initial}$ [a], and of the recovered crystalline solutes in equilibrium with the saturated solution, $T_{mp,equilibrated}$ [a] at 101 kPa [a].

Chemical Solute	$T_{mp,initial}$/K	$T_{mp,equilibrated}$/K	$T_{mp,literatuare}$/K
o-Acetoacetanisidide	359.7 ± 0.5	359.4 ± 0.5	359.2 [b] 355.2–357.2 [33]
Anthracene	490.7 ± 0.5	491.1 ± 0.4	488.9–491.3 [34]
Benzoin	408.6 ± 0.4	408.3 ± 0.5	408.2 [34]
Diphenyl sulfone	398.2 ± 0.4	398.0 ± 0.5	398.2 [34]
2-Ethylanthraquinone	383.8 ± 0.5	383.5 ± 0.5	382.9 [35] 384.2 [36]
Salicylamide	414.1 ± 0.5	413.9 ± 0.4	411.9–414.9 [37]
Thioxanthen-9-one	488.2 ± 0.5	488.0 ± 0.5	486.6–487.9 [34]
Xanthene	374.4 ± 0.5	374.1 ± 0.5	373.3–374.6 [34]
4-*tert*-Butylbenzoic acid	439.3 ± 0.4	439.1 ± 0.5	440 [34]
3-Chlorobenzoic acid	429.6 ± 0.5	429.2 ± 0.5	427.4–429.9 [37]
4-Chlorobenzoic acid	512.5 ± 0.3	512.7 ± 0.4	512.3–513.5 [37]
2-Chloro-5-nitrobenzoic acid	440.1 ± 0.5	440.4 ± 0.5	437.2–438.2 [38] 438.2–439.2 [39] 437–441 [b]
4-Chloro-3-nitrobenzoic acid	456.3 ± 0.5	456.5 ± 0.4	453.2–455.2 [39] 453–456 [b]
3,4-Dichlorobenzoic acid	480.7 ± 0.5	480.3 ± 0.5	478–481 [b]
2,3-Dimethoxybenzoic acid	394.9 ± 0.5	395.3 ± 0.5	393.1 [40] 394–397 [b]
3,4-Dimethoxybenzoic acid	452.9 ± 0.5	452.7 ± 0.5	453.1 [37]
3,5-Dimethoxybenzoic acid	455.9 ± 0.5	455.6 ± 0.5	454–458 [b]
3,5-Dinitrobenzoic acid	481.2 ± 0.5	481.0 ± 0.5	479.2–481.7 [37]
2-Methoxybenzoic acid	375.1 ± 0.5	374.9 ± 0.4	374.7 [37]
4-Methoxybenzoic acid	456.3 ± 0.5	456.5 ± 0.5	453–455.8 [37]
2-Methylbenzoic acid	377.2 ± 0.5	376.9 ± 0.5	376.5–376.9 [37]
3-Methylbenzoic acid	382.4 ± 0.4	382.8 ± 0.4	381.9 [36]
2-Methyl-3-nitrobenzoic acid	455.9 ± 0.5	455.7 ± 0.5	456.6 [41]
3-Methyl-4-nitrobenzoic acid	489.7 ± 0.5	489.4 ± 0.5	489.1 [41]
4-Methyl-3-nitrobenzoic acid	459.6 ± 0.5	459.2 ± 0.5	459.8 [41]
2-Naphthoxyacetic acid	424.6 ± 0.5	424.1 ± 0.5	424.3 [42]

Table 3. *Cont.*

Chemical Solute	$T_{mp,initial}$/K	$T_{mp,equilibrated}$/K	$T_{mp,literatuare}$/K
3-Nitrobenzoic acid	414.7 ± 0.4	414.5 ± 0.4	408.7–414.3 [37]
4-Nitrobenzoic acid	512.2 ± 0.5	511.9 ± 0.5	512.4 [37]
3,4,5-Trimethoxybenzoic acid	443.9 ± 0.5	444.2 ± 0.5	444.5 [43]

[a] standard uncertainties, u, $u(T_{mp})$ = 0.3 to 0.5 K, as stated, and $u(P)$ = 3 kPa. [b] experimental melting point temperatures available on ChemSpider, Royal Society of Chemistry. database, accessed 2 May 2018.

3. Results and Discussion

The experimental mole fraction solubilities, $X_{S,organic}$, of the 29 different crystalline organic solutes dissolved in isobutyl acetate are tabulated in the second and fourth columns of Table 4. The numerical values represent the average of 6 to 10 independent experimental determinations, including the follow-up measurements that were performed after the initial three-day equilibration time. The follow-up studies confirmed that, in all cases, equilibrium was achieved after three days. The tabulated $X_{S,organic}$ values were reproducible to within ±2.5% (relative error). We were not able to find in the published chemical and engineering literature solubility data for these organic solutes in isobutyl acetate that we could compare our experimental values against. This was not unexpected, in that it was only until recently that researchers started performing solubility measurements in isobutyl acetate.

Table 4. Mole fraction solubilities, $X_{S,organic}$, of select crystalline nonelectrolyte organic compounds dissolved in isobutyl acetate at a temperature of 298.15 K and an ambient atmospheric pressure of 101 kPa [a].

Chemical Name	$X_{S,organic}$	Chemical Name	$X_{S,organic}$
o-Acetoacetanisidide	0.0464	3,4-Dimethoxybenzoic acid	0.00461
Anthracene	0.00506	3,5-Dimethoxybenzoic acid	0.00432
Benzoin	0.00882	3,5-Dinitrobenzoic acid	0.0309
Diphenyl sulfone	0.0302	2-Methoxybenzoic acid	0.0353
2-Ethylanthraquinone	0.0341	4-Methoxybenzoic acid	0.00789
Salicylamide	0.0500	2-Methylbenzoic acid	0.1222
Thioxanthen-9-one	0.00306	3-Methylbenzoic acid	0.1265
Xanthene	0.1057	2-Methyl-3-nitrobenzoic acid	0.0247
4-*tert*-Butylbenzoic acid	0.0750	3-Methyl-4-nitrobenzoic acid	0.0111
3-Chlorobenzoic acid	0.0541	4-Methyl-3-nitrobenzoic acid	0.0183
4-Chlorobenzoic acid	0.00825	2-Naphthoxyacetic acid	0.0148
2-Chloro-5-nitrobenzoic acid	0.0517	3-Nitrobenzoic acid	0.0967
4-Chloro-3-nitrobenzoic acid	0.0246	4-Nitrobenzoic acid	0.00682
3,4-Dichlorobenzoic acid	0.0133	3,4,5-Trimethoxybenzoic acid	0.00825
2,3-Dimethoxybenzoic acid	0.0206		

[a] Standard uncertainties and relative uncertainties are: $u(T)$ = 0.05 K; $u(p)$ = 5 kPa; and $u_r(x)$ = 0.025.

The published experimental solubility data that we found were for the following: hydrogen [13], nitrogen [13], carbon dioxide [13], carbon monoxide [13], naphthalene [14], fluorene [15], benzoic acid [16], acetylsalicylic acid [17], isophthalic acid [16], forchlorfenuron [18], metronidiazole benzoate [19], iminostilbene [20], isovanillin [21], lansoprazole [22], 2,2′-bis(2-hydroxyethoxy)-1,1′-binaphthalene [23], lovastatin [24], etodolac [25], simvastatin [26,27], 18β-glycyrrhetinic acid [28], and exo-5,6-dehydronorcantharidin [29] dissolved in isobutyl acetate. Our Abraham model solute descriptor database contains solute descriptors for 13 of the 20 aforementioned solutes. As part of the current study, descriptor values were determined for forchlorfenuron, metronidiazole benzoate, iminostilbene, 2,2′-*bis*(2-hydroxyethoxy)-1,1′-binaphthalene, etodolac, and 18β-glycyrrhetinic acid using the published solubility data taken from the published chemical literature [18–20,23,25,28]. In calculating the solute descriptors of etodolac we eliminated the solubility data reported by Rathi and Deshpande [44] from our computations. The authors reported the solubility of etodolac in 1,4-dioxane and in water in terms of mole fraction and molar solubilities. The reported mole fraction solubilities are not internally consistent with the solubility data expressed in units of molarity [45].

As an informational note, we attempted to calculate the solute descriptors for the remaining two solutes without success. The database for exo-5,6-dehydronorcantharidin contained only a single solvent capable of acting as hydrogen-bond donors (no alcohol solvents), which prevented the calculation of a meaningful set of solute descriptor values. The two lone pairs of electrons on each of the four oxygen atoms on exo-5,6-dehydronorcantharidin can serve as H-bond acceptor sites. In calculating the molecule's solute descriptors, it is imperative that the database contains solvent molecules capable of acting as H-bond donors. In Table 5, we have compiled our newly calculated solute descriptors, as well as the numerical values for all of the solutes that will be considered in the current study. Readers interested in learning about how the Abraham model solute descriptors are calculated from published solubility data can refer to earlier papers [12,46,47] that describe the computational method in detail.

Table 5. Solute descriptors of the compounds used in the regression analysis for determining the Abraham model correlations for isobutyl acetate.

Solute	E	S	A	B	L	V
Hydrogen	0.000	0.000	0.000	0.000	−1.200	0.1086
Nitrogen	0.000	0.000	0.000	0.000	−0.978	0.2222
Carbon monoxide	0.000	0.000	0.000	0.040	−0.836	0.2220
Carbon dioxide	0.000	0.280	0.050	0.100	0.058	0.2809
Isobutyl acetate	0.520	0.570	0.000	0.470	3.161	1.0284
Naphthalene	1.340	0.920	0.000	0.200	5.161	1.0854
Anthracene	2.290	1.340	0.000	0.280	7.568	1.4544
Xanthene	1.502	1.070	0.000	0.230	7.153	1.4152
Fluorene	1.588	1.060	0.000	0.250	6.922	1.3565
Benzoic acid	0.730	0.900	0.590	0.400	4.657	0.9317
4-*tert*-Butylbenzoic acid	0.730	1.111	0.551	0.443	6.547	1.4953
3-Chlorobenzoic acid	0.840	0.950	0.630	0.320	5.197	1.0541
4-Chlorobenzoic acid	0.840	1.020	0.630	0.270	4.947	1.0541
3,4-Dichlorobenzoic acid	0.950	0.920	0.670	0.260	5.623	1.1766
2,3-Dimethoxybenzoic acid	0.890	1.636	0.564	0.703	6.612	1.3309
3,4-Dimethoxybenzoic acid	0.950	1.646	0.570	0.755	6.746	1.3309
3,5-Dimethoxybenzoic acid	0.950	1.531	0.684	0.564	6.699	1.3309
3,4,5-Trimethoxybenzoic acid	1.001	1.760	0.603	0.850	7.711	1.5309
2-Methoxybenzoic acid	0.899	1.410	0.450	0.620	5.636	1.1313
4-Methoxybenzoic acid	0.899	1.250	0.620	0.520	5.741	1.1313
2-Methylbenzoic acid	0.730	0.840	0.420	0.440	4.677	1.0726
3-Methylbenzoic acid	0.730	0.890	0.600	0.400	4.819	1.0726
2-Methyl-3-nitrobenzoic acid	1.040	1.396	0.541	0.532	6.332	1.2468
3-Methyl-4-nitrobenzoic acid	1.040	1.336	0.525	0.500	6.266	1.2468
4-Methyl-3-nitrobenzoic acid	1.040	1.461	0.659	0.521	6.434	1.2468
3-Nitrobenzoic acid	0.990	1.180	0.730	0.520	5.601	1.1059
4-Nitrobenzoic acid	0.990	1.520	0.680	0.400	5.770	1.1059
3,5-Dinitrobenzoic acid	1.250	1.630	0.700	0.590	6.984	1.2801
2-Chloro-5-nitrobenzoic acid	1.250	1.400	0.670	0.460	6.513	1.2283
4-Chloro-3-nitrobenzoic acid	1.250	1.470	0.700	0.440	6.685	1.2283
Isophthalic acid	1.100	1.360	1.055	0.585	6.144	1.1470
2-Naphthoxyacetic acid	1.610	1.940	0.690	0.764	8.553	1.5003
2-Ethylanthraquinone	1.410	1.545	0.000	0.557	8.781	1.8106
Thioxanthen-9-one	1.940	1.441	0.000	0.557	8.436	1.5357
Diphenylsulfone	1.570	2.150	0.000	0.700	8.902	1.6051
Acetylsalicylic acid	0.781	1.690	0.710	0.670	6.279	1.2879
o-Acetoacetanisidide	1.190	2.333	0.264	1.025	8.563	1.6108

Table 5. *Cont.*

Solute	E	S	A	B	L	V
Salicylamide	1.160	1.650	0.630	0.480	5.910	1.0315
Benzoin	1.587	2.115	0.196	0.847	9.159	1.6804
Forchlorfenuron	1.870	1.928	1.186	0.702	9.986	1.7617
Metronidazole benzoate	1.430	2.452	0.000	1.077	10.171	1.9563
Iminostilbene	2.000	1.801	0.246	0.401	9.013	1.5542
Isovanillin	1.040	1.477	0.308	0.681	5.868	1.1313
Lansoprazole	2.300	2.652	0.607	1.514	13.187	2.3700
2,2′-*bis*(2-Hydroxyethoxy)-1,1′-binaphthalene	3.610	3.377	0.491	1.633	17.398	2.8606
Lovastatin	1.230	2.730	0.310	1.760	15.459	3.2853
18β-Glycyrrhetinic acid	1.800	2.809	0.823	2.220	18.246	3.8984
Etodolac	1.610	1.860	0.455	1.170	11.019	2.2390

The Abraham model correlations that we have derived thus far for describing the solubilizing properties of organic mono-solvents have used the two molar solubility ratios, log ($C_{S,organic}/C_{S,water}$) and log ($C_{S,organic}/C_{S,gas}$), as the independent solute property that appears on the left-hand side of Equations (1) and (2), respectively. Published solubility data are often reported in the published literature as mole fraction solubilities, and this is how we have elected to report our measured solubility data in Table 3 for the 29 crystalline organic compounds for which solubility measurements were performed. The conversion of mole fraction solubilities to molar solubilities is relatively straightforward and involves dividing $X_{S,organic}$ by the ideal molar volume of the saturated solution, as follows:

$$C_{S,organic} \approx X_{S,organic}/[X_{S,organic}\, V_{Solute} + (1 - X_{S,organic})\, V_{Solvent}] \quad (3)$$

A numerical value of $V_{solvent}$ = 0.1341 liter mol^{-1} was used for the molar volume of the isobutyl acetate solvent. Mole fraction solubility data taken from the published chemical literature [14–29] were converted into molar solubilities in a similar fashion. The estimated values of V_{solute}, taken from the Chemspider website [48], were used for the solid organic solutes. In Table 6, we have assembled the calculated molar solubility ratios for the 44 solid organic compounds that will be used in the regression analyses for obtaining Abraham model correlations for isobutyl acetate. Also included in Table 6 are the logarithms of the gas-to-isobutyl acetate partition coefficients, log *K*, and logarithms of the water-to-isobutyl acetate transfer coefficients, log *P*, for the four inorganic gases (hydrogen, nitrogen, carbon monoxide, and carbon dioxide) and for isobutyl acetate, as well as the log $C_{S,gas}$ and log $C_{S,water}$ values used in calculating the individual molar solubility ratios.

Table 6. Experimental logarithms of molar solubility ratios, log ($C_{S,organic}/C_{S,gas}$) and log ($C_{S,organic}/C_{S,water}$), for solutes dissolved in isobutyl acetate at 298.15 K, as well as the logarithms of solute's molar gas phase concentration, log $C_{S,gas}$, and the solute's molar solubility in water, log $C_{S,water}$, at 298.15 K.

Solute	log ($C_{S,organic}/C_{S,gas}$)	log ($C_{S,organic}/C_{S,water}$)	log $C_{S,gas}$	log $C_{S,water}$
Hydrogen	−1.073 [a]	0.647 [b]		
Nitrogen	−0.761 [a]	1.039 [b]		
Carbon monoxide	−0.628 [a]	0.992 [b]		
Carbon dioxide	0.660 [a]	0.740 [b]		
Isobutyl acetate	3.880 [a]	2.161 [b]		
Naphthalene	5.709	3.979	−5.35	−3.62
Anthracene	8.036	5.006	−9.46	−6.43
Xanthene	7.601	5.101	−7.71	−5.21
Fluorene	7.280	4.830	−7.45	−5.00
Benzoic acid	6.706	1.566	−6.69	−1.55

Table 6. *Cont.*

Solute	$\log (C_{S,organic}/C_{S,gas})$	$\log (C_{S,organic}/C_{S,water})$	$\log C_{S,gas}$	$\log C_{S,water}$
4-*tert*-Butylbenzoic acid	8.862	3.638	−9.123	−3.899
3-Chlorobenzoic acid	7.408	2.258	−7.80	−2.65
4-Chlorobenzoic acid	7.149	2.349	−8.36	−3.56
3,4-Dichlorobenzoic acid	7.715	2.975	−8.72	−3.98
2,3-Dimethoxybenzoic acid	9.432	1.347	−10.245	−2.16
3,4-Dimethoxybenzoic acid	9.478	1.031	−10.942	−2.495
3,5-Dimethoxybenzoic acid	9.678	1.984	−11.170	−3.476
3,4,5-Trimethoxybenzoic acid	10.593	1.338	−11.805	−2.55
2-Methoxybenzoic acid	7.776	0.976	−8.354	−1.554
4-Methoxybenzoic acid	8.270	1.570	−9.50	−2.80
2-Methylbenzoic acid	6.325	2.025	−6.36	−2.06
3-Methylbenzoic acid	7.100	2.120	−7.12	−2.14
2-Methyl-3-nitrobenzoic acid	8.713	1.976	−9.447	−2.71
3-Methyl-4-nitrobenzoic acid	8.510	2.146	−9.594	−3.23
4-Methyl-3-nitrobenzoic acid	9.104	1.819	−9.97	−2.685
3-Nitrobenzoic acid	8.473	1.543	−8.61	−1.68
4-Nitrobenzoic acid	8.587	1.687	−9.88	−2.98
3,5-Dinitrobenzoic acid	10.080	1.780	−10.717	−2.417
2-Chloro-5-nitrobenzoic acid	9.125	2.175	−9.538	−2.588
4-Chloro-3-nitrobenzoic acid	9.474	2.264	−10.21	−3.00
Isophthalic acid	9.457	0.484	−12.083	−3.11
2-Naphthoxyacetic acid	11.584	1.665	−12.543	−2.624
2-Ethylanthraquinone	9.744	4.930	−10.344	−5.530
Thioxanthen-9-one	8.965	3.897	−10.608	−5.54
Diphenyl sulfone	10.380	2.990	−11.03	−3.64
Acetylsalicylic acid	9.411	0.911	−10.18	−1.68
o-Acetoacetanisidide	11.307	1.073	−11.775	−1.541
Salicylamide	9.011	1.326	−9.439	−1.754
Benzoin	11.218	2.487	−12.401	−3.67
Forchlorfenuron	14.174	2.755	−15.369	−3.95
Metronidazole benzoate	12.143	2.521	−12.697	−3.075
Iminostilbene	10.670	4.389	−11.592	−5.311
Isovanillin	7.661	0.823	−8.390	−1.552
Lansoprazole	16.370	1.677	−18.364	−3.671
2,2′-bis(2-Hydroxyethoxy)-2,2′-binaphthalene	20.307	3.298	−21.965	−4.956
Lovastatin	18.159	4.177	−19.597	−5.585
Simvastatin	17.932	4.282	−18.68	−5.03
18β-Glycyrrhetinic acid	22.179	4.349	−23.917	−6.087
Etodolac	13.451	3.267	−13.790	−3.606

[a] Tabulated value pertains to the logarithm of the solute's gas-to-isobutyl acetate partition coefficient. [b] Tabulated value pertains to the logarithm of the solute's water-to-isobutyl acetate transfer coefficient.

While most of the experimental data points compiled in Table 6 pertain to solid organic compounds, the range of values spanned by the individual solute descriptors are as follows: from E = 0.000 to E = 3.610; from **S** = 0.000 to **S** = 3.377; from **A** = 0.000 to **A** = 1.186; from **B** = 0.000 to **B** = 1.860; from **V** = 0.1086 to **V** = 3.4628; and **L**= −1.200 to **L** = 17.398, which is sufficiently large enough to allow for the development of meaningful Abraham model correlations for predicting the molar solubilities of many organic compounds commonly encountered in industrial manufacturing processes. The range of solute descriptors covered determines the area of predictive chemical space over which the correlation can be used. An analysis of the experimental data points in Table 6 yielded the following two Abraham model expressions:

$$\log P \text{ or } \log (C_{S,organic}/C_{S,water}) = 0.234(0.055) + 0.351(0.053)\,\mathbf{E} - 0.471(0.080)\,\mathbf{S} - 1.050(0.060)\,\mathbf{A} - 4.982(0.142)\,\mathbf{B} + 4.212(0.081)\,\mathbf{V} \quad (4)$$

$$(\text{with } N = 49,\ SD = 0.112,\ SEE = 0.119;\ R^2 = 0.992,\ F = 1099)$$

$$\log K \text{ or } \log (C_{S,organic}/C_{S,gas}) = 0.173(0.044) - 0.353(0.047)\,\mathbf{E} + 1.183(0.073)\,\mathbf{S} + 2.463(0.058)\,\mathbf{A} + 0.936(0.015)\,\mathbf{L} \quad (5)$$
$$(\text{with } N = 49,\ SD = 0.112,\ SEE = 0.117,\ R^2 = 0.999,\ F = 15753)$$

where the numerical value enclosed within the parentheses following each coefficient represents the standard error in the respective coefficient. The statistical information associated with correlation includes the following: the number of experimental data points used in determining the equation coefficients, N; the standard deviation, SD; the squared correlation coefficient, R^2; the standard error of the estimate, SEE; and the Fisher F-statistic, F. As an informational note, the $b \times \mathbf{B}$ term was not included in the Equation (5) regression analysis because isobutyl acetate lacks acidic hydrogen, and thus cannot act as an H-bond donor. Both correlations were obtained using the IBM SPSS Statistical 22 commercial software.

Careful examination of the statistical information reveals that both mathematical expressions provide a reasonably accurate description of the observed log $(C_{S,organic}/C_{S,water})$ and log $(C_{S,organic}/C_{S,gas})$ values, as documented by the small standard deviations (SD = 0.112 log units for both equations) and small standard errors of the estimate (SEE = 0.117 and SEE = 0.119 log units for Equations (4) and (5), respectively). The descriptive ability of both equations is depicted graphically in Figures 1 and 2, which compare the logarithms of the observed molar solubility ratios to the back-calculated values based on our derived Abraham model correlations. As expected from the near-unity squared correlation coefficients, the back-calculated and observed values are in very good agreement.

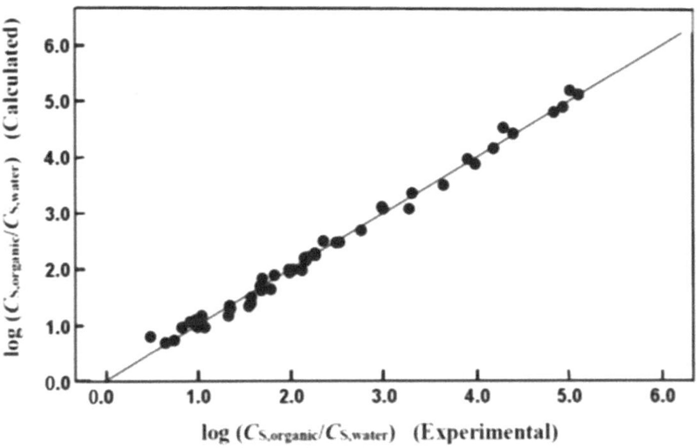

Figure 1. Comparison of experimental log $(C_{S,organic}/C_{S,water})$ data versus back-calculated values based on Equation (4) for organic compounds and inorganic gases dissolved in isobutyl acetate.

Abraham model correlations have now been determined for eleven alkyl alkanoate mono-solvents. In Table 7, we have summarized the equation coefficients for the following: methyl acetate, ethyl acetate, propyl acetate, isopropyl acetate, butyl acetate, isobutyl acetate, *tert*-butyl acetate, pentyl acetate, methyl butanoate, isopropyl myristate, and dimethyl adipate [49–52]. The tabulated equation coefficients pertain to the 'dry, anhydrous' organic solvents. The words 'dry, anhydrous' indicate that the organic solvent was not in direct contact with water, as would be the case for practical partitioning processes involving the removal of the solute from water with, for example, either ethyl acetate or butyl acetate as the extracting organic solvent. Abraham model correlations have been published for 'wet' ethyl acetate and 'wet' butyl acetate in an earlier paper [53]; however, there has not been sufficient practical partition coefficient data for the other biphasic aqueous-alkyl alkanoate extraction systems to develop meaningful Abraham model correlations.

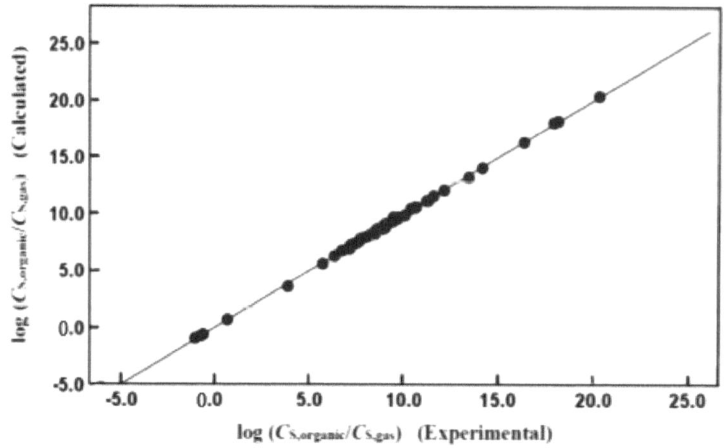

Figure 2. Comparison of experimental log ($C_{S,organic}/C_{S,gas}$) data versus back-calculated values based on Equation (5) for organic compounds and inorganic gases dissolved in isobutyl acetate.

Table 7. Abraham model correlations for predicting log ($C_{S,organic}/C_{S,water}$) and log ($C_{S,organic}/C_{S,gas}$) molar solubility ratios into select alkyl alkanoate mono-solvents.

Log (P or $C_{S,org}/C_{S,water}$)	c_p	e_p	s_p	a_p	b_p	v_p
Methyl acetate	0.351	0.223	−0.150	−1.035	−4.527	3.972
Ethyl acetate	0.328	0.314	−0.348	−0.847	−4.899	4.142
Propyl acetate	0.362	0.280	−0.390	−0.975	−4.928	4.183
Isopropyl acetate	0.307	0.314	−0.481	−0.952	−4.779	4.159
Butyl acetate	0.289	0.336	−0.501	−0.913	−4.964	4.262
Isobutyl acetate	0.234	0.351	−0.471	−1.050	−4.982	4.212
tert-Butyl acetate	0.456	0.324	−0.661	−1.068	−4.680	4.101
Pentyl acetate	0.182	0.261	−0.474	−1.017	−4.952	4.388
Methyl butyrate	0.238	0.368	−0.538	−1.031	−4.623	4.253
Isopropyl myristate	−0.605	0.930	−1.153	−1.682	−4.093	4.249
Dimethyl adipate	0.128	0.546	−0.404	−1.001	−4.481	3.987
Log (K or $C_{S,org}/C_{S,gas}$)	c_k	e_k	s_k	a_k	b_k	l_k
Methyl acetate	0.134	−0.477	1.749	2.678	0.000	0.876
Ethyl acetate	0.171	−0.403	1.428	2.726	0.000	0.914
Propyl acetate	0.246	−0.346	1.318	2.537	0.000	0.916
Isopropyl acetate	0.233	−0.495	1.324	2.550	0.000	0.928
Butyl acetate	0.154	−0.439	1.223	2.586	0.000	0.953
Isobutyl acetate	0.173	−0.353	1.183	2.463	0.000	0.936
tert-Butyl acetate	0.178	−0.444	1.045	2.522	0.000	0.964
Pentyl acetate	0.154	−0.424	1.172	2.506	0.000	0.962
Methyl butyrate	0.201	−0.502	1.290	2.469	0.000	0.958
Dimethyl adipate	0.051	−0.248	1.579	2.513	0.000	0.877

An examination of the numerical entries in Table 7 reveals that the equation coefficients for isobutyl acetate are very similar to those of butyl acetate. There should be very little difference in the solubilizing characteristics of these two alkyl acetate solvents. The sign and the magnitude of the c_p, e_p, s_p, a_p, b_p, and v_p coefficients determine whether or not a given solute–solvent interaction increases or decreases the solute's solubility in the organic solvent relative to the solute's solubility in water. For example, the negative a_p and b_p equation coefficients for isobutyl acetate, when substituted into Equation (1), would result in negative $a_p \times \mathbf{A}$ and $b_p \times \mathbf{B}$ terms, and thus decrease the calculated log ($C_{S,organic}/C_{S,water}$) value. In other words, a solute capable of hydrogen-bond formation is expected to be more soluble in water than in isobutyl acetate when all other solute–solvent interactions are ignored. Large solute molecules, on the other hand, would have a positive $v_p \times \mathbf{V}$ term, which would increase the calculated log ($C_{S,organic}/C_{S,water}$) value. Large solute molecules would tend to be more soluble in isobutyl acetate when all other solute–solvent interactions are ignored. It is the sum of the five interaction terms, however, that determines the molar solubility ratio.

4. Summary

Mathematical equations based on the Abraham solvation parameter model have been determined by performing multi-linear regression analyses on a chemically diverse set containing 49 organic solutes dissolved in isobutyl acetate. The derived Abraham model expressions were found to accurately predict the observed solubility data to within an overall standard deviation of 0.12 log units or less. Based on prior experience using the Abraham model, we expect that the derived correlations reported in the current study will provide accurate solubility predictions for additional organic compounds dissolved in isobutyl acetate, provided that one does not venture outside of the range of the predictive area of chemical space defined by the 49 solute data sets used in deriving Equations (4) and (5). Moreover, the derived Abraham model correlations will allow us to use experimental solubility data determined in isobutyl acetate in our future solute descriptor calculations. As noted previously in the manuscript, several research groups have started measuring the solubility of drug molecules dissolved in isobutyl acetate.

Author Contributions: Conceptualization and writing—original draft preparation, W.E.A.J.; data curation and analysis, R.M., T.K., J.F., J.C., E.Y., S.M., A.C., D.K. and N.S.; writing—review and editing, R.M., T.K., J.F., J.C., E.Y., S.M., A.C., D.K. and N.S. All authors have read and agreed to the published version of the manuscript.

Funding: This research received no external funding.

Data Availability Statement: Data are contained within the article.

Conflicts of Interest: The authors declare no conflicts of interest.

References

1. Abraham, M.H. Scales of solute hydrogen-bonding: Their construction and application to physicochemical and biochemical processes. *Chem. Soc. Rev.* **1993**, *22*, 73–83. [CrossRef]
2. Abraham, M.H.; Ibrahim, A.; Zissimos, A.M. Determination of sets of solute descriptors from chromatographic measurements. *J. Chromatogr. A* **2004**, *1037*, 29–47. [CrossRef]
3. Abraham, M.H.; Smith, R.E.; Luchtefeld, R.; Boorem, A.J.; Luo, R.; Acree, W.E., Jr. Prediction of solubility of drugs and other compounds in organic solvents. *J. Pharm. Sci.* **2010**, *99*, 1500–1515. [CrossRef] [PubMed]
4. Jalan, A.; Ashcraft, R.W.; West, R.H.; Green, W.H. Predicting solvation energies for kinetic modeling. *Annu. Rep. Prog. Chem. Sec. C Phys. Chem.* **2010**, *106*, 211–258. [CrossRef]
5. Clarke, E.D. Beyond physical properties–application of Abraham descriptors and LFER analysis in agrochemical research. *Bioorg. Med. Chem.* **2009**, *17*, 4153–4159. [CrossRef] [PubMed]
6. Poole, C.F.; Ariyasena, T.C.; Lenca, N. Estimation of the environmental properties of compounds from chromatographic properties and the solvation parameter method. *J. Chromatogr. A* **2013**, *1317*, 85–104. [CrossRef]
7. Poole, C.F.; Atapattu, S.N. Recent advances for estimating environmental properties for small molecules from chromato-graphic measurements and the solvation parameter model. *J. Chromatogr. A* **2022**, *1687*, 463682. [CrossRef]
8. Endo, S.; Goss, K.-U. Applications of polyparameter linear free energy relationships in environmental chemistry. *Environ. Sci. Technol.* **2014**, *48*, 12477–12491. [CrossRef]
9. Poole, C.F. Assessment of liquid-liquid partition for the assignment of descriptors for the solvation parameter model. *J. Chromatogr. A* **2024**, *1721*, 464850. [CrossRef]
10. Poole, C.F. Determination of solvation parameter model compound descriptors by gas chromatography. *J. Chromatogr. A* **2024**, *1717*, 464711. [CrossRef]
11. Dayani, M.T.; Ambagaspitiya, A.W.T.D.; Atapattu, S.N.; Ariyasena, T.C. Determination of experimental solute descriptor values for safrole by liquid-liquid partitioning and gas chromatography. *Phys. Chem. Liq.* 2024, *in press*. [CrossRef]
12. Liu, X.; Acree, W.E., Jr.; Abraham, M.H. Descriptors for some compounds with pharmacological activity; calculation of properties. *Int. J. Pharm.* **2022**, *617*, 121597. [CrossRef] [PubMed]

13. Orlov, A.A.; Demenko, D.Y.; Bignaud, C.; Valtz, A.; Marcou, G.; Horvath, D.; Coquelet, C.; Varnek, A.; de Meyer, F. Chemoinformatics-driven design of new physical solvents for selective CO_2 absorption. *Environ. Sci. Technol.* **2021**, *55*, 15542–15553. [CrossRef] [PubMed]
14. Li, Q.-S.; Yi, Z.-M.; Su, M.-G.; Sun, X.-F. Solubility of naphthalene in isobutyl acetate, n-butyric acid, ethyl acetate, *N*-methyl pyrrolidone, *N,N*-dimethylformamide, and tetrahydrofuran. *J. Chem. Eng. Data* **2008**, *53*, 2701–2703. [CrossRef]
15. Sun, M.; Xiao, Y.; Zhou, Y.; Zhang, B.; Cui, P.; Zhou, L.; Yin, Q. Solubility measurement and thermodynamic properties of fluorene in 14 pure solvents at temperatures from 278.15 to 318.15 K. *J. Chem. Eng. Data* **2023**, *68*, 2491–2499. [CrossRef]
16. Zhao, S.; Chen, X.; Dai, Q.; Wang, L. Solubilities of benzene carboxylic acids in isobutyl acetate from (299.73 to 353.15) K. *J. Chem. Eng. Data* **2011**, *56*, 2399–2402. [CrossRef]
17. Tully, G.; Hou, G.; Glennon, B. Solubility of benzoic acid and aspirin in pure solvents using focused beam reflective measurement. *J. Chem. Eng. Data* **2016**, *61*, 594–601. [CrossRef]
18. Wang, Y.; Wu, Y.; Li, H.; Li, Y.; Li, T.; Ren, B. Solubility measurement and Hansen solubility parameter of forchlorfenuron in 12 organic pure solvents at various temperatures. *J. Chem. Eng. Data* **2022**, *67*, 3278–3289. [CrossRef]
19. Yu, F.; Wang, F.; Zhang, L.; Zhang, M.; Zhou, L.; Xie, C.; Bao, Y.; Chen, W.; Gong, J. Uncovering dissolution behavior and thermodynamic properties of metronidazole benzoate in twelve mono-solvents by experiments and molecular simulation. *J. Mol. Liq.* **2024**, *393*, 123539. [CrossRef]
20. Wan, Y.; Gao, X.; Wang, R.; Li, F.; Li, Y.; He, H. Research on solubility behavior of iminostilbene in twelve mono-solvents: Measurement, modeling, molecular simulation and thermodynamic properties. *J. Chem. Thermodyn.* **2024**, *188*, 107182. [CrossRef]
21. Zhang, S.; Wang, H.; Dai, J.; Niu, Y.; Yin, Q.; Zhou, L. Solubility determination, model evaluation and solution thermodynamics of isovanillin in 15 pure solvents and 4 binary solvents. *J. Mol. Liq.* **2021**, *340*, 116847. [CrossRef]
22. Hong, M.; Xu, L.; Ren, G.; Chen, J.; Qi, M. Solubility of lansoprazole in different solvents. *Fluid Phase Equilib.* **2012**, *331*, 18–25. [CrossRef]
23. Liu, Y.; Cui, P.; Zhou, L.; Yin, Q. Measurement and correlation of solubility and thermodynamic properties of 2,2′-bis(2-hydroxyethoxy)-1,1′-binaphthalene form B in twelve pure solvents. *J. Chem. Eng. Data* **2022**, *67*, 1541–1549. [CrossRef]
24. Nti-Gyabaah, J.; Chiew, Y.C. Solubility of lovastatin in ethyl acetate, propyl acetate, isopropyl acetate, butyl acetate, *sec*-butyl acetate, isobutyl acetate, *tert*-butyl acetate, and 2-butanone, between (285 and 313) K. *J. Chem. Eng. Data* **2008**, *53*, 2060–2065. [CrossRef]
25. Hu, X.; Gong, Y.; Cao, Z.; Huang, Z.; Sha, J.; Li, Y.; Li, T.; Ren, B. Solubility, Hansen solubility parameter and thermodynamic properties of etodolac in twelve organic pure solvents at different temperatures. *J. Mol. Liq.* **2020**, *316*, 113779. [CrossRef]
26. Yan, J.; Yin, Q.; Jiang, C.; Gong, J.; Zhang, M.; Wang, Y.; Hou, B.; Hao, H. Solution thermodynamics of simvastatin in pure solvents and binary solvent mixtures. *Fluid Phase Equilib.* **2015**, *406*, 77–90. [CrossRef]
27. Nti-Gyabaah, J.; Chan, V.; Chiew, Y.C. Solubility and limiting activity coefficient of simvastatin in different organic solvents. *Fluid Phase Equilib.* **2009**, *280*, 35–41. [CrossRef]
28. Shen, Y.; Li, R.; Zhao, P.; Liu, W.; Yang, X.; Zhang, Z. Equilibrium solubility of 18β-glycyrrhetinic acid in 12 pure solvents: Determination, correlation, and Hansen solubility parameter. *J. Chem. Eng. Data* **2022**, *67*, 3243–3251. [CrossRef]
29. He, H.; Wan, Y.; Sha, J.; Sun, R.; Jiang, G.; Li, Y.; Li, T.; Ren, B. Equilibrium solubility of exo-5,6-dehydronorcantharidin in thirteen pure solvents: Determination, correlation, Hansen solubility parameter and thermodynamic properties. *J. Mol. Liq.* **2020**, *312*, 113384. [CrossRef]
30. Monton, J.B.; Munoz, R.; Burguet, M.C.; de la Torre, J. Isobaric vapor-liquid equilibria for the binary systems isobutyl alcohol + isobutyl acetate and *tert*-butyl alcohol + *tert*-butyl acetate at 20 and 101.3 kPa. *Fluid Phase Equilib.* **2005**, *227*, 19–25. [CrossRef]
31. Acree, W.E., Jr. IUPAC-NIST Solubility Data Series. Volume 99. Solubility of Benzoic Acid and Substituted Benzoic Acids in Both Neat Organic Solvents and Organic Solvent Mixtures. *J. Phys. Chem. Ref. Data* **2013**, *42*, 033103/1–033103/525. [CrossRef]
32. Fritz, J.S.; Lisicki, N.M. Titration of acids in nonaqueous solvents. *Anal. Chem.* **1951**, *23*, 589–591. [CrossRef]
33. Maiti, A.; Reddy, P.V.N.; Sturdy, M.; Marler, L.; Pegan, S.D.; Mesecar, A.D.; Pezzuto, J.M.; Cushman, M. Synthesis of casimiroin and optimization of its quinone reductase 2 and aromatase inhibitory activities. *J. Med. Chem.* **2009**, *52*, 1873–1884. [CrossRef]
34. Acree, W.; Chickos, J.S. Phase Transition Enthalpy Measurements of Organic and Organometallic Compounds and Ionic Liquids. Sublimation, Vaporization, and Fusion Enthalpies from 1880 to 2015. Part 2. C_{11}-C_{192}. *J. Phys. Chem. Ref. Data* **2017**, *46*, 013104/1–013104/532. [CrossRef]
35. Bai, H.; Fang, X.; Peng, C.; Liu, Q.; Xie, W.; Jia, L.; Song, Z. Dissolution behavior of 2-ethylanthraquinone in various neat solvents: Solubility, correlation and thermodynamic analysis. *J. Mol. Liq.* **2019**, *289*, 111122. [CrossRef]
36. Liu, C.X. Solubility Measurement and Intrinsic Kinetics Study on Hydrogenation of Alkylanthraquinone. Ph.D. Thesis, Tianjin University, Tianjin, China, 2005.
37. Acree, W.; Chickos, J.S. Phase transition enthalpy measurements of organic and organometallic compounds. sublimation, vaporization and fusion enthalpies from 1880 to 2015. Part 1. C_1–C_{10}. *J. Phys. Chem. Ref. Data* **2016**, *45*, 033101/1–033101/565. [CrossRef]
38. Dippy, J.F.J.; Hughes, S.R.C.; Laxton, J.W. Chemical constitution and the dissociation constants of monocarboxylic acids. XV. Steric effects in substituted nitrobenzoic acids. *J. Chem. Soc.* **1956**, 2995–3000. [CrossRef]
39. Le Fave, G.M. Some reactions of the trifluoromethyl group in the benzotrifluoride series. I. Hydrolysis. *J. Am. Chem. Soc.* **1949**, *71*, 4148–4149. Available online: https://pubs.acs.org/doi/epdf/10.1021/ja01180a507 (accessed on 10 May 2024). [CrossRef]

40. Wan, Y.; He, H.; Gao, X.; Guo, X.; Li, F.; Li, Y. Solid-liquid equilibrium of 2,3-dimethoxybenzoic acid in fifteen mono-solvents: Determination, correlation, Hansen solubility parameter, molecular dynamic simulation and thermodynamic analysis. *J. Mol. Liq.* **2022**, *348*, 118029. [CrossRef]
41. Monte, M.J.S.; Hillesheim, D.M. Thermodynamic study on the sublimation of six methylnitrobenzoic acids. *J. Chem. Thermodyn.* **2001**, *33*, 103–112. [CrossRef]
42. Li, H.; Wang, Y.; Wu, Y.; Li, Y.; Li, T.; Ren, B. Solubility and thermodynamic properties of 2-naphthoxyacetic acid in twelve mono-solvents. *J. Chem. Thermodyn.* **2023**, *177*, 106934. [CrossRef]
43. Huang, C.-Y.; Tang, M.; Chen, Y.-P. Solid-liquid equilibria for 3-aminophenol + 2-aminophenol, 4-(acetylamino)toluene + 2-(acetylamino)toluene, and 3,4,5-trimethoxybenzoic acid + 2,4,5-trimethoxybenzoic acid. *J. Chem. Eng. Data* **2005**, *50*, 40–43. [CrossRef]
44. Rathi, P.B.; Deshpande, K.V. Extended Hildebrand approach: An empirical model for solubility prediction of etodolac in 1,4-dioxane and water mixtures. *J. Solut. Chem.* **2014**, *43*, 1886–1903. [CrossRef]
45. Acree, W.E., Jr. Commentary on Extended Hildebrand approach: An empirical model for solubility prediction of etodolac in 1,4-dioxane and water mixtures. *J. Solut. Chem.* **2017**, *46*, 2130–2133. [CrossRef]
46. Charlton, A.K.; Daniels, C.R.; Acree, W.E., Jr.; Abraham, M.H. Solubility of crystalline nonelectrolyte solutes in organic solvents: Mathematical correlation of acetylsalicylic acid solubilities with the Abraham general solvation model. *J. Solut. Chem.* **2003**, *32*, 1087–1102. [CrossRef]
47. Hart, E.; Lee, G.; Qian, E.; Jodray, M.; Barrera, M.; Fischer, R.; Che, M.; Liu, Y.; Zha, O.; Woods, D.; et al. Determination of Abraham model solute descriptors for 4-*tert*-butylbenzoic acid from experimental solubility data in organic mono-solvents. *Phys. Chem. Liq.* **2019**, *57*, 445–452. [CrossRef]
48. ChemSpider Database, The Royal Society of Chemistry. Available online: http://www.chemspider.com/ (accessed on 14 May 2024).
49. Sinha, S.; Yang, C.; Wu, E.; Acree, W.E., Jr. Abraham solvation parameter model: Examination of possible intramolecular hydrogen-bonding using calculated solute descriptors. *Liquids* **2022**, *2*, 131–146. [CrossRef]
50. Shanmugam, N.; Zhou, A.; Motati, R.; Yao, E.; Kandi, T.; Longacre, L.; Benavides, D.; Motati, S.; Acree, W.E., Jr. Development of Abraham model correlations for dimethyl adipate from measured solubility data of nonelectrolyte organic compounds. *Phys. Chem. Liq.* **2023**, *61*, 328–339. [CrossRef]
51. Longacre, L.; Wu, E.; Yang, C.; Zhang, M.; Sinha, S.; Varadharajan, A.; Acree, W.E., Jr. Development of Abraham model correlations for solute transfer into the *tert*-butyl acetate mono-solvent and updated equations for both ethyl acetate and butyl acetate. *Liquids* **2022**, *2*, 258–288. [CrossRef]
52. Sedov, I.A.; Salikov, T.M.; Khaibrakhmanova, D.R.; Wadawadigi, A.; Zha, O.; Qian, E.; Hart, E.; Barrera, M.; Acree, W.E., Jr.; Abraham, M.H. Determination of Abraham model correlations for solute transfer into propyl acetate based on experimental activity coefficient and solubility data. *J. Solut. Chem.* **2018**, *47*, 634–653. [CrossRef]
53. Sprunger, L.M.; Proctor, A.; Acree, W.E., Jr.; Abraham, M.H.; Benjelloun-Dakhama, N. Correlation and prediction of partition coefficient between the gas phase and water, and the solvents dry methyl acetate, dry and wet ethyl acetate, and dry and wet butyl acetate. *Fluid Phase Equilib.* **2008**, *270*, 30–44. [CrossRef]

liquids

MDPI

Article

Unprecedented High Probe-Reported Polarity of Deep Eutectic Solvents Composed of Lanthanide Salts and Urea

Anushis Patra, Vaishali Khokhar and Siddharth Pandey *

Department of Chemistry, Indian Institute of Technology Delhi, Hauz Khas, New Delhi 110016, India; anuchemistry2019@gmail.com (A.P.); vaishalikhokhar57@gmail.com (V.K.)

* Correspondence: sipandey@chemistry.iitd.ac.in; Tel.: +91-11-26596503; Fax: +91-11-26581102

Abstract: Deep eutectic solvents (DESs) have emerged as viable alternatives to toxic organic solvents. The most intriguing aspect of these solvents is perhaps the widely varying physicochemical properties emerging from the changes in the constituents that form DESs along with their composition. Based on the constituents, a DES can be hydrophilic/polar or hydrophobic/non-polar, rendering a vastly varying spectrum of polarity a possibility. DESs formed by mixing urea (U) with hydrated lanthanide salts, lanthanum nitrate hexahydrate (La : U), cerium nitrate hexahydrate (Ce : U), and gadolinium nitrate hexahydrate (Gd : U), respectively, exhibit very high polarity as manifested via the probe-reported empirical parameters of dipolarity/polarizability (π^*). The highest π^* of 1.70 exhibited by the DES (Gd : U) in a 1 : 2 molar ratio is unprecedented. The π^* ranges from 1.50 to 1.70 for these DESs, which is almost the highest reported for any solvent system. The π^* decreases with an increasing amount of urea in the DES; however, the anomalous trends in H-bond donating acidity (α) and H-bond accepting basicity (β) appear to be due to the hydrated water of the lanthanide salt. The emission band maxima of the fluorescence probe of the "effective" dielectric constant (ε_{eff}) of the solubilizing media, pyrene-1-carboxaldehyde (PyCHO), in salt-rich DESs reflect higher cybotactic region dipolarity than that offered by water. Probe Nile red aggregates readily in these DESs to form non-fluorescent H-aggregates, which is a characteristic of highly polar solvents. The behavior of probe pyranine also corroborates these outcomes as the (lanthanide salt : urea) DES system supports the formation of the deprotonated form of the probe in the excited state. The (lanthanide salt : urea) DES system offers solubilizing media of exceptionally high polarity, which is bound to expand their application potential.

Keywords: deep eutectic solvent; hydrated lanthanide salts; urea; dipolarity/polarizability (π^*); pyrene-1-carboxaldehyde; Nile red; aggregates; pyranine

1. Introduction

Currently, the most emerged and prominent solvent class belonging to the area of green chemistry arguably comprises deep eutectic solvents (DESs). DESs are usually obtained by simply mixing two or more constituents in a particular molar ratio followed by gentle heating. In DESs, the thermodynamics of mixing ensures significant lowering of the normal freezing point compared to those of individual constituents [1–5]. While the application potential of DESs lies in their environmentally benign nature with negligible toxicity along with the low costs of the constituents and preparation, there is an escalated interest in academia due to the fact that a wide variety of physicochemical properties of these neoteric media are possible based on the identities of the constituents and their composition used to prepare these solvents [6–10]. Based on the chemical nature of the constituents, DESs are divided into several classes. In the formation of DESs, usually one component behaves as an H-bond donor (HBD), and the other behaves as an H-bond acceptor (HBA), leading to extensive H-bonding and other interactions (van der Waals, electrostatic, etc.) resulting in a liquid state under wide ranges of temperature and pressure [1–5].

DESs have been extensively investigated not only for their applications, but also for their physicochemical properties. Consequently, among other classifications, DESs may also be divided into "hydrophilic" and "hydrophobic" DESs, with "hydrophilic" DESs offering complete water miscibility and relatively high polarities [11,12]. DESs composed of a mixture of quaternary ammonium salt as an HBA with a suitable HBD have shown to possess relatively high polarity (as reported by several solvatochromic probes) in comparison to common solvents [13]. Extensive research has been carried out to characterize metal-based DESs by understanding their physicochemical properties with variations in the identity and composition of the constituents [14,15]. Kamlet-Taft parameters, in this context, have afforded the understanding of the H-bonding interactions that occur inside hydrated metal salt-based DESs [16–19]. Herein, we report our findings of exceptionally high probe-reported polarity exhibited by a set of type IV DESs constituting hydrated lanthanide metal salts as HBAs and urea as prototypical HBDs [14]. These DESs have recently been shown to support probe aggregation at unprecedented low concentrations [20]. Specifically, three (3) HBD urea-based DESs containing the HBA metal salts lanthanum nitrate hexahydrate (La : U), cerium nitrate hexahydrate (Ce : U), and gadolinium nitrate hexahydrate (Gd : U), respectively, are found to exhibit some of the highest probe-reported polarities in terms of the empirical Kamlet-Taft parameters of dipolarity/polarizability (π^*) [21] and HBD acidity (α) [22] along with E_T values [22]. The highly dipolar nature of the solubilization microenvironment offered by these DESs was further emphasized by the cybotactic region dipolarity of several fluorescence probes.

2. Materials and Methods

Hydrated metal salts [lanthanum nitrate hexahydrate (La), cerium nitrate hexahydrate (Ce), and gadolinium nitrate hexahydrate (Gd)] and urea were chosen as the DES constituents and were obtained with the highest purity $\geq$ 99% from Sigma-Aldrich and SRL Enterprises, respectively. *N,N*-Diethyl-4-nitroaniline (DENA) and 4-nitroaniline (NA) (Figure 1) were used as absorbance probes in this work. DENA and NA were obtained from Frinton Laboratories and Spectrochem Co. Ltd.(Hainesport, NJ, USA), respectively, while Pyridin-*N*-oxide (PyO, 95%), Pyrene-1-carboxyldhyed (PyCHO, 99%), 9-diethylamino-5-benzo[a]phenoxazinone (Nile red, $\geq$98%, HPLC), and 8-hydroxypyrene-1,3,6-trisulfonic acid (pyranine, 99%) (Figure 1) were purchased from Sigma-Aldrich (St. Louis, MO, USA). These were used without any further purification.

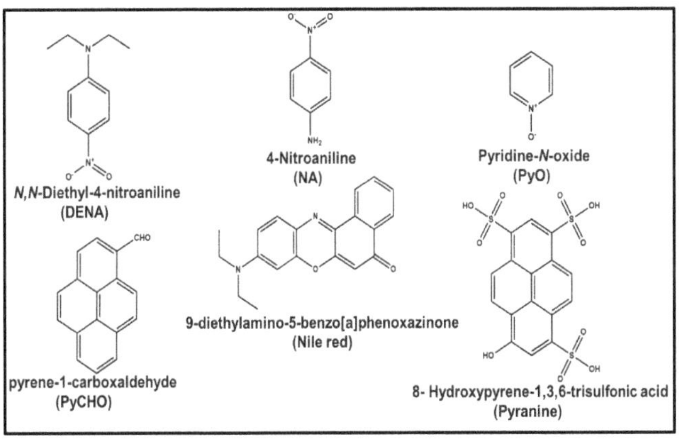

Figure 1. The molecular structures of the solvatochromic probes used in this work.

DESs with mixtures of (La : U), (Ce : U), and (Gd : U) were prepared by simply mixing the lanthanide salts (all $\geq$ 99+% purity) with urea (>99.9% purity) at permissible (in the context of DES formation) molar composition ranges (1 : 3 to 1 : 7 for La : U, 1 : 3.5 to 1 : 7

for Ce : U, and 1 : 2 to 1 : 7 for Gd : U). At several compositions, simply mixing the two solid constituents resulted in liquid-state DESs under ambient conditions (no heating was required). All UV-vis molecular absorbance and fluorescence as well as NMR spectroscopic probes were dissolved in the DESs, and the probe responses were acquired using adequate sampling and data collection techniques using appropriate instrumentation.

Stock solutions of probes were prepared by dissolving a precalculated amount of the respective probe in ethanol and storing it at 4 ± 1 °C in pre-cleaned brown-amber glass vials. Mettler Toledo AB104-S balance was used to weigh the proper amount of probe with an accuracy of ± 0.0001 g. A prerequisite amount of the prepared stock solution was taken in a 1 cm^2 quartz cuvette. Grade 1 nitrogen gas was used to evaporate the ethanol. The UV-visible absorbance spectra of all samples were collected by using Lambda 35 double-beam spectrophotometer (purchased from PerkinElmer) containing a photomultiplier tube (PMT) detector with a flexible deuterium and a Xe lamp as a light source. The required amount of DESs under investigation was transferred to the cuvette. Dissolution of the probes in DESs was confirmed using the linearity of the absorbance versus the concentration plot. The ^{13}C of 250 mM PyO dissolved in La : U-based DESs was obtained on Brucker 500 MHz (DPX-500) NMR spectrometers(Billerica, Massachusetts, USA). $CDCl_3$ was used as internal reference, and the sample was taken in the sealed capillary inside the same NMR tube (consists of $CDCl_3$). The steady-state fluorescence spectra were obtained using an FLS1000 spectrometer(Livingston, West Lothian, UK) with emission monochromators (STGM325-M) and grating excitation (STGM325-X) containing a 450 W xenon arc lamp as the radiation source, a red PMT detector, and a temperature controller (a Quantum Northwest Luma 40). All parts of the instrument are product of Quantum Northwest, Inc. (Libert Lake, WA, USA). The spectrometer was bought from Edinburgh Instruments, Ltd. Appropriate blank was subtracted from the emission and excitation spectra before analysis. Data analysis was accomplished using SigmaPlot v14.5. A Horiba-Jobin Yvon Fluorocube time-correlated single-photon counting (TCSPC) fluorimeter was used to measure the fluorescence lifetime of the selected probe. Samples were also characterized through Attenuated Total Reflectance-Fourier Transform Infrared (ATR-FTIR) absorbance data, which were acquired on a double-beam spectrophotometer (Agilent Technologies Cary 660 ATR, Santa Clara, CA, USA) from 4000 to 400 cm^{-1}.

3. Results and Discussion

Well-established UV–vis molecular absorbance probes, *N,N*-diethyl-4-nitroaniline (DENA) [21] and 4-nitroaniline (NA) [23] (Figure 1), were employed to obtain the dipolarity/polarizability (π*) and H-bond accepting (HBA) basicity (β) of the three DESs at several different compositions using the following expressions:

$$\pi* = 8.649 - 0.314\bar{\upsilon}_{DENA} \quad (1)$$

$$\beta = -0.357\bar{\upsilon}_{NA} - 1.176\pi * + 11.12 \quad (2)$$

where $\bar{\upsilon}_{DENA}$ and $\bar{\upsilon}_{NA}$ are the lowest energy absorbance band maxima of the probes DENA and NA, respectively, in kK units (1000 cm^{-1} = 1 kK). Our attempts to obtain the E_T values [that subsequently afford HBD acidity (α)] using Reichardt's betaine dyes 30 and 33, the most used probes for this purpose, were futile as both of these dyes were insoluble in our DESs or due to the protonation of –O^-; these dyes did not exhibit lowest-energy intramolecular charge-transfer absorbance bands. Schneider et al. [24] used pyridine-*N*-oxide (PyO) (Figure 1) as a probe to characterize the HBD acidity (α) of several solvents using ^{13}C NMR chemical shifts with the following equations:

$$\alpha_{24} = 2.32 - 0.15 \times d_{24} \quad (3)$$

where d_{24} is the difference (in ppm) of the ^{13}C NMR chemical shifts (δ) of carbon 2 with respect to that of carbon 4 of PyO. The correlation between α was thus estimated, and the α from Reichardt's dye (α_{RD}) was established by Freire's [22] group recently as follows:

$$\alpha_{24} = 0.88 \times \alpha_{RD} \tag{4}$$

supporting the suitability of PyO as an alternative probe to characterize the HBD acidity of the solvents. It was further shown that the $E_T(30)$ [22] and E_T^N [22] single-parameter polarity scales can be subsequently obtained using the following equations:

$$E_T(30) = (\alpha_{RD} + 2.03 + 0.72\pi*)/0.0649 \tag{5}$$

and

$$E_T^N = (E_T(30) - 30.7)/32.4 \tag{6}$$

The absorbance spectra and the respective absorbance maxima of DENA and NA in all three DESs at all investigated compositions are presented in Figure S1 and Table S1, respectively (the representative spectra within DESs (Gd : U) at 1 : 2 and 1 : 7 molar ratios are shown in Figure 2). With the increase in the amount of urea in all three DES systems, a significant hypsochromic shift in the DENA wavelength maxima is clearly observed. While similar to DENA, the NA wavelength maxima depends, to some extent, on the identity of the lanthanide metal, interestingly, and contrary to DENA, the band maxima of NA are statistically the same irrespective of the amount of urea in the DES system. The ^{13}C NMR scans of PyO within (La : U) at 1 : 3 and 1 : 7 molar ratios are presented in Figure 3 as representatives (others ratios are shown in Figure S2). It is to be noted that (Ce : U) based DESs appear to react with PyO, and due to the high number of unpaired electrons in Gd, the ^{13}C NMR of PyO in (Ce : U) and (Gd : U) DESs could not be obtained. The values of the empirical parameters π^*, β, α, and E_T^N, estimated from Equations (1)–(6) for the applicable DES systems, are presented in Tables 1 and 2. A careful examination of the empirical parameters listed in Tables 1 and 2 reveals that π^*, E_T^N, and α decrease, while β increases monotonically as the relative amount of urea is increased within a DES system.

Table 1. Dipolarity/polarizability (π^*) and HBA basicity (β) of (Ln : U)-based DESs at 298.15 K.

DESs	Ratio	π^*	β
La : U	1 : 3	1.64	0.07
	1 : 4	1.62	0.10
	1 : 5	1.59	0.13
	1 : 6	1.54	0.19
	1 : 7	1.53	0.20
Ce : U	1 : 3.5	1.61	0.30
	1 : 4	1.61	0.30
	1 : 5	1.58	0.33
	1 : 6	1.54	0.38
	1 : 7	1.50	0.43
Gd : U	1 : 2	1.70	−0.01
	1 : 3	1.62	0.08
	1 : 4	1.59	0.12
	1 : 5	1.58	0.13
	1 : 6	1.54	0.18
	1 : 7	1.53	0.19

Table 2. HBD acidity values (α's), E_T^N and $E_T(30)$, of (La : U) DESs at 298.15K.

DES	Ratio	α_{24}	α_{RD}	E_T^N	$E_T(30)$
	1 : 3	1.55	1.77	1.42	76.67
	1 : 4	1.51	1.71	1.39	75.64
La : U	1 : 5	1.46	1.66	1.35	74.54
	1 : 6	1.39	1.58	1.29	72.65
	1 : 7	1.38	1.56	1.29	72.35

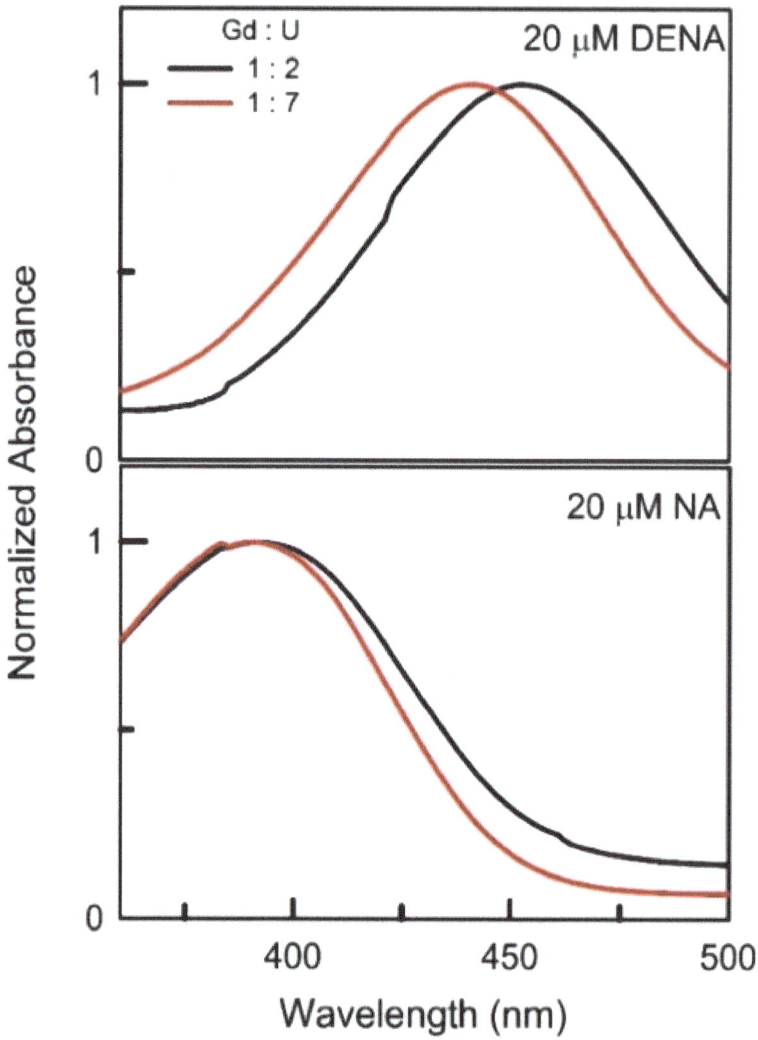

Figure 2. Representative absorbance spectra of 20 ± 2 μM DENA and 20 ± 2 μM NA within DESs (Gd : U) at 1 : 2 and 1 : 7 molar ratios.

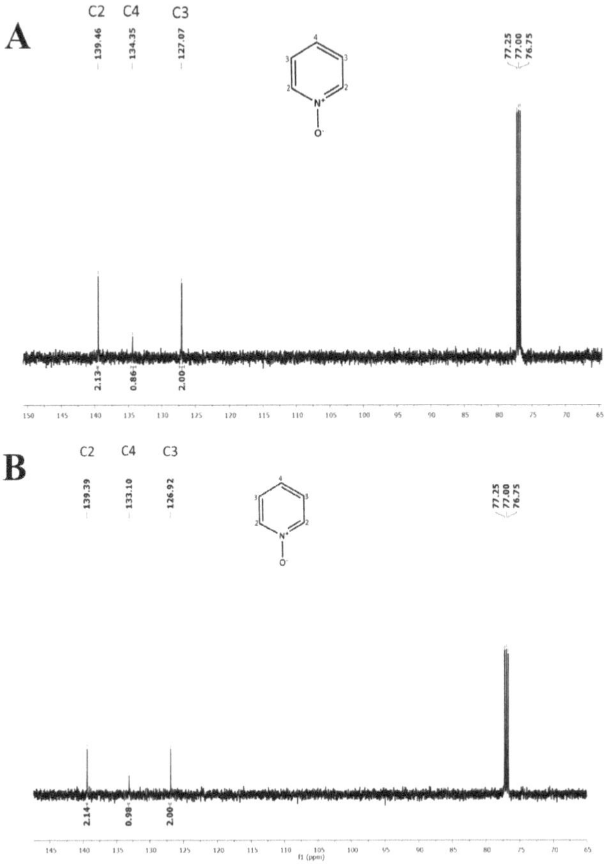

Figure 3. Representative ^{13}C NMR spectra of 250 ± 5 mM PyO within DESs (La : U) at 1 : 3 (**A**) and 1 : 7 (**B**) molar ratios.

There are two highly unusual aspects associated with these probe-reported empirical parameters that clearly emerge. First and foremost, the values of the polarity indicators, π^* and E_T^N, are exceptionally high—they are almost the highest reported for any solvent system (π^* ranges from 1.50 to 1.70, and E_T^N ranges from 1.29 to 1.42). In comparison, for choline chloride (ChCl)-based DESs, the reported π^* is 1.23 for (ChCl : U :: 1 : 2) [11], 1.373 for (ChCl : citric acid :: 1 : 1) [25], and 1.406 for (ChCl : $ZnCl_2$ :: 1 : 2) [25]. The highest π^* for any DES that we could find in the literature is 1.671 for (betaine : citric acid :: 1 : 1) [25] under ambient conditions. Clearly, these (lanthanide salt : urea) based DESs offer exceptionally high dipolarity/polarizability, especially at compositions with a high metal salt content. This further manifests in the unprecedented high E_T^N values [E_T^N of 1.42 for (La : U :: 1 : 3) is the highest, and we could not find comparable E_T^N values for any solvent system in the literature]. The exceptionally high probe-reported polarity of these (lanthanide salt : urea) DESs can be effectively utilized in chemical synthesis and separation along with many other applications.

The second unusual aspect of the data is as follows: The decreases in π^* and E_T^N with an increasing relative amount of urea is easy to comprehend as urea, the neutral constituent, is increased, and lanthanide salt, the ionic constituent, is decreased. However, surprisingly, α shows a decrease and β shows an increase as the HBD constituent urea is increased (and HBA salt is decreased) within the DES system. This is tentatively attributed to the presence of hydrated water in the system, which contributes significantly to the HBD acidity, and as the relative amount of hydrated metal salt decreases, the α also decreases (while most of the urea is involved in the H-bonding interaction). A slight increase in β as the relative amount of urea is increased in the DES system appears to be a complex interplay of the interactions involving salt, hydrated water, and urea. Further experimentation in our research labs is ongoing to obtain insights into this observation. It is also noteworthy that for these DESs, the α values are fairly high [e.g., 1.77 for (La : U :: 1 : 3)], if not the highest [>2.00 for DESs constituted of (ChCl : D-sorbital :: 1 : 2), (ChCl : citric acid :: 1 : 1), (ChCl : lactic acid :: 1 : 1), (betaine : glycerol :: 1 : 2), (betaine : citric acid :: 1 : 1), and (betaine : urea :: 1 : 2)] [25]. Surprisingly, the β values of these lanthanide : urea DESs are significantly lower in comparison to those of other DES systems [among the reported β values of DESs, only (ChCl : urea : ethylene glycol :: 1 : 1 : 1) (β = 0.000), (ChCl : ethylene glycol : formamide :: 1 : 1 : 1) (β = 0.000), and (betaine : citric acid :: 1 : 1) (β = 0.012) exhibit such low H-bond accepting basicity] [25].

In order to corroborate these unusually high probe-reported polarities of lanthanide salt/urea DESs, we explored the behavior of judiciously selected fluorescence probes within these DESs. First, a wavelength-shift fluorescence probe pyrene-1-carboxaldehyde (PyCHO), which provides information about the "effective" dielectric constant (ε_{eff}) of solubilizing media [26], is used to assess the polarity of (lanthanide salt : urea) DESs. PyCHO exhibits a well-structured emission band in non-polar solvents, which becomes broad and structureless as polarity is increased [27]. The strong dependency of the PyCHO emission spectra on ε_{eff} is mirrored by a gradual bathochromic shift with an increase in ε_{eff} in the medium [26]. As expected, the emission spectra and band maxima of PyCHO in (lanthanide salt : urea) DESs presented in Figure 4 (the emission spectra for other DESs at all of the investigated compositions are presented in Figure S3) show a broad structureless band with a gradual bathochromic shift with a decrease in the amount of urea, suggesting an increasing polarity (or "effective" dielectric constant). More importantly, the emission band maxima of PyCHO in (lanthanide salt : urea) DES systems are in the range of 472–486 nm (Table 3), while for other hydrophilic ChCl-based DESs, the values lie in the 452–460 nm range [28]. The emission maxima for these DESs, particularly at a lower (Ln : U) molar ratio, are even higher than those reported in water (*ca.* 475 nm) [28], implying that the ε_{eff} value of these DES systems are higher than those of water. These outcomes clearly demonstrate the unusually high polarity surrounding the probe cybotactic region for PyCHO when dissolved in these type IV (Ln : U) DESs and corroborate the outcomes of the empirical polarity parameters discussed above.

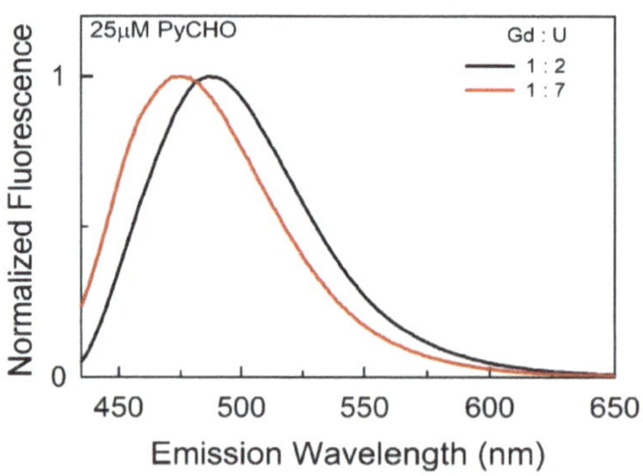

Figure 4. Representative emission spectra of 25 ± 2 μM PyCHO (λ_{ex} = 406 nm) within DESs (Gd : U) at 1 : 2 and 1 : 7 molar ratios.

Table 3. Emission wavelength maxima (λ_{max}^{em}/nm) of 25 ± 2 μM PyCHO in (Ln : U)-based DESs at 298.15 K.

Ratio	La : U	Ce : U	Gd : U
1 : 7	479	474	472
1 : 6	482	476	477
1 : 5	482	477	477
1 : 4	483	479	483
1 : 3.5		482	
1 : 3	486		484
1 : 2			486

Next, we used the popular fluorescent twisted intramolecular charge-transfer (TICT) probe 9-diethylamino-5-benzo[a]phenoxazinone, commonly called Nile red [29], which, due to its planer structure and poor solubility, is known to form aggregates in an aqueous (and highly polar) medium [29]. The fluorescence spectral response of Nile red reveals unique solute–solute interactions when solubilized in (lanthanide salt : urea) DESs. Most interestingly, the absorbance and excitation spectra of Nile red are different from each other; the excitation maxima of the probe, particularly for (Ce : U) and (Gd : U) DESs, are significantly hypsochromically shifted compared to the absorbance maxima (Figure S4 and Table 4), clearly implying probe aggregation within these DESs. Furthermore, the spectral maxima strongly depend on the composition of a given DES system (highlighted in Figure 5) compared to (Gd : U :: 1 : 2), and the absorbance and excitation maxima of Nile red in a (Gd : U :: 1 : 7) are significantly blue-shifted (~28 and ~25 nm, respectively). Overall, the unusual behavior of Nile red within these metal DESs indicates the formation of H-aggregates. The hypothesis of the formation of H-aggregates in DESs is further supported by the absence of the characteristic sharp bathochromically shifted emission spectra for the J-aggregates, indicating the Nile red aggregates formed within these DESs to be non-emissive H-aggregates [30]. The presence of aggregates was further confirmed using resonance light scattering (RLS). It was observed that with the increase in the concentration of Nile red, the intensity of the band at *ca.* 642 nm is increased in (Ln : U :: 1 : 7) DESs (Figure S5). The hydrophobic nature of Nile red and its poor solubility in water combined with π-π stacking interactions of the planar structure contribute to its aggregation to form non-emissive H-aggregates [29,31]. We believe the exceptionally high polarity of these DESs leads to the aggregation of Nile red in lanthanide salt + urea DESs. Finally, it is noteworthy that the emission maxima of Nile red vary from 658 nm to 677 nm in these DESs (Table 4), which is considerably bathochromic compared to other DESs, ionic liquids, and polar solvents (emission maxima $\leq$ 580 nm) [28,32]. This enormous difference in emission maxima could be a combined effect of the very high polarity of the (lanthanide salt : urea) DESs (leading to lower energy emission bands) and the fact that emission originates from the aggregates of the probe.

Table 4. Absorbance (λ_{max}^{ab}, 10 ± 1 μM), excitation (λ_{max}^{ex}, 25 ± 2 μM), and emission (λ_{max}^{em}, 25 ± 2 μM) wavelength maxima of Nile red in (Ln : U)-based DESs at 298.15 K.

Ratio	La : U			Ce : U			Gd : U		
	λ_{max}^{ab}/nm	λ_{max}^{ex}/nm	λ_{max}^{em}/nm	λ_{max}^{ab}/nm	λ_{max}^{ex}/nm	λ_{max}^{em}/nm	λ_{max}^{ab}/nm	λ_{max}^{ex}/nm	λ_{max}^{em}/nm
1 : 7	616	617	672	617	590	658	618	607	670
1 : 6	620	618	672	622	592	661	621	608	670
1 : 5	623	619	674	625	602	663	628	612	671
1 : 4	632	627	674	628	607	663	630	618	672
1 : 3.5				634	614	664			
1 : 3	632	631	675				636	619	675
1 : 2							646	632	677

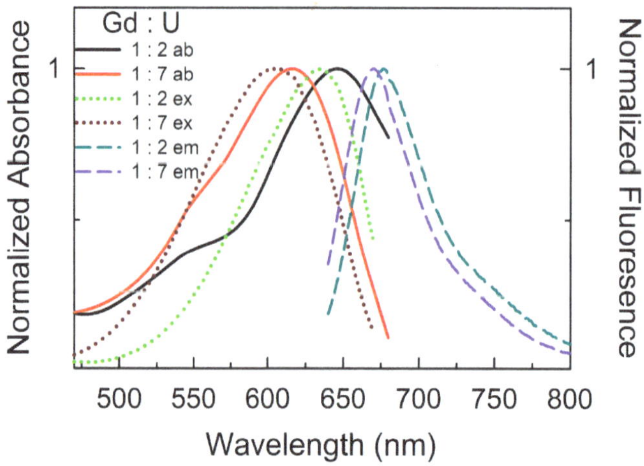

Figure 5. Representative absorbance (ab, 10 ± 1 μM), excitation (ex, 25 ± 2 μM), and emission (em, 25 ± 2 μM) spectra of Nile red dissolved in DESs (Gd : U) at 1 : 2 and 1 : 7 molar ratios.

Finally, we used a popular amphiphilic fluorescence probe 8-hydroxypyrene-1,3,6-trisulfonic acid (pyranine) [33], an arylsulfonate with an −OH group, and a planer aromatic ring that undergoes excited-state proton transfer (ESPT) depending on the environment of media [33]. This probe is highly sensitive to its cybotactic region, and depending on the polarity of the surrounding milieu, it can exist in two different forms, protonated (ROH) and deprotonated (RO^-) [34,35]. The emission spectra of pyranine in three (lanthanide salt : urea) DES systems are presented in Figure S6; representative spectra are shown in Figure 6. Based on the literature, the emission band with maxima appearing at 450 (±5) nm characterizing the excited protonated form (ROH*) is the dominant form when the relative amount of urea in the DES is smaller. However, increasing the amount of urea in our DES system results in the appearance of a shoulder at a long wavelength (*ca.* 515 nm), which is due to the emergence of the deprotonated form of pyranine (RO^{-*}). The ESPT to the DES at higher relative urea amounts is further supported by the excited-state intensity decay [I(t)] measurements. The I(t) acquired at both 450 nm and 515 nm in all the investigated (Ln : U) DESs and the associated global fit decay parameters are reported in Table S2. Importantly, it is observed that at low urea concentrations, the decay fits best to a single exponential function at both wavelengths, implying the presence of only the ROH*, whereas at higher urea concentrations, the decay starts to show a better fit to a double exponential decay function, with one of the recovered pre-exponential factors being negative, implying that the process takes place after excitation. It has been reported earlier that pyranine readily undergoes ESPT in a methanol–urea mixture; as the urea amount is increased, more and more RO^{-*} is formed [36]. The formation of UH^+ is proposed to facilitate the deprotonation of pyranine in its excited state [36]. We believe an increase in the HBA basicity (β) of the media with an increasing urea is responsible for the enhanced deprotonation of pyranine as the urea is increased in the DES system. Clearly, the complex interplay of high polarity and the HBA basicity of (lanthanide salt : urea) DES systems controls the photophysical behavior of the probe pyranine.

In order to obtain insights into the interactions between Ln salt and urea as the DES is formed in the absence of any external probe, the FTIR absorbance spectral signature of a representative DES (Gd : U :: 1 : 2) is compared with that of $Gd(NO_3)_3.6H_2O$ and urea, respectively (Figure 7 and Table S3 [37,38]). Significant shifts in the relevant band maxima of the Gd salt as well as urea upon the formation of DES corroborate the presence of strong interspecies interactions within the DES. These interactions appear to be responsible in affording the significantly high polarity to these DESs. In a seminal work, Edler's group reported very high surface tension and density along with low viscosity and glass transition temperatures associated with similar DESs [39]. The existence of strongly bonded but fluxional oligomeric polyanions and polycations was put forth. It was stated that the excess of the molecular component in the DESs resulted in an intercalating H-bonded nanostructure possessing water and urea. It may be proposed that such structuring may result in a highly polar cybotactic region as exhibited by the optical probes.

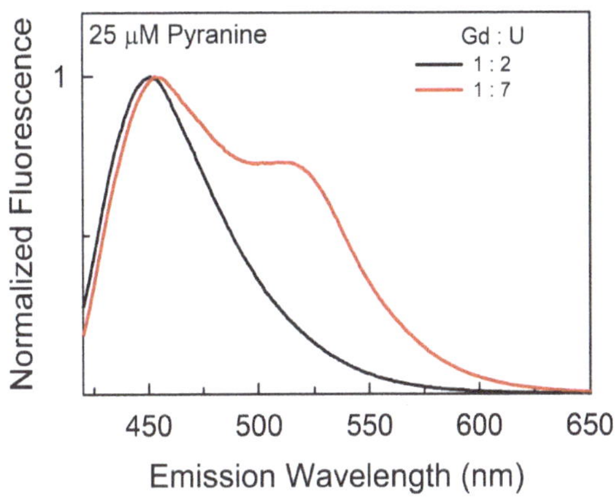

Figure 6. Representative emission spectra of 25 ± 2 μM pyranine ($\lambda_{ex} = 406$ nm) dissolved in DESs (Gd : U) at 1 : 2 and 1 : 7 molar ratios.

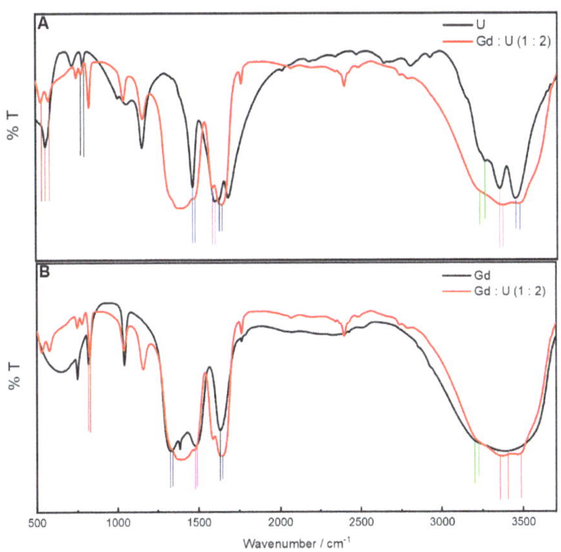

Figure 7. FTIR absorbance spectra of urea (U, panel **A**), $Gd(NO_3)_3.6H_2O$ (Gd, panel **B**), and DES (Gd : U :: at 1 : 2) (panels **A** and **B**), respectively, under ambient conditions. Vertical lines represent shift in maxima when going from neat constituent to DES.

4. Conclusions

We reported our findings that DESs composed of lanthanide salt + urea exhibit polarity (via empirical dipolarity/polarizability, π^*, and/or E_T^N parameters, emission maxima of PyCHO, aggregation by Nile red, and prototropism by pyranine) that is higher than that of any reported solvent system in the literature, especially for lanthanide salt-rich DES systems. The unusually high polarity of these DESs is attributed to lanthanide salt as well as the hydrated water of the salt. The polarity can be effectively tuned by varying the relative amount of lanthanide salt and urea within the DES. The hydrated water of the salt, obtained via controlling H-bonding, plays a decisive role in controlling the HBD acidity, α, and HBA basicity, β, as the composition of the DES, i.e., the lanthanide salt-to-urea ratio, is changed. These DES systems open new avenues of applications where high cybotactic region polarity is desired.

Supplementary Materials: The following supporting information can be downloaded at https://www.mdpi.com/article/10.3390/liquids4030028/s1, Figure S1: Absorbance spectra of 20 $\pm$ 2 μM DENA and 20 $\pm$ 2 μM NA dissolved in Ln:U-based DESs at room temperature; Figure S2: ^{13}C NMR spectra of 250 $\pm$ 5 mM PyO within La : U at 1 : 4 (Panel A), 1 : 5 (Panel B), and 1 : 6 (Panel C) molar ratios; Figure S3: Fluorescence emission spectra of 25 $\pm$ 2 μm PyCHO (λ_{ex} = 406 nm) dissolved in Ln : U-based DESs at room temperature; Figure S4: Absorbance spectra of 10 $\pm$ 1 μM Nile red and fluorescence excitation spectra of 25 $\pm$ 2 μM Nile red dissolved in Ce : U-based (panel A) and Gd : U-based (panel B) DESs at room temperature; Figure S5: RLS spectra of Nile red with variation in concentration dissolved in Ln : U-based (1 : 7) DESs at room temperature. Figure S6: Emission spectra of 25 $\pm$ 2 μM pyranine (λ_{ex} = 406 nm) dissolved in Ln : U-based DESs at room temperature; Table S1: Absorbance wavelength (λ^{ab}_{max}/nm) of 20 $\pm$ 2 μM DENA and 20 $\pm$ 2 μM NA dissolved in Ln : U-based deep eutectic solvent at room temperature. Table S2: Recovered excited-state intensity decay parameters for pyranine (λ_{em} = 450 nm, 515 nm; excitation with 404 nm violet laser diode) dissolved in Ln : U-based DES at room temperature by using global fit; Table S3: FTIR absorbance band maxima of urea, $Gd(NO_3)_3.6H_2O$, and DES (Gd : U) at 1 : 2 molar ratios.

Author Contributions: Conceptualization, Methodology, Visualization, Writing—Original Draft, and Data Collection, A.P.; Data Collection, V.K.; Writing—Review and Editing, Supervision, Resources, and Project Administration, S.P. All authors have read and agreed to the published version of the manuscript.

Funding: This work was generously funded by the Council of Scientific & Industrial Research, EMR-II (CSIR-EMR-II), GoI through a grant to Siddharth Pandey [grant number 01(3043)/21/ EMR-II]. The authors gratefully acknowledge the Science and Engineering Research Board (SERB), GoI for providing funding [grant no. CRG/2021/000602] for the UV-vis spectrophotometer. We would also like to thank the Central Research Facility at IIT Delhi for providing the NMR measurements.

Data Availability Statement: The data presented in this study are available in the Supplementary Materials.

Acknowledgments: Anushis Patra would like to acknowledge the Indian Institute of Technology, Delhi Government of India for his Senior Research Fellowship (SRF).

Conflicts of Interest: There are no conflicts of interest to declare.

References

1. Hansen, B.B.; Spittle, S.; Chen, B.; Poe, D.; Zhang, Y.; Klein, J.M.; Horton, A.; Adhikari, L.; Zelovich, T.; Doherty, B.W.; et al. Deep Eutectic Solvents: A Review of Fundamentals and Applications. *Chem. Rev.* **2021**, *121*, 1232–1285. [CrossRef] [PubMed]
2. Wagle, D.V.; Zhao, H.; Baker, G.A. Deep Eutectic Solvents: Sustainable Media for Nanoscale and Functional Materials. *Acc. Chem. Res.* **2014**, *47*, 2299–2308. [CrossRef] [PubMed]
3. Liu, Y.; Friesen, J.B.; McAlpine, J.B.; Lankin, D.C.; Chen, S.-N.; Pauli, G.F. Natural Deep Eutectic Solvents: Properties, Applications, and Perspectives. *J. Nat. Prod.* **2018**, *81*, 679–690. [CrossRef] [PubMed]
4. Mbous, Y.P.; Hayyan, M.; Hayyan, A.; Wong, W.F.; Hashim, M.A.; Looi, C.Y. Applications of Deep Eutectic Solvents in Biotechnology and Bioengineering—Promises and Challenges. *Biotechnol. Adv.* **2017**, *35*, 105–134. [CrossRef] [PubMed]
5. Zhang, Q.; De Oliveira Vigier, K.; Royer, S.; Jérôme, F. Deep Eutectic Solvents: Syntheses, Properties and Applications. *Chem. Soc. Rev.* **2012**, *41*, 7108–7146. [CrossRef] [PubMed]
6. Paiva, A.; Craveiro, R.; Aroso, I.; Martins, M.; Reis, R.L.; Duarte, A.R.C. Natural Deep Eutectic Solvents—Solvents for the 21st Century. *ACS Sustain. Chem. Eng.* **2014**, *2*, 1063–1071. [CrossRef]
7. Zainal-Abidin, M.H.; Hayyan, M.; Ngoh, G.C.; Wong, W.F. Doxorubicin Loading on Functional Graphene as a Promising Nanocarrier Using Ternary Deep Eutectic Solvent Systems. *ACS Omega* **2020**, *5*, 1656–1668. [CrossRef] [PubMed]

8. Ruesgas-Ramón, M.; Figueroa-Espinoza, M.C.; Durand, E. Application of Deep Eutectic Solvents (DES) for Phenolic Compounds Extraction: Overview, Challenges, and Opportunities. *J. Agric. Food Chem.* **2017**, *65*, 3591–3601. [CrossRef]
9. van Osch, D.J.G.P.; Dietz, C.H.J.T.; Warrag, S.E.E.; Kroon, M.C. The Curious Case of Hydrophobic Deep Eutectic Solvents: A Story on the Discovery, Design, and Applications. *ACS Sustain. Chem. Eng.* **2020**, *8*, 10591–10612. [CrossRef]
10. Tomé, L.C.; Mecerreyes, D. Emerging Ionic Soft Materials Based on Deep Eutectic Solvents. *J. Phys. Chem. B* **2020**, *124*, 8465–8478. [CrossRef]
11. Pandey, A.; Pandey, S. Solvatochromic Probe Behavior within Choline Chloride-Based Deep Eutectic Solvents: Effect of Temperature and Water. *J. Phys. Chem. B* **2014**, *118*, 14652–14661. [CrossRef] [PubMed]
12. Cao, J.; Su, E. Hydrophobic Deep Eutectic Solvents: The New Generation of Green Solvents for Diversified and Colorful Applications in Green Chemistry. *J. Clean. Prod.* **2021**, *314*, 127965. [CrossRef]
13. Farooq, M.Q.; Abbasi, N.M.; Anderson, J.L. Deep Eutectic Solvents in Separations: Methods of Preparation, Polarity, and Applications in Extractions and Capillary Electrochromatography. *J. Chromatogr. A* **2020**, *1633*, 461613. [CrossRef] [PubMed]
14. Khokhar, V.; Dhingra, D.; Pandey, S. Effect of Temperature and Composition on Density and Dynamic Viscosity of (Lanthanide Metal Salts + Urea) Deep Eutectic Solvents. *J. Mol. Liq.* **2022**, *360*, 119396. [CrossRef]
15. Khalid, A.; Tahir, S.; Khalid, A.R.; Hanif, M.A.; Abbas, Q.; Zahid, M. Breaking New Grounds: Metal Salts Based-Deep Eutectic Solvents and Their Applications- a Comprehensive Review. *Green Chem.* **2024**, *26*, 2421–2453. [CrossRef]
16. Xia, Q.; Liu, Y.; Meng, J.; Cheng, W.; Chen, W.; Liu, S.; Liu, Y.; Li, J.; Yu, H. Multiple Hydrogen Bond Coordination in Three-Constituent Deep Eutectic Solvents Enhances Lignin Fractionation from Biomass. *Green Chem.* **2018**, *20*, 2711–2721. [CrossRef]
17. Tong, Z.; Meng, J.; Liu, S.; Liu, Y.; Zeng, S.; Wang, L.; Xia, Q.; Yu, H. Room Temperature Dissolving Cellulose with a Metal Salt Hydrate-Based Deep Eutectic Solvent. *Carbohydr. Polym.* **2021**, *272*, 118473. [CrossRef] [PubMed]
18. Huo, D.; Sun, Y.; Yang, Q.; Zhang, F.; Fang, G.; Zhu, H.; Liu, Y. Selective Degradation of Hemicellulose and Lignin for Improving Enzymolysis Efficiency via Pretreatment Using Deep Eutectic Solvents. *Bioresour. Technol.* **2023**, *376*, 128937. [CrossRef] [PubMed]
19. Chen, H.; Sun, C.; Hu, Y.; Xia, C.; Sun, F.; Zhang, Z. Reaction Characteristics of Metal-Salt Coordinated Deep Eutectic Solvents during Lignocellulosic Pretreatment. *J. Environ. Chem. Eng.* **2023**, *11*, 109531. [CrossRef]
20. Khokhar, V.; Kumar, M.; Pandey, S. Pyrene Aggregation at Unprecedented Low Concentrations in (Lanthanide Metal Salt + Urea) Deep Eutectic Solvents. *Phys. Chem. Chem. Phys.* **2023**, *25*, 64–68. [CrossRef]
21. Kamlet, M.J.; Abboud, J.L.; Taft, R.W. The Solvatochromic Comparison Method. 6. The π^* Scale of Solvent Polarities. *J. Am. Chem. Soc.* **1977**, *99*, 6027–6038. [CrossRef]
22. Madeira, P.P.; Passos, H.; Gomes, J.; Coutinho, J.A.P.; Freire, M.G. Alternative Probe for the Determination of the Hydrogen-Bond Acidity of Ionic Liquids and Their Aqueous Solutions. *Phys. Chem. Chem. Phys.* **2017**, *19*, 11011–11016. [CrossRef] [PubMed]
23. Kamlet, M.J.; Taft, R.W. The Solvatochromic Comparison Method. I. The β-Scale of Solvent Hydrogen-Bond Acceptor (HBA) Basicities. *J. Am. Chem. Soc.* **1976**, *98*, 377–383. [CrossRef]
24. Schneider, H.D.; Badrieh, Y.; Migron, Y.; Marcus, Y. Hydrogen Bond Donation Properties of Organic Solvents and Their Aqueous Mixtures from 13C NMR Data of Pyridine-N-Oxide. *Z. Phys. Chem.* **1992**, *177*, 143–156. [CrossRef]
25. Wojeicchowski, J.P.; Abranches, D.O.; Ferreira, A.M.; Mafra, M.R.; Coutinho, J.A.P. Using COSMO-RS to Predict Solvatochromic Parameters for Deep Eutectic Solvents. *ACS Sustain. Chem. Eng.* **2021**, *9*, 10240–10249. [CrossRef]
26. Kalyanasundaram, K.; Thomas, J.K. Solvent-Dependent Fluorescence of Pyrene-3-carboxaldehyde and Its Applications in the Estimation of Polarity at Micelle-Water Interfaces. *J. Phys. Chem.* **1977**, *81*, 2176–2180. [CrossRef]
27. Street, K.W.; Acree, W.E. Estimation of the Effective Dielectric Constant of Cyclodextrin Cavities Based on the Fluorescence Properties of Pyrene-3-Carboxaldehyde. *Appl Spectrosc* **1988**, *42*, 1315–1318. [CrossRef]
28. Pandey, A.; Rai, R.; Pal, M.; Pandey, S. How Polar Are Choline Chloride-Based Deep Eutectic Solvents? *Phys. Chem. Chem. Phys.* **2014**, *16*, 1559–1568. [CrossRef]
29. Dutta, A.K.; Kamada, K.; Ohta, K. Spectroscopic Studies of Nile Red in Organic Solvents and Polymers. *J. Photochem. Photobiol. A Chem.* **1996**, *93*, 57–64. [CrossRef]
30. Kurniasih, I.N.; Liang, H.; Mohr, P.C.; Khot, G.; Rabe, J.P.; Mohr, A. Nile Red Dye in Aqueous Surfactant and Micellar Solution. *Langmuir* **2015**, *31*, 2639–2648. [CrossRef]
31. Mishra, A.; Behera, R.K.; Behera, P.K.; Mishra, B.K.; Behera, G.B. Cyanines during the 1990s: A Review. *Chem. Rev.* **2000**, *100*, 1973–2012. [CrossRef]
32. Greenspan, P.; Fowler, S.D. Spectrofluorometric Studies of the Lipid Probe, Nile Red. *J. Lipid Res.* **1985**, *26*, 781–789. [CrossRef]
33. Tolbert, L.M.; Solntsev, K.M. Excited-State Proton Transfer: From Constrained Systems to "Super" Photoacids to Superfast Proton Transfer. *Acc. Chem. Res.* **2002**, *35*, 19–27. [CrossRef]
34. Ireland, J.F.; Wyatt, P.A.H. Acid-Base Properties of Electronically Excited States of Organic Molecules. *Adv. Phys. Org. Chem.* **1976**, *12*, 131–221.
35. Förster, T. Primary Photophysical Processes. *Pure Appl. Chem.* **1973**, *34*, 225–234. [CrossRef]
36. Htun, T. Excited-State Proton Transfer in Nonaqueous Solvent. *J. Fluoresc.* **2003**, *13*, 323–329. [CrossRef]
37. Grdadolnik, J.; Maréchal, Y. Urea and Urea–Water Solutions—An Infrared Study. *J. Mol. Struct.* **2002**, *615*, 177–189. [CrossRef]

38. Nelson, D.L.; Irish, D.E. Interactions in Lanthanide Systems. I. A Raman and Infrared Study of Aqueous Gadolinium Nitrate. *J. Chem. Phys.* **1971**, *54*, 4479–4489. [CrossRef]
39. Hammond, O.S.; Bowron, D.T.; Edler, K.J. Structure and Properties of "Type IV" Lanthanide Nitrate Hydrate:Urea Deep Eutectic Solvents. *ACS Sustain. Chem. Eng.* **2019**, *7*, 4932–4940. [CrossRef]

liquids

Article

AbraLlama: Predicting Abraham Model Solute Descriptors and Modified Solvent Parameters Using Llama

Andrew S. I. D. Lang * and Youngmin Lee

Department of Computing & Mathematics, Oral Roberts University, Tulsa, OK 74171, USA; ttmn_ym@oru.edu
* Correspondence: alang@oru.edu

Abstract: This study explores the application of fine-tuned large language models for predicting physicochemical properties, specifically focusing on Abraham model solute descriptors (E, S, A, B, V) and modified solvent parameters (e_0, s_0, a_0, b_0, v_0). By leveraging ChemLLaMA, a specialized version of the LLaMA model for cheminformatics tasks, we developed the AbraLlama-Solvent and AbraLlama-Solute models using curated datasets of experimentally derived solute descriptors and solvent parameters. Our findings demonstrate that AbraLlama-Solvent and AbraLlama-Solute predict modified solvent parameters and solute descriptors with high accuracy, comparable to existing methods. The AbraLlama-Solvent model shows varying prediction accuracy across different solvents, influenced by their position within the chemical space, while the AbraLlama-Solute model consistently predicts solute descriptors with high accuracy. Both models are available as applications on Hugging Face, facilitating easy predictions from SMILES strings. This research highlights the potential of LLMs in chemistry applications, offering practical tools for solvent comparison and expanding the applicability of Abraham solvation equations to a broader range of organic solvents.

Keywords: Abraham general solvation model; Abraham solute descriptors; Abraham solvent parameters; solvent comparison; solvent replacement; ChemLLaMA

1. Introduction

The use of Artificial Intelligence (AI), including using fine-tuned large language models (LLMs), for predicting molecular properties has garnered significant attention in recent years [1,2]. LLMs are advanced AI systems that utilize deep learning techniques, specifically transformer architectures, to process and understand large amounts of textual data. These models are initially pre-trained on extensive corpora to capture general language patterns and semantic relationships, which are then fine-tuned on domain-specific datasets for specialized applications, such as cheminformatics; see the recent article by Luong and Singh for a review of how transformer models are being used in cheminformatics [3]. Building on this interest, our study leverages the capabilities of ChemLLaMA [4], a specialized version of the LLaMA (Large Language Model Meta AI) tailored for cheminformatics tasks, to predict Abraham model solute descriptors (E, S, A, B, V) and modified Abraham solvent parameters (e_0, s_0, a_0, b_0, v_0) [5]. This aligns with recent advancements in machine learning models for predicting solute parameters and solvation energies, which have shown promising results in both accuracy and applicability, such as the SoluteGC, SoluteML, and DirectML models [6].

The Abraham general solvation model is integral to predicting solubility and partition coefficients across a wide range of solvation systems. The model is based on linear free energy relationships (LFERs) between solute descriptors and solvent parameters [7–10]. This model is expressed as

$$\log P = c + e \cdot E + s \cdot S + a \cdot A + b \cdot B + v \cdot V, \quad (1)$$

where log P represents the solvent/water partition coefficient. Under typical conditions, this model can also predict the solubility of organic compounds in organic solvents [9] as follows:

$$\log S_s = \log S_w + c + e \cdot E + s \cdot S + a \cdot A + b \cdot B + v \cdot V, \quad (2)$$

where S_s is the molar concentration of the solute in the organic solvent, and S_w is the molar concentration of the solute in water. The parameters c, e, s, a, and b are the standard Abraham solvent parameters, and E, S, A, B, and V are the Abraham solute descriptors: E is the solute excess molar refractivity in units of $(cm^3/mol)/10$, S is the solute dipolarity/polarizability, A and B are the overall or summation hydrogen bond acidity and basicity, and V is the McGowan characteristic volume in units of $(cm^3/mol)/100$.

To determine the Abraham solvent parameters, linear regression is typically performed on the partition coefficients and solubilities of solutes with known Abraham descriptors, which have been experimentally measured in the solvent. The intercept c (see Equation (1)) is traditionally allowed to float, ideally capturing information not characterized by the other solvent–solute interaction terms. However, this approach complicates direct comparisons of different solvents. To address this, we utilized modified Abraham solvent parameters (e_0, s_0, a_0, b_0, v_0), derived by regressing with the intercept set to zero, as described by Bradley et al. [5], to facilitate more straightforward comparisons.

Identifying alternative solvents for various processes is essential due to the significant environmental and health risks associated with some traditional solvents. Increased regulatory demands; the push for sustainability; reactivity; volatility; and cost considerations are also critical factors in solvent selection. While several solvent replacement tools and lists exist [11], this work aims to provide an open tool for comparing solvents based on their modified Abraham solvent parameters. Solvents with closely matching parameters are likely to exhibit similar solvation properties [5].

Thus, the objectives of this study are twofold: firstly, to demonstrate how large language models (LLMs) can predict physicochemical properties, specifically, modified Abraham solvent parameters and Abraham solute descriptors, and secondly, to provide open models (AbraLlama-Solvent and AbraLlama-Solute) that facilitate the prediction of these parameters and descriptors. This extends the applicability of the Abraham solvation equations to a broader range of organic solvents and aids in identifying solvents with comparable solvation properties.

Our findings demonstrate that the models predict modified Abraham solvent parameters and Abraham solute descriptors with accuracy comparable to existing methods. Furthermore, by making these models available as apps on Hugging Face, a platform that provides open-source tools and models for the easy deployment and sharing of machine learning models, we provide a valuable service to the community, enabling easy and accessible calculations without requiring users to have expertise in AI model deployment [12]. This research highlights the significant potential of LLMs in chemistry applications and offers a practical tool for solvent comparison.

2. Materials and Methods

2.1. Datasets

Experimentally determined Abraham solute descriptors (E, S, A, B, V) were obtained from the UFZ-LSER database (version 3.2.1) using the search parameter "%C%" and with the radio button "experimental descriptors" selected [13]. Standard Abraham solvent parameters (c, e, s, a, b, v) were provided by Dr. William E. Acree Jr., who compiled them from the literature. We have made this solvent dataset, which contains both Abraham model log P and log K (not used in this work) equation coefficients for both pure solvents and mixtures, available under a CC0 license on Figshare [14], where the subscripts 3 and 4 in the dataset refer to equations involving log P and log K respectively [15].

2.2. Data Preprocessing

The Abraham solute descriptor dataset contained multiple records for many compounds. To ensure consistency, we filtered the data and kept only entries labeled as "LSER Dataset for CompTox users (2017)". This resulted in a final dataset of $N = 6852$ compounds with experimentally derived Abraham solute descriptors. We excluded mixtures from the Abraham solvent parameter dataset and kept only pure solvents ($N = 122$) with experimentally derived solvent parameters. Since the c-parameter can complicate direct comparisons between solvents, we calculated modified Abraham solvent parameters (e_0, s_0, a_0, b_0, v_0) for all solvents using the method described by Bradley et al. [5]. This process involved calculating log P values for all solutes in each solvent using Equation (1). We then determined the modified Abraham solvent parameters by regressing these log P values with a linear equation with zero intercept: $\log P = e_0 \cdot E + s_0 \cdot S + a_0 \cdot A + b_0 \cdot B + v_0 \cdot V$. Both the original and modified Abraham solvent parameters for the modeling dataset are available under a CC0 license from Figshare [16].

2.3. Model Development

We leveraged ChemLLaMA, a fine-tuned version of the LLaMA transformer model, which was developed to predict molecular properties from SMILES strings [4]. This study utilized the 30-million-parameter ChemLLaMA model [17] and further fine-tuned it to predict both Abraham solute descriptors and modified Abraham solvent parameters. Note that we did not freeze the pre-trained model for fine-tuning for this research. We refer the reader to the literature for a more detailed description of ChemLLaMA, including the source code [4,17].

To create the AbraLlama-Solvent and AbraLlama-Solute models presented here, we fine-tuned ChemLLaMA using the curated datasets mentioned above. Training occurred on a single GPU (NVIDIA A30) using PyTorch, specifically PyTorch-Lightning (v. 1.9.5) [18] and Lightning-Bolts (v. 0.7.0) [19]. Both models underwent 20 epochs of training with a learning rate of 0.0001 with a 'Linear Warmup Cosine Annealing Learning Rate' scheduler which reaches the peak learning rate at the second epoch. The training setup for the AbraLlama-Solute model used 5-fold cross-validation and a batch size pair of (64, 64). Due to the smaller dataset size, 10-fold cross-validation and a batch size pair of (10, 10) were employed for training the AbraLlama-Solvent model. Cross-validation allowed us to report test-set-equivalent statistics.

Both models were retrained on the complete, unpartitioned datasets to maximize predictive accuracy and overall utility. These AbraLlama models are available as applications on Hugging Face, allowing users to predict modified solvent parameters and solute descriptors from SMILES string inputs [12].

Our methodology involved pre-training transformer models on extensive datasets, followed by fine-tuning them on task-specific datasets, ensuring consistent training environments for reliable comparisons. The key steps included using uniform tokenizers, standardized training data, and identical training configurations across models. Advanced optimization techniques and learning rate schedulers were employed to ensure effective model convergence. Additionally, rigorous cross-validation and repeated fine-tuning iterations were used to mitigate statistical anomalies and enhance robustness. This structured approach, which encompasses dataset preparation, model pre-training, fine-tuning, and thorough evaluation, provides a replicable framework adaptable for model comparison and fine-tuning in various domains. For more details, refer to the ChemLLaMA paper [4] and the Python source code for ChemLLaMA [17] and AbraLlama-Solvent and AbraLlama-Solute available on GitHub [20].

3. Results and Discussion

The ChemLLaMA model [4], originally fine-tuned from the LLaMA large language model (LLM) for cheminformatics applications, was further refined to develop models capable of predicting Abraham model solute descriptors (E, S, A, B, V) and modified

Abraham solvent parameters (e_0, s_0, a_0, b_0, v_0). The performance of these fine-tuned ChemLLaMA models, designated as AbraLlama-Solvent and AbraLlama-Solute, was evaluated using cross-validation. Table 1 presents the aggregated cross-validated statistics from all folds, demonstrating the models' ability to predict new molecules with accuracy comparable to other established methods [5,13].

Table 1. Cross-validated statistics for the AbraLlama-Solvent and AbraLlama-Solute models.

Modified Solvent Parameters				Solute Descriptors			
Parameter	*N*	RMSE	R^2	Descriptor	*N*	RMSE	R^2
e_0	122	0.163	0.32	E	6852	0.132	0.97
s_0	122	0.353	0.66	S	6852	0.240	0.90
a_0	122	0.655	0.81	A	6852	0.135	0.85
b_0	122	0.480	0.40	B	6852	0.123	0.96
v_0	122	0.318	0.49	V	6852	0.097	0.98

3.1. *AbraLlama-Solvent*

The AbraLlama-Solvent model predicts modified Abraham solvent parameters (e_0, s_0, a_0, b_0, v_0) with varying degrees of accuracy depending on the solvent's position within the chemical space, as determined using principal component analysis (PCA). The region around the origin—the center of the chemical space—is populated by acyclic alcohols, esters, and ethers consisting solely of carbon, oxygen, and hydrogen atoms (C, O, H). In contrast, the outer regions include compounds with rings and those containing nitrogen, sulfur, and fluorine (N, S, F).

The prediction error (the Euclidean distance from the predicted to measured values) is correlated with the distance from the center of the chemical space ($p < 0.0001$), as shown in Figure 1. Four clear outliers—carbon disulfide, trifluoroethanol, pyridine, and dimethyl sulfoxide—exhibit significant deviations from the model's predictions. When these outliers are excluded, the model's cross-validated statistics improve markedly: e_0 (RMSE = 0.154, R^2 = 0.364), s_0 (RMSE = 0.330, R^2 = 0.698), a_0 (RMSE = 0.512, R^2 = 0.883), b_0 (RMSE = 0.404, R^2 = 0.458), v_0 (RMSE = 0.289, R^2 = 0.546).

3.2. *AbraLlama-Solute*

The AbraLlama-Solute model predicts Abraham solvent descriptors (E, S, A, B, V) with high accuracy, achieving R^2-values over 0.9 for all descriptors except A, which has a cross-validated R^2-value of 0.85 (see Table 1). To investigate outliers, we compare the predicted log P values using the 1-octanol solvent parameters (c = 0.088, e = 0.562, s = −1.054, a = 0.034, b = −3.460, v = 3.814) calculated from the measured and predicted solute descriptors. The high density of points along the 45-degree line (see Figure 2) demonstrates the effectiveness of the solute descriptor prediction models and helps identify outliers.

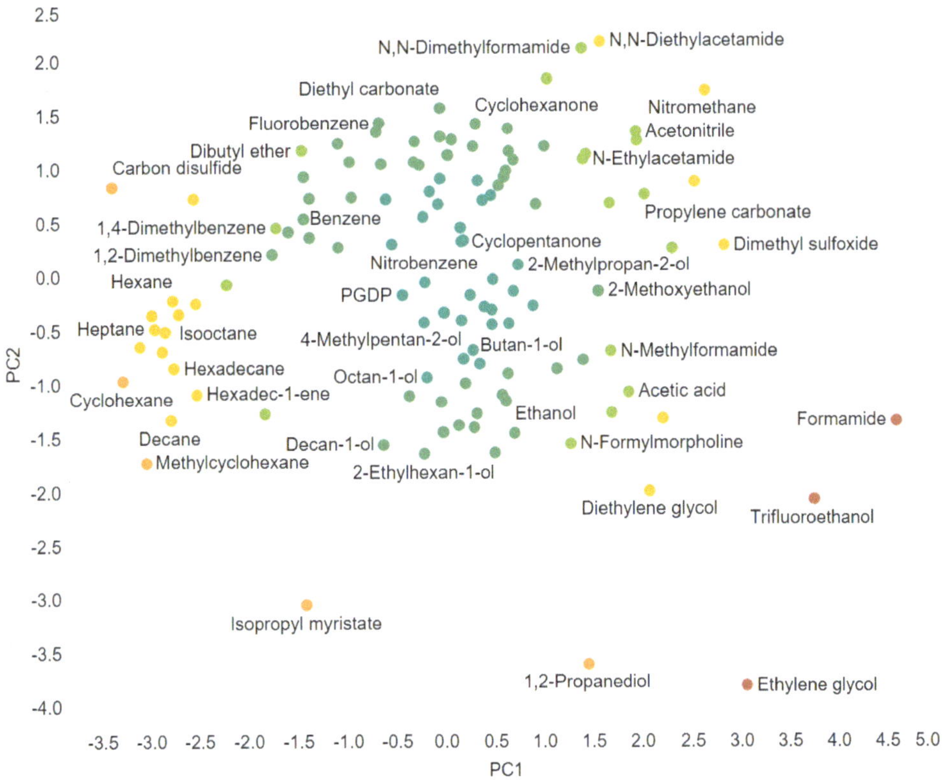

Figure 1. Solvent chemical space (PCA) colored by prediction error: green indicates small errors (Euclidean distance < 1), yellow-orange indicates medium errors ($1 \leq$ Euclidean distance ≤ 1.45), and red indicates large errors (Euclidean distance > 1.45).

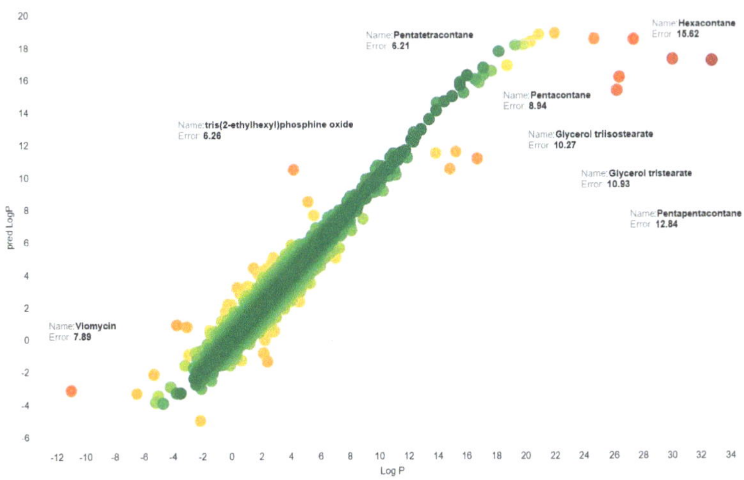

Figure 2. Calculated log P values ($N = 6852$) using the solvent parameters for 1-octanol, comparing measured vs. predicted Abraham solute descriptors. The data points are colored by absolute error.

4. Conclusions

This study successfully demonstrates the application of fine-tuned large language models (LLMs) in predicting molecular properties, specifically the prediction of Abraham model solute descriptors (E, S, A, B, V) and modified Abraham solvent parameters (e_0, s_0, a_0, b_0, v_0). By leveraging the capabilities of ChemLLaMA [4], we developed the AbraLlama-Solvent and AbraLlama-Solute models, which predict these critical parameters with accuracy consistent with previous studies [5,13].

The AbraLlama-Solvent model predicts modified solvent parameters across a wide range of solvents, facilitating the comparison of solvents with similar solvation properties. The PCA highlighted that prediction errors correlate with the distance from the center of the chemical space, indicating a need for further refinement in regions containing complex solvent structures and for the considered use of Hugging Face apps [12]. Similarly, the AbraLlama-Solute model achieves high accuracy in predicting Abraham solute descriptors, demonstrating its robustness and reliability.

Future work will focus on expanding the solvent dataset to include a broader range of solvents, enhancing the model's accuracy in predicting solvent parameters, and developing a user-friendly solvent selection tool.

Author Contributions: Conceptualization, A.S.I.D.L.; methodology, A.S.I.D.L. and Y.L.; Hugging Face apps, Y.L.; data curation, A.S.I.D.L.; writing, A.S.I.D.L. and Y.L. All authors have read and agreed to the published version of the manuscript.

Funding: This research received no external funding.

Data Availability Statement: Solute data are available for download online [13]. Solvent data are available under a CC0 license from Figshare [14,16].

Acknowledgments: We would like to extend our sincere gratitude to William E. Acree, Jr. for generously providing the solvent data used in this analysis and for his many years of invaluable collaboration and support. His contributions have been instrumental in advancing our research and in fostering an open and collaborative scientific community.

Conflicts of Interest: The authors declare no conflicts of interest.

References

1. Deng, J.; Yang, Z.; Wang, H.; Ojima, I.; Samaras, D.; Wang, F. A systematic study of key elements underlying molecular property prediction. *Nat. Commun.* **2023**, *14*, 6395. [CrossRef] [PubMed]
2. Lang, A.S.I.D.; Chong, W.K.; Wörner, J.H. Fine-Tuning ChemBERTa-2 for Aqueous Solubility Prediction. *Ann. Chem. Sci. Res.* **2023**, *4*, 1–3. [CrossRef]
3. Luong, K.-D.; Singh, A. Application of Transformers in Cheminformatics. *J. Chem. Inf. Model.* **2024**, *64*, 4392–4409. [CrossRef] [PubMed]
4. Lee, Y.; Lang, A.S.I.D.; Cai, D.; Wheat, S.R. The Role of Model Architecture and Scale in Predicting Molecular Properties: Insights from Fine-Tuning RoBERTa, BART, and LLaMA. *arXiv* **2024**. [CrossRef]
5. Bradley, J.-C.; Abraham, M.H.; Acree, W.E.; Lang, A.S. Predicting Abraham model solvent coefficients. *Chem. Cent. J.* **2015**, *9*, 12. [CrossRef] [PubMed]
6. Chung, Y.; Vermeire, F.H.; Wu, H.; Walker, P.J.; Abraham, M.H.; Green, W.H. Group Contribution and Machine Learning Approaches to Predict Abraham Solute Parameters, Solvation Free Energy, and Solvation Enthalpy. *J. Chem. Inf. Model.* **2022**, *62*, 433–446. [CrossRef] [PubMed]
7. Abraham, M.H.; Zissimos, A.M.; Acree, W.E. Partition of solutes into wet and dry ethers; an LFER analysis. *New J. Chem.* **2003**, *27*, 1041–1044. [CrossRef]
8. Abraham, M.H.; Acree, W.E. Comparison of solubility of gases and vapours in wet and dry alcohols, especially octan-1-ol. *J. Phys. Org. Chem.* **2008**, *21*, 823–832. [CrossRef]
9. Abraham, M.H.; Smith, R.E.; Luchtefeld, R.; Boorem, A.J.; Luo, R.; Acree, W.E. Prediction of solubility of drugs and other compounds in organic solvents. *J. Pharm. Sci.* **2010**, *99*, 1500–1515. [CrossRef] [PubMed]
10. Jouyban, A.; Acree, W.E., Jr. Michael H. Abraham and his developed parameters: Various applications in medicine, chemistry and biology. *Pharm. Sci.* **2022**, *28*, 170–173. [CrossRef]
11. Lee, J.L.; Chong, G.H.; Ota, M.; Guo, H.; Smith, R.L. Solvent Replacement Strategies for Processing Pharmaceuticals and Bio-Related Compounds—A Review. *Liquids* **2024**, *4*, 352–381. [CrossRef]

12. Lang, A.S.I.D.; Lee, Y. AbraLlama Hugging Face App: Predicting Abraham Model Solute Descriptors and Modified Solvent Parameters Using Llama. Hugging Face. 2024. Available online: https://huggingface.co/spaces/ttmn/AbraLlama (accessed on 24 May 2024).
13. Ulrich, N.; Endo, S.; Brown, T.N.; Watanabe, N.; Bronner, G.; Abraham, M.H.; Goss, K.-U. *UFZ-LSER Database v 3.2.1*; Helmholtz Centre for Environmental Research-UFZ: Leipzig, Germany, 2017; Available online: http://www.ufz.de/lserd (accessed on 24 May 2024).
14. Acree, W.E., Jr.; Land, A.S.I.D.; Lee, Y. Dataset: Abraham model Log P and Log K equation coefficients. *Figshare* **2024**. [CrossRef]
15. Sinha, S.; Yang, C.; Wu, E.; Acree, W.E., Jr. Abraham Solvation Parameter Model: Examination of Possible Intramolecular Hydrogen-Bonding Using Calculated Solute Descriptors. *Liquids* **2022**, *2*, 131–146. [CrossRef]
16. Lang, A.S.I.D.; Lee, Y. Dataset: AbraLlama: Predicting Abraham Model Solute Descriptors and Modified Solvent Parameters Using Llama. *Figshare* **2024**. [CrossRef]
17. Lee, Y.; Lang, A.S.I.D.; Cai, D.; Wheat, S.R. Transformers and Chemistry. Available online: https://github.com/BrightBlueCheese/transformers_and_chemistry (accessed on 24 May 2024).
18. Falcon, W. The PyTorch Lightning Team. PyTorch Lightning (Version 1.9.5). 2024. Available online: https://github.com/Lightning-AI/pytorch-lightning/ (accessed on 24 May 2024).
19. The PyTorch Lightning Bolts Team. PyTorch Lightning Bolts (Version 0.7.0). 2024. Available online: https://github.com/Lightning-Universe/lightning-bolts (accessed on 24 May 2024).
20. Lee, Y.; Lang, A.S.I.D. AbraLLaMA Source Code. Available online: https://github.com/BrightBlueCheese/AbraLLaMA (accessed on 24 May 2024).

liquids

MDPI

Article

Solvation Enthalpies and Free Energies for Organic Solvents through a Dense Neural Network: A Generalized-Born Approach †

Sergei F. Vyboishchikov

Institut de Química Computacional i Catàlisi and Departament de Química, Universitat de Girona, Carrer Maria Aurèlia Capmany 69, 17003 Girona, Spain; vyboishchikov@googlemail.com

† Dedicated to Professor William E. Acree, Jr.

Abstract: A dense artificial neural network, ESE-ΔH-DNN, with two hidden layers for calculating both solvation free energies ΔG°_{solv} and enthalpies ΔH°_{solv} for neutral solutes in organic solvents is proposed. The input features are generalized-Born-type monatomic and pair electrostatic terms, the molecular volume, and atomic surface areas of the solute, as well as five easily available properties of the solvent. ESE-ΔH-DNN is quite accurate for ΔG°_{solv}, with an RMSE (root mean square error) below 0.6 kcal/mol and an MAE (mean absolute error) well below 0.4 kcal/mol. It performs particularly well for alkane, aromatic, ester, and ketone solvents. ESE-ΔH-DNN also exhibits a fairly good accuracy for ΔH°_{solv} prediction, with an RMSE below 1 kcal/mol and an MAE of about 0.6 kcal/mol.

Keywords: solvation; solvation free energy; solvation enthalpy; artificial neural networks; generalized-Born method

1. Introduction

Solvation is a major effect in chemistry and has to be accounted for in chemical computations. Typically, in high demand is the solvation *free energy* ΔG°_{solv}, since it makes a significant contribution to the total free energy of chemical reactions in solution and is also crucially important for the estimation of partition coefficients. Nevertheless, solvation *enthalpy* ΔH°_{solv} also plays a significant role, as it gives a direct measure of the heat of solvation and provides, in conjunction with ΔG°_{solv}, access to the solvation entropy ΔS°_{solv}. Hence, the development of reliable and efficient computational tools for prediction ΔG°_{solv} and ΔH°_{solv} is an important task in computational chemistry.

While highly accurate explicit methods for solvation energy evaluation, such as umbrella sampling [1,2] and free energy perturbation [3], are available, they are quite costly. Consequently, most practical calculations of solvation energy rely on the continuum solvation (CS) model, in which the electrostatic E_{elst} and nonelectrostatic contributions to ΔG°_{solv} are evaluated separately. The two most prominent CS approaches are the polarizable continuum model (PCM) [4–15] and the generalized Born (GB) method [16,17], including SMx [18–21]. In the PCM family of methods, the solute placed in a cavity interacts with the solvent, represented by a continuum. The cavity surface needs to be constructed and induced charges placed on it have to be computed.

In contrast, in the GB method, E_{elst} is calculated directly from atomic charges $\{Q_I\}$:

$$E_{elst}^{GB} = -\frac{1}{2}\sum_I E_I^{self} - \sum_{I<J} E_{IJ}^{pair} = -\frac{1}{2}\left(1-\frac{1}{\varepsilon}\right)\sum_I \frac{Q_I^2}{R_I} - \left(1-\frac{1}{\varepsilon}\right)\sum_{I<J}\frac{Q_I Q_J}{f_{IJ}} \quad (1)$$

where ε is the dielectric constant of the solvent and f_{IJ} is a function of the atomic radii R_I (in this context referred to as the Born radii) and the interatomic distance r_{IJ}. The monatomic terms (self-terms) $E_I^{self} = (1 - 1/\varepsilon)Q_I^2/R_I$ in Equation (1) correspond to the solvation energy in the Born theory [22] for spherical ions. The function f_{IJ} in the pair term E_{IJ}^{pair} = (1

$- 1/\varepsilon)Q_IQ_J/f_{IJ}$ is designed to interpolate between the Coulombic limit $(1 - 1/\varepsilon)Q_IQ_J/r_{IJ}$ at large r_{IJ} and the Born or Onsager limits at small r_{IJ}, when the diatomic fragment IJ collapses either into an single ion of charge $Q_I + Q_J$ or into a neutral.

The choice of effective Born radii R_I in the traditional GB methods is crucial for achieving an acceptable accuracy [23]. A typical form of f_{IJ} is the following [16]:

$$f_{IJ} = \sqrt{r_{IJ}^2 + R_I R_J \exp\left(-r_{IJ}^2/(4R_I R_J)\right)}, \tag{2}$$

although alternative expressions for f_{IJ} are also in use [18,23–26]. The GB approach has been implemented within the framework of numerous solvation energy schemes [17–19,21]. The GB-type methods are generally more efficient than the PCM-type schemes, since they avoid an explicit cavity construction and induced-charge calculation. However, the GB methods are in general less accurate than PCM.

Nonelectrostatic correction is necessary for both PCM- and GB-type methods in order to achieve reasonably accurate results. Typically, it is made of linear or nonlinear terms depending on atomic surfaces S_I, with the simplest form of the correction being $\sum_I \kappa_I S_I$ [16]. In addition, the correction term often involves the molecular volume, induced charges, and possibly other characteristics of the solute and typically contains some adjustable parameters that must be fitted on a suitable database. Thus, the CS methods inevitably have some degree of empiricity.

In our previous works [27–32], we developed a family of non-iterative methods for evaluating $\Delta G^\circ_{\mathrm{solv}}$ (uESE—*universal Easy Solvation Energy*). The electrostatics is included by means of COSMO (*COnductor-like Screening MOdel*) [14,15], while the correction term depends on induced charges and surfaces. The atomic charges necessary for the COSMO calculation can be evaluated by means of various techniques [27,33–36], including ab initio or DFT-based, semiempirical, and even electronegativity-equalization (EE) [31] charges. However, changing the charge scheme requires re-parameterization of the correction term. Although the semiempirical and EE charges provide $\Delta G^\circ_{\mathrm{solv}}$ estimates of reasonable quality, the accuracy of the EE-based method (ESE-EE) [31] is not as high as that of the uESE method [29] employing DFT charges. In order to improve the quality of an EE-based scheme, artificial neural networks (ANN) were introduced as a computational framework [37,38]. In the ESE-EE-DNN method [37,39], the ANN input features are the COSMO electrostatic energy, atomic cavity surface areas, total cavity volume, and induced surface charges. In contrast, the ESE-GB-DNN method [38,40] is based on GB-style input features. In particular, the number of atoms in the solute molecule, the total charge, the molecular volume, the atomic surface areas as well as GB-type self-terms and pair terms ($E_I{}^{\mathrm{self}}$ and $E_{IJ}{}^{\mathrm{pair}}$ summed over the elements) are used to represent the solute, while the solvent is described by just three parameters: the dielectric constant ε, the boiling point (BP), and the number of non-hydrogen atoms. The full list of solvents and their properties is given in the Supporting Information (Table S1). Since no cavity construction is needed, the ESE-GB-DNN method is substantially faster than ESE-EE-DNN (and dramatically more efficient than any DFT-based approach), allowing for a virtually instantaneous $\Delta G^\circ_{\mathrm{solv}}$ calculation for solutes up to 100 atoms in size.

In the present paper, I further develop the GB-based approach proposed in [38]. The purpose of the present work is to create a consistent ANN-based predictor of both $\Delta G^\circ_{\mathrm{solv}}$ and $\Delta H^\circ_{\mathrm{solv}}$ at 298 K for organic solvents.

Neural networks for $\Delta G^\circ_{\mathrm{solv}}$ evaluation were developed previously. Chen et al. [41] presented a graph ANN with atomistic embedding. Vermeire and Green's approach [42] employed a directed message-passing ANN with SMILES (*Simplified Molecular-Input Line-Entry System*) and InChI (*International Chemical Identifier*) as input features. Another successful, but sophisticated, message-passing ANN for neutral solutes was presented by Low et al. [43]. Lim and Jung [44] developed a recurrent ANN and a graph convolutional ANN based on atomic vectors. Alibakhshi and Hartke [45] achieved quite accurate results within their ANN that uses a self-consistent C-PCM input. The works by Bernazzani et al. [46],

Hutchinson and Kobayashi [47], Wang et al. [48], Jaquis et al. [49], and Chung et al. [50] are also worth mentioning. The latter two address approaches that target solvation enthalpy in addition to Gibbs free energy.

2. Methods

The atomic charges used for the calculations are computed using an EE-based scheme described in detail in my previous paper [38]. This is similar but not identical to the version by Svobodová Vařeková et al. [51].

The original input features used to describe a solute closely follow the approach used in my previous paper [38]. They are as follows:

- (1) the number of atoms in the solute molecule N;
- (2) the molecular volume V_{tot}: $V_{tot}= \sum_I V_I$;
- (3) the total surface area S_{tot} composed of atomic surfaces: $S_{tot}= \sum_I S_I$;
- (4–12) atomic surface areas summed over all the atoms of a given element L: $S_L = \sum_{I \epsilon L} S_I$ for L = H, C, N, O, F, S, Cl, Br, and I. The atomic volumes V_I and surfaces S_I are efficiently calculated by simple formulas based on geometric considerations. The details are given in [38];
- (13–21) the Born-type self-terms, also summed over all the atoms of a given element L:

$$E_1{}^{\text{Born}}(L) = \sum_{I \epsilon L} E_I{}^{\text{self}} = (1 - 1/\varepsilon) \sum_{I \epsilon L} Q_I{}^2/R_I \tag{3}$$

for L = H, C, N, O, F, S, Cl, Br, I calculated from the EE charges Q_I;

- (22–51) the Born-type pair terms:

$$E_2{}^{\text{Born}}(L_1,L_2) = \sum_{I \epsilon L_1} \sum_{J \epsilon L_2} E_{IJ}{}^{\text{pair}} = (1 - 1/\varepsilon) \sum_{I \epsilon L_1} \sum_{J \epsilon L_2} Q_I Q_J / f_{IJ} \tag{4}$$

The thirty L_1–L_2 pair terms are as follows: H–H, C–C, C–H, N–N, N–H, N–C, O–O, O–H, O–C, O–N, F–F, F–H, F–C, F–O, S–H, S–C, S–N, S–O, Cl–H, Cl–C, Cl–N, Cl–O, Cl–F, Br–H, Br–C, Br–N, Br–O, Br–Cl, I–H, and I–C.

The radii R_I used in Equations (2)–(4) are unmodified Bondi [52] radii;

- (52–56) five solvent-related input features: in addition to the dielectric constant, boiling point, and the number of nonhydrogen atoms employed in my previous work [38], in this paper, the molar volume and the number of hydrogen-bond centers (the sum of the donor and acceptor centers) are also used.

Thus, the initial input feature set consistes of 56 parameters. Some of them exhibit a strong correlation on the training dataset used (vide infra). For instance, V_{tot} and S_{tot} have a correlation coefficient of 0.99; S_{tot} and N have a correlation coefficient of 0.97; $E_1{}^{\text{Born}}$(Cl) and $E_2{}^{\text{Born}}$(Cl,H) have a correlation coefficient of −0.99, etc. In total, 16 pairs of features with correlation coefficients greater than 0.96 were identified. To reduce the number of ANN parameters, principal component analysis using the sklearn.decomposition.PCA class from the Python Scikit-learn package [53] was applied. This process allowed truncating the 16 most correlated features, resulting in a 56 × 40 transformation matrix. Consequently, a vector of 40 input features is produced and fed into the ANN.

I employed a dense ANN with 40 input neurons and 2 hidden layers with 14 and 6 neurons, respectively, and an output layer with two neurons (corresponding to $\Delta G^\circ_{\text{solv}}$ and $\Delta H^\circ_{\text{solv}}$). The *ReLU* (*Rectified Linear Unit*) activation function for the hidden layers and the linear activation for the output layer were used. Other ANN configurations were also tested, including an ANN with a single hidden layer, but the abovementioned 40 × 14 × 6 × 2 network (Figure 1) turned out to be the most accurate. It contains 678 adjustable parameters (40 × 14 + 14 × 6 + 6 × 2 = 656 weights plus 14 + 6 + 2 = 22 biases). The input data are min-max scaled and fed into the dense ANN described above. The ANN fitting was performed on a suitable database (vide infra) using the Nesterov-accelerated [54] Adaptive Moment Estimation algorithm [55] as implemented in the tensorflow.keras.optimizers.Nadam class [56], with mean squared error as the loss function and L_2 regularization with a strength λ = 0.01.

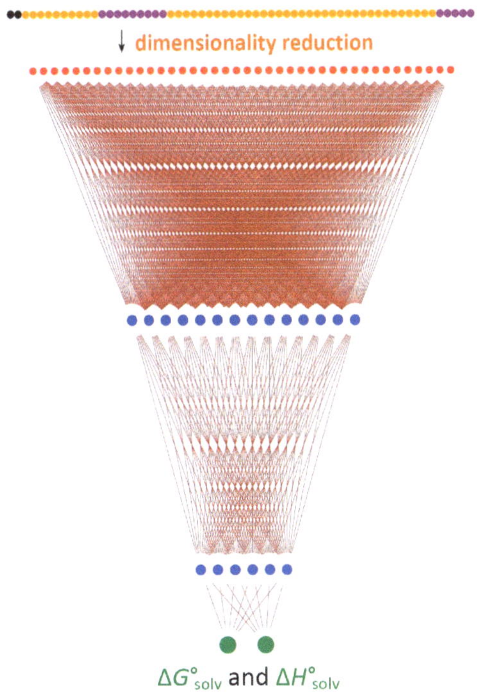

Figure 1. DNN architecture used in the present work. The first row (various colors) denotes the original 56 input features. The dimensionality reduction is achieved via a 56 × 40 linear transformation. The second row (40 red circles) represents the DNN input layer (40 linear combinations of the 56 initial features). The following blue circles denote two hidden layers (14 and 6 neurons, respectively). The green circles at the bottom are neurons in the output layer, corresponding to $\Delta G^{\circ}{}_{\text{solv}}$ and $\Delta H^{\circ}{}_{\text{solv}}$.

3. Results and Discussion

3.1. Database

The datasets used for the training and testing of ESE-ΔH-DNN are based on the database of $\Delta H^{\circ}{}_{\text{solv}}$ and partition coefficients at T = 298 K by Prof. W. E. Acree [57,58]. The logarithms of partition coefficients $\log_{10}p$ were converted to solvation free energies: $\Delta G^{\circ}{}_{\text{solv}} = -RT\ln(10)\log_{10}p$. Subsequently, chemical names of the solutes were automatically translated into SMILES codes using the website of the National Institutes of Health [59]. The resulting SMILES codes were then automatically converted to Cartesian coordinates, and the geometries were optimized with MMFF94 (*Merck Molecular Force Field*) [60] using the rdkit.Chem.AllChem module [61]. Subsequently, the geometries were manually controlled and corrected when necessary. Atomic charges, volumes, and surfaces were computed based on these geometries. The solvent parameters (boiling points, dielectric constants, molar volume, and the number of hydrogen-bond centers) were retrieved semi-automatically from other public databases [62,63]. Finally, the resulting database contained 5201 $\Delta H^{\circ}{}_{\text{solv}}$ values and 3789 $\Delta G^{\circ}{}_{\text{solv}}$ values. It was randomly divided into a training/validation set (80%) and testing set (20%). The training/validation set was further split into a training set and a validation set, with 20% assigned for validation. Once trained, the optimized ESE-ΔH-DNN parameters (neuron weights and biases) were incorporated into a user-friendly Fortran code that reads the molecular geometry, computes the EE charges and input features, and finally evaluates $\Delta G^{\circ}{}_{\text{solv}}$ and $\Delta H^{\circ}{}_{\text{solv}}$ via ESE-ΔH-DNN.

3.2. Training

The Cartesian coordinates of the solutes obtained as described above were used to calculate the EE charges and atomic surfaces and volumes as explained in the Section 2. Along with the five solvent-related features (52–56), they were submitted first to the dimension-reducing 56 × 40 linear transformation and subsequently to the ANN training. The mean-square error was used as the loss function. The learning rate was initially set to 1×10^{-3} mol^2/kcal^2 and then reduced to 1×10^{-4} mol^2/kcal^2 within the same training run. Since ESE-ΔH-DNN was trained simultaneously for both $\Delta H°_{solv}$ and $\Delta G°_{solv}$, the loss function was defined as the sum of two residual sums of squares calculated from the predicted and reference $\Delta H°_{solv}$ and $\Delta G°_{solv}$ values. After a number of training runs, ESE-ΔH-DNN with the fitted parameters was evaluated on the testing sets that include various classes of solvents.

3.3. Performance of ESE-ΔH-DNN

The statistical data for predicted $\Delta H°_{solv}$ and $\Delta G°_{solv}$ values for various classes of solvents are reported in Tables 1 and 2. Table 1 shows errors in predicted values, while Table 2 contains information about the correlation of predicted and reference values. Individual data for the training, validation, and testing sets are given in the Supporting Information (Tables S2 and S3).

Table 1. Mean signed error (MSE), mean absolute error (MAE), and root mean square error (RMSE) of the predicted solvation free energy and solvation enthalpy for various testing subsets of ESE-ΔH-DNN.

	$\Delta G°_{solv}$, kcal/mol			$\Delta H°_{solv}$, kcal/mol		
Solvent Class [a]	**MSE**	**MAE**	**RMSE**	**MSE**	**MAE**	**RMSE**
Alkanes (245/186)	−0.10	0.24	0.34	−0.27	0.52	0.75
Alkoxyalcohols (44/7)	−0.13	1.03	1.55	−0.02	0.40	0.43
Aromatic (81/71)	0.04	0.19	0.25	0.11	0.61	0.85
Amides (56/39)	−0.17	0.47	0.62	0.19	0.68	1.06
Ethers (64/65)	−0.21	0.34	0.70	0.22	0.58	0.80
Esters (45/8)	0.02	0.25	0.33	−0.54	0.59	0.80
Haloalkanes (123/85)	0.18	0.41	0.55	0.16	0.60	0.80
Haloaromatic (50/38)	−0.17	0.48	0.69	−0.31	0.65	0.93
Ketones (60/41)	0.05	0.26	0.36	0.41	0.73	0.90
Miscellaneous (154/472)	0.01	0.39	0.57	−0.09	0.63	1.00
ALL (922/1036)	−0.03	0.36	0.59	−0.04	0.62	0.96

[a] The numbers in parentheses are the number of entries in the $\Delta G°_{solv}$ and $\Delta H°_{solv}$ databases, correspondingly.

Table 2. Slope, intercept (in kcal/mol), and coefficient of determination R^2 of the predicted solvation free energy and solvation enthalpy for various testing subsets of the ESE-ΔH-DNN.

	$\Delta G°_{solv}$			$\Delta H°_{solv}$		
Solvent Class [a]	**Slope**	**Intercept**	R^2	**Slope**	**Intercept**	R^2
Alkanes (245/186)	0.98	−0.16	0.978	0.99	−0.33	0.972
Alkoxyalcohols (44/7)	0.55	−1.61	0.384	1.21	1.85	0.972
Aromatic (81/71)	0.96	−0.07	0.989	0.91	0.03	0.965
Amides (56/39)	0.84	−0.67	0.900	0.88	−1.13	0.969
Ethers (64/65)	0.94	−0.42	0.933	0.97	−0.09	0.976
Esters (45/8)	0.92	−0.27	0.976	0.81	−2.33	0.853
Haloalkanes (123/85)	0.92	−0.24	0.956	0.97	−0.13	0.962
Haloaromatic (50/38)	1.02	−0.11	0.944	1.08	0.46	0.949
Ketones (60/41)	0.94	−0.13	0.978	0.96	0.07	0.979
Miscellaneous (154/472)	0.89	−0.36	0.926	0.99	−0.23	0.953
ALL (922/1036)	0.93	−0.29	0.939	0.98	−0.29	0.960

[a] The numbers in parentheses are the number of entries in the $\Delta G°_{solv}$ and $\Delta H°_{solv}$ databases, correspondingly.

3.3.1. Performance for Solvation Gibbs Free Energies

The overall quality of ESE-ΔH-DNN for ΔG°_{solv} over the entire testing set is characterized by an RMSE (root mean square error) below 0.6 kcal/mol and an MAE (mean absolute error) well below 0.4 kcal/mol. This performance is very convincing: even DFT-based methods often yield an RMSE between 0.6 and 1.5 kcal/mol. For instance, for nonaqueous solvents, our uESE [29] yields an RMSE between 0.6 and 1 kcal/mol depending on the solvent class, while the SCCS model by Hille et al. [12] yields an average RMSE of about 0.8 kcal/mol. The recent methods from W. H. Green's group [50] exhibit an RMSE range from about 0.7 to 1.5 kcal/mol. It should be noted, however, that the databases used are not identical and the comparison is therefore illustrative only. Considering the ESE-ΔH-DNN results separately for various solvent classes (Table 1), very good results (with RMSE < 0.5 kcal/mol, MAE < 0.3 kcal/mol) were obtained for nonpolar solvents (alkanes and aromatic solvents), as well as for esters and ketone solvents. For other solvent classes, the results are also compelling, with an RMSE mostly below 0.7 kcal/mol. The only clearly unfavorable exception is alkoxyalcohols, for which a significant error (RMSE > 1.5 kcal/mol) is detected. The failure with alkoxyalcohols is also reflected by poor slope and R^2 values (Table 2).

The ESE-ΔH-DNN results for the entire testing set are illustrated in Figure 2a. There are 420 entries (46%) with an error below 0.2 kcal/mol and 564 entries (61%) with an error below 0.3 kcal/mol, thus confirming the overall good performance of ESE-ΔH-DNN. Still, there are some 60 (6.5%) outliers with a deviation $\Delta\Delta G^{\circ}_{solv}$ > 1 kcal/mol. The worst failure is benzene dissolved in 2-butoxyethanol and pentane in anisole, for which ESE-ΔH-DNN fails to reproduce experimental positive ΔG°_{solv}. For nitromethane in triethylene glycol and SO_2F_2 in tributyl phosphate, the predicted ΔG°_{solv} values are much too negative. On the contrary, for C_3F_8 dissolved in hexafluorobenzene, a too positive ΔG°_{solv} is incorrectly predicted.

The results for amide solvents (Figure 2b) may also appear somewhat disturbing based on a slope value of 0.84. Indeed, ESE-ΔH-DNN tends to slightly overestimate $|\Delta G^{\circ}_{solv}|$ for amides, but in fact the accuracy is still quite good: even for the worst-case amide solvent, methylformamide, the RMSE is just 0.81 kcal/mol and there are only 9 (16%) outliers ($\Delta\Delta G^{\circ}_{solv}$ > 1 kcal/mol) out of 56 entries. Other examples shown in Figure 2 are alkane and aromatic solvents. Figure 2c,d demonstrate a very good agreement between the predicted and reference data for these solvent classes. There are only five (6%) outliers in each case.

The statistical ΔG°_{solv} data for all the solvents tested are summarized in Table 3. ESE-ΔH-DNN performs consistently well for all the alkane, aromatic, and ketone solvents. The results for the ethers are also very encouraging, except for anisole, where RMSE is about 1.2 kcal/mol. Ether and amide solvents are discussed above. The results for haloalkanes are also very good. Only chloroform and methylene iodide exhibit slightly larger deviations (RMSE about 0.7 kcal/mol). The haloaromatics are mostly very good (RMSE around 0.5 kcal/mol), with the only exception being perfluorobenzene. The problematic solvents within the alkoxyalcohol class are butoxyethanol, diethylene glycol, and triethylene glycol, the former being the worst case with an RMSE of almost 2.9 kcal/mol. The broad class of various solvents that are grouped together in Table 3 as *Miscellaneous* demonstrate a reliable accuracy, with an RMSE ranging from 0.2 to 1 kcal/mol. Eleven of the 24 solvents exhibit an RMSE below 0.4 kcal/mol, and eight more exhibit an RMSE below 0.8 kcal/mol. Only tributyl phosphate is slightly problematic, with an RSME of about 1 kcal/mol, which is still an acceptable accuracy for many applications.

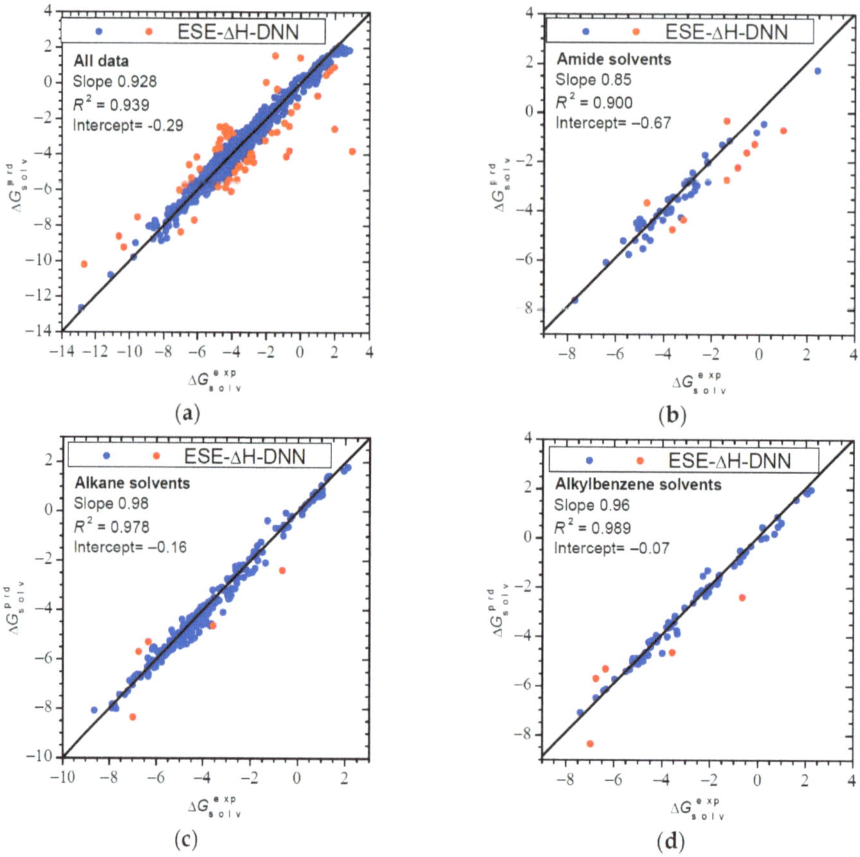

Figure 2. Solvation free energies (in kcal/mol) calculated by ESE-ΔH-DNN: (**a**) the entire testing set (922 entries); (**b**) amide solvents (56 entries); (**c**) alkane solvents (245 entries); (**d**) aromatic solvents (81 entries) versus reference values. Red points denote outliers with a deviation greater than 1 kcal/mol. The slanting line represents the identity line.

Table 3. MSE, MAE, and RMSE of the predicted solvation free energy and solvation enthalpy for various solvents (testing subsets).

Solvent [a]	**ΔG°_{solv}, kcal/mol**			**ΔH°_{solv}, kcal/mol**		
	MSE	**MAE**	**RMSE**	**MSE**	**MAE**	**RMSE**
Alkane solvents:						
Pentane (6/3)	−0.03	0.13	0.20	−0.04	0.41	0.43
Hexane (27/32)	−0.15	0.33	0.49	−0.08	0.49	0.61
Heptane (27/49)	−0.02	0.27	0.36	−0.31	0.54	0.80
Octane (27/12)	−0.08	0.16	0.23	0.01	0.31	0.38
Nonane (14/2)	−0.08	0.20	0.25	−0.36	0.36	0.46
Decane (15/11)	0.08	0.17	0.22	−0.28	0.50	0.71
Undecane (10/2)	−0.20	0.32	0.42	−0.33	0.34	0.48
Dodecane (11/8)	−0.07	0.17	0.19	−0.29	0.41	0.57
Hexadecane (65/17)	−0.06	0.22	0.31	−0.25	0.35	0.42
Tetradecane (2/3)	−0.12	0.24	0.27	0.50	0.68	0.87
Pentadecane (1/0)	−0.32	0.32	0.32			

Table 3. *Cont.*

Solvent [a]	ΔG°_{solv}, kcal/mol			ΔH°_{solv}, kcal/mol		
	MSE	MAE	RMSE	MSE	MAE	RMSE
Methylcyclohexane (5/0)	−0.29	0.29	0.50			
Cyclohexane (33/51)	−0.23	0.32	0.42	−0.48	0.67	0.96
Cyclooctane (2/0)	−0.10	0.10	0.11			
Alkoxyalcohol solvents:						
2-methoxyethanol (6/7)	0.23	0.27	0.36	−0.02	0.40	0.43
2-ethoxyethanol (4/0)	0.55	0.55	0.59			
2-butoxyethanol (6/0)	−1.16	1.39	2.79			
diethylene glycol (17/0)	−0.42	0.99	1.16			
triethylene glycol (11/0)	0.43	1.49	1.74			
Aromatic solvents:						
Benzene (13/46)	−0.09	0.17	0.23	−0.14	0.59	0.84
Toluene (25/25)	−0.03	0.21	0.29	0.56	0.65	0.86
Ethylbenzene (12/0)	0.09	0.19	0.20			
o-xylene (8/0)	0.08	0.12	0.15			
m-xylene (11/0)	0.14	0.18	0.20			
p-xylene (12/0)	0.17	0.24	0.31			
Amide solvents:						
Formamide (15/18)	−0.53	0.63	0.73	0.39	0.91	2.33
Methylformamide (7/8)	−0.40	0.60	0.81	0.36	1.10	1.42
N-methylacetamide (15/0)	−0.24	0.42	0.56			
N-methyl-2-pyrrolidone (19/14)	0.25	0.35	0.46	0.53	0.80	1.30
Ether solvents:						
diethyl ether (7/7)	−0.28	0.30	0.51	0.09	0.22	0.29
dipropyl ether (5/0)	−0.14	0.18	0.28			
diisopropyl ether (4/0)	−0.23	0.44	0.47			
dibutyl ether (6/20)	−0.17	0.24	0.41	0.22	0.53	0.72
methyl *tert*-butyl ether (5/0)	−0.24	0.39	0.45			
bis(2-ethoxyethyl) ether (1/0)	0.51	0.51	0.51			
Tetrahydrofuran (17/32)	−0.02	0.25	0.43	0.19	0.69	0.93
Tetrahydropyran (4/0)	−0.04	0.16	0.18			
anisole(15/0)	−0.52	0.56	1.22			
Ester solvents:						
methyl acetate (10/8)	−0.04	0.37	0.47	−0.54	0.58	0.80
ethyl acetate (13/24)	0.00	0.27	0.36	−0.15	0.56	0.74
propyl acetate (6/0)	0.11	0.16	0.17			
butyl acetate (8/0)	0.09	0.17	0.21			
pentyl acetate (7/0)	−0.02	0.23	0.26			
hexyl acetate (1/0)	−0.08	0.08	0.08			
Haloalkane solvents:						
Dichloromethane (9/24)	0.02	0.26	0.29	−0.59	0.68	0.87
Chloroform (48/21)	0.34	0.51	0.67	0.77	0.86	1.08
carbon tetrachloride (45/40)	0.21	0.30	0.40	0.29	0.42	0.54
1-chlorobutane (8/0)	0.02	0.35	0.39			
Dibromomethane (2/0)	0.32	0.42	0.52			
Bromoethane (2/0)	0.13	0.39	0.41			
methylene iodide (9/0)	−0.57	0.63	0.71			
Haloaromatic solvents:						
Fluorobenzene (3/0)	−0.54	0.54	0.58			
Chlorobenzene (21/30)	−0.30	0.41	0.53	−0.28	0.64	0.96
Bromobenzene (14/0)	−0.35	0.44	0.53			
Iodobenzene (8/0)	−0.08	0.30	0.45			
Hexafluorobenzene (4/0)	1.24	1.24	1.69			

Table 3. *Cont.*

Solvent [a]	ΔG°_{solv}, kcal/mol			ΔH°_{solv}, kcal/mol		
	MSE	MAE	RMSE	MSE	MAE	RMSE
Ketone solvents:						
Acetone (20/27)	0.01	0.35	0.50	0.32	0.78	0.95
2-butanone (10/7)	0.18	0.21	0.28	0.35	0.41	0.52
3-pentanone (1/0)	0.21	0.21	0.21			
2-hexanone (2/0)	0.23	0.23	0.24			
4-methyl-2-pentanone (1/0)	0.29	0.29	0.29			
Cyclohexanone (13/7)	0.05	0.20	0.25	0.78	0.85	1.00
Acetophenone (8/0)	0.10	0.19	0.23			
Cyclopentanone (1/0)	−0.06	0.06	0.06			
2-methylcyclohexanone (4/0)	−0.36	0.36	0.39			
Miscellaneous solvents:						
Benzonitrile (9/0)	0.08	0.21	0.27			
tributyl phosphate (16/0)	−0.14	0.57	1.01			
propylene carbonate (12/17)	0.04	0.20	0.30	−0.19	0.61	0.74
carbon disulfide (5/0)	−0.36	0.43	0.51			
Triethylamine (4/10)	0.18	0.19	0.35	0.65	0.93	1.58
Ethoxybenzene (3/0)	−0.14	0.66	0.77			
2-methylpyridine (4/0)	0.40	0.41	0.47			
benzyl ether (1/0)	−0.57	0.57	0.57			
3-methylphenol (1/0)	1.87	1.87	1.87			
acetic acid (10/14)	−0.08	0.48	0.60	0.12	0.66	0.94
Nitroethane (1/0)	0.60	0.60	0.60			
benzyl alcohol (2/0)	−0.30	0.30	0.35			
Butyronitrile (9/0)	−0.11	0.16	0.20			
Aniline (11/0)	−0.44	0.61	0.69			
Nitromethane (4/9)	0.09	0.21	0.31	−0.49	0.55	0.66
Nitrobenzene (3/0)	0.68	0.68	0.76			
dimethyl sulfoxide (11/37)	0.22	0.65	0.79	−0.58	0.75	1.08
Propionitrile (4/0)	−0.03	0.34	0.39			
Acetonitrile (6/41)	0.15	0.25	0.33	−0.23	0.89	1.23
ethyl benzoate (3/0)	−0.03	0.24	0.26			
Sulfolane (18/0)	0.24	0.36	0.43			
Pyridine (12/14)	0.05	0.22	0.29	0.21	1.37	1.97
diethyl carbonate (5/15)	−0.12	0.33	0.51	0.10	0.29	0.33

[a] The numbers in parentheses are the number of entries in the ΔG°_{solv} and ΔH°_{solv} databases, correspondingly.

3.3.2. Performance for Solvation Enthalpies

Tables 1–3 also provide statistical data for predicted solvation enthalpies ΔH°_{solv}. Overall, ESE-ΔH-DNN yields somewhat larger errors for ΔH°_{solv} than for ΔG°_{solv}, with a global RMSE of 0.96 kcal/mol and an MAE of 0.62 kcal/mol. While it is difficult to pinpoint the exact reasons of the lower absolute accuracy of ESE-ΔH-DNN for ΔH°_{solv}, it is tempting to assume that a large total range of ΔH°_{solv} contributes to it. In fact, ΔH°_{solv} is virtually always more negative than ΔG°_{solv} due to negative solvation entropy. An inspection of Figure 3 shows that larger deviations often occur for more negative ΔH°_{solv}. The worst result is obtained for benzo-15-crown-5 in propanol, with $|\Delta H^\circ_{solv}|$ overestimated by 6.7 kcal/mol and for glycerol in methanol and *tert*-butanol, for which $|\Delta H^\circ_{solv}|$ is underestimated by 5.5 and 4.9 kcal/mol, respectively. Nevertheless, it is encouraging that 859 of 1036 values exhibit a deviation within 1 kcal/mol and 603 within 0.5 kcal/mol.

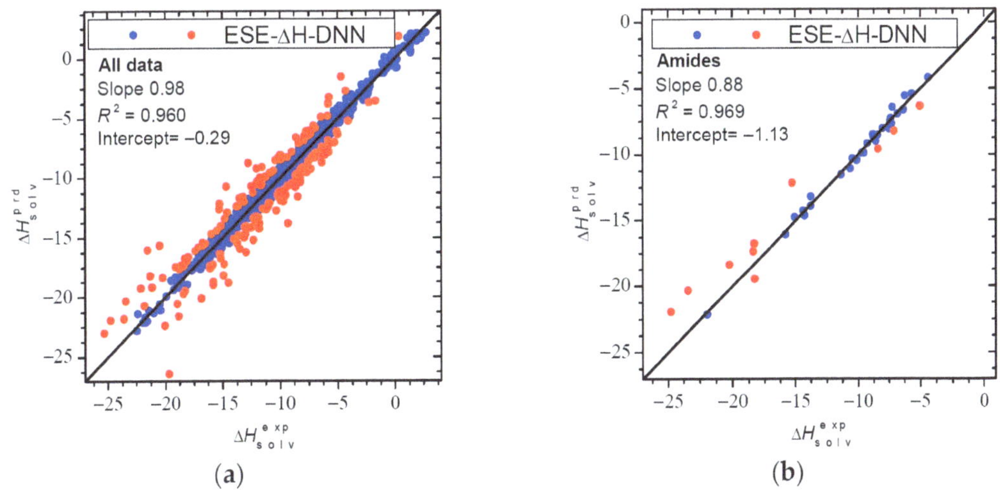

Figure 3. Solvation enthalpies (in kcal/mol) calculated by ESE-ΔH-DNN: (**a**) the entire testing set (1036 entries); (**b**) amide solvents (39 entries) versus reference values. Red points denote outliers with a deviation greater than 1 kcal/mol. The slanting line represents the identity line.

Various classes of solvent demonstrate a fairly homogeneous distribution of errors for ΔH°_{solv}, with RSME ranging from approximately 0.4 kcal/mol for alkoxyalkohols to about 1.1 kcal/mol for amides (see the rightmost column of Table 1). There are no solvent classes with notably large errors. It is worth noting that the performance of ESE-ΔH-DNN for ΔH°_{solv} does not correlate with that for ΔG°_{solv}: the lowest ΔH°_{solv} RSME (0.43 kcal/mol) is observed for alkoxyalkohols, which is the most problematic in terms of ΔG°_{solv}. The largest RMSE observed is for amides, but it does not remarkably deviate from the global average.

As for individual solvents, relatively large errors are detected for some amides and amines: formamide, pyridine, triethylamine, methylformamide, and *N*-methylpyrrolidone. While methylformamide is also problematic for ΔG°_{solv}, pyridine and *N*-methylpyrrolidone show quite accurate results in terms of ΔG°_{solv},

In addition to the tests conducted on our randomly selected testing dataset, I performed evaluations using the structures retrieved from the Minnesota Solvation Database (MNSol) [64]. The statistical results for nonpolar solvents are presented in Table 4. Individual data are given in the Supporting Information (Table S4). Overall, these data confirm a good performance of ESE-ΔH-DNN, with a global RMSE of about 0.7 kcal/mol. An inferior performance was observed only for ethyl acetate, tributyl phosphate, and diethyl ether solvents.

Table 4. MSE, MAE, and RMSE of the solvation free energy predicted by ESE-ΔH-DNN for various nonpolar solvents (MNSol database).

Solvent [a]	**ΔG°_{solv}, kcal/mol**			**Solvent** [a]	**ΔG°_{solv}, kcal/mol**		
	MSE	**MAE**	**RMSE**		**MSE**	**MAE**	**RMSE**
acetic acid (7)	0.43	0.78	0.87	fluoroctane (6)	−0.11	0.16	0.19
aniline (9)	0.00	0.53	0.63	heptane (69)	−0.08	0.36	0.59
anisole (8)	−0.06	0.22	0.28	hexadecane (198)	−0.12	0.29	0.56
benzene (74)	−0.04	0.42	0.75	hexadecyl iodide (9)	0.44	0.44	0.51
bromobenzene (27)	−0.47	0.50	0.61	hexane (59)	−0.23	0.36	0.56
bromoform (12)	0.10	0.26	0.32	iodobenzene (20)	−0.31	0.41	0.54
bromooctane (5)	0.25	0.25	0.28	isooctane (32)	−0.37	0.38	0.46

Table 4. *Cont.*

Solvent [a]	ΔG°_{solv}, kcal/mol			Solvent [a]	ΔG°_{solv}, kcal/mol		
	MSE	MAE	RMSE		MSE	MAE	RMSE
butyl acetate (21)	0.10	0.38	0.50	isopropylbenzene (19)	0.03	0.22	0.26
butylbenzene (10)	0.42	0.42	0.47	isopropyltoluene (6)	0.25	0.25	0.30
carbon disulfide (14)	−0.51	0.58	0.67	mesitylene (7)	0.31	0.31	0.37
carbon tetrachloride (78)	0.01	0.25	0.41	nonane (26)	−0.11	0.20	0.25
chlorobenzene (38)	−0.59	0.60	0.72	nonanol (10)	0.27	0.42	0.50
chloroform (108)	0.09	0.65	0.98	octane (38)	−0.23	0.28	0.37
chlorohexane (11)	−0.07	0.17	0.20	pentadecane (9)	0.12	0.16	0.20
cyclohexane (92)	−0.49	0.53	0.81	pentane (26)	−0.41	0.43	0.47
decalin (27)	−0.26	0.36	0.56	perfluorobenzene (15)	0.94	0.94	1.02
decane (39)	−0.11	0.23	0.34	phenyl ether (6)	−0.55	0.55	0.62
decanol (11)	0.38	0.50	0.58	*sec*–butylbenzene (5)	0.31	0.31	0.35
dibromoethane (10)	−0.17	0.26	0.31	*tert*–butylbenzene (14)	0.12	0.18	0.21
dibutyl ether (14)	−0.22	0.47	0.65	tetrachloroethene (10)	0.04	0.23	0.30
dichloromethane (11)	−0.26	0.39	0.55	tetrahydrofuran (7)	0.18	0.27	0.30
diethyl ether (71)	−0.03	0.74	1.30	tetralin (9)	−0.92	0.92	1.15
diisopropyl ether (21)	−0.10	0.45	0.66	toluene (50)	0.01	0.20	0.31
dimethylpyridine (6)	−0.16	0.34	0.54	tributyl phosphate (16)	1.43	1.49	1.79
dodecane (8)	−0.29	0.35	0.45	triethylamine (7)	0.25	0.26	0.38
ethoxybenzene (7)	−0.08	0.22	0.30	trimethylbenzene (11)	0.27	0.27	0.29
ethyl acetate (23)	0.72	0.95	1.99	undecane (13)	−0.09	0.29	0.40
ethylbenzene (29)	−0.06	0.24	0.33	xylene (48)	0.08	0.25	0.34
fluorobenzene (7)	−0.66	0.66	0.79	ALL (1543)	−0.08	0.42	0.71

[a] The numbers in parentheses are the number of entries in the MNSol database.

4. Conclusions

In this paper, I proposed ESE-ΔH-DNN, a dense neural network for calculating the ΔG°_{solv} and ΔH°_{solv} based on physically sound input features such as generalized-Born-type electrostatic terms, the molecular volume, and the atomic surface area of the solute as well as five characteristics of the solvent. ESE-ΔH-DNN is defined for neutral solutes in a wide range of organic solvents. It requires only the molecular geometry as the input and provides accurate results, with an average RMSE below 0.6 kcal/mol. It is particularly accurate for alkane, aromatic, ester, and ketone solvents. The limitations (which also present opportunities for further development) are that ESE-ΔH-DNN is trained only on neutral solutes and is thus not automatically applicable to ions. It also has lower accuracy for specific alkoxyalcohol solvents (butoxyethanol, triethylene glycol, and diethylene glycol), as well as for hexafluorobenzene.

ESE-ΔH-DNN also demonstrates a fairly good accuracy for ΔH°_{solv} prediction, although the errors are slightly larger, with an average RMSE still below 1 kcal/mol. One positive aspect is that the error is quite homogeneous, with the RMSE showing small variance among diverse solvent classes.

Supplementary Materials: The following supporting information can be downloaded at: https://www.mdpi.com/article/10.3390/liquids4030030/s1: Part 1: ESE-ΔH-DNN neural network buildup: 56 initial input vector components; 56 × 40 dimensionality reduction linear transformation; ESE-ΔH-DNN weights and biases; Part 2: Solvents and their properties. Part 3: Predicted and reference solvation free energies and solvation enthalpies for the training, validation, and testing sets; predicted and reference solvation free energies for nonpolar solvents from MNSol database. Reference [64] appear in the Supplementary material.

Funding: This research received no external funding.

Data Availability Statement: The executable ESE-ΔH-DNN program and a user guide are openly available for download: https://github.com/vyboishchikov/ESE-DeltaH-DNN. Accessed on 3 July 2024.

Acknowledgments: I am very grateful to W. E. Acree for kindly providing me with the large database of experimental ΔG°_{solv} and ΔH°_{solv} used in this work.

Conflicts of Interest: The author declares no conflicts of interest.

References

1. Torrie, G.; Valleau, J. Nonphysical sampling distributions in Monte Carlo free-energy estimation: Umbrella sampling. *J. Comput. Phys.* **1977**, *23*, 187–199. [CrossRef]
2. You, W.; Tang, Z.; Chang, C.A. Potential mean force from umbrella sampling simulations: What can we learn and what is missed? *J. Chem. Theory Comput.* **2019**, *15*, 2433–2443. [CrossRef] [PubMed]
3. Chipot, C.; Pohorille, A. Calculating free energy differences using perturbation theory. In *Free Energy Calculations: Theory and Applications in Chemistry and Biology*; Chipot, C., Pohorille, A., Eds.; Springer Series in Chemical Physics; Springer: Berlin/Heidelberg, Germany; New York, NY, USA, 2007; Volume 86, p. 33. ISBN 540-38447-2.
4. Tomasi, J.; Mennucci, B.; Cammi, R. Quantum mechanical continuum solvation models. *Chem. Rev.* **2005**, *105*, 2999–3094. [CrossRef] [PubMed]
5. Skyner, R.E.; McDonagh, J.L.; Groom, C.R.; van Mourik, T.; Mitchell, J.B.O. A review of methods for the calculation of solution free energies and the modelling of systems in solution. *Phys. Chem. Chem. Phys.* **2015**, *17*, 6174–6191. [CrossRef] [PubMed]
6. Barone, V.; Cossi, M.; Tomasi, J. A new definition of cavities for the computation of solvation free energies by the polarizable continuum model. *J. Chem. Phys.* **1997**, *107*, 3210–3221. [CrossRef]
7. Cancès, E.; Mennucci, B.; Tomasi, J. A new integral equation formalism for the polarizable continuum model: Theoretical background and applications to isotropic and anistropic dielectrics. *J. Chem. Phys.* **1997**, *107*, 3032–3041. [CrossRef]
8. Mennucci, B.; Cancès, E.; Tomasi, J. Evaluation of solvent effects in isotropic and anisotropic dielectrics, and in ionic solutions with a unified integral equation method: Theoretical bases, computational implementation and numerical applications. *J. Phys. Chem. B* **1997**, *101*, 10506–10517. [CrossRef]
9. Cossi, M.; Barone, V.; Robb, M.A. A direct procedure for the evaluation of solvent effects in MC-SCF calculations. *J. Chem. Phys.* **1999**, *111*, 5295–5302. [CrossRef]
10. Tomasi, J.; Mennucci, B.; Cancès, E. The IEF version of the PCM solvation method: An overview of a new method addressed to study molecular solutes at the QM ab initio level. *J. Mol. Struct. (Theochem)* **1999**, *464*, 211–226. [CrossRef]
11. Marenich, A.V.; Cramer, C.J.; Truhlar, D.G. Universal solvation model based on solute electron density and on a continuum model of the solvent defined by the bulk dielectric constant and atomic surface tensions. *J. Phys. Chem. B* **2009**, *113*, 6378–6396. [CrossRef]
12. Pomogaeva, A.; Chipman, D.M. Hydration energy from a composite method for implicit representation of solvent. *J. Chem. Theory Comput.* **2014**, *10*, 211–219. [CrossRef]
13. Hille, C.; Ringe, S.; Deimel, M.; Kunkel, C.; Acree, W.E.; Reuter, K.; Oberhofer, H. Generalized molecular solvation in non-aqueous solutions by a single parameter implicit solvation scheme. *J. Chem. Phys.* **2019**, *150*, 041710. [CrossRef]
14. Klamt, A.; Schüürmann, G. COSMO: A new approach to dielectric screening in solvents with explicit expressions for the screening energy and its gradient. *J. Chem. Soc. Perkin Trans.* **1993**, *2*, 799–805. [CrossRef]
15. Klamt, A. The COSMO and COSMO-RS solvation models. *WIREs Comput. Mol. Sci.* **2011**, *1*, 699–709. [CrossRef]
16. Still, W.C.; Tempczyk, A.; Hawley, R.C.; Hendrickson, T.J. Semianalytical treatment of solvation for molecular mechanics and dynamics. *J. Am. Chem. Soc.* **1990**, *112*, 6127–6129. [CrossRef]
17. Marenich, A.V.; Cramer, C.J.; Truhlar, D.G. Generalized Born solvation model SM12. *J. Chem. Theory Comput.* **2013**, *9*, 609–620. [CrossRef] [PubMed]
18. Cramer, C.J.; Truhlar, D.G. General parameterized SCF model for free energies of solvation in aqueous solution. *J. Am. Chem. Soc.* **1991**, *113*, 8305–8311. [CrossRef]
19. Cramer, C.J.; Truhlar, D.G. An SCF solvation model for the hydrophobic effect and absolute free energies of aqueous solvation. *Science* **1992**, *256*, 213–217. [CrossRef]
20. Cramer, C.J.; Truhlar, D.G. A universal approach to solvation modeling. *Acc. Chem. Res.* **2008**, *41*, 760–768. [CrossRef]
21. Marenich, A.V.; Cramer, C.J.; Truhlar, D.G. Uniform treatment of solute–solvent dispersion in the ground and excited electronic states of the solute based on a solvation model with state-specific polarizability. *J. Chem. Theory Comput.* **2013**, *9*, 3649–3659. [CrossRef]
22. Born, M. Volumen und Hydratationswärme der Ionen. *Z. Physik* **1920**, *1*, 45–48. (In German) [CrossRef]
23. Onufriev, A.V.; Case, D.A. Generalized Born implicit solvent models for biomolecules. *Annu. Rev. Biophys.* **2019**, *48*, 275–296. [CrossRef] [PubMed]
24. Grycuk, T. Deficiency of the Coulomb-field approximation in the generalized Born model: An improved formula for Born radii evaluation. *J. Chem. Phys.* **2003**, *119*, 4817–4826. [CrossRef]
25. Grant, J.; Pickup, B.; Sykes, M.; Kitchen, C.; Nicholls, A. The Gaussian Generalized Born model: Application to small molecules. *Phys. Chem. Chem. Phys.* **2007**, *9*, 4913–4922. [CrossRef] [PubMed]
26. Lange, A.W.; Herbert, J.M. Improving Generalized Born models by exploiting connections to Polarizable Continuum Models. I. An improved effective Coulomb operator. *J. Chem. Theory Comput.* **2012**, *8*, 1999–2011. [CrossRef] [PubMed]

27. Voityuk, A.A.; Vyboishchikov, S.F. A simple COSMO-based method for calculation of hydration energies of neutral molecules. *Phys. Chem. Chem. Phys.* **2019**, *21*, 18706–18713. [CrossRef] [PubMed]
28. Voityuk, A.A.; Vyboishchikov, S.F. Fast and accurate calculation of hydration energies of molecules and ions. *Phys. Chem. Chem. Phys.* **2020**, *22*, 14591–14598. [CrossRef]
29. Vyboishchikov, S.F.; Voityuk, A.A. Fast non-iterative calculation of solvation energies for water and nonaqueous solvents. *J. Comput. Chem.* **2021**, *42*, 1184–1194. [CrossRef] [PubMed]
30. Vyboishchikov, S.F.; Voityuk, A.A. Solvation free energies for aqueous and nonaqueous solutions computed using PM7 atomic charges. *J. Chem. Inf. Model.* **2021**, *61*, 4544–4553. [CrossRef]
31. Vyboishchikov, S.F. A quick solvation energy estimator based on electronegativity equalization. *J. Comput. Chem.* **2023**, *44*, 307–318. [CrossRef]
32. Vyboishchikov, S.F.; Voityuk, A.A. Noniterative solvation energy method based on atomic charges. In *Chemical Reactivity: Approaches and applications*; Kaya, S., von Szentpály, L., Serdaroğlu, G., Guo, K., Eds.; Elsevier: Amsterdam, The Netherlands, 2023; Volume 2, pp. 399–427. [CrossRef]
33. Voityuk, A.A.; Stasyuk, A.J.; Vyboishchikov, S.F. A simple model for calculating atomic charges in molecules. *Phys. Chem. Chem. Phys.* **2018**, *20*, 23328–23337. [CrossRef] [PubMed]
34. Vyboishchikov, S.F.; Voityuk, A.A. Iterative atomic-charge partitioning of valence electron density. *J. Comp. Chem.* **2019**, *40*, 875–884. [CrossRef] [PubMed]
35. Marenich, A.V.; Jerome, S.V.; Cramer, C.J.; Truhlar, D.G. Charge Model 5: An extension of Hirshfeld population analysis for the accurate description of molecular interactions in gaseous and condensed phases. *J. Chem. Theory Comput.* **2012**, *8*, 527–541. [CrossRef] [PubMed]
36. Kříž, K.; Řezáč, J. Reparametrization of the COSMO solvent model for semiempirical methods PM6 and PM7. *J. Chem. Inf. Model.* **2019**, *59*, 229–235. [CrossRef] [PubMed]
37. Vyboishchikov, S.F. Dense neural network for calculating solvation free energies from electronegativity-equalization atomic charges. *J. Chem. Inf. Model.* **2023**, *63*, 6283–6292. [CrossRef] [PubMed]
38. Vyboishchikov, S.F. Predicting solvation free energies using electronegativity-equalization atomic charges and a dense neural network: A generalized-Born approach. *J. Chem. Theory Comput.* **2023**, *19*, 8340–8350. [CrossRef]
39. The Program Executable Is Available Free of Charge. Available online: https://github.com/vyboishchikov/ESE-EE-DNN (accessed on 3 July 2024).
40. The Program Executable Is Available Free of Charge. Available online: https://github.com/vyboishchikov/ESE-GB-DNN (accessed on 3 July 2024).
41. Chen, Y.; Krämer, A.; Charron, N.E.; Husic, B.E.; Clementi, C.; Noé, F. Machine learning implicit solvation for molecular dynamics. *J. Chem. Phys.* **2021**, *155*, 084101. [CrossRef]
42. Vermeire, F.H.; Green, W.H. Transfer learning for solvation free energies: From quantum chemistry to experiments. *Chem. Engin. J.* **2021**, *418*, 129307. [CrossRef]
43. Low, K.; Coote, M.L.; Izgorodina, E.I. Explainable solvation free energy prediction combining graph neural networks with chemical intuition. *J. Chem. Inf. Model.* **2022**, *62*, 5457–5470. [CrossRef]
44. Lim, H.; Jung, I. MLSolvA: Solvation free energy prediction from pairwise atomistic interactions by machine learning. *J. Cheminform.* **2021**, *13*, 56. [CrossRef]
45. Alibakhshi, A.; Hartke, B. Improved prediction of solvation free energies by machine-learning polarizable continuum solvation model. *Nat. Commun.* **2021**, *12*, 3584. [CrossRef] [PubMed]
46. Bernazzani, L.; Duce, C.; Micheli, A.; Mollica, V.; Tiné, M.R. Quantitative structure–property relationship (QSPR) prediction of solvation Gibbs energy of bifunctional compounds by recursive neural networks. *J. Chem. Eng. Data* **2010**, *55*, 5425–5428. [CrossRef]
47. Hutchinson, S.T.; Kobayashi, R. Solvent-specific featurization for predicting free energies of solvation through machine learning. *J. Chem. Inf. Model.* **2019**, *59*, 1338–1346. [CrossRef] [PubMed]
48. Wang, B.; Wang, C.; Wu, K.; Wei, G.W. Breaking the polar–non-polar division in solvation free energy prediction. *J. Comput. Chem.* **2018**, *39*, 217–233. [CrossRef] [PubMed]
49. Jaquis, B.J.; Li, A.; Monnier, N.D.; Sisk, R.G.; Acree, W.E.; Lang, A.S.I.D. Using machine learning to predict enthalpy of solvation. *J. Solut. Chem.* **2019**, *48*, 564–573. [CrossRef]
50. Chung, Y.; Vermeire, F.H.; Wu, H.; Walker, P.J.; Abraham, M.H.; Green, W.H. Group contribution and machine learning approaches to predict Abraham solute parameters, solvation free energy, and solvation enthalpy. *J. Chem. Inf. Model.* **2022**, *62*, 433–446. [CrossRef] [PubMed]
51. Svobodová Vařeková, R.; Jiroušková, Z.; Vaněk, J.; Suchomel, Š.; Koča, J. Electronegativity equalization method: Parameterization and validation for large sets of organic, organohalogene and organometal molecule. *Int. J. Mol. Sci.* **2007**, *8*, 572–582. [CrossRef]
52. Bondi, A. Van der Waals volumes and radii. *J. Phys. Chem.* **1964**, *68*, 441–451. [CrossRef]
53. Pedregosa, F.; Varoquaux, G.; Gramfort, A.; Michel, V.; Thirion, B.; Grisel, O.; Blondel, M.; Prettenhofer, P.; Weiss, R.; Dubourg, V.; et al. Scikit-learn: Machine Learning in Python. *J. Mach. Learn. Res.* **2011**, *12*, 2825–2830. Available online: https://scikit-learn.org/stable (accessed on 1 March 2024).

54. Nesterov, Y. A method of solving a convex programming problem with convergence rate $O(1/k^2)$. *Sov. Math. Dokl.* **1983**, *27*, 372–376.
55. Dozat, T. Incorporating Nesterov Momentum into Adam. In International Conference on Learning Representations. 2016. Available online: https://openreview.net/pdf/OM0jvwB8jIp57ZJjtNEZ.pdf (accessed on 8 November 2023).
56. Martín, A.; Agarwal, A.; Barham, P.; Brevdo, E.; Chen, Z.; Citro, C.; Corrado, G.S.; Davis, A.; Dean, J.; Devin, M.; et al. TensorFlow: Large-Scale Machine Learning on Heterogeneous Systems. 2015. Available online: http://tensorflow.org (accessed on 30 April 2023).
57. Lu, J.Z.; Acree, W.E.; Abraham, M.H. Updated Abraham model correlations for enthalpies of solvation of organic solutes dissolved in benzene and acetonitrile. *Phys. Chem. Liquids* **2019**, *57*, 84–99. [CrossRef]
58. Acree, W.E., Jr.; (University of North Texas, Denton, TX, USA). Personal communication, 2024.
59. Available online: http://cactus.nci.nih.gov/chemical/structure (accessed on 1 March 2024).
60. Halgren, T.A. Merck molecular force field. I. Basis, form, scope, parameterization, and performance of MMFF94. *J. Comput. Chem.* **1996**, *17*, 490–519. [CrossRef]
61. Landrum, G. Rdkit Documentation, Release 1 September 2019. 2019. Available online: https://buildmedia.readthedocs.org/media/pdf/rdkit/latest/rdkit.pdf (accessed on 1 March 2024).
62. Available online: https://webbook.nist.gov (accessed on 1 March 2024).
63. Available online: https://www.stenutz.eu/chem (accessed on 1 March 2024).
64. Marenich, A.V.; Kelly, C.P.; Thompson, J.D.; Hawkins, G.D.; Chambers, C.C.; Giesen, D.J.; Winget, P.; Cramer, C.J.; Truhlar, D.G. Minnesota Solvation Database—Version 2012, University of Minnesota, 26 November 2012. Available online: https://conservancy.umn.edu/bitstream/handle/11299/213300/MNSolDatabase_v2012.zip (accessed on 10 June 2024).

liquids

MDPI

Article

Vaporisation Thermodynamics: Are Triazolium Ionic Liquids a Real Alternative to Popular Imidazolium-Based Ionic Liquids?

Sergey P. Verevkin [1,2,*] and Dzmitry H. Zaitsau [3,4]

[1] Competence Centre CALOR, Department of "Science and Technology of Life, Light and Matter", Faculty of Interdisciplinary Research, University of Rostock, Albert-Einstein-Str. 27, 18059 Rostock, Germany
[2] Department of Physical Chemistry, Kazan Federal University, Kremlevskaya Str. 18, 420008 Kazan, Russia
[3] Institute of Technical Thermodynamics, University of Rostock, Albert-Einstein-Str. 2, 18059 Rostock, Germany; zaitsaudz@gmail.com
[4] FVTR GmbH, Joachim-Jungius-Str. 9, 18059 Rostock, Germany
* Correspondence: sergey.verevkin@uni-rostock.de

Abstract: New experimental vapour pressures and vaporisation enthalpies of the *ionic liquids* [2,4-dimethyl-1,2,4-triazolium][NTf_2], [2-methyl-4-ethyl-1,2,4-triazolium][NTf_2], and [2-ethyl-4-methyl-1,2,4-triazolium][NTf_2] were measured using the Langmuir method in combination with the quartz crystal microbalance. New experimental vapour pressures and vaporisation enthalpies of the *molecular liquids* 1H-1,2,4-triazole, 1-methyl-1,2,4-triazole, 1-ethyl-1,2,4-triazole, and 1H-1,2,3-triazole were measured using the transpiration method. Structure–property relationships between molecular and ionic liquids were studied. These results will facilitate chemical engineering calculations of processes involving ILs.

Keywords: ionic liquid; vapour pressure; enthalpy of vaporisation; structure–property relationships

1. Introduction

Triazolium-based ionic liquids (ILs) are considered a new class of solvents for cellulose. The cations of these ILs are structurally similar to the popular imidazolium cations, but the 1,2,4- and 1,2,3-triazolium cations lack the isolated ring proton, which leads to a lower formation of N-heterocyclic carbenes and thus to a lower reactivity and fewer undesired side reactions [1–3]. Triazolium ILs can be easily produced in high purity and without contamination by halides, can be a good alternative to "classical" imidazolium ILs for a number of applications, and can even enable new applications [4].

Since the ILs are generally regarded as neoteric solvents, the solution to practical questions concerning miscibility, separation of mixtures, volatilisation, vapour pressure, thermal stability, etc., requires knowledge of thermodynamic data. Such data are not yet available for the new triazolium ILs. This work extends our previous studies on ILs containing fluorinated anions [5–8] and deals with the vaporisation thermodynamics of 1,2,4-triazolium-based ILs with the [NTf_2] anion (see Figure 1).

The experimental data on the vapour pressures and vaporisation enthalpies of the three ionic liquids [di-a1kyl-1,2,4-triazolium][NTf_2] provide only limited information on the thermal properties of this new class of solvents. It is therefore important to compare these new data with those for "classical" imidazolium-based ILs in order to qualitatively assess the advantages and disadvantages of the new materials. In this context, we will refer to our previous results on the vaporisation thermodynamics of the [$C_n$mim][NTf_2] family [5] for comparison.

In fact, the experimental investigation of the vapour pressures of extremely low volatile ILs is a very challenging task, which is why a sufficient amount of thermal data for triazolium-based ILs can hardly be expected in the near future. For this reason, the development of reliable methods for predicting vaporisation thermodynamics could be important for chemical engineering calculations. The experience gained in our recent studies through a systematic comparison of "*molecular* liquids versus *ionic* liquids" [9–11] could be used in this work to predict the vaporisation enthalpies of 1,2,4-triazolium and 1,2,3-triazolium ILs paired with the [NTf_2] anion. For this purpose, the vapour pressures of the *molecular liquids* (triazoles) shown in Figure 2 were measured, and their vaporisation enthalpies were involved in structure–property correlations with the corresponding triazolium-based *ionic liquids*.

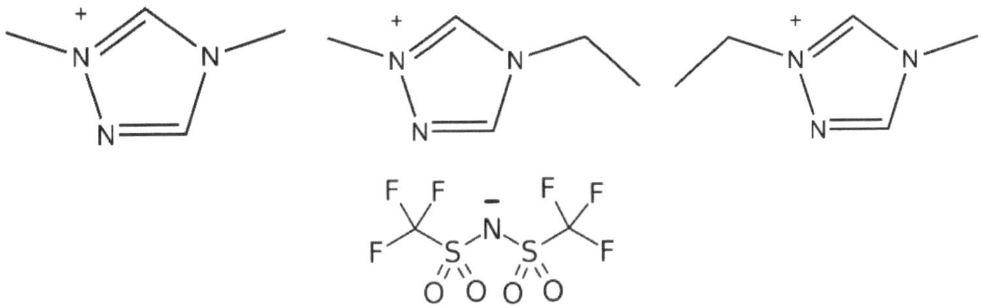

Figure 1. *Ionic liquids* under study in this work: [2,4-dimethyl-1,2,4-triazolium][NTf_2], [2-methyl-4-ethyl-1,2,4-triazolium][NTf_2], and [2-ethyl-4-methyl-1,2,4-triazolium][NTf_2], with the anion [NTf_2] = (trifluoromethylsulfonyl)imide. The abbreviations [2-C_n-4-C_m-1,2,4-T][NTf_2] with n, m = 1, 2 are helpful for the presentation of the data in the tables in this paper.

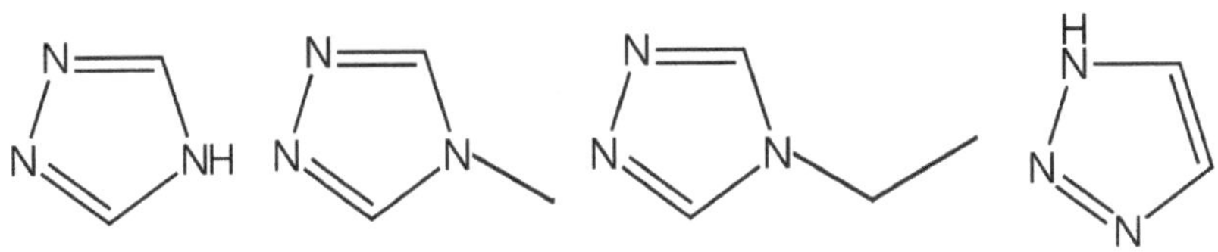

288-88-0 6086-21-1 16778-70-4 288-36-8

Figure 2. *Molecular liquids* under study in this work: 1H-1,2,4-triazole, 1-methyl-1,2,4-triazole, 1-ethyl-1,2,4-triazole, and 1H-1,2,3-triazole.

2. Experimental Section

2.1. Materials

The *ionic liquid* samples [2,4-dimethyl-1,2,4-triazolium][NTf_2], [2-methyl-4-ethyl-1,2,4-triazolium][NTf_2], and [2-ethyl-4-methyl-1,2,4-triazolium][NTf_2] used in this work were prepared and purified by Dr. hab. Andrzej Skrzypczak, Poznań University of Technology, Institute of Chemical Technology and Engineering, Poznań, Poland, who passed away in 2023. The degree of purity of >99% was determined using the HPLC method.

Prior to the experiments, the samples were vacuum-evaporated for more than 24 h at 413 K and 10^{-5} Pa to reduce possible traces of solvents and moisture. The samples used for the vaporisation studies using a quartz crystal microbalance (QCM) were additionally conditioned within 12 h in the vacuum chamber at the highest temperature of the experiment. This additional purification allowed the removal of residual traces of water and volatile impurities as well as the collection of a sufficient amount of vaporised IL on the quartz sensor required for FTIR analysis.

The *molecular liquid* samples 1H-1,2,4-triazole, 1-methyl-1,2,4-triazole, 1-ethyl-1,2,4-triazole, and 1H-1,2,3-triazole were of commercial origin. Provenance and purities are listed in Table S1 (Supplemental Materials). Before starting the vapour pressure measurements with the transpiration method, the samples were preconditioned in the saturator (see below for details). The purity of samples was analysed with a Hewlett Packard 5890 II series gas chromatograph equipped with a flame ionisation detector and an HP-5 capillary column (length, inside diameter, and film thickness of 25 m × 0.32 mm × 0.25 μm). The GC temperature programme started at T = 323 K, followed by heating at a rate of 0.167 $K \cdot s^{-1}$ to T = 573 K.

2.2. Absolute Vapour Pressure Measurements

The vapour pressures and standard molar enthalpies of vaporisation of the ILs were determined using the QCM method [5,12]. In this method, a sample of an IL is placed in an open measuring cavity within the thermostat block and exposed to a vacuum (10^{-5} Pa) with the entire open surface (Langmuir evaporation). The sensor is located directly above the cavity containing the sample. During evaporation in a vacuum, a certain amount of the sample is deposited on the quartz crystal. The change in the vibrational frequency of the crystal is directly related to the mass of the compound deposited on the crystal in a given time. The molar standard enthalpies of vaporisation were derived from the temperature dependencies of the experimentally measured change in the vibration frequency of the quartz crystal. The essential experimental details are provided in the Supplemental Materials.

The vapour pressures and standard molar enthalpies of vaporisation of the molecular liquids (1H-1,2,4-triazole, 1-methyl-1,2,4-triazole, 1-ethyl-1,2,4-triazole, and 1H-1,2,3-triazole) were determined using the transpiration method [13]. In this method, a stream of nitrogen was passed through the saturator filled with the sample at a precisely defined flow rate at a constant temperature (±0.1 K). After reaching saturation equilibrium in the saturator, the material transported by the gas flow was collected in a cold trap within a defined time and its mass was determined by GC. The standard molar enthalpies and entropies of vaporisation were derived from the temperature dependence of the absolute vapour pressures. The necessary details can be found in the Supplemental Materials.

3. Results and Discussion

3.1. Absolute Vapour Pressures and Vaporisation Thermodynamics of Ionic Liquids

Admittedly, one of the most important attributes of ionic liquids is that they have a negligible vapour pressure (compared to the common molecular solvents). However, this statement applies more to ambient temperatures, and it is obvious that at higher temperatures, which are relevant for chemical engineering, the vapour pressure already assumes a considerable size and should be quantified for the development of future technologies. For example, reliable knowledge of the temperature dependence of the vapour pressure of ILs is essential for the optimisation of modern catalytic processes such as Solid Catalyst with Ionic Liquid Layer (SCILL) [14] or Supported Ionic Liquid Phase (SILP) [15] in order to assess a possible long-term uptake of ILs at any practical temperature. The experimental vapour pressures at different temperatures for the triazolium ILs measured in this work are collected in Table S2. The equations used to approximate the experimental data allow the vapour pressure to be calculated at any practically relevant temperature. The 373 K and 473 K temperatures appear to be reasonable choices for many practical applications, and the extremely low values of $[NTf_2]$-based IL vapour pressures at these temperatures (see Table 1) indicate that the negligible mass uptake of IL in various catalytic or separation applications can also be expected at elevated temperatures.

Table 1. Absolute vapour pressures of ILs $[2\text{-}C_n\text{-}4\text{-}C_m\text{-}1,2,4\text{-}T][NTf_2]$ with n, m = 1, 2 at 373 K and at 473 K.

Cation	$p_{sat} \times 10^6$ [Pa]	$p_{sat} \times 10^3$ [Pa]
	373 K	473 K
$[2\text{-}C_1\text{-}4\text{-}C_1\text{-}1,2,4\text{-}T]$	20	70.8
$[2\text{-}C_1\text{-}4\text{-}C_2\text{-}1,2,4\text{-}T]$	36	134.8
$[2\text{-}C_2\text{-}4\text{-}C_1\text{-}1,2,4\text{-}T]$	35	121.5

The absolute vapour pressure data of ILs are still rarely available in the literature, and it is interesting to compare the general vapour pressure values within the families of imidazolium-based ILs with fluorinated anions $[CF_3SO_3]^-$, $[PF_6]^-$, $[BF_4]^-$, and $[NTF_2]^-$ (see Figure 3).

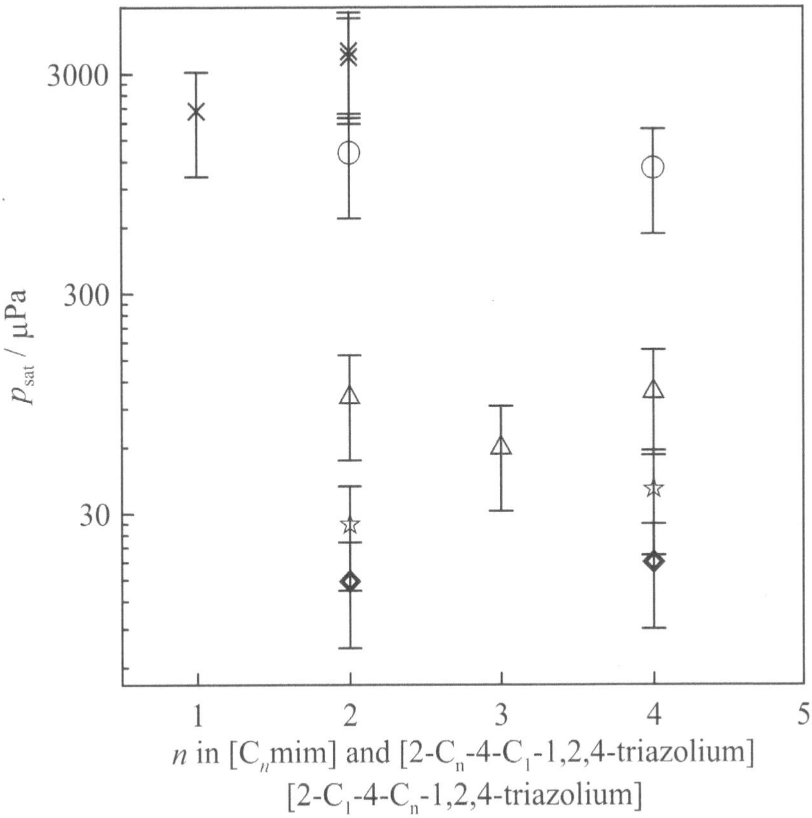

Figure 3. Vapour pressures of typical ILs. Chain-length dependence of absolute vapour pressures at T = 423.15 K for homologous series: (Δ)—[$C_n$mim][CF_3SO_3] from [16], (○)—[$C_n$mim][NTf_2] [5], (◊)—[$C_n$mim][PF_6] from [7], (☆)—[$C_n$mim][BF_4] from [6], and (x)—[2-C_n-4-C_m-1,2,4-triazolium][NTf_2] from this work.

Figure 3 shows that the vapour pressures of [di-alkyl-1,2,4-triazolium][NTf_2] are several times higher than those of ILs based on the [BF_4] and [PF_6] anions. The values of the absolute vapour pressures of the [di-alkyl-1,2,4-triazolium][NTf_2] and [$C_n$mim][NTf_2] series do not differ significantly. The apparent decrease in vapour pressure for ILs with the [C_2mim] cation compared to ILs with the [C_4mim] cation appears to be common for this series of ILs with [PF_6] and [BF_4] anions and is not as pronounced for ILs with the same cation but [NTf_2] and [CF_3SO_3] anions. In the case of [di-alkyl-1,2,4-triazolium][NTf_2], replacing one of the CH_3 substituents with C_2H_5 leads to an almost twofold increase in vapour pressure. Such an increase can be interpreted as an even–odd effect in the vaporisation properties for ILs with a short alkyl chain. A similar dependence is observed for [$C_n$mim][CF_3SO_3] when going from n = 3 to 4 [16].

The standard molar enthalpies of vaporisation of ionic liquids (see Table 2) were derived from the temperature dependence of the absolute vapour pressures given in Table S2 (see for details in Supplemental Materials).

Table 2. Thermodynamics of vaporisation of [2-C_n-4-C_m-1,2,4-T][NTf_2] with n, m = 1, 2 derived from experimental results [a].

IL Triazole Cation	T-Range	T_{av}	$\Delta_l^g H_m^o(T_{av})$ [b]	$\Delta_l^g G_m^o(T_{av})$ [c]	$\Delta_l^g C_{p,m}^o$ [d]	$\Delta_l^g H_m^o$(298 K) [e]
	K		kJ·mol^{-1}		J·K^{-1}·mol^{-1}	kJ·mol^{-1}
1	2	3	4	5	6	7
[2-C_1-4-C_1-1,2,4-T]	348-395	371.0	122.2 ± 1.0	69.5 ± 1.0	−49	125.7 ± 1.2
[2-C_1-4-C_2-1,2,4-T]	345-393	368.4	123.7 ± 1.0	68.1 ± 1.0	−57	127.7 ± 1.3
[2-C_2-4-C_1-1,2,4-T]	343-390	365.9	122.4 ± 1.0	68.5 ± 1.0	−57	126.3 ± 1.3

[a] Uncertainties of thermodynamic functions are expressed as the standard uncertainty. [b] The standard molar enthalpy of vaporisation at the average temperature, T_{av}, of the QCM experiment. [c] The standard Gibbs energies of vaporisation at the average temperature, T_{av}, of the QCM experiment. [d] From Table S3. [e] Adjusted to 298 K according to Kirchhoff's Law using the $\Delta_l^g C_{p,m}^o$–values from column 6. The experimental QCM uncertainties were extended with uncertainty of the heat capacity difference assessed to be of 20 J·K^{-1}·mol^{-1}.

The vaporisation enthalpies of triazolium-based ionic liquids were determined for the first time. However, it is interesting to compare these results with those known for imidazolium-based ionic liquids that are paired with the same [NTf_2] anion and have a comparable number of C atoms in the alkyl chains attached to the cation. It has been found that the $\Delta_l^g H_m^o$(298 K) = 122.7 ± 1.1 kJ·mol^{-1} [5] of [C_1mim][NTf_2] is slightly lower than that of [2-C_1-4-C_1-1,2,4-T][NTf_2] from this work. The vaporisation enthalpy $\Delta_l^g H_m^o$(298 K) = 126.4 ± 1.1 kJ·mol^{-1} [5] of [C_2mim][NTf_2] is almost the same as that of the corresponding triazolium-based ILs [2-C_1-4-C_2-1,2,4-T][NTf_2] and [2-C_2-4-C_1-1,2,4-T][NTf_2] (see Table 2). In general, however, the results for the triazolium- and imidazolium-based ILs can be considered to agree within their combined uncertainties. This similarity opens up the possibility of establishing structure–property correlations between the imidazolium-based ILs (for which a considerable amount of vaporisation thermodynamics data are available) and the triazolium-based ILs, whose data are only limited by the results of the current study. This correlation will be discussed in Section 3.4.

3.2. *Absolute Vapour Pressures and Vaporisation Thermodynamics of Molecular Liquids: 1,2,4-Triazoles*

The primary vapour pressure–temperature dependencies for 1H-1,2,4-triazole, 1-methyl-1,2,4-triazole, 1-ethyl-1,2,4-triazole, and 1H-1,2,3-triazole measured in this work (see Table S4) were fitted uniformly using a three-parametric equation (see Supplemental Materials for details). From these data, the standard molar enthalpies of vaporisation/sublimation at the respective temperatures T were derived (see Table S4). The final results at the reference temperature T = 298 K, $\Delta_{l,cr}^g H_m^o$(298 K), are summarised and compared in Table 3.

As can be seen from Table 3, the available experimental data for 1H-1,2,4-triazole are confusing. Our complementary transpiration measurements for this compound have helped to resolve inconsistencies and determine the level of enthalpy of sublimation for this compound. The weighted average $\Delta_{cr}^g H_m^o$(298 K) = 84.3 ± 0.5 kJ·mol^{-1} was recommended for thermochemical calculations. The latter value was combined with the fusion enthalpy for this compound (see Table S5) to derive the vaporisation enthalpy $\Delta_l^g H_m^o$(298 K) = 71.2 ± 1.0 kJ·mol^{-1} of 1H-1,2,4-triazole (see Table 3), which is essential as "*molecular liquid*" for the purpose of this work. The vapour pressures and vaporisation enthalpies for 1-methyl-1,2,4-triazole, 1-ethyl-1,2,4-triazole, and 1H-1,2,3-triazole were measured for the first time in this work.

Table 3. Compilation of available enthalpies of sublimation/vaporisation $\Delta_{cr,l}^{g}H_{m}^{o}$ for triazoles.

Compound/CAS	M [a]	*T*-Range	$\Delta_{cr,l}^{g}H_{m}^{o}(T_{av})$	$\Delta_{cr,l}^{g}H_{m}^{o}$(298 K) [b]	Ref.
		K	kJ·mol^{-1}	kJ·mol^{-1}	
1H-1,2,4-triazole (cr)	K	322–350	84.1 ± 1.0	84.6 ± 1.0	[17]
	n/a			(80.6 ± 0.5)	[18]
	K	281.5–295.7	84.1 ± 1.3	84.0 ± 1.3	[19]
	C	306.0	80.5 ± 0.9	(80.6 ± 0.9)	[20]
	T	322.2–367.5	83.7 ± 0.6	84.3 ± 0.7	Table S4
				84.3 ± 0.5 [c]	average
1H-1,2,4-triazole (liq)				71.2 ± 1.0	Table S5
1-methyl-1,2,4-triazole (liq)	T	275.0–337.8	53.1 ± 0.3	53.4 ± 0.4	Table S4
1-ethyl-1,2,4-triazole (liq)	T	280.6–313.1	57.0 ± 0.4	56.8 ± 0.5	Table S4
1H-1,2,3-triazole (liq)	T	297.2–348.7	59.9 ± 0.4	59.9 ± 0.4	Table S4
1-methyl-1,2,3-triazole (liq)				44.0 ± 1.5	Figure 4
1-ethyl-1,2,3-triazole (liq)				47.0 ± 1.5	Figure 5

[a] Techniques: K = Knudsen-effusion method; T = transpiration method; C = calorimetry; n/a = not available. [b] Uncertainty of the sublimation/vaporisation enthalpy $U(\Delta_{cr,l}^{g}H_{m}^{o})$ is the expanded uncertainty (0.95 level of confidence, k = 2) calculated according to a procedure described elsewhere [21,22]. It includes uncertainties from the experimental conditions, uncertainties of vapour pressure, uncertainties from the fitting equation, and uncertainties from temperature adjustment to T = 298 K. [c] Weighted mean value (the uncertainties were used as a weighting factor). Values in parentheses were excluded from the calculation of the mean. Value in bold is recommended for further thermochemical calculations.

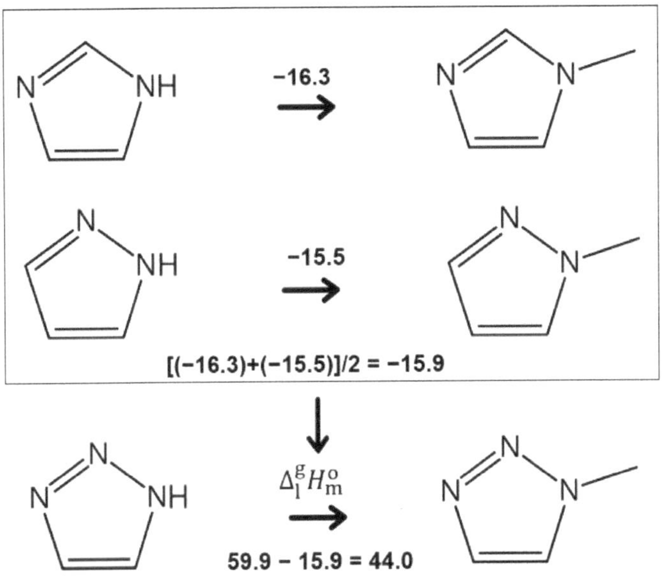

Figure 4. Evaluation of the vaporisation enthalpy, $\Delta_{l}^{g}H_{m}^{o}$(298 K), for 1-methyl-1,2,3-triazole using structure–property relationships. Experimental data are listed in Table 3 and Table S6. (All data are given in kJ·mol^{-1}).

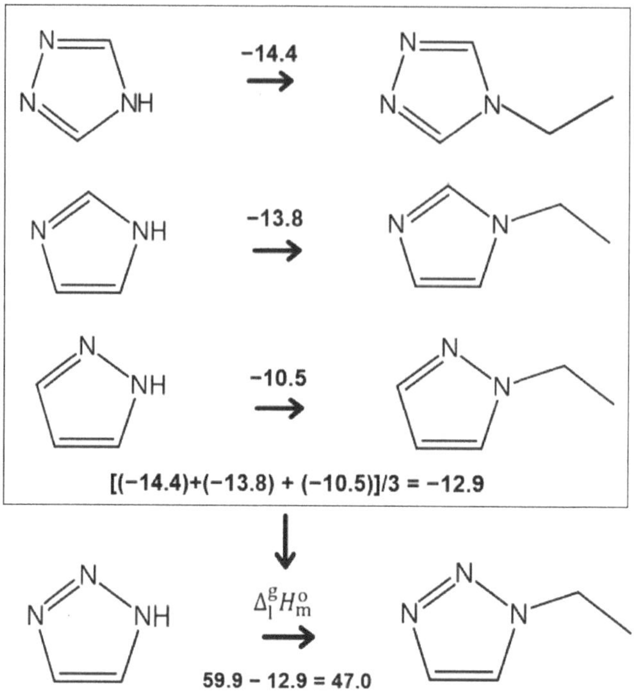

Figure 5. Evaluation of the vaporisation enthalpy, $\Delta_l^g H_m^o$(298 K), for 1-ethyl-1,2,3-triazole using structure–property relationships. Experimental data are listed in Table 3 and Table S6. (All data are given in kJ·mol^{-1}).

3.3. Vaporisation Thermodynamics of Molecular Liquids: 1,2,3-Triazoles

From the 1,2,3-triazole family, only 1H-1,2,3-triazole was commercially available for thermodynamic studies. For the purposes of this work, the vaporisation enthalpies for 1-methyl-1,2,3-triazole and 1-ethyl-1,2,3-triazole are required for the structure–property correlations with the corresponding ionic liquids. To assess these vaporisation enthalpies, we used reliable thermodynamic data on imidazole and alkylimidazoles as well as on pyrazole and alkylpyrazoles from the literature (see Table S6). The idea of evaluation is shown in Figure 4.

It has turned out that the replacement of the hydrogen bonded to the nitrogen atom in both 1-H-imidazole and 1H-pyrazole by the CH_3 group lowers the vaporisation enthalpies for both molecules by ≈16 kJ·mol^{-1} (see Figure 4). It can therefore be assumed that the same decrease can also be expected for 1H-1,2,3-triazole. With the known vaporisation enthalpy $\Delta_l^g H_m^o$(298 K) = 59.9 ± 0.4 kJ·mol^{-1} of 1H-1,2,3-triazole (see Table 3) and the averaged contribution H → CH_3 = −15.9 kJ·mol^{-1} (see Figure 4), the required vaporisation enthalpy $\Delta_l^g H_m^o$(298 K) = 44.0 ± 1.5 kJ·mol^{-1} of 1-methyl-1,2,3-triazole was derived (see Table 3).

A similar idea was realised for 1-methyl-1,2,3-triazole, where the averaged contribution H → CH_3CH_2 = −12.9 kJ·mol^{-1} (see Figure 5) was added to $\Delta_l^g H_m^o$(298 K) = 59.9 ± 0.4 kJ·mol^{-1} of 1H-1,2,3-triazole (see Table 3) to obtain the required vaporisation enthalpy $\Delta_l^g H_m^o$(298 K) = 47.0 ± 1.5 kJ·mol^{-1} of 1-ethyl-1,2,3-triazole (see Table 3).

3.4. Molecular versus Ionic Liquids: Structure–Property Relationships for Predicting Vaporisation Thermodynamics

It is noticeable that the vaporisation enthalpies of heterocycles with two or three nitrogen atoms have very similar vaporisation enthalpies (see Figure 6).

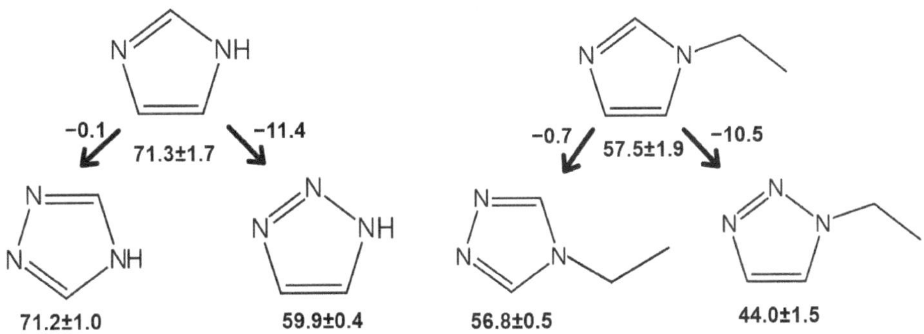

Figure 6. Comparison of the vaporisation enthalpies, $\Delta_l^g H_m^o$(298 K), of 1-H-imidazole with those of 1H-1,2,4-triazole and 1H-1,2,3-triazole (**left**). Comparison of the vaporisation enthalpies, $\Delta_l^g H_m^o$(298 K), of 1-ethyl-imidazole with those of 1-ethyl-1,2,4-triazole and 1-ethyl-1,2,3-triazole (**right**). Experimental data are from Table 3 and Table S6. (All data are given in kJ·mol^{-1}).

Indeed, $\Delta_l^g H_m^o$(298 K) = 71.2 ± 1.0 kJ·mol^{-1} of 1H-1,2,4-triazole (see Table 3) is indistinguishable from $\Delta_l^g H_m^o$(298 K) = 71.3 ± 1.7 kJ·mol^{-1} of 1H-imidazole (see Table S6). Also, $\Delta_l^g H_m^o$(298 K) = 56.8 ± 0.5 kJ·mol^{-1} of 1-ethyl-1,2,4-triazole (see Table 3) is almost the same as the $\Delta_l^g H_m^o$(298 K) = 57.5 ± 1.9 kJ·mol^{-1} of 1-ethyl-imidazole (see Table S6). A similar trend was already observed in Section 3.1 for the triazolium- and imidazolium-based ILs.

As a matter of fact, the vaporisation thermodynamics of the [$C_n$mim][NTf_2] *ionic liquids* family with alkyl chain length n = 1 to 18 are well established [5], and the reliable vaporisation enthalpies, $\Delta_l^g H_m^o$(298 K), are compiled in Table S7. The reliable vaporisation enthalpies, $\Delta_l^g H_m^o$(298 K), of the corresponding *molecular liquids* of the 1-alkyl-imidazole family with alkyl chain length n = 1 to 18 are also known and are summarised in Table S7. The correlation of the $\Delta_l^g H_m^o$(298 K)-values of molecular liquids (axis X) and ionic liquids (axis Y) is shown in Figure 7.

As can be seen from this plot, there is a remarkable linear correlation between the $\Delta_l^g H_m^o$(298 K)-values for the [$C_n$mim][NTf_2] family and the 1-alkyl-imidazole family:

$$\Delta_l^g H_m^o([C_n\text{mim}][NTf_2]) = 0.9781 \times \Delta_l^g H_m^o(\text{1-alkyl-imidazole}) + 69.1 \text{ with } R^2 = 0.9989$$

The very high correlation coefficient R^2 can be seen as proof of the good consistency of the experimental data for each family.

It is now interesting to know how the 1,2,4-triazolium-based ILs are associated with this correlation. The experimental vaporisation enthalpies for these *ionic liquids* from Table 2 and their corresponding *molecular liquids*, 1-methyl-1,2,4-triazole and 1-ethyl-1,2,4-triazole from Table 3, are plotted in Figure 7, and the points are highlighted in red. Figure 7 shows that the point for [2-C_1-4-C_1-1,2,4-T][NTf_2] is slightly above the line, but this outlier is to be expected since such behaviour of the first representatives in the homologous series is typical for *ionic* and *molecular* liquids [5]. However, the points for [2-C_1-4-C_2-1,2,4-T][NTf_2] and [2-C_2-4-C_1-1,2,4-T][NTf_2] lie practically (within the uncertainties) on the correlation of the imidazole derivatives. This observation suggests that the vaporisation enthalpies of ILs based on 1,2,4-triazolium with a longer than ethyl chain length are likely to be very similar to those of the corresponding [$C_n$mim][NTf_2] family. This assumption is important for chemical engineering calculations with the alkyl-1,2,4-triazolium-based ILs, which are promising for various practical applications [23].

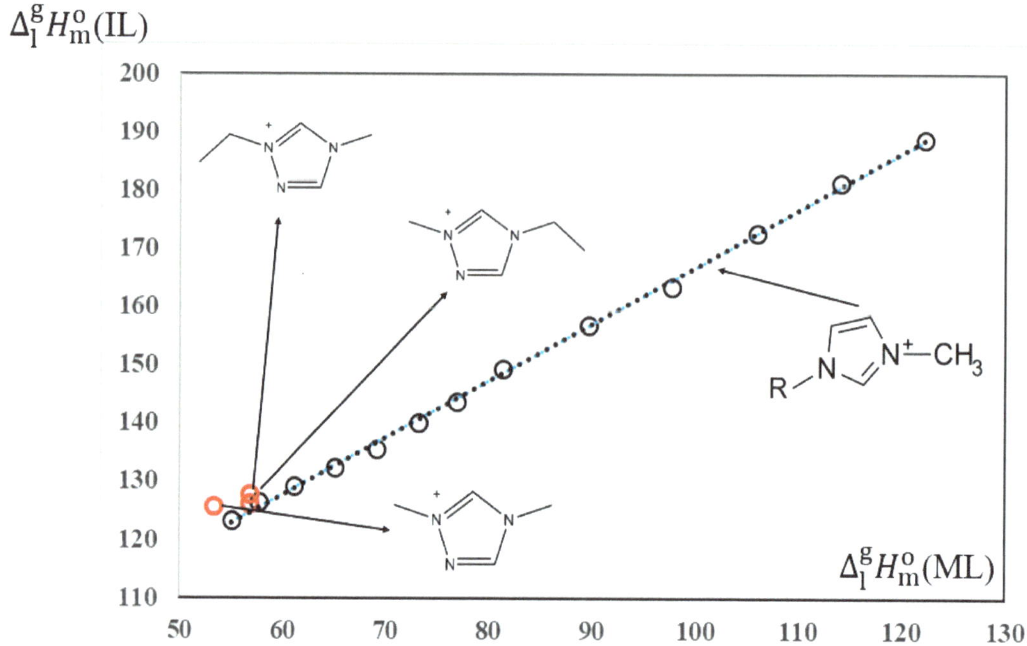

Figure 7. The correlation of the $\Delta_l^g H_m^o$(298 K)-values of *molecular liquids* (axis *X*) versus *ionic liquids* (axis *Y*). Experimental data are listed in Table 3, Tables S6 and S7. (All data are given in kJ·mol^{-1}).

What about the vaporisation enthalpies, $\Delta_l^g H_m^o$(298 K), of the 1,2,3-triazolium-based ILs which are even more promising than 1,2,4-triazolium-based ILs [24,25]? To approach the answer to this question, we compare the experimental data for heterocycles in Figure 6. While the derivatives of 1-H-imidazole and 1-H-1,2,4-triazole have very close $\Delta_l^g H_m^o$(298 K)-values, the derivatives of 1-H-1,2,3-triazole, in contrast, are obviously and systematically more volatile by ≈11 kJ·mol^{-1}. These systematics, in conjunction with the structure–property observations specific to imidazolium and triazolium ILs evident in Figure 7, suggest that the vaporisation enthalpies for the [alkyl-1,2,3-triazolium][NTf_2] family could be reliably estimated from the data for the [$C_n$mim][NTf_2] family by subtracting 11 kJ·mol^{-1} from imidazolium-based ILs of the corresponding chain length. These estimates can be reliably used for chemical engineering calculations with the alkyl-1,2,3-triazolium-based ILs required for various practical applications.

4. Conclusions

Structure–property correlations between similarly shaped *ionic* and *molecular* liquids have proven to be a valuable tool to determine the consistency of experimental data within homologous series and to reveal general trends in vaporisation thermodynamics. These trends allow an assessment of the vaporisation enthalpies, $\Delta_l^g H_m^o$(298 K), of new ILs with the help of the well-established data for imidazolium-based *ionic liquids* combined with the evaluated data for *molecular liquids*.

Supplementary Materials: The following supporting information can be downloaded at https://www.mdpi.com/article/10.3390/liquids4030032/s1, Table S1: Provenance and purity of ILs studied in this work. Table S2: Details on vapor pressure measurements with Quartz-Crystal Microbalance. Results for vapor pressure temperature dependence and vaporization enthalpies determined by QCM. Table S3: Details on vapour pressure measurements of molecular liquids using the transpiration method. Compilation of data on molar heat capacities heat capacity differences. Table S4: Results of transpiration method: absolute vapour pressures, standard molar vaporisation/sublimation enthalpies and standard molar vaporisation/sublimation entropies. Table S5: Thermodynamics of phase transitions of 1H-1,2,4-triazole. Table S6: Auxiliary data for vaporisation enthalpies of nitrogen containing heterocycles available from the literature. Table S7: Compilation of enthalpies of vaporization for [Cnmim][NTf2] and for alkyl-imidazole. Figure S1: The scheme of the QCM experimental setup. Figures S2–S4: The IR spectra for ionic liquids under study. Figure S5: The scheme of the transpiration experimental setup. References [5,7,12,13,21,22,26–37] are cited in the Supplementary Materials.

Author Contributions: Methodology, S.P.V.; validation, S.P.V.; formal analysis, S.P.V. and D.H.Z.; investigation, D.H.Z.; data curation, D.H.Z.; writing—original draft, S.P.V.; writing—review and editing, D.H.Z.; project administration, S.P.V.; funding acquisition, S.P.V. All authors have read and agreed to the published version of the manuscript.

Funding: This work has been supported by the German Research Foundation (DFG) in the frame of the priority programme SPP 1807 "Control of London Dispersion Interactions in Molecular Chemistry" (grants VE 265-9/1 and VE 265-9/2), as well as of the priority programme SPP 1708 "Material Synthesis Near Room Temperature" (grant VE 265/14-1). DHZ acknowledges the financial support from DFG, grant ZA 872/3-1, 407078203.

Data Availability Statement: The data supporting the results of this study are available in the manuscript text and in the electronic Supplementary Information of this article.

Acknowledgments: The work is carried out in accordance with the Strategic Academic Leadership Program "Priority 2030" of the Kazan Federal University of the Government of the Russian Federation.

Conflicts of Interest: Although the author Dzmitry H. Zaitsau was employed by the company FVTR GmbH, Zaitsau and Verevkin declare that the research was conducted in the absence of any commercial or financial relationships that could be construed as a potential conflict of interest. The company had no role in the design of the study; in the collection, analyses, or interpretation of data; in the writing of the manuscript, or in the decision to publish the results.

References

1. Brehm, M.; Pulst, M.; Kressler, J.; Sebastiani, D. Triazolium-Based Ionic Liquids: A Novel Class of Cellulose Solvents. *J. Phys. Chem. B* **2019**, *123*, 3994–4003. [CrossRef] [PubMed]
2. Daily, L.A.; Miller, K.M. Correlating structure with thermal properties for a series of 1-alkyl-4-methyl-1,2,4-triazolium ionic liquids. *J. Org. Chem.* **2013**, *78*, 4196–4201. [CrossRef] [PubMed]
3. Fei, Y.; Chen, Z.; Zhang, J.; Yu, M.; Kong, J.; Wu, Z.; Cao, J.; Zhang, J. Thiazolium-based ionic liquids: Synthesis, characterization and physicochemical properties. *J. Mol. Liq.* **2021**, *342*, 117553. [CrossRef]
4. Chand, D.; Wilk-Kozubek, M.; Smetana, V.; Mudring, A.-V. Alternative to the Popular Imidazolium Ionic Liquids: 1,2,4-Triazolium Ionic Liquids with Enhanced Thermal and Chemical Stability. *ACS Sustain. Chem. Eng.* **2019**, *7*, 15995–16006. [CrossRef]
5. Verevkin, S.P.; Zaitsau, D.H.; Emel'yanenko, V.; Yermalayeu, A.V.; Schick, C.; Liu, H.; Maginn, E.J.; Bulut, S.; Krossing, I.; Kalb, R.; et al. Making sense of enthalpy of vaporization trends for ionic liquids: New experimental and simulation data show a simple linear relationship and help reconcile previous data. *J. Phys. Chem. B* **2013**, *117*, 6473–6486. [CrossRef] [PubMed]
6. Zaitsau, D.H.; Yermalayeu, A.V.; Schubert, T.J.S.; Verevkin, S.P. Alkyl-imidazolium tetrafluoroborates: Vapor pressure, thermodynamics of vaporization, and enthalpies of formation. *J. Mol. Liq.* **2017**, *242*, 951–957. [CrossRef]
7. Zaitsau, D.H.; Yermalayeu, A.V.; Emel'yanenko, V.N.; Butler, S.; Schubert, T.; Verevkin, S.P. Thermodynamics of imidazolium-based ionic liquids containing PF6 anions. *J. Phys. Chem. B* **2016**, *120*, 7949–7957. [CrossRef]
8. Zaitsau, D.H.D.H.; Verevkin, S.P.S.P. Imidazolium-based ionic liquids containing FAP anion: Thermodynamic study. *J. Mol. Liq.* **2019**, *287*, 110959. [CrossRef]
9. Verevkin, S.P.; Zaitsau, D.H.; Zherikova, K.V. Molecular liquids versus ionic liquids: Thermodynamic insights into the interplay between inter-molecular and intra-molecular hydrogen bonding. *J. Mol. Liq.* **2023**, *382*, 121938. [CrossRef]
10. Verevkin, S.P.; Zaitsau, D.H.; Ludwig, R. Molecular Liquids versus Ionic Liquids: The Interplay between Inter-Molecular and Intra-Molecular Hydrogen Bonding as Seen by Vaporisation Thermodynamics. *Molecules* **2023**, *28*, 539. [CrossRef]
11. Verevkin, S.P.; Zherikova, K.V.; Martynenko, E.A. Molecular versus ionic liquids: Development of a thermodynamic framework for predicting vaporization thermodynamics. *J. Mol. Liq.* **2022**, *350*, 118576. [CrossRef]
12. Verevkin, S.P.; Zaitsau, D.H.; Emel'yanenko, V.; Heintz, A.; Emel'yanenko, V.N.; Heintz, A. A new method for the determination of vaporization enthalpies of ionic liquids at low temperatures. *J. Phys. Chem. B* **2011**, *115*, 12889–12895. [CrossRef]
13. Verevkin, S.P.; Emel'yanenko, V.N. Transpiration method: Vapor pressures and enthalpies of vaporization of some low-boiling esters. *Fluid Phase Equilib.* **2008**, *266*, 64–75. [CrossRef]

14. Kernchen, U.; Etzold, B.; Korth, W.; Jess, A. Solid catalyst with ionic liquid layer (SCILL)—A new concept to improve selectivity illustrated by hydrogenation of cyclooctadiene. *Chem. Eng. Technol.* **2007**, *30*, 985–994. [CrossRef]
15. Fehrmann, R.; Riisager, A.; Haumann, M. *Supported Ionic Liquids: Fundamentals and Applications*; Fehrmann, R., Riisager, A., Haumann, M., Eds.; WILEY-VCH Verlag: Weinheim, Germany, 2014; ISBN 9783527324293.
16. Zaitsau, D.H.; Yermalayeu, A.V.; Emel'yanenko, V.N.; Verevkin, S.P. Thermodynamics of Imidazolium-Based Ionic Liquids Containing the Trifluoromethanesulfonate Anion. *Chem. Eng. Technol.* **2018**, *41*, 1604–1612. [CrossRef]
17. Zimmermann, H.; Geisenfelder, H. Über die Mesomerieenergie von Azolen. *Z. Elektrochem. Berichte Bunsenges. Phys. Chem.* **1961**, *65*, 368–371. [CrossRef]
18. Faour, M.; Akasheh, T.S. Heat of combustion of some N-heterocycle compounds. Part 1. *J. Chem. Soc. Perkin Trans. 2* **1985**, *6*, 811–813. [CrossRef]
19. Jiménez, P.; Roux, M.V.; Turrión, C. Thermochemical properties of N-heterocyclic compounds II. Enthalpies of combustion, vapour pressures, enthalpies of sublimation, and enthalpies of formation of 1,2,4-triazole and benzotriazole. *J. Chem. Thermodyn.* **1989**, *21*, 759–764. [CrossRef]
20. Sabbah, R.; Perez, L. Energetics of Intramolecular Bonds in 1H-1,2,4-Triazole and 1H-Benzotriazole. *Aust. J. Chem.* **1999**, *52*, 235. [CrossRef]
21. Verevkin, S.P.; Sazonova, A.Y.; Emel'yanenko, V.N.; Zaitsau, D.H.; Varfolomeev, M.A.; Solomonov, B.N.; Zherikova, K. V Thermochemistry of halogen-substituted methylbenzenes. *J. Chem. Eng. Data* **2015**, *60*, 89–103. [CrossRef]
22. Emel'yanenko, V.N.; Verevkin, S.P. Benchmark thermodynamic properties of 1,3-propanediol: Comprehensive experimental and theoretical study. *J. Chem. Thermodyn.* **2015**, *85*, 111–119. [CrossRef]
23. Rodriguez Castillo, A.S.; Guihéneuf, S.; Biard, P.; Paquin, L.; Amrane, A.; Couvert, A. Physicochemical properties of some hydrophobic room-temperature ionic liquids applied to volatile organic compounds biodegradation processes. *J. Chem. Technol. Biotechnol.* **2018**, *93*, 215–223. [CrossRef]
24. Yacob, Z.; Liebscher, J. 1,2,3-Triazolium Salts as a Versatile New Class of Ionic Liquids. In *Ionic Liquids—Classes and Properties*; InTech: London, UK, 2011.
25. Khan, S.S.; Hanelt, S.; Liebscher, J. Versatile synthesis of 1,2,3-triazolium-based ionic liquids. *Arkivoc* **2009**, *2009*, 193–208. [CrossRef]
26. Sauerbrey, G. Verwendung von Schwingquarzen zur Wägung dünner Schichten und zur Mikrowägung. *Z. Phys.* **1959**, *155*, 206–222. [CrossRef]
27. Clarke, E.C.W.; Glew, D.N. Evaluation of thermodynamic functions from equilibrium constants. *Trans. Faraday Soc.* **1966**, *62*, 539–547. [CrossRef]
28. Kulikov, D.; Verevkin, S.P.; Heintz, A. Determination of vapor pressures and vaporization enthalpies of the aliphatic branched C5 and C6 alcohols. *J. Chem. Eng. Data* **2001**, *46*, 1593–1600. [CrossRef]
29. Chickos, J.S.; Hosseini, S.; Hesse, D.G.; Liebman, J.F. Heat capacity corrections to a standard state: A comparison of new and some literature methods for organic liquids and solids. *Struct. Chem.* **1993**, *4*, 271–278. [CrossRef]
30. Acree, W.E., Jr.; Chickos, J.S. Phase transition enthalpy measurements of organic and organometallic compounds and ionic liquids. Sublimation, vaporization, and fusion enthalpies from 1880 to 2015. Part 2. C11–C192. *J. Phys. Chem. Ref. Data* **2017**, *46*, 13104. [CrossRef]
31. Ahmadi, A.; Haghbakhsh, R.; Raeissi, S.; Hemmati, V. A simple group contribution correlation for the prediction of ionic liquid heat capacities at different temperatures. *Fluid Phase Equilib.* **2015**, *403*, 95–103. [CrossRef]
32. Ribeiro da Silva, M.A.; Ribeiro da Silva, M.d.D.M.; Matos, M.A.R.; Jiménez, P.; Victoria Roux, M.; Martin-Luengo, M.; Elguero, J.; Claramunt, R.; Cabildo, P. Enthalpies of combustion, heat capacities, and enthalpies of vaporisation of 1-phenylimidazole and 1-phenylpyrazole. *J. Chem. Thermodyn.* **2000**, *32*, 237–245. [CrossRef]
33. Almeida, A.R.R.P.; Monte, M.J.S. Thermodynamic study of phase transitions of imidazoles and 1-methylimidazoles. *J. Chem. Thermodyn.* **2012**, *44*, 163–168. [CrossRef]
34. Safronov, S.P.; Nagrimanov, R.N.; Samatov, A.A.; Emel'yanenko, V.N.; Zaitsau, D.H.; Pimerzin, A.A.; Skrzypczak, A.; Verevkin, S.P. Benchmark properties of pyrazole derivatives as a potential liquid organic hydrogen carrier: Evaluation of thermochemical data with complementary experimental and computational methods. *J. Chem. Thermodyn.* **2019**, *128*, 173–186. [CrossRef]
35. Verevkin, S.S.P.; Zaitsau, D.H.D.; Emel'yanenko, V.N.; Paulechka, Y.U.; Blokhin, A.V.A.; Bazyleva, A.A.B.; Kabo, G.G.J.; Emel'yanenko, V.; Paulechka, E.; Blokhin, A.V.A.; et al. Thermodynamics of ionic liquids precursors: 1-methylimidazole. *J. Phys. Chem. B* **2011**, *115*, 4404–4411. [CrossRef] [PubMed]
36. Vitorino, J.; Agapito, F.; Bernardes, C.E.S.; Minas da Piedade, M.E. Thermochemistry of 1-alkylimidazoles. *J. Chem. Thermodyn.* **2015**, *80*, 59–64. [CrossRef]
37. Emel'yanenko, V.N.; Portnova, S.V.; Verevkin, S.P.; Skrzypczak, A.; Schubert, T. Building blocks for ionic liquids: Vapor pressures and vaporization enthalpies of 1-(n-alkyl)-imidazoles. *J. Chem. Thermodyn.* **2011**, *43*, 1500–1505. [CrossRef]

liquids

MDPI

Article

Calculation of Hydrogen Bonding Enthalpy Using the Two-Parameter Abraham Equation

Boris N. Solomonov, Mansur B. Khisamiev and Mikhail I. Yagofarov *

Department of Physical Chemistry, Kazan Federal University, Kremlevskaya Str. 18, Kazan 420008, Russia; boris.solomonov@kpfu.ru (B.N.S.); mansur.khisamiev@mail.ru (M.B.K.)
* Correspondence: miiyagofarov@kpfu.ru

Abstract: In this work, an approach to the calculation of hydrogen bonding enthalpies is proposed. It employs the correlation proposed by M.H. Abraham, establishing the connection between the equilibrium constant (K_{HB}) and acidity (α_2^H) and basicity (β_2^H) parameters: $\log K_{HB} = 7.354 \cdot \alpha_2^H \cdot \beta_2^H - 1.099$. Hydrogen bonding enthalpy ($\Delta_{HB}H$) is found using the compensation relationship with Gibbs energy ($\Delta_{HB}G$): $\Delta_{HB}G = 0.66 \cdot \Delta_{HB}H + 2.5$ kJ·mol^{-1}. This relationship enables the calculation of the enthalpy, Gibbs energy and entropy of hydrogen bonding. The validity of this approach was tested against 122 experimental hydrogen bonding enthalpies values available from the literature. The root mean square deviation and average deviation equaled 1.6 kJ·mol^{-1} and 0.5 kJ·mol^{-1}, respectively.

Keywords: hydrogen bonding enthalpy; complexation equilibrium; compensation relationship

1. Introduction

There is a continuing interest in the hydrogen bonding phenomenon [1–4]. This is evidenced by the growing number of publications using hydrogen bonding (HB) as a keyword. However, it is worth highlighting that the number of direct experimental studies on the thermodynamic parameters of hydrogen bond formation has been declining over the last 3 decades. Among the reasons for this is the technically complex and time-consuming measurement procedure. Several approaches for the simple prediction of the thermodynamic functions of HB have been proposed [5–8]. These works complement the existing data on the Gibbs energies ($\Delta_{HB}G$) and enthalpies ($\Delta_{HB}H$) of acid-base complexation. However, the use of these methods is limited by the available constants characterizing the proton acceptor and proton donor abilities.

Recently [9,10], we developed an approach for the calculation of complexation enthalpies (protic acid-base and charge-transfer complexes) using the experimental values of the equilibrium constants at 298.15 K. The determination of complexation enthalpies is a far more difficult procedure, compared with that of the Gibbs energies. This is evidenced by the better reproducibility of the Gibbs energies [1,9,10], and it naturally follows from the fact that most binding enthalpies values are found from the temperature dependences of the equilibrium constants (K_{HB}) [1,4].

The purpose of this work is to test the predictive capability of a successful method for K_{HB} prediction developed by Abraham et al. combined with the approach for $\Delta_{HB}H$ calculation using K_{HB} values.

2. Methodology

Below, the complex formation equilibrium (1) between a proton donor, A-H, and proton acceptor, B, is considered:

$$\mathrm{A-H+B} \overset{K_{HB}}{\rightleftarrows} \mathrm{A-H\cdots B} \quad (1)$$

The Gibbs energy of hydrogen bonding is related to the equilibrium constant of reaction (1) as follows:

$$\Delta_{HB}G = -RT \ln K_{HB} \quad (2)$$

The hydrogen bonding enthalpy can be determined from direct calorimetric experiments [11–13] or from the temperature dependence of K_{HB}, along with the entropy ($\Delta_{HB}S$) [1,4]:

$$\ln K_{HB} = -\frac{\Delta_{HB}H}{R} \cdot \frac{1}{T} + \frac{\Delta_{HB}S}{R} \quad (3)$$

K_{HB} depends on the proton acceptor ability of B and proton donor ability of A-H. Since the 1980s, several empirical two-parameter approaches have been proposed for predicting K_{HB} by multiplying the descriptors characterizing A-H and B, particularly by Raevsky et al. [6,7] and Abraham et al. [8]. M.H. Abraham and co-workers developed the acidity and basicity scales based on the analysis of 1:1 complexation constants in tetrachloromethane of (*a*) a set of acids with reference bases [14] and (*b*) a set of bases with reference acids [15]. In order to construct the scales, (1) two arbitrary acids (set *a*) or bases (set *b*) were chosen with the fixed acidity ($\log K_A^H$) and basicity constants ($\log K_B^H$), (2) the $\log K_{HB}$ values were correlated with the latter fixed $\log K_A^H$ and $\log K_B^H$ parameters to characterize the reference bases (set *a*) and acids (set *b*), and (3) the $\log K_A^H$ and $\log K_B^H$ values were obtained for the rest of the proton acceptors and proton donors from the correlations with $\log K_{HB}$. As a result, the logarithm of the concentration constant of hydrogen bonding ($K_{HB,c}$) in tetrachloromethane can be found using Equation (4):

$$\log K_{HB,c} = (7.354 \pm 0.019) \cdot \alpha_2^H \cdot \beta_2^H - (1.094 \pm 0.007) \quad (4)$$

where $\alpha_2^H = (\log K_A^H + 1.1)/4.636$ and $\beta_2^H = (\log K_B^H + 1.1)/4.636$ are re-scaled acidity and basicity parameters. For 1312 acid-base systems, the root mean square deviation of Equation (4) was 0.09 log units [8]. The acidity (α_2^H) and basicity (β_2^H) parameters have been derived for a wide range of organic compounds [14–16], which makes Equation (4) quite effective in the estimation of K_{HB} values. This methodology was later developed and fruitfully applied in the studies of hydrogen bonding and fluid phase equilibria by M.H. Abraham and his colleagues, particularly by W.E. Acree, Jr. [17–19], to whom the present issue is devoted.

In recent works [20,21], we studied the compensation relationship between the Gibbs energies and enthalpies of solvation and vaporization of organic non-electrolytes at 298.15 K and distinguished four types of solute–solvent systems, exhibiting different types of the Gibbs energy–enthalpy relationship, depending on their hydrogen bonding capabilities. Aromatic and short-chained aliphatic compounds incapable of hydrogen bonding dissolved in non-hydrogen-bonded solvents followed a general linear trend. Whenever hydrogen bonds were formed between a solute and non-hydrogen-bonded solvent, the Gibbs energies of solvation appeared to be systematically more positive; the slope of the relationship stayed the same. Based on these findings, we derived Equation (5), connecting the Gibbs energies and enthalpies of hydrogen bonding [9]:

$$\Delta_{HB}G/\left(\text{kJ} \cdot \text{mol}^{-1}\right) = 0.660 \cdot \Delta_{HB}H/\left(\text{kJ} \cdot \text{mol}^{-1}\right) + 2.5 \cdot n \quad (5)$$

where n is a number of hydrogen bonds in acid-base complex. Equation (5) was also applied to charge-transfer complexes with iodine and interhalogens [10]. The validity of Equation (5) was confirmed by comparing the complex formation enthalpies calculated using Equation (6) with the literature data on 293 acid-base systems [9] and 152 charge-transfer complexes:

$$\Delta_{HB}H/(\text{kJ} \cdot \text{mol}^{-1}) = \frac{\Delta_{HB}G/(\text{kJ} \cdot \text{mol}^{-1}) - 2.5 \cdot n}{0.660} \quad (6)$$

The root mean square deviation was 1.2 kJ·mol^{-1} for the hydrogen-bonded complexes and 1.4 kJ·mol^{-1} for the complexes with iodine and interhalogens.

It is worth highlighting that, in our previous studies of solvation thermodynamics [20, 21], the mole-fraction scale was adopted, so $\Delta_{\text{HB}}G$ in Equations (5) and (6) is also a mole-fraction-scale-based quantity. The relationships between mole-fraction and concentration constants and the respective Gibbs energies of hydrogen bonding under infinite dilution conditions are given by Equations (6) and (7):

$$K_{\text{HB}} = K_{\text{HB,c}}/V_{\text{m}}^{\text{S}} \tag{7}$$

$$\Delta_{\text{HB}}G = \Delta_{\text{HB}}G_c + RT\ln\left(\frac{V_{\text{m}}^{\text{S}}}{\text{L}\cdot\text{mol}^{-1}}\right) \tag{8}$$

where V_{m}^{S} is the molar volume of solvent at 298.15 K.

In this work, Equation (6) was combined with Equation (4) to obtain the hydrogen bonding enthalpies in tetrachloromethane directly from the acidity and basicity parameters of the proton donor and proton acceptor. Taking into account the molar volume of tetrachloromethane of 0.0965 L·mol^{-1} and combining Equations (4) and (6), one comes to a two-parameter Equation (9) for the calculation of $\Delta_{\text{HB}}H$:

$$\Delta_{\text{HB}}H/(\text{kJ}\cdot\text{mol}^{-1}) = -63.61\cdot\alpha_2^{\text{H}}\cdot\beta_2^{\text{H}} + 3.03 \tag{9}$$

It was tested against 122 literature values of the hydrogen bonded enthalpies determined in tetrachloromethane medium. The quality of the prediction was judged based on the root mean square deviation (*RMS*) and average deviation (*AD*) between the calculated and literature values given by Equations (10) and (11):

$$RMS = \sqrt{\frac{\Sigma[\Delta_{\text{HB}}H(\text{calc}) - \Delta_{\text{HB}}H(\text{lit})]^2}{N-1}} \tag{10}$$

$$AD = \frac{\Sigma[\Delta_{\text{HB}}H(\text{calc}) - \Delta_{\text{HB}}H(\text{lit})]}{N} \tag{11}$$

3. Results

The experimental data on the Δ_{HB}H values were collected from comprehensive books by Joesten and Schaad [1] and Laurence and Gal [3], as well as original research papers. For some compounds, the specific interaction enthalpies derived using the data on solvation enthalpies in the inert solvents and the pure base according to Ref. [22] are provided. The comparison between the calculated and literature data, along with the Δ_{HB}G found using Equations (4) and (8), is shown in Table 1.

Table 1. The Gibbs energies of hydrogen bonding at 298.15 K calculated according to Equation (4), the enthalpies of hydrogen bonding calculated according to Equation (9), and the literature values.

Proton Donor	Proton Acceptor	$-\Delta_{\text{HB}}G$(Eq. 4) [a] kJ·mol^{-1}	$-\Delta_{\text{HB}}H$(Eq.9) [a] kJ·mol^{-1}	$-\Delta_{\text{HB}}H$(lit.) kJ·mol^{-1}	Δ [b] kJ·mol^{-1}	Ref.
Phenol	Benzene	3.1	8.5	6.5	2.0	[1]
	Toluene	3.1	8.5	6.9	1.6	[1]
	m-Xylene	3.6	9.2	8.7	0.5	[1]
	p-Xylene	3.6	9.2	9.0	0.2	[1]
	Mesitylene	4.3	10.4	9.2	1.2	[1]
	Acetophenone	11.6	21.4	19.7	1.7	[1]
		11.6	21.4	21.1	0.3	[1]

Table 1. *Cont.*

Proton Donor	Proton Acceptor	$-\Delta_{HB}G$(Eq. 4) [a] kJ·mol^{-1}	$-\Delta_{HB}H$(Eq.9) [a] kJ·mol^{-1}	$-\Delta_{HB}H$(lit.) kJ·mol^{-1}	Δ [b] kJ·mol^{-1}	Ref.
	Cyclohexanone	13.7	24.5	22.7	1.8	[22]
	Butanone	12.4	22.6	21.8	0.8	[1]
	Acetone	11.9	21.8	19.7	2.1	[23]
		11.9	21.8	22.3	−0.5	[22]
		11.9	21.8	21.3	0.5	[24]
	Methyl acetate	10.9	20.3	18.8	1.5	[1]
	Ethyl acetate	10.9	20.3	21.8	−1.5	[1]
	Butyrolactone	11.9	21.8	20.5	1.3	[1]
	Dimethylformamide	18.2	31.4	28.7	2.7	[1]
		18.2	31.4	29.1	2.3	[22]
	Propanal	10.9	20.3	18.0	2.3	[1]
	Dimethylacetamide	19.2	32.9	30.8	2.1	[1]
		19.2	32.9	30.3	2.6	[22]
	Pyridine	15.2	26.8	27.2	−0.4	[1]
		15.2	26.8	26.8	0.0	[22]
	Diethyl ether	10.9	20.3	20.1	0.2	[1]
	Benzonitrile	7.9	15.7	13.8	1.9	[1]
		7.9	15.7	19.3	−3.6	[1]
	Tetrahydrofuran	11.6	21.4	22.1	−0.7	[1]
		11.6	21.4	23.0	−1.6	[1]
	Chlorocyclohexane	2.1	6.9	8.2	−1.3	[1]
	1-Bromohexane	2.6	7.7	6.7	1.0	[1]
	1-Chlorobutane	2.1	6.9	7.2	−0.2	[1]
		2.1	6.9	9.3	−2.4	[1]
	1-Bromobutane	2.6	7.7	7.2	0.5	[1]
	1-Iodobutane	3.3	8.8	5.4	3.4	[1]
	1-Iodohexane	3.3	8.8	7.3	1.5	[1]
	Diethyl sulfide	7.6	15.3	15.1	0.2	[1]
	Di-*n*-butyl sulfide	7.6	15.3	14.2	1.1	[1]
	Benzaldehyde	9.4	18.0	18.0	0.0	[1]
4-Fluorophenol	Benzene	3.3	8.7	7.3	1.4	[3]
		3.3	8.7	7.7	1.0	[3]
	Toluene	3.3	8.7	7.9	0.8	[3]
	p-Xylene	3.8	9.5	8.1	1.4	[3]
	Mesitylene	4.6	10.7	8.8	1.9	[3]
	Pentamethylbenzene	4.8	11.1	10.3	0.8	[3]
	Hexamethylbenzene	5.1	11.5	10.8	0.7	[3]
	1-Heptene	1.4	5.9	6.8	−0.9	[3]
	Diethyl ether	11.5	21.1	24.1	−3.0	[3]
		11.5	21.1	22.7	−1.6	[3]
	Acetophenone	12.2	22.3	20.8	1.5	[3]
	Acetone	12.5	22.7	22.4	0.3	[3]
	Butanone	13.0	23.6	21.1	2.5	[3]
		13.0	23.6	22.9	0.7	[3]
	Cyclohexanone	14.4	25.6	24.3	1.3	[3]
	Benzaldehyde	9.9	18.7	18.6	0.1	[3]
	Methyl formate	9.6	18.3	18.0	0.3	[3]
	Ethyl formate	9.6	18.3	18.0	0.3	[3]
	Ethyl acetate	11.5	21.1	20.8	0.3	[3]
	Methyl acetate	11.5	21.1	20.8	0.3	[3]
	Nitrobenzene	7.0	14.3	11.5	2.8	[3]
	Chlorocyclohexane	2.2	7.1	8.7	−1.6	[1]
	Bromocyclohexane	3.8	9.5	8.2	1.3	[1]
	1-Chlorobutane	2.2	7.1	8.1	−1.0	[1]
	1-Bromobutane	2.7	7.9	7.6	0.3	[1]
	1-Iodobutane	3.5	9.1	6.5	2.6	[1]

Table 1. *Cont.*

Proton Donor	Proton Acceptor	$\frac{-\Delta_{HB}G(\text{Eq. 4})}{\text{kJ·mol}^{-1}}$ a	$\frac{-\Delta_{HB}H(\text{Eq.9})}{\text{kJ·mol}^{-1}}$ a	$\frac{-\Delta_{HB}H(\text{lit.})}{\text{kJ·mol}^{-1}}$	$\frac{\Delta}{\text{kJ·mol}^{-1}}$ b	Ref.
	Benzonitrile	8.3	16.3	17.5	−1.2	[3]
	Dimethyl sulfide	7.2	14.7	13.0	1.7	[3]
	Ethanethiol	5.9	12.7	10.4	2.3	[3]
	Diethyl sulfide	8.0	15.9	14.7	1.2	[3]
	Dibutyl sulfide	8.0	15.9	15.5	0.4	[3]
	Pyridine	16.0	28.0	29.6	−1.6	[3]
Chloroform	Pyridine	3.5	9.0	10.0	−1.0	[1]
	Tetrahydrofuran	2.6	7.7	9.2	−1.5	[22]
	Triethylamine	4.5	10.7	13.5	−2.8	[22]
	Benzene	0.4	4.4	7.1	−2.7	[1]
	Diethyl ether	2.4	7.4	7.2	0.2	[22]
	Acetone	2.6	7.8	7.3	0.5	[22]
	Dimethylformamide	4.2	10.2	11.8	−1.6	[22]
	Dimethylacetamide	4.5	10.6	13.0	−2.4	[1]
	Dimethyl sulfoxide	4.3	10.4	10.5	−0.1	[22]
	Ethyl acetate	2.4	7.4	7.7	−0.3	[1]
		2.4	7.4	7.4	0.0	[22]
	Diethyl sulfide	1.6	6.2	7.1	−0.9	[1]
	Acetonitrile	1.5	6.1	4.9	1.2	[22]
	Nitromethane	1.5	6.1	3.7	2.4	[22]
Pyrrole	Acetonitrile	4.0	9.8	10.3	−0.5	[25]
	Anisole	3.7	9.4	7.4	2.0	[25]
	Benzene	1.6	6.1	5.7	0.4	[25]
	Benzonitrile	4.3	10.3	10.3	0.0	[25]
	Chlorobenzene	0.6	4.6	4.1	0.5	[25]
	Cyclohexanone	7.5	15.2	13.7	1.5	[25]
	Dimethylformamide	10.1	19.1	16.6	2.5	[25]
	Dimethyl sulfoxide	10.4	19.5	18.0	1.5	[25]
	Ethyl acetate	6.0	12.8	13.1	−0.3	[25]
	HMPA c	13.8	24.7	24.0	0.7	[25]
	Nitrobenzene	3.6	9.2	8.8	0.4	[25]
	Nitromethane	4.0	9.8	8.8	1.0	[25]
	Pyridine	8.4	16.5	16.2	0.3	[25]
	Tetrahydrofuran	6.4	13.5	15.2	−1.7	[25]
	Toluene	1.6	6.1	6.1	0.0	[25]
N-methylaniline	Dimethylformamide	4.8	11.1	9.9	1.2	[22]
	Dimethyl sulfoxide	5.0	11.3	9.9	1.4	[22]
	Ethyl acetate	2.8	8.0	5.4	2.6	[22]
	Pyridine	4.0	9.8	8.5	1.3	[22]
Methanol	Benzylamine	12.6	22.8	22.3	0.5	[22]
	Triethylamine	13.8	24.7	24.1	0.6	[22]
Butan-1-ol	Dimethylformamide	11.0	20.5	19.2	1.3	[1]
	Dimethylacetamide	11.7	21.5	18.7	2.8	[1]
	Triethylamine	11.8	21.7	23.1	−1.4	[1]
	Pyridine	9.2	17.7	18.4	−0.7	[1]
	Diethyl ether	6.5	13.7	16.3	−2.6	[1]
		6.5	13.7	13.6	0.1	[22]
	Anisole	4.1	9.9	7.1	2.8	[1]
	Tetrahydrofuran	7.0	14.4	12.8	1.6	[1]
	Acetophenone	7.0	14.4	15.9	−1.5	[1]
	Butanone	7.5	15.1	12.8	2.3	[1]
	Cyclohexanone	8.3	16.3	17.2	−0.9	[1]

Table 1. *Cont.*

Proton Donor	Proton Acceptor	$-\Delta_{HB}G$(Eq. 4) [a] kJ·mol^{-1}	$-\Delta_{HB}H$(Eq.9) [a] kJ·mol^{-1}	$-\Delta_{HB}H$(lit.) kJ·mol^{-1}	Δ [b] kJ·mol^{-1}	Ref.
2-methylbutan-2-ol	Dimethylformamide	9.19	17.7	15.1	2.6	[22]
	Dimethyl sulfoxide	9.45	18.1	15.0	3.1	[22]
	Ethyl acetate	5.41	12.0	9.3	2.7	[22]
	Triethylamine	9.84	18.7	19.7	−1.0	[22]
Hexanol	Benzylamine	10.7	20.1	21.3	−1.2	[22]
	Pyridine	9.2	17.7	15.6	2.1	[22]
	Tetrahydrofuran	7.0	14.4	12.3	2.1	[22]

[a] The α_2^H and β_2^H values [16,26] used for the calculation are listed in Table S1 of Supplementary Materials. [b] The difference between columns 4 and 5. [c] Hexamethylphosphoramide.

The data on 122 acid-base pairs were collected in total. Among 50 proton acceptors, aromatic and aliphatic amines, amides, ethers, esters, ketones, nitriles, aromatic hydrocarbons, nitro-compounds, halohydrocarbons, sulfides, and hexamethylphosphoramide were considered. The proton donors included phenol and 4-fluorophenol, pyrrole, methanol, butan-1-ol, hexan-1-ol, 2-methylbutan-2-ol, *N*-methylaniline, and chloroform. When collecting the enthalpies values, we attempted to cover more systems (pyrrole, *N*-methylaniline, 2-methylbutan-2-ol as acids; amines—alcohols pairs) that were not included in our recent work [9] devoted to the calculation of hydrogen bonding enthalpies from the experimental binding constants.

4. Discussion

For the 122 acid-base systems considered, *RMS* equaled 1.6 kJ·mol^{-1} and *AD* was 0.5 kJ·mol^{-1}. The deviations greater than 3 kJ·mol^{-1} are rather inconsistent with other available data. Although the maximum positive deviation of 3.4 kJ·mol^{-1} can be seen for the 1-iodobutane–phenol pair, excellent agreement was observed for 1-bromobutane and 1-iodohexane. One of the values from the Ref. available for the benzonitrile–phenol pair [1], for which the deviation is minimal and equals −3.6 kJ·mol^{-1}, deviates from another value by 5.2 kJ·mol^{-1}.

When analyzing the deviations between the calculated and experimental values, one should consider the possible sources of uncertainties for each magnitude. First, the expected uncertainty of Equation (4) of 0.09 log units contributes ±0.8 kJ·mol^{-1} to $\Delta_{HB}H$ found using Equation (6) or (9). The uncertainty of Equation (6) itself was earlier estimated as 2 kJ·mol^{-1} [9]. On the other hand, the uncertainty of the $\Delta_{HB}H$ measurement (usually due to K_{HB} variation with temperature) depends on the error in K_{HB}. It was earlier evaluated at the level of ca. 0.05–0.1 log units [15]. Then, one can estimate the uncertainty in $\Delta_{HB}H$ determination from its temperature dependence, similarly to Refs. [27,28]. Assuming that the temperature uncertainty is 0.01 K and three measurements at 288–308 K are performed, the respective standard deviation of $\Delta_{HB}H$ would equal 1–2 kJ·mol^{-1}. The issues with maintaining infinite dilution conditions and temperature effects on extinction coefficients during spectroscopic measurement may lead to far greater errors in $\Delta_{HB}H$, which is reflected in poor agreement between data on some acid-base systems studied multiple times, e.g., phenol-pyridine. In this case, standard deviation of the literature $\Delta_{HB}H$ values exceeds 4 kJ·mol^{-1}. Based on these estimates, the agreement within 3 kJ·mol^{-1} for most compounds listed in Table 1 and a *RMS* of 1.6 kJ·mol^{-1} can be considered excellent.

Thus, the comparison between the hydrogen bonding enthalpies calculated according to Equation (9) and the experimental data confirms the consistency between the compensation relationship given by Equation (5) and Abraham's two-parameter equation (4). The acidity and basicity parameters of Equation (4) are readily available for more than 600 compounds [26], or $3.6 \cdot 10^5$ acid-base pairs. It is reasonable to expect that Equation (9) can be a useful tool for estimating $\Delta_{HB}H$ in numerous acid-base systems not studied before without measuring K_{HB}. Hydrogen bonding entropies can also be estimated from the calculated Gibbs energies and enthalpies.

One can recall the two- or multi-parameter models [6,7,29,30] for estimating complex formation enthalpies. They rely on the empirical acidity and basicity parameters correlated to the experimental enthalpic data. An important difference of the present study is that Equations (4), (5) and (9) do not follow from the experimental hydrogen bonding enthalpies at all, and their consistency with $\Delta_{HB}H$ in the literature is demonstrated independently.

Supplementary Materials: The following supporting information can be downloaded at: https://www.mdpi.com/article/10.3390/liquids4030034/s1, Table S1: Abraham's acidity parameters of the proton donors and basicity parameters of proton acceptors used in this research.

Author Contributions: Conceptualization, B.N.S.; methodology, B.N.S. and M.I.Y.; investigation, B.N.S., M.B.K. and M.I.Y.; writing—original draft preparation, B.N.S. and M.I.Y.; writing—review and editing, B.N.S. and M.I.Y. All authors have read and agreed to the published version of the manuscript.

Funding: The work was carried out in accordance with the Strategic Academic Leadership Program "Priority 2030" of the Kazan Federal University of the Government of the Russian Federation.

Data Availability Statement: Data are contained within the article.

Conflicts of Interest: The authors declare no conflicts of interest.

References

1. Joesten, M.D.; Schaad, L. *Hydrogen Bonding*; Marcel Dekker: New York, NY, USA, 1974; p. 599.
2. Laurence, C.; Graton, J.; Berthelot, M.; Besseau, F.; Le Questel, J.-Y.; Luçon, M.; Ouvrard, C.; Planchat, A.; Renault, E. An enthalpic scale of hydrogen-bond basicity. 4. Carbon π bases, oxygen bases, and miscellaneous second-row, third-row, and fourth-row bases and a survey of the 4-fluorophenol affinity scale. *J. Org. Chem.* **2010**, *75*, 4105–4123. [CrossRef]
3. Laurence, C.; Gal, J. Thermodynamic and spectroscopic scales of hydrogen-bond basicity and affinity. In *Lewis Basicity and Affinity Scales: Data and Measurement*; John Wiley & Sons, Ltd.: New York, NY, USA, 2009; pp. 111–227.
4. Pimentel, G.C.; McClellan, A. Hydrogen bonding. *Annu. Rev. Phys. Chem.* **1971**, *22*, 347–385. [CrossRef]
5. Drago, R.S.; O'Bryan, N.; Vogel, G.C. A frequency shift-enthalpy correlation for a given donor with various hydrogen-bonding acids. *J. Am. Chem. Soc.* **1970**, *92*, 3924–3929. [CrossRef]
6. Raevsky, O.A.; Novikov, V.P. Unification of the donor-acceptor interaction characteristics to reveal structure-activity-relationships. *Pharm. Chem. J.* **1982**, *16*, 583–586.
7. Raevsky, O.A.; Grigor'ev, V.Y.; Kireev, D.B.; Zefirov, N.S. Complete thermodynamic description of H-bonding in the framework of multiplicative approach. *QSAR* **1992**, *11*, 49–63. [CrossRef]
8. Abraham, M.H.; Grellier, P.L.; Prior, D.V.; Taft, R.W.; Morris, J.J.; Taylor, P.J.; Laurence, C.; Berthelot, M.; Doherty, R.M. A general treatment of hydrogen bond complexation constants in tetrachloromethane. *J. Am. Chem. Soc.* **1988**, *110*, 8534–8536. [CrossRef]
9. Solomonov, B.N.; Yagofarov, M.I. Can the hydrogen bonding enthalpy be calculated from the binding constant at 298.15 K? *J. Mol. Liq.* **2024**, *409*, 125353. [CrossRef]
10. Solomonov, B.N.; Khisamiev, M.B.; Yagofarov, M.I. Calculation of the formation enthalpies of charge-transfer complexes with iodine from the binding constants at 298.15 K. *J. Mol. Liq.* **2024**, *411*, 125690. [CrossRef]
11. Arnett, E.M.; Joris, L.; Mitchell, E.; Murty, T.; Gorrie, T.; Schleyer, P.v.R. Hydrogen-bonded complex formation. III. Thermodynamics of complexing by infrared spectroscopy and calorimetry. *J. Am. Chem. Soc.* **1970**, *92*, 2365–2377. [CrossRef]
12. Epley, T.D.; Drago, R.S. Calorimetric studies on some hydrogen-bonded adducts. *J. Am. Chem. Soc.* **1967**, *89*, 5770–5773. [CrossRef]
13. Spencer, J.; Gleim, J.E.; Blevins, C.H.; Garrett, R.C.; Mayer, F.J. Enthalpies of solution and transfer enthalpies. An analysis of the pure base calorimetric method for the determination of hydrogen bond enthalpies. *J. Phys. Chem.* **1979**, *83*, 1249–1255. [CrossRef]
14. Abraham, M.H.; Grellier, P.L.; Prior, D.V.; Duce, P.P.; Morris, J.J.; Taylor, P.J. Hydrogen bonding. Part 7. A scale of solute hydrogen-bond acidity based on log K values for complexation in tetrachloromethane. *J. Chem. Soc. Perkin Trans. 2* **1989**, *6*, 699–711. [CrossRef]
15. Abraham, M.H.; Grellier, P.L.; Prior, D.V.; Morris, J.J.; Taylor, P.J. Hydrogen bonding. Part 10. A scale of solute hydrogen-bond basicity using log K values for complexation in tetrachloromethane. *J. Chem. Soc. Perkin Trans. 2* **1990**, *4*, 521–529. [CrossRef]
16. Abraham, M.H. Scales of solute hydrogen-bonding: Their construction and application to physicochemical and biochemical processes. *Chem. Soc. Rev.* **1993**, *22*, 73–83. [CrossRef]
17. Abraham, M.H.; Gola, J.M.R.; Cometto-Muñiz, J.E.; Acree, W.E., Jr. Hydrogen Bonding between Solutes in Solvents Octan-1-ol and Water. *J. Org. Chem.* **2010**, *75*, 7651–7658. [CrossRef]
18. Abraham, M.H.; Acree, W.E.; Earp, C.E.; Vladimirova, A.; Whaley, W.L. Studies on the hydrogen bond acidity, and other descriptors and properties for hydroxyflavones and hydroxyisoflavones. *J. Mol. Liq.* **2015**, *208*, 363–372. [CrossRef]

19. Abraham, M.H.; Acree, W.E. Descriptors for the hydrogen halides, their solution properties and hydrogen- bonding acidity and basicity: Comparison of the latter with gas phase data. *J. Mol. Liq.* **2019**, *275*, 667–673. [CrossRef]
20. Solomonov, B.N.; Yagofarov, M.I. Compensation relationship in Thermodynamics of solvation and vaporization: Features and applications. I. Non-hydrogen-bonded systems. *J. Mol. Liq.* **2022**, *368*, 120762. [CrossRef]
21. Solomonov, B.N.; Yagofarov, M.I. Compensation relationship in thermodynamics of solvation and vaporization: Features and applications. II. Hydrogen-bonded systems. *J. Mol. Liq.* **2023**, *372*, 121205. [CrossRef]
22. Solomonov, B.N.; Novikov, V.B.; Varfolomeev, M.A.; Mileshko, N.M. A new method for the extraction of specific interaction enthalpy from the enthalpy of solvation. *J. Phys. Org. Chem.* **2005**, *18*, 49–61. [CrossRef]
23. Gramstad, T. Studies of hydrogen bonding—Part VII: Hydrogen-bond association of phenol and pentachlorophenol with carbonyl compounds and ethers. *Spectrochim. Acta* **1963**, *19*, 497–508. [CrossRef]
24. Perelygin, I.S.; Ahunov, T.F. IR spectra and hydrogen bonds of hydroxyls of chlorine-substituted phenols. I. Complexes with acetonitrile. *Opt. Spektrosk.* **1971**, *30*, 679–683, Chem. Abst. 75: 27627.
25. Catalan, J.; Gomez, J.; Couto, A.; Laynez, J. Toward a solvent basicity scale: The calorimetry of the pyrrole probe. *J. Am. Chem. Soc.* **1990**, *112*, 1678–1681. [CrossRef]
26. Abraham, M.H.; Chadha, H.S.; Whiting, G.S.; Mitchell, R.C. Hydrogen bonding. 32. An analysis of water-octanol and water-alkane partitioning and the Δlog P parameter of Seiler. *J. Pharm. Sci.* **1994**, *83*, 1085–1100. [CrossRef]
27. Cebe, P.; Thomas, D.; Merfeld, J.; Partlow, B.P.; Kaplan, D.L.; Alamo, R.G.; Wurm, A.; Zhuravlev, E.; Schick, C. Heat of fusion of polymer crystals by fast scanning calorimetry. *Polymer* **2017**, *126*, 240–247. [CrossRef]
28. Bolmatenkov, D.N.; Yagofarov, M.I.; Valiakhmetov, T.F.; Rodionov, N.O.; Solomonov, B.N. Vaporization enthalpies of benzanthrone, 1-nitropyrene, and 4-methoxy-1-naphthonitrile: Prediction and experiment. *J. Chem. Thermodyn.* **2022**, *168*, 106744. [CrossRef]
29. Drago, R.S.; Wayland, B.B. A Double-Scale Equation for Correlating Enthalpies of Lewis Acid-Base Interactions. *J. Am. Chem. Soc.* **1965**, *87*, 3571–3577. [CrossRef]
30. Vogel, G.C.; Drago, R.S. The ECW Model. *J. Chem. Educ.* **1996**, *73*, 701. [CrossRef]

liquids

MDPI

Article

Nanoheterogeneity in Protic and Aprotic Alkylimidazolium Bistriflimide Ionic Liquids

Timur I. Magsumov and Igor A. Sedov *

Chemical Institute, Kazan Federal University, Kremlevskaya Str., 18, Kazan 420008, Russia; timomax@mail.ru
* Correspondence: igor_sedov@inbox.ru

Abstract: Many ionic liquids, including alkylimidazolium salts, form a nanoheterogeneous structure with polar and apolar domains in their liquid phase. Using molecular dynamics simulations, the influence of the structure of the cations of a series of aprotic ($[C_nC_1Im][TFSI]$, $[C_nC_nIm][TFSI]$) and protic ($[HC_nIm][TFSI]$) alkylimidazolium bistrilimides on the domain structure of their liquid phase was studied. The characteristic sizes of domains and the extent of domain segregation in different liquids have been compared. It has been shown that the latter, but not the former, is a key factor determining the magnitude of the Gibbs free energy of cavity formation in nanostructured ionic liquids, which in turn governs their solvation properties.

Keywords: ionic liquids; domain structure; nanoheterogeneity; molecular dynamics

1. Introduction

Ionic liquids are salts that exist in a liquid state at relatively low temperatures. The vast number of potential combinations of cation and anion allows for the tailoring of ionic liquids for various tasks. Changes in the structure of the cation or anion affect the supramolecular organization of the liquid phase of these solvents. Ionic liquids containing apolar hydrocarbon chains in their cations often exhibit nanoscale segregation into polar and apolar domains [1–3]. Nanoheterogeneity affects various properties of ionic liquids including solvation behavior, conductivity, diffusion coefficients, viscosity, spectral properties, and so on.

Small- or wide-angle X-ray (SAXS, WAXS) or small-angle neutron (SANS) scattering methods can be used to obtain information about the nanostructure of ionic liquids [2,4,5]. Scattering data have been reported for quite a large number of ionic liquids, with a particular focus on the intensity peaks observed at values of the scattering vector modulus q below 1 $Å^{-1}$. These peaks are indicative of the nanosegregated structure of ionic liquids. An increase in the first peak intensity and decrease in the value of q corresponding to its maximum is observed with increasing length of alkyl chain in cations [2,6–8] due to the growth of the distance between the charged groups separated by apolar domains. The introduction of polar groups, such as hydroxyls or ether moieties, into the cation results in a deterioration of hydrophobic interactions between alkyl residues, which in turn leads to a strong decrease or complete disruption of the structural heterogeneity [2,9]. These observations are consistent with other known properties of these liquids. For example, an increase in the alkyl size in alkylammonium ionic liquids leads to the growth of the solubility of hydrocarbons, which are preferentially solvated by apolar domains [10]. Conversely, the hydrocarbon solubility in 2-hydroxyethylammonium nitrate lacking domain segregation is much lower than in propyl- or ethylammonium nitrate [11].

Interesting comparisons of the structure of symmetric and asymmetric dialkylimidazolium bis(trifluoromethanesulfonyl)imides (bistriflimides) were performed by Xiao et al. [12] and Rocha et al. [6]. As follows from their scattering data, the size of the heterogeneous domains in these liquids is mainly determined by the longest alkyl chain. For example, it is similar in 1,3-dibutylimidazolium and 1-butyl-3-methylimidazolium bistriflimides but much larger in 1-methyl-3-heptylimidazolium bistriflimide. It was suggested that long alkyl chains interdigitate in apolar domains.

Another comparison that is important to understand the relationship between the nature of ions and the structure of the liquid phase is contrasting aprotic (AIL) and protic (PIL) ionic liquids. The presence of an acidic hydrogen can strengthen the interactions in polar domains between cation and anion due to the formation of hydrogen bonds, which influences the structure of the liquid phase. Martinelli et al. showed that protic 1-alkylimidazolium bistriflimides [7,13] show similar X-ray scattering patterns to the aprotic imidazolium bistriflimides [8] with the same length of an alkyl chain, namely similar peak positions but higher peak intensities in the case of PILs. This enhancement was attributed to the increase in the domain segregation in 1-alkylimidazolium PILs in comparison with 1,3-dialkylimidazolium AILs.

Computer simulations have been successfully used to study the nanoheterogeneous structure of ionic liquids. Molecular dynamics have been carried out in order to provide insights into the structure of alkylimidazolium [14,15], alkylammonium [10,16], alkylpyrrolidinium [17], and alkylphosphonium [18–20] ionic liquids. In addition to the scattering patterns, various computational approaches can be used to assess the spatial heterogeneity using molecular dynamics snapshots. The most common of them are calculation of radial distribution functions, cluster analysis, and Voronoi tessellation. In our recent work [16], we used Ripley's functions, finite-volume integrals of radial distribution functions, and local atom density variance to identify the existence of a domain nanostructure in ionic liquids and characterize the domain length scale. In addition, we have shown that the Gibbs free energy of cavity formation in the liquid is sensitive to the presence of domain segregation and has much lower values in apolar than in polar domains [10,16]. This quantity is important to understand the influence of nanoheterogeneity on the solvation properties of a liquid.

In the present work, we study the structure of the liquid phase of 21 alkylimidazolium bistriflimides using molecular dynamics simulations. Three groups of bistriflimides are considered (Figure 1): aprotic symmetric ([C_nC_nIm][TFSI]) and asymmetric ([C_nC_1Im][TFSI]) 1,3-dialkylimidazolium salts, and protic 1-alkylimidazolium salts ([HC_nIm][TFSI]). The characteristic size of segregated domains in the studied ionic liquids is assessed using different methods. The relationships between the nature of substituents, the presence and size of domains, and the cavity formation thermodynamics are analyzed.

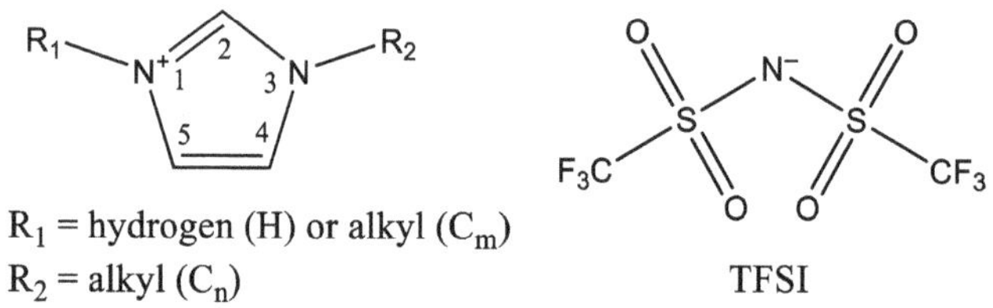

Figure 1. Structures of cations (HC_nIm and C_nC_mIm) and anion (TFSI) of the studied ILs.

2. Methodology

2.1. Molecular Dynamics Simulations

Molecular dynamics simulations were performed using the GROMACS 2022.5 package [21]. Topologies of cations and anions were created using the fftool program [22]. For all ions, the models developed by Canongia Lopes et al. were used [23–26]. The partial charges from these models were scaled by 0.8, which is the value used in a number of previous studies [27–29] in order to achieve the best agreement of simulation with the experimental properties of 1-alkyl-3-methylimidazolium salts. Charge scaling represents a computationally cheaper alternative for models with explicit atomic level polarizability which lead to more accurate results [30–32]. The cutoff distances for van der Waals and short-range electrostatic interactions were set to 1.2 nm. For long-range electrostatic interactions, the particle mesh Ewald method with an interpolation order of 4 and Fourier grid spacing of 0.12 nm was used. The covalent bonds with hydrogen atoms were constrained using the LINCS algorithm. The leapfrog integration algorithm with time step 1 fs was used.

Cubic simulation cells with periodic boundary conditions contained 500 pairs of cation and anion of IL. The v-rescale algorithm with a temperature coupling constant of 1 ps was used for temperature control. A constant pressure (1 bar) was maintained by the Parrinello–Rahman barostat with a pressure coupling constant of 5 ps. Before the production run, the simulation cell was energy minimized and equilibrated at 298 K. After this, five cycles of heating to high temperatures at constant volume and cooling to 298 K in NPT conditions were carried out to ensure the random configuration of ions and the absence of crystalline structures. Then, the cell was additionally equilibrated for 10 ns in NPT conditions before starting the production run. The 100 ns long production runs were recorded in NPT conditions at 298 K. The instantaneous configurations recorded every 1 ps were analyzed using programs written by authors as well as the standard GROMACS tools.

In simulations, we did not observe crystallization or the systematic drift of any properties calculated for different instantaneous configurations or blocks. The densities of ionic liquids from simulations agree well with the experimental values (Table 1).

Table 1. Comparison of calculated and experimental values of the density (ρ) of ILs at 298 K.

Ionic Liquid	ρ/(g·mL^{-1})	
	MD	Experimental
1-methylimidazolium bis(trifluoromethanesulfonyl)imide ([HC_1Im][TFSI])	1.607	1.614 [33]
1-ethylimidazolium bis(trifluoromethanesulfonyl)imide ([HC_2Im][TFSI])	1.553	1.56913 [7], 1.563 [34]
1-butylimidazolium bis(trifluoromethanesulfonyl)imide ([HC_4Im][TFSI])	1.467	1.47194 [7]
1-hexylimidazolium bis(trifluoromethanesulfonyl)imide ([HC_6Im][TFSI])	1.399	1.40242 [7]
1-octylimidazolium bis(trifluoromethanesulfonyl)imide ([HC_8Im][TFSI])	1.345	1.34633 [7]
1-decylimidazolium bis(trifluoromethanesulfonyl)imide ([$HC_{10}Im$][TFSI])	1.301	1.28754 [7]
1-dodecylimidazolium bis(trifluoromethanesulfonyl)imide ([$HC_{12}Im$][TFSI])	1.266	1.26068 [7]
1,3-dimethylimidazolium bis(trifluoromethanesulfonyl)imide ([C_1C_1Im][TFSI])	1.525	1.5692 [6], 1.56729 [35], 1.557 [33]
1-ethyl-3-methylimidazolium bis(trifluoromethanesulfonyl)imide ([C_2C_1Im][TFSI])	1.482	1.5195 [36], 1.51845 [37], 1.51859 [38]
1-propyl-3-methylimidazolium bis(trifluoromethanesulfonyl)imide ([C_3C_1Im][TFSI])	1.445	1.475 [39], 1475.7 [40], 1.4756 [41]

Table 1. *Cont.*

Ionic Liquid	ρ/(g·mL^{-1})	
	MD	Experimental
1-butyl-3-methylimidazolium bis(trifluoromethanesulfonyl)imide ($[C_4C_1Im][TFSI]$)	1.412	1.43704 [42], 1.43635 [39], 1.4369 [41]
1-pentyl-3-methylimidazolium bis(trifluoromethanesulfonyl)imide ($[C_5C_1Im][TFSI]$)	1.382	1.40221 [39], 1.4045 [40], 1.4036 [41]
1-hexyl-3-methylimidazolium bis(trifluoromethanesulfonyl)imide ($[C_6C_1Im][TFSI]$)	1.355	1.3719 [39], 1.3727 [41], 1.37081 [42]
1-heptyl-3-methylimidazolium bis(trifluoromethanesulfonyl)imide ($[C_7C_1Im][TFSI]$)	1.33	1.3454 [41], 1.3446 [6], 1.34413 [43]
1-octyl-3-methylimidazolium bis(trifluoromethanesulfonyl)imide ($[C_8C_1Im][TFSI]$)	1.308	1.32032 [38], 1.32054 [44], 1.32081 [45]
1-decyl-3-methylimidazolium bis(trifluoromethanesulfonyl)imide ($[C_{10}C_1Im][TFSI]$)	1.269	1.27259 [45], 1.27001 [35], 1.2784 [46]
1-dodecyl-3-methylimidazolium bis(trifluoromethanesulfonyl)imide ($[C_{12}C_1Im][TFSI]$)	1.239	1.2447 [47], 1.2423 [48], 1.24355 [35]
1,3-diethylimidazolium bis(trifluoromethanesulfonyl)imide ($[C_2C_2Im][TFSI]$)	1.444	1.4632 [49], 1.4749 [6]
1,3-dipropylimidazolium bis(trifluoromethanesulfonyl)imide ($[C_3C_3Im][TFSI]$)	1.381	1.399 [6]
1,3-dibutylimidazolium bis(trifluoromethanesulfonyl)imide ($[C_4C_4Im][TFSI]$)	1.329	1.3428 [6], 1.340 [50]
1,3-dihexylimidazolium bis(trifluoromethanesulfonyl)imide ($[C_6C_6Im][TFSI]$)	1.249	1.2550 [6]

2.2. Functions Characterizing Nanoheterogeneity

Various approaches can be used to analyze the structure of the liquid phase of ionic liquids. First of all, we compared the positions of peak maxima in the calculated and experimental SAXS curves. The calculation was carried out using the gmx saxs program from the GROMACS package, which uses the Cromer method [51].

The nanoheterogeneity of the liquid phase was also characterized using several functions discussed previously [16]. One of them is the finite-volume Kirkwood–Buff-like integral [16,52] of the radial distribution function $g(r)$ for a given set of atoms with average number density ρ:

$$G(R) = 1/\rho + \int_0^R (g(r) - 1)4\pi r^2 dr \quad (1)$$

This function is calculated separately for the sets of all "polar" and "apolar" heavy atoms without distinguishing the atom or element type within the set. For a completely random (Poisson) distribution of points in space, the function $G(R)$ equals zero at any R. The positive or negative values reflect the increased or decreased probability of finding atoms of the considered set at a given distance from another atom. Hence, clustering behavior results in positive values of $G(R)$ at the length scales of the cluster diameter.

Another homogeneity-related function, $\Gamma(R)$, equals the difference in the density-normalized average number of atoms belonging to two different sets (domains), A and B, within radius R from an atom that belongs to one of these sets (A). The reference atom itself should be counted in order to expect a zero value in the case of a homogeneous spatial distribution of atoms in both sets. $\Gamma(R)$ can be expressed through pairwise radial distribution functions:

$$\Gamma(R) = 1/\rho_A + \int_0^R (g_{AA}(r) - g_{AB}(r))4\pi r^2 dr \tag{2}$$

If reference A comprises apolar fragments, this function shows the prevalence of apolar atoms in the vicinity of other apolar atoms. Otherwise, if A refers to the polar fragments, it shows the prevalence of polar atoms in the vicinity of other polar atoms.

The third function we use to characterize structural (in)homogeneity is the local atom number density variance $\frac{\sigma^2(R)}{\langle N\rangle(R)}$ inside the balls of radius R centered at random points.

Here,

$$\langle N\rangle(R) = 4/3\pi R^3 \rho \tag{3}$$

is the average number of atom centers inside these balls, and $\sigma^2(R)$ is the variance of $\langle N\rangle(R)$. This function has a large positive value if atoms tend to form clusters at the given length scale. It is calculated separately for polar and apolar atom sets.

2.3. The Gibbs Free Energy of Cavity Formation

The Gibbs free energy of the formation of small cavities in the solvent, which describes its solvation properties, can be calculated using a variation of the Widom test particle insertion method [53]. The formation of a cavity is considered as the insertion of a hard-sphere particle with an infinitely large interaction potential. This particle excludes all centers of solvent heavy (non-hydrogen) atoms within its radius R. Molecular dynamics is used to obtain instantaneous solvent configurations. A large number of insertions is performed into the random points of space for each configuration. Neglecting small fluctuations in the volume of the system, the Gibbs free energy of cavity formation can be related to the probability $p(R)$ of successful insertion, i.e., the fraction of insertion attempts resulting in the absence of heavy atom centers within radius R:

$$\Delta_{cav}G = -kT \ln p \tag{4}$$

The accuracy increases with the number of configurations and insertions, which becomes increasingly important for large cavity radii R. Hence, we produced long (100 ns) trajectories with instantaneous configurations recorded each 1 ps (10^5 configurations in total) and performed 10^5 insertions for each configuration, which results in 10^{10} insertions per liquid. This allows us to obtain $\Delta_{cav}G$ values for $R < 0.4$ nm with the standard deviation below 0.5 kJ·mol^{-1} according to block analysis.

3. Results and Discussion

3.1. Scattering Curves Calculated from Simulations

X-ray scattering curves calculated from the MD trajectory snapshots are presented in Figure 2. As in experiments [6–8,12], up to three peaks are observed in these curves. Peak I corresponds to the smallest values of q and characterizes the separation distance between charged groups by the alkyl chain, i.e., the size of apolar domains. Its intensity and position depend on the length of the alkyl substituent. Peak I can appear as a shoulder in the case of short alkyl chains and is unnoticeable for substituents with fewer than four carbon atoms. Peak II observed around 9 nm^{-1} was reported [54] to describe the distance between polar groups of the same charge not separated by alkyl chains. The positions of the maxima of peaks I and II for calculated and experimental scattering curves are given in Table 2. Peak III is positioned near 14 nm^{-1} and is caused by ions located at close distances (about 0.45 nm). Its position and intensity are almost the same for different bistriflimides except for protic 1-methyl and 1-ethylimidazolium salts showing slightly lower q values of the peak maximum, in agreement with experimental observations [6,7].

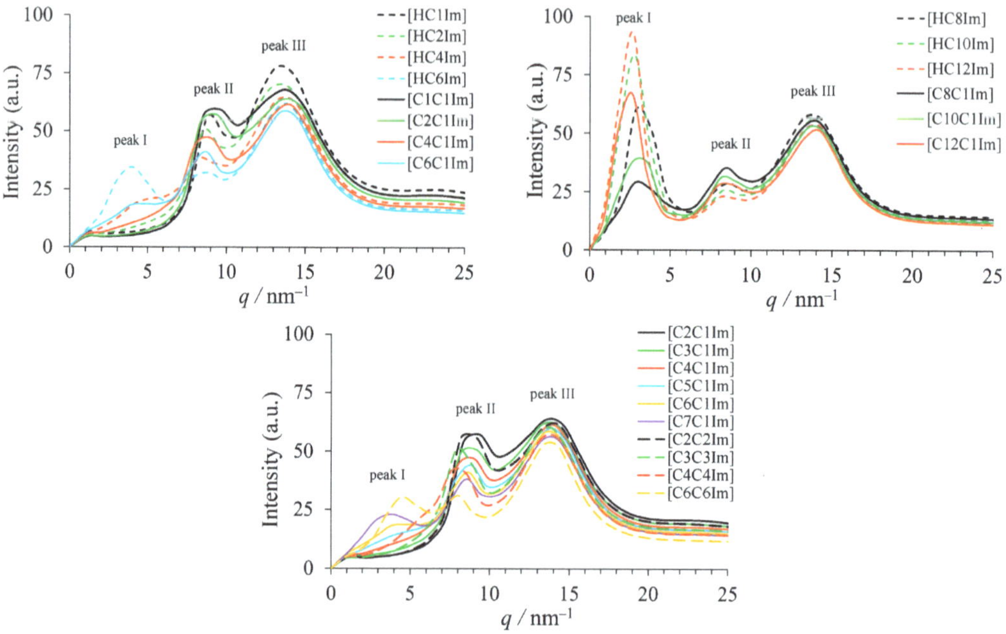

Figure 2. Calculated X-ray scattering curves for the studied ionic liquids.

The data in Table 2 indicate a general agreement between calculated and experimental [6–8,12] positions of peaks I and II as well. The values of q corresponding to the maximum of peak I significantly decrease with increasing alkyl chain length. Larger alkyl substituents also cause an increase in the intensity of peak I. This peak is not observed for [HC_1Im][TFSI], [HC_2Im][TFSI], [C_1C_1Im][TFSI], [C_2C_1Im][TFSI], [C_3C_1Im][TFSI], [C_2C_2Im][TFSI], and [C_3C_3Im][TFSI], indicating that there is no significant nanoheterogeneity in these ionic liquids. For [C_4C_1Im][TFSI] and [C_4C_4Im][TFSI], a small shoulder is observed. Protic ionic liquids ([HC_nIm][TFSI]) have higher peak I intensity compared to their aprotic counterparts ([C_nC_1Im][TFSI]). This can be explained by the stabilization of domain structure by hydrogen bonds in the polar domain between the anion and NH group of the cation. A comparison of SAXS plots for ionic liquids with the same total number of carbon atoms in positions 1 and 3 of the imidazolium ring of the cation (e.g., [C_4C_4Im][TFSI] and [C_7C_1Im][TFSI]) shows that ionic liquids with long alkyl groups form larger domains corresponding to lower q and produce much more intensive scattering peaks. Asymmetric and symmetric ionic liquids with the same length of the alkyl chain (e.g., [C_6C_1Im][TFSI] and [C_6C_6Im][TFSI]) produce peaks with much closer positions corresponding to a similar domain size as pointed out previously [12].

Peak II undergoes a small shift to lower q with alkyl chain length growing from C_2 to C_4. This can also be explained by the emergence of the domain structure with the clusters of polar groups. Further growth of alkyl chain length does not change significantly the position of peak II.

Table 2. Positions of the maxima of peak I in calculated and experimental X-ray scattering curves.

Ionic Liquid	Peak I (nm^{-1})		Peak II (nm^{-1})	
	Calc.	Exp.	Calc.	Exp.
$[HC_1Im][TFSI]$	no peak	–	8.9	–
$[HC_2Im][TFSI]$	no peak	no peak [7]	8.7	8.7 [7]
$[HC_4Im][TFSI]$	4.9 (weak peak on shoulder)	5.1 (weak peak) [7]	8.3	8.6 [7]
$[HC_6Im][TFSI]$	3.9	4.1 [7]	8.7	8.4 [7]
$[HC_8Im][TFSI]$	3.1	3.4 [7]	8.5	8.3 [7]
$[HC_{10}Im][TFSI]$	2.8	2.9 [7]	8.4	8.3 [7]
$[HC_{12}Im][TFSI]$	2.6	2.6 [7]	8.3	8.3 [7]
$[C_1C_1Im][TFSI]$	no peak	no peak [6]	9.1	9.1 [6]
$[C_2C_1Im][TFSI]$	no peak	no peak [6,8,12]	9.0	9.0 [8], 9.0 [6], 8.9 [12]
$[C_3C_1Im][TFSI]$	no peak	no peak [6,12]	8.7	8.7 [12], 8.7 [6]
$[C_4C_1Im][TFSI]$	no peak	shoulder [6,8,12]	8.7	8.6 [8], 8.5 [6], 8.5 [12]
$[C_5C_1Im][TFSI]$	4.4 (weak)	4.4 (weak) [12], 4.9 (weak) [6]	8.8	8.5 [12], 8.5 [6]
$[C_6C_1Im][TFSI]$	3.9	4.1 (weak) [8], 4.3 [6]	8.7	8.5 [8], 8.4 [6]
$[C_7C_1Im][TFSI]$	3.8	3.6 [6]	8.6	8.4 [6]
$[C_8C_1Im][TFSI]$	3.0	3.5 [8], 3.4 [6]	8.5	8.5 [8], 8.4 [6]
$[C_{10}C_1Im][TFSI]$	3.0	2.8 [8], 2.9 [6]	8.4	8.5 [8], 8.5 [6]
$[C_{12}C_1Im][TFSI]$	2.6	2.4 [8], 2.5 [6]	8.3	8.5 [8], 8.4 [6]
$[C_2C_2Im][TFSI]$	no peak	no peak [6,12]	8.6	8.6 [12], 8.7 [6]
$[C_3C_3Im][TFSI]$	no peak	shoulder [6,12]	8.1	8.1 [12], 8.1 [6]
$[C_4C_4Im][TFSI]$	shoulder	shoulder [6,12]	7.9	7.9 [12], 8.0 [6]
$[C_6C_6Im][TFSI]$	4.5	4.6 [6]	8.0	7.8 [6]

3.2. Heterogeneity of Polar/Apolar Atom Distribution

For the analysis using functions described in Section 2.2, only the carbon atoms of the side chain of the imidazolium cation were considered as belonging to the apolar domain (also called apolar atoms below). The remaining heavy atoms, namely all the atoms in the anion and carbon and nitrogen atoms in the imidazolium ring of the cation, were attributed to the polar domain.

As stated above, RDF integrals $G(R)$ and $\Gamma(R)$, and local density fluctuations $\frac{\sigma^2(R)}{\langle N\rangle(R)}$ can be calculated separately for the polar and apolar domains. This always resulted in a pair of curves sharing similar features. Most importantly, $G(R)$ and $\Gamma(R)$ had their first zero, and $\frac{\sigma^2(R)}{\langle N\rangle(R)}$ had its maximum at the close values of radii R for calculations with polar and apolar atoms. The same result was observed in the analysis of the structure of alkylammonium salts [16]. Hence, these functions characterize a general length scale of heterogeneities in ionic liquids. Below, only the results of calculations of $G(R)$ and $\frac{\sigma^2(R)}{\langle N\rangle(R)}$ for the polar domain and $\Gamma(R)$ for the polar atoms as reference are shown.

Figures 3–5 show the dependences of the calculated functions on radius R.

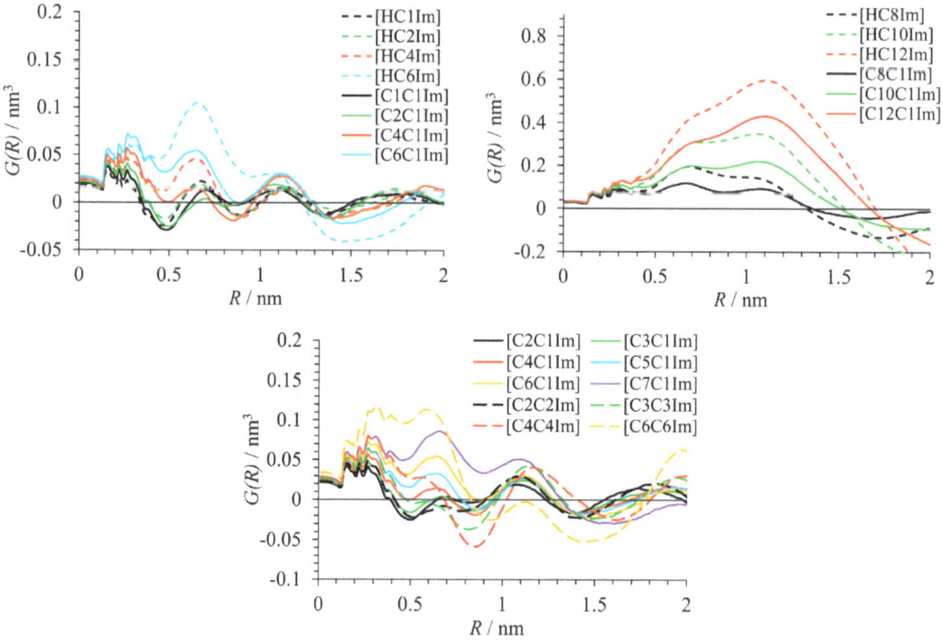

Figure 3. Function $G(R)$ characterizing inhomogeneity of heavy atom distribution in polar fragments of ILs.

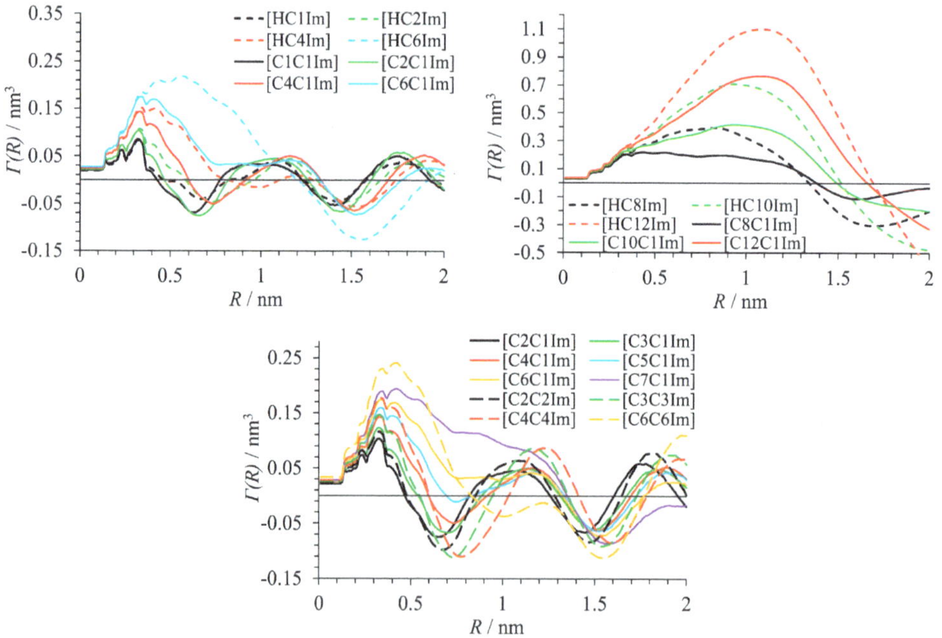

Figure 4. Function $\Gamma(R)$ characterizing inhomogeneity of heavy atom distribution in ILs around polar atoms.

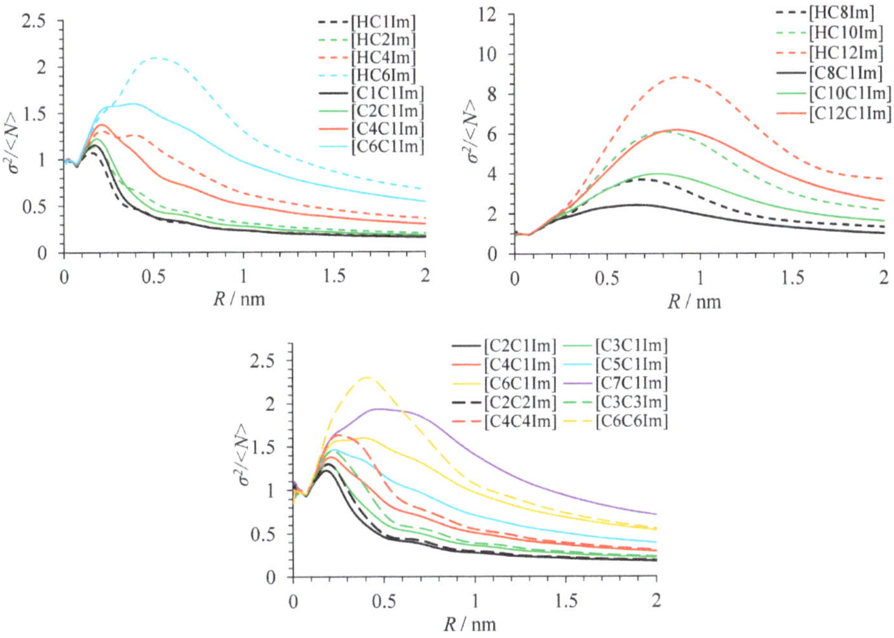

Figure 5. The variance of the number density $\frac{\sigma^2}{\langle N \rangle}$ of polar atoms in ionic liquids.

$G(R)$ and $\Gamma(R)$ show the oscillating behavior typical for domainless liquids [16] in the case of bistriflimides with fewer than four carbon atoms in the alkyl substituent. For the liquids with seven or more carbon atoms in the substituent, large-period oscillations caused by well-defined domains are present. In the liquids from C_4 to C_6, progressive degradation of the first minimum of oscillations corresponding to the emergence and strengthening of domain structure is observed. From the curves in Figures 3 and 4, it is clear that heterogeneous structure is more pronounced in protic than in aprotic ionic liquids with the same alkyl substituent. Analysis of the radial distribution function between the center of mass of the anion and N3 atom of the imidazolium ring shows that its first maximum is observed at the lower distance R for PILs such as [$HC_{10}C_1Im$][TFSI] than for AILs like [$C_{10}C_1Im$][TFSI] (Figure 6). A shoulder in the RDF for [$HC_{10}C_1Im$][TFSI] corresponds to the part of cations not involved in hydrogen bonding with anion.

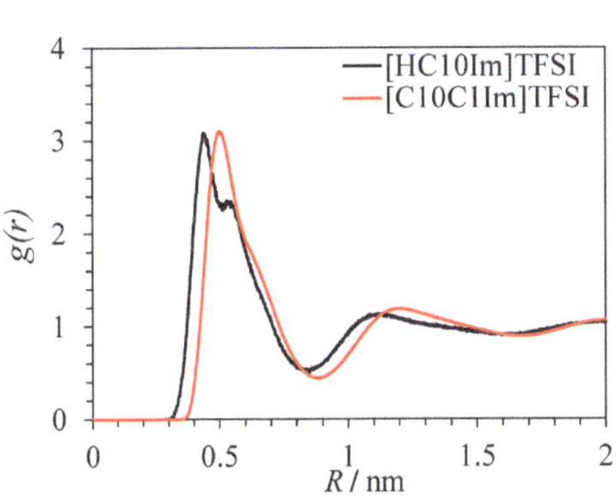

Figure 6. Radial distribution function $g(r)$ between the center of mass of the anion and N3 atom of imidazolium ring for [$HC_{10}Im$][TFSI] and [$C_{10}C_1Im$][TFSI].

The above conclusions are fully supported by the curves of the variance of atom density shown in Figure 5. C_1–C_3 liquids have no maximum or even shoulder caused by the difference in the atom density inside and outside domains. Other liquids show signs of nanoheterogeneity. The variance is also larger for protic than for aprotic liquids. Symmetric ionic liquids $[C_nC_nIm][TFSI]$ exhibit better segregation of polar and apolar domains than asymmetric liquids $[C_nC_1Im][TFSI]$, which is also visible in the $G(R)$ and $\Gamma(R)$ curves.

The radii corresponding to the first zero of $G(R)$ and $\Gamma(R)$ as well as double of the radius of maximum of $\frac{\sigma^2(R)}{\langle N\rangle(R)}$ function for domain-forming bistriflimides are given in Table 3. All these quantities generally reflect the size of the domains. The values of R_0 for $G(R)$ and $\Gamma(R)$ almost always coincide. They correspond to the radius of the sphere at which polar and apolar atoms become on average represented proportionally to their total number in the liquid, i.e., the sphere around a polar atom includes both polar and apolar domains. The value of maximum atom number variance R_{max} is always much smaller. It would correspond to the radius of domains if they were spherical. The values of $2R_{max}$ are usually closer to the values of R_0.

Table 3. Quantitative characteristics of domains in the studied ionic liquids: R_0 is the radius corresponding to the first zero of $G(R)$ or $\Gamma(R)$, $2R_{max}$ is twice the radius corresponding to the maximum of $\frac{\sigma^2}{\langle N\rangle}$.

Ionic Liquid	$G(R)$	$\Gamma(R)$	$\frac{\sigma^2}{\langle N\rangle}$
	R_0/nm	R_0/nm	$2R_{max}$/nm
$[HC_4Im][TFSI]$	0.8	0.82	0.80
$[HC_6Im][TFSI]$	1.23	1.21	1.06
$[HC_8Im][TFSI]$	1.34	1.35	1.40
$[HC_{10}Im][TFSI]$	1.54	1.52	1.60
$[HC_{12}Im][TFSI]$	1.71	1.70	1.80
$[C_4C_1Im][TFSI]$	–	0.60	–
$[C_5C_1Im][TFSI]$	0.78	0.70	–
$[C_6C_1Im][TFSI]$	0.87	1.33	0.76
$[C_7C_1Im][TFSI]$	1.32	1.36	0.96
$[C_8C_1Im][TFSI]$	1.37	1.39	1.30
$[C_{10}C_1Im][TFSI]$	1.52	1.50	1.56
$[C_{12}C_1Im][TFSI]$	1.68	1.67	1.76
$[C_4C_4Im][TFSI]$	0.69	0.61	–
$[C_6C_6Im][TFSI]$	0.83	0.84	0.80

The obtained results indicate that the size of domains rapidly grows in the liquids from C_4 to C_6. The values of R_0 and $2R_{max}$ can diverge for these liquids due to the presence of double peaks in the calculated curves, which is linked to the loose domain segregation. For the liquids with longer alkyl substituents, single sharp peaks and agreeing domain size characteristics were obtained, evidencing that they have well-defined nanostructure. The results obtained for $[C_6C_1Im][TFSI]$ and $[C_6C_6Im][TFSI]$ allow us to suggest that they form domains with a similar size. The size of domains coincides for protic and aprotic bistriflimides with the same alkyl group.

3.3. Gibbs Free Energy of Cavity Formation

The concept of cavity formation is widely used in studies of solvation processes. Solvation of any molecule can be considered as its accommodation in the cavity of the molecular size that has been pre-formed in a solvent. The free energy cost of cavity formation is determined by the solvent–solvent interactions. In ionic liquids with domain segregation, cavities predominantly form in apolar domains. This should lead to lower free energy of solvation and higher hydrocarbon solubility in such liquids.

Calculated values of the Gibbs free energy of cavity formation $\Delta_{cav}G$ in the studied liquids are shown in Figure 7. $\Delta_{cav}G$ always decreases with increasing length of the alkyl substituent. A rapid decrease is observed for C_1–C_6 liquids due to the emergence and strengthening of domain segregation. In contrast, only a small decrease occurs for C_8–C_{12} liquids, in which well-defined domains are present, and cavities form predominantly in apolar domains with similar free energy cost for all liquids. At the same time, these liquids have a significantly different domain length scale, which has no huge effect on $\Delta_{cav}G$.

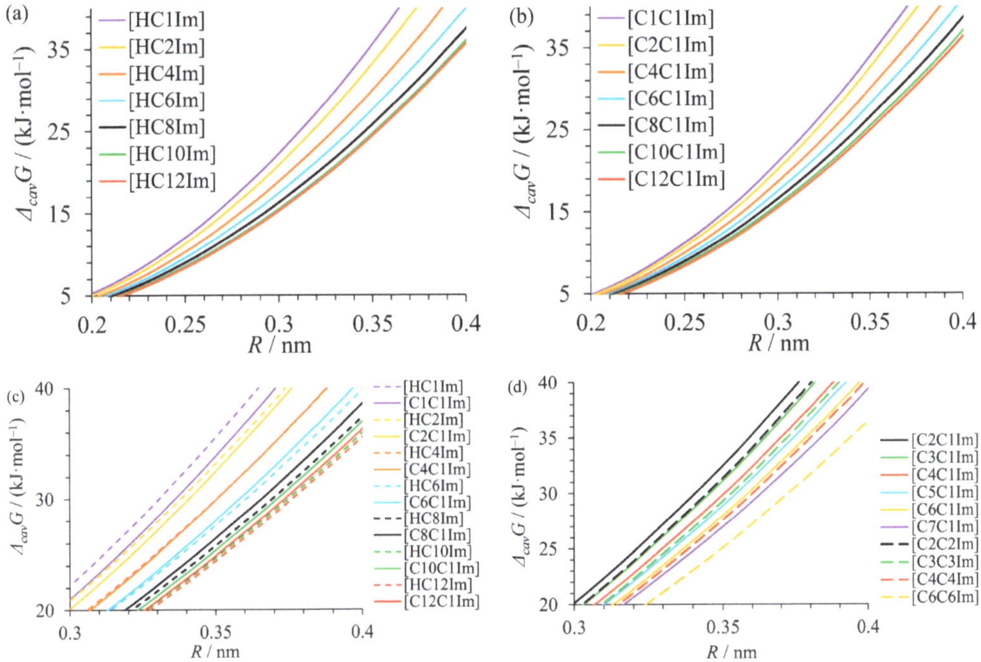

Figure 7. The Gibbs free energy of cavity formation in alkylimidazolium bistriflimides at 298 K for spherical cavities with various radii R: (**a**,**b**) dependence on the alkyl substituent chain length; (**c**) comparison of protic and aprotic ionic liquids; (**d**) comparison of liquids with symmetrically and asymmetrically alkylated cations.

An interesting trend is observed for the difference in the cavity formation Gibbs free energy between protic and aprotic ionic liquids (Figure 7c). For liquids with fewer than four carbon atoms in the alkyl chain, $\Delta_{cav}G$ is slightly higher in protic liquids. In the absence of a domain structure, hydrogen bonds formed between cations and anions in protic liquids impede cavity formation, similar to what was observed for the hydrogen-bonded molecular solvents [55]. For [HC_4Im][TFSI] and [C_4C_1Im][TFSI], the values of $\Delta_{cav}G$ become almost the same. Protic liquids with larger alkyl chains have lower $\Delta_{cav}G$ values than aprotic liquids. In this case, hydrogen bonds are isolated in polar domains and improve domain segregation, which facilitates cavity formation in apolar domains.

A comparison of aprotic ionic liquids with symmetric and asymmetric cations (Figure 7d) indicates a similar cost of cavity formation for ionic liquids with the same total number of carbon atoms (e.g., [C_3C_3Im][TFSI] and [C_5C_1Im][TFSI]) but not for liquids with the same length of the alkyl groups ([C_6C_6Im][TFSI] and [C_6C_1Im][TFSI]) which were shown above to have a similar domain size. This is in part a result of a much higher fraction of liquid volume occupied by the apolar domains in ionic liquids with two alkyl groups than in those with a single group. A rough approximation with a two times higher fraction of apolar domain in $[C_6C_6Im][TFSI]$ than in $[C_6C_1Im][TFSI]$ leads to the difference $\Delta\Delta_{cav}G = RT\ln 2 = 1.7$ kJ·mol^{-1} if only the cavities in the apolar domain are considered. The observed difference is even larger due to the better segregation of polar and apolar domains in the liquids in symmetrical cations, which has been shown above. Hence, the solvation properties of such pairs of liquids will differ significantly despite the same domain length scale.

4. Conclusions

A nanoheterogeneous structure comprising polar and apolar domains is typical for ionic liquids with sufficiently long alkyl substituents. It has previously been shown that in the case of alkylammonium nitrates, even ethylammonium nitrate shows signs of nanoheterogeneity. In the case of alkylimidazolium bistriflimides ($[HC_nIm][TFSI]$ and $[C_nC_mIm][TFSI]$), our analysis of molecular dynamics snapshots as well as previously reported scattering data indicate domain segregation in the presence of butyl or longer alkyl substituents. As the length of the alkyl substituent increases, larger domain sizes and better segregation are observed. At the same time, ionic liquids with symmetric and asymmetric cations with the same alkyl substituent ($[C_nC_nIm][TFSI]$ and $[C_nC_1Im][TFSI]$) have similar domain length scales, but better segregation is observed for symmetric cations. Protic alkylimidazolium bistriflimides $[HC_nIm][TFSI]$ have more pronounced nanoheterogeneity than aprotic salts $[C_nC_1Im][TFSI]$ but do not differ from them in domain size. This is explained by hydrogen bonding between the acidic hydrogen in the cation and the anion.

The nanoheterogeneous structure can dramatically affect the solvation properties of ionic liquids. For the first time, we have analyzed the dependence of the Gibbs free energy of cavity formation in ionic liquids on both the domain length scale and the extent of domain segregation. We have shown that the latter, but not the former, plays a crucial role in determining the solvation properties of nanostructured ionic liquids. For example, a rather small decrease of the Gibbs free energy of cavity formation is observed from 1-octyl- to 1-undecyl-3-methylimidazolium bistriflimide despite an apparent increase in domain size. At the same time, increasing the alkyl substituent chain length by a single methylene group from C_1 (no domain segregation) to C_6 (clear evidence of nanoheterogeneity) liquids leads to a much more pronounced decrease in $\Delta_{cav}G$. In addition, cation–anion hydrogen bonding in aprotic ionic liquids strengthens the domain structure if it is present, and thus results in a higher propensity of cavity formation in $[HC_nIm][TFSI]$ liquids than in $[C_nC_1Im][TFSI]$ liquids with $n > 4$. For the liquids with smaller alkyl substituents and no domain segregation, the situation is reversed: cation–anion hydrogen bonds increase the energy cost of cavity formation and result in larger $\Delta_{cav}G$ values in AILs.

The most intriguing result of the work comes from the comparison of ionic liquids with symmetrically and asymmetrically substituted cations ($[C_nC_nIm][TFSI]$, $[C_nC_1Im][TFSI]$, and $[C_{2n-1}C_1Im][TFSI]$). The Gibbs free energies of cavity formation were found to be governed by the total number of carbon atoms in the substituents. The length scale of the domains again does not play a significant role, while the fraction of the volume occupied by the apolar domains and the extent of domain segregation are important.

Thus, it has been shown that the structure of the ionic liquid cation has complex relationships with the bulk phase properties such as cavity formation thermodynamics even for a series of structurally similar alkylimidazolium bistriflimides. This study once again highlights the links between the structure of the ions of different ionic liquids, the nanostructure of their liquid phase, and their solvation properties.

Author Contributions: Conceptualization, T.I.M. and I.A.S.; software, I.A.S.; investigation, T.I.M. and I.A.S.; data curation, T.I.M.; writing—original draft preparation, T.I.M. and I.A.S. All authors have read and agreed to the published version of the manuscript.

Funding: The research was funded by the Russian Science Foundation, project 24-13-00062.

Data Availability Statement: Data are contained within the article. The raw data are available on request from the authors.

Conflicts of Interest: The authors declare no conflicts of interest.

References

1. Hayes, R.; Warr, G.G.; Atkin, R. Structure and Nanostructure in Ionic Liquids. *Chem. Rev.* **2015**, *115*, 6357–6426. [CrossRef] [PubMed]
2. Greaves, T.L.; Kennedy, D.F.; Mudie, S.T.; Drummond, C.J. Diversity Observed in the Nanostructure of Protic Ionic Liquids. *J. Phys. Chem. B* **2010**, *114*, 10022–10031. [CrossRef] [PubMed]
3. Wang, Y.-L.; Li, B.; Sarman, S.; Mocci, F.; Lu, Z.-Y.; Yuan, J.; Laaksonen, A.; Fayer, M.D. Microstructural and Dynamical Heterogeneities in Ionic Liquids. *Chem. Rev.* **2020**, *120*, 5798–5877. [CrossRef] [PubMed]
4. Triolo, A.; Russina, O.; Bleif, H.-J.; Di Cola, E. Nanoscale Segregation in Room Temperature Ionic Liquids. *J. Phys. Chem. B* **2007**, *111*, 4641–4644. [CrossRef] [PubMed]
5. Atkin, R.; Warr, G.G. The Smallest Amphiphiles: Nanostructure in Protic Room-Temperature Ionic Liquids with Short Alkyl Groups. *J. Phys. Chem. B* **2008**, *112*, 4164–4166. [CrossRef]
6. Rocha, M.A.A.; Neves, C.M.S.S.; Freire, M.G.; Russina, O.; Triolo, A.; Coutinho, J.A.P.; Santos, L.M.N.B.F. Alkylimidazolium Based Ionic Liquids: Impact of Cation Symmetry on Their Nanoscale Structural Organization. *J. Phys. Chem. B* **2013**, *117*, 10889–10897. [CrossRef]
7. Abdurrokhman, I.; Elamin, K.; Danyliv, O.; Hasani, M.; Swenson, J.; Martinelli, A. Protic Ionic Liquids Based on the Alkyl-Imidazolium Cation: Effect of the Alkyl Chain Length on Structure and Dynamics. *J. Phys. Chem. B* **2019**, *123*, 4044–4054. [CrossRef]
8. Martinelli, A.; Maréchal, M.; Östlund, Å.; Cambedouzou, J. Insights into the Interplay between Molecular Structure and Diffusional Motion in 1-Alkyl-3-Methylimidazolium Ionic Liquids: A Combined PFG NMR and X-Ray Scattering Study. *Phys. Chem. Chem. Phys.* **2013**, *15*, 5510. [CrossRef]
9. Triolo, A.; Russina, O.; Caminiti, R.; Shirota, H.; Lee, H.Y.; Santos, C.S.; Murthy, N.S.; Castner, E.W., Jr. Comparing Intermediate Range Order for Alkyl- vs. Ether-Substituted Cations in Ionic Liquids. *Chem. Commun.* **2012**, *48*, 4959. [CrossRef] [PubMed]
10. Sedov, I.A.; Magsumov, T.I.; Salikov, T.M.; Solomonov, B.N. Solvation of Apolar Compounds in Protic Ionic Liquids: The Non-Synergistic Effect of Electrostatic Interactions and Hydrogen Bonds. *Phys. Chem. Chem. Phys.* **2017**, *19*, 25352–25359. [CrossRef]
11. Sedov, I.A.; Salikov, T.M.; Solomonov, B.N. Contrasting the Solvation Properties of Protic Ionic Liquids with Different Nanoscale Structure. *J. Mol. Liq.* **2019**, *290*, 111361. [CrossRef]
12. Xiao, D.; Hines, L.G.; Li, S.; Bartsch, R.A.; Quitevis, E.L.; Russina, O.; Triolo, A. Effect of Cation Symmetry and Alkyl Chain Length on the Structure and Intermolecular Dynamics of 1,3-Dialkylimidazolium Bis(Trifluoromethanesulfonyl)Amide Ionic Liquids. *J. Phys. Chem. B* **2009**, *113*, 6426–6433. [CrossRef]
13. Garaga, M.N.; Dracopoulos, V.; Werner-Zwanziger, U.; Zwanziger, J.W.; Maréchal, M.; Persson, M.; Nordstierna, L.; Martinelli, A. A Long-Chain Protic Ionic Liquid inside Silica Nanopores: Enhanced Proton Mobility Due to Efficient Self-Assembly and Decoupled Proton Transport. *Nanoscale* **2018**, *10*, 12337–12348. [CrossRef]
14. Ghatee, M.H.; Zolghadr, A.R.; Moosavi, F.; Ansari, Y. Studies of Structural, Dynamical, and Interfacial Properties of 1-Alkyl-3-Methylimidazolium Iodide Ionic Liquids by Molecular Dynamics Simulation. *J. Chem. Phys.* **2012**, *136*, 124706. [CrossRef] [PubMed]
15. Dudariev, D.; Koverga, V.; Kalugin, O.; Miannay, F.-A.; Polok, K.; Takamuku, T.; Jedlovszky, P.; Idrissi, A. Insight to the Local Structure of Mixtures of Imidazolium-Based Ionic Liquids and Molecular Solvents from Molecular Dynamics Simulations and Voronoi Analysis. *J. Phys. Chem. B* **2023**, *127*, 2534–2545. [CrossRef] [PubMed]
16. Sedov, I.A.; Magsumov, T.I. Highlighting the Difference in Nanostructure between Domain-Forming and Domainless Protic Ionic Liquids. *Phys. Chem. Chem. Phys.* **2022**, *24*, 21477–21494. [CrossRef]
17. Kashyap, H.K.; Santos, C.S.; Murthy, N.S.; Hettige, J.J.; Kerr, K.; Ramati, S.; Gwon, J.; Gohdo, M.; Lall-Ramnarine, S.I.; Wishart, J.F.; et al. Structure of 1-Alkyl-1-Methylpyrrolidinium Bis(Trifluoromethylsulfonyl)Amide Ionic Liquids with Linear, Branched, and Cyclic Alkyl Groups. *J. Phys. Chem. B* **2013**, *117*, 15328–15337. [CrossRef]
18. Gontrani, L.; Russina, O.; Lo Celso, F.; Caminiti, R.; Annat, G.; Triolo, A. Liquid Structure of Trihexyltetradecylphosphonium Chloride at Ambient Temperature: An X-Ray Scattering and Simulation Study. *J. Phys. Chem. B* **2009**, *113*, 9235–9240. [CrossRef]
19. Wang, Y.-L.; Li, B.; Sarman, S.; Laaksonen, A. Microstructures and Dynamics of Tetraalkylphosphonium Chloride Ionic Liquids. *J. Chem. Phys.* **2017**, *147*, 224502. [CrossRef]
20. Liu, X.; Zhao, Y.; Zhang, X.; Zhou, G.; Zhang, S. Microstructures and Interaction Analyses of Phosphonium-Based Ionic Liquids: A Simulation Study. *J. Phys. Chem. B* **2012**, *116*, 4934–4942. [CrossRef]
21. Abraham, M.J.; Murtola, T.; Schulz, R.; Páll, S.; Smith, J.C.; Hess, B.; Lindahl, E. GROMACS: High Performance Molecular Simulations through Multi-Level Parallelism from Laptops to Supercomputers. *SoftwareX* **2015**, *1–2*, 19–25. [CrossRef]
22. Padua, A. Fftool v1.0.0. 2015. Available online: https://zenodo.org/records/18618 (accessed on 15 May 2024).

23. Canongia Lopes, J.N.; Pádua, A.A.H. Molecular Force Field for Ionic Liquids III: Imidazolium, Pyridinium, and Phosphonium Cations; Chloride, Bromide, and Dicyanamide Anions. *J. Phys. Chem. B* **2006**, *110*, 19586–19592. [CrossRef] [PubMed]
24. Canongia Lopes, J.N.; Deschamps, J.; Pádua, A.A.H. Modeling Ionic Liquids Using a Systematic All-Atom Force Field. *J. Phys. Chem. B* **2004**, *108*, 2038–2047. [CrossRef]
25. Canongia Lopes, J.N.; Pádua, A.A.H. Molecular Force Field for Ionic Liquids Composed of Triflate or Bistriflylimide Anions. *J. Phys. Chem. B* **2004**, *108*, 16893–16898. [CrossRef]
26. Gouveia, A.S.L.; Bernardes, C.E.S.; Tomé, L.C.; Lozinskaya, E.I.; Vygodskii, Y.S.; Shaplov, A.S.; Lopes, J.N.C.; Marrucho, I.M. Ionic Liquids with Anions Based on Fluorosulfonyl Derivatives: From Asymmetrical Substitutions to a Consistent Force Field Model. *Phys. Chem. Chem. Phys.* **2017**, *19*, 29617–29624. [CrossRef]
27. Bhargava, B.L.; Balasubramanian, S. Refined Potential Model for Atomistic Simulations of Ionic Liquid [Bmim][PF6]. *J. Chem. Phys.* **2007**, *127*, 114510. [CrossRef] [PubMed]
28. Chaban, V. Polarizability versus Mobility: Atomistic Force Field for Ionic Liquids. *Phys. Chem. Chem. Phys.* **2011**, *13*, 16055. [CrossRef]
29. Doherty, B.; Zhong, X.; Gathiaka, S.; Li, B.; Acevedo, O. Revisiting OPLS Force Field Parameters for Ionic Liquid Simulations. *J. Chem. Theory Comput.* **2017**, *13*, 6131–6145. [CrossRef]
30. Schröder, C. Comparing Reduced Partial Charge Models with Polarizable Simulations of Ionic Liquids. *Phys. Chem. Chem. Phys.* **2012**, *14*, 3089. [CrossRef]
31. Son, C.Y.; McDaniel, J.G.; Schmidt, J.R.; Cui, Q.; Yethiraj, A. First-Principles United Atom Force Field for the Ionic Liquid $BMIM^+$ BF_4^-: An Alternative to Charge Scaling. *J. Phys. Chem. B* **2016**, *120*, 3560–3568. [CrossRef]
32. McDaniel, J.G.; Yethiraj, A. Influence of Electronic Polarization on the Structure of Ionic Liquids. *J. Phys. Chem. Lett.* **2018**, *9*, 4765–4770. [CrossRef] [PubMed]
33. Watanabe, H.; Doi, H.; Saito, S.; Matsugami, M.; Fujii, K.; Kanzaki, R.; Kameda, Y.; Umebayashi, Y. Hydrogen Bond in Imidazolium Based Protic and Aprotic Ionic Liquids. *J. Mol. Liq.* **2016**, *217*, 35–42. [CrossRef]
34. Rodrigues, A.S.M.C.; Rocha, M.A.A.; Almeida, H.F.D.; Neves, C.M.S.S.; Lopes-da-Silva, J.A.; Freire, M.G.; Coutinho, J.A.P.; Santos, L.M.N.B.F. Effect of the Methylation and N–H Acidic Group on the Physicochemical Properties of Imidazolium-Based Ionic Liquids. *J. Phys. Chem. B* **2015**, *119*, 8781–8792. [CrossRef]
35. Kolbeck, C.; Lehmann, J.; Lovelock, K.R.J.; Cremer, T.; Paape, N.; Wasserscheid, P.; Fröba, A.P.; Maier, F.; Steinrück, H.-P. Density and Surface Tension of Ionic Liquids. *J. Phys. Chem. B* **2010**, *114*, 17025–17036. [CrossRef]
36. Součková, M.; Klomfar, J.; Pátek, J. Measurements and Group Contribution Analysis of 0.1MPa Densities for Still Poorly Studied Ionic Liquids with the [PF6] and [NTf2] Anions. *J. Chem. Thermodyn.* **2014**, *77*, 31–39. [CrossRef]
37. Fröba, A.P.; Kremer, H.; Leipertz, A. Density, Refractive Index, Interfacial Tension, and Viscosity of Ionic Liquids [EMIM][$EtSO_4$], [EMIM][NTf_2], [EMIM][$N(CN)_2$], and [OMA][NTf_2] in Dependence on Temperature at Atmospheric Pressure. *J. Phys. Chem. B* **2008**, *112*, 12420–12430. [CrossRef]
38. Zorębski, E.; Geppert-Rybczyńska, M.; Zorębski, M. Acoustics as a Tool for Better Characterization of Ionic Liquids: A Comparative Study of 1-Alkyl-3-Methylimidazolium Bis[(Trifluoromethyl)Sulfonyl]Imide Room-Temperature Ionic Liquids. *J. Phys. Chem. B* **2013**, *117*, 3867–3876. [CrossRef]
39. Zorębski, M.; Zorębski, E.; Dzida, M.; Skowronek, J.; Jężak, S.; Goodrich, P.; Jacquemin, J. Ultrasonic Relaxation Study of 1-Alkyl-3-Methylimidazolium-Based Room-Temperature Ionic Liquids: Probing the Role of Alkyl Chain Length in the Cation. *J. Phys. Chem. B* **2016**, *120*, 3569–3581. [CrossRef] [PubMed]
40. Esperança, J.M.S.S.; Visak, Z.P.; Plechkova, N.V.; Seddon, K.R.; Guedes, H.J.R.; Rebelo, L.P.N. Density, Speed of Sound, and Derived Thermodynamic Properties of Ionic Liquids over an Extended Pressure Range. 4. [C_3Mim][NTf_2] and [C_5Mim][NTf_2]. *J. Chem. Eng. Data* **2006**, *51*, 2009–2015. [CrossRef]
41. Xue, L.; Gurung, E.; Tamas, G.; Koh, Y.P.; Shadeck, M.; Simon, S.L.; Maroncelli, M.; Quitevis, E.L. Effect of Alkyl Chain Branching on Physicochemical Properties of Imidazolium-Based Ionic Liquids. *J. Chem. Eng. Data* **2016**, *61*, 1078–1091. [CrossRef]
42. Gomes De Azevedo, R.; Esperança, J.M.S.S.; Szydlowski, J.; Visak, Z.P.; Pires, P.F.; Guedes, H.J.R.; Rebelo, L.P.N. Thermophysical and Thermodynamic Properties of Ionic Liquids over an Extended Pressure Range: [Bmim][NTf2] and [Hmim][NTf2]. *J. Chem. Thermodyn.* **2005**, *37*, 888–899. [CrossRef]
43. Zorębski, E.; Zorębski, M.; Dzida, M.; Goodrich, P.; Jacquemin, J. Isobaric and Isochoric Heat Capacities of Imidazolium-Based and Pyrrolidinium-Based Ionic Liquids as a Function of Temperature: Modeling of Isobaric Heat Capacity. *Ind. Eng. Chem. Res.* **2017**, *56*, 2592–2606. [CrossRef]
44. Santos, D.; Santos, M.; Franceschi, E.; Dariva, C.; Barison, A.; Mattedi, S. Experimental Density of Ionic Liquids and Thermodynamic Modeling with Group Contribution Equation of State Based on the Lattice Fluid Theory. *J. Chem. Eng. Data* **2016**, *61*, 348–353. [CrossRef]
45. Khalil, R.; Chaabene, N.; Azar, M.; Malham, I.B.; Turmine, M. Effect of the Chain Lengthening on Transport Properties of Imidazolium-Based Ionic Liquids. *Fluid Phase Equilibria* **2020**, *503*, 112316. [CrossRef]
46. Lago, S.; Rodríguez, H.; Soto, A.; Arce, A. Deterpenation of Citrus Essential Oil by Liquid−Liquid Extraction with 1-Alkyl-3-Methylimidazolium Bis(Trifluoromethylsulfonyl)Amide Ionic Liquids. *J. Chem. Eng. Data* **2011**, *56*, 1273–1281. [CrossRef]

47. Tariq, M.; Forte, P.A.S.; Gomes, M.F.C.; Lopes, J.N.C.; Rebelo, L.P.N. Densities and Refractive Indices of Imidazolium- and Phosphonium-Based Ionic Liquids: Effect of Temperature, Alkyl Chain Length, and Anion. *J. Chem. Thermodyn.* **2009**, *41*, 790–798. [CrossRef]
48. Domańska, U.; Królikowski, M.; Wlazło, M.; Więckowski, M. Phase Equilibrium Investigation on 2-Phenylethanol in Binary and Ternary Systems: Influence of High Pressure on Density and Solid–Liquid Phase Equilibrium. *J. Phys. Chem. B* **2018**, *122*, 6188–6197. [CrossRef]
49. Tomé, L.I.N.; Carvalho, P.J.; Freire, M.G.; Marrucho, I.M.; Fonseca, I.M.A.; Ferreira, A.G.M.; Coutinho, J.A.P.; Gardas, R.L. Measurements and Correlation of High-Pressure Densities of Imidazolium-Based Ionic Liquids. *J. Chem. Eng. Data* **2008**, *53*, 1914–1921. [CrossRef]
50. Hasse, B.; Lehmann, J.; Assenbaum, D.; Wasserscheid, P.; Leipertz, A.; Fröba, A.P. Viscosity, Interfacial Tension, Density, and Refractive Index of Ionic Liquids [EMIM][$MeSO_3$], [EMIM][$MeOHPO_2$], [EMIM][$OcSO_4$], and [BBIM][NTf_2] in Dependence on Temperature at Atmospheric Pressure. *J. Chem. Eng. Data* **2009**, *54*, 2576–2583. [CrossRef]
51. Cromer, D.T.; Mann, J.B. X-Ray Scattering Factors Computed from Numerical Hartree–Fock Wave Functions. *Acta Crystallogr. A* **1968**, *24*, 321–324. [CrossRef]
52. Ben-Naim, A. The Kirkwood–Buff Integrals for One-Component Liquids. *J. Chem. Phys.* **2008**, *128*, 234501. [CrossRef] [PubMed]
53. Widom, B. Some Topics in the Theory of Fluids. *J. Chem. Phys.* **1963**, *39*, 2808–2812. [CrossRef]
54. Annapureddy, H.V.R.; Kashyap, H.K.; De Biase, P.M.; Margulis, C.J. What Is the Origin of the Prepeak in the X-Ray Scattering of Imidazolium-Based Room-Temperature Ionic Liquids? *J. Phys. Chem. B* **2010**, *114*, 16838–16846. [CrossRef]
55. Sedov, I.; Magsumov, T. The Gibbs Free Energy of Cavity Formation in a Diverse Set of Solvents. *J. Chem. Phys.* **2020**, *153*, 134501. [CrossRef] [PubMed]

liquids

MDPI

Article

Effect of Intramolecular Hydrogen Bond Formation on the Abraham Model Solute Descriptors for Oxybenzone

Jocelyn Chen, Audrey Chen, Yixuan Yang and William E. Acree *

Department of Chemistry, University of North Texas, Denton, TX 76203, USA
* Correspondence: bill.acree@unt.edu

Abstract: Solute descriptors derived from experimental solubility data for oxybenzone dissolved in 21 different organic solvents indicate that the hydrogen atom on the hydroxyl functional group forms an intramolecular hydrogen bond with the lone electron pair on the oxygen atom of the neighboring >C=O functional group. Group contribution methods developed for estimating the Abraham model solute descriptors from the molecule's Canonical SMILES code significantly overestimate the Abraham model's hydrogen bond acidity solute descriptor of oxybenzone. An informed user-modified Canonical SMILES code is proposed to identify which hydrogen atoms are involved in intramolecular H-bond formation. The identified hydrogen atom(s) can be used to define a new functional/fragment group and numerical group contribution value.

Keywords: intramolecular hydrogen-bond formation; oxybenzone solubility; Abraham model solute descriptors; group contribution methods; machine learning

1. Introduction

The chemical manufacturing sector relies heavily on predictive methods to estimate the thermodynamic and physical properties of fluid mixtures, as experimental measurements cannot keep pace with the large number of new chemical compounds synthesized or discovered each year. Currently, experimental quantities are available for only a very small fraction of the multicomponent mixtures commonly encountered in industrial processes. Quantitative structure property/activity relationships (QSPRs/QSARs), linear free energy relationships (LFERs), and functional/fragment group contribution methods have been widely used by research scientists and design engineers to predict properties of both neat organic compounds and fluid mixtures based on the molecular structure of each individual compound or mixture component. The simpler predictive methods combine structural information, and easily measure property data to serve as required input parameters in mathematical correlations, while the more sophisticated predictive approaches utilize input parameters (such as atomic charges, atom–atom polarizabilities, molecular orbital energies, and frontier orbital densities) deduced from more theoretical quantum-mechanical computation methods. Binary interaction terms have even been incorporated in some methods to represent the various types of molecular interactions believed to be present. The introduction of machine learning and deep learning models has increased the applicability and predictive accuracy of the various estimation methods. Physical and thermodynamic properties that have been reasonably predicted by QSPR/QSAR, LFER, and group contribution approaches include vapor pressures [1,2], critical temperature [3], flash point temperatures [4,5], water-to-organic solvent partition coefficients [2,6–8], Gibbs energies of solvation and Ostwald solubility coefficients [9–15], liquid viscosities [16–19], surface tensions [20], enthalpies of combustion and enthalpies of formation [21–24], enthalpies of solvation [25,26], isobaric solid, liquid and gas molar heat capacities [27–32], enthalpies of vaporization and sublimation [33–37], total solid-to-liquid phase change entropies [38,39], and activity coefficients [40–44].

The focus of the current study is limited in scope to the group contribution and machine learning approaches that have been developed for estimating Abraham model solute descriptors based on inputting the solute molecule's Canonical SMILES. Our recent experience has indicated that such methods often do not properly account for intramolecular hydrogen bond formation. For example, both the UFZ-LSER and MIT group contribution methods significantly overestimate the Abraham model hydrogen bond acidity solute descriptor. In the case of 1,4-dihydroxyanthraquinone (see Figure 1 for the molecular structure), the UFZ-LSER [45] and MIT [14,46] predictive software yielded values of **A** = 0.82 and **A** = 0.862, respectively. The two estimated values are significantly larger than the experiment-based value of **A** = 0.00 obtained by fitting the measured molar solubility of 1,4-dihydroxyanthraquinone in accordance with expressions derived from the Abraham model using measured solubility data for the organic solute dissolved in acetic acid, acetone, acetonitrile, 1-butanol and toluene [47].

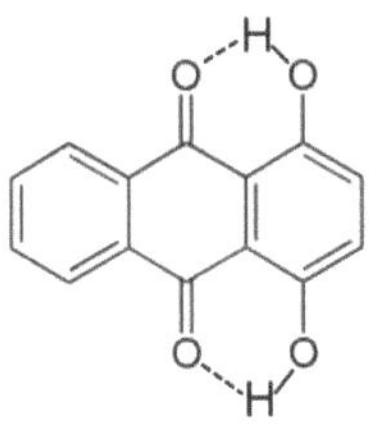

Figure 1. Molecular structure of 1,4-dihydroxyanthraquinone showing the intramolecular hydrogen bonds (dashed lines).

In this study, we identify several molecules in our Abraham model solute descriptor database that are known to engage in intramolecular H-bond formation. We propose a relatively simple modification of the molecule's Canonical SMILES code to indicate which -OH hydrogen atoms cannot act as H-bond donors in their interactions with surrounding solvent molecules. As part of this study, we also measured the mole fraction solubility of oxybenzone (more formally named 2-hydroxy-4-methoxybenzophenone) dissolved in hexane, heptane, octane, dodecane, cyclohexane, methylcyclohexane, 2,2,4-trimethylpentane, diisopropyl ether, propan-1-ol, butan-1-ol, pentan-1-ol, heptan-1-ol, octan-1-ol, 2-methylpropan-1-ol, butan-2-ol, 2-methylbutan-1-ol, 3-methylbutan-1-ol, 4-methylpentan-2-ol, 2-ethylhexan-1-ol, cyclopentanol, and ethylene glycol at 298.15 K. Oxybenzone was selected because we wished to report experiment-based solute descriptors for one additional organic compound that exhibits intramolecular H-bond formation. Published infrared [48,49], X-ray crystallographic [50,51], proton nuclear magnetic resonance chemical shift [48,52], and computational studies [53,54] have revealed that intramolecular H-bond formation occurs between the hydroxyl hydrogen atom in 2-hydroxybenzophenone derivatives and the lone electron pairs on the oxygen atom of the neighboring (>C=O) functional group, as depicted in Figure 2. Intramolecular H-bond formation does not prevent the second lone electron pair on the oxygen >C=O from forming an intermolecular hydrogen bond with protic solvents [55]. Several amphoteric alcohol solvents were included in the current study because of their ability to act both as a hydrogen bond donor and a hydrogen bond acceptor. Published studies [55–58] have suggested the possible formation of bifurcated three-center hydrogen bonds in several organic compounds that exhibit intramolecular H-bond formation, which is where the hydroxyl hydrogen atom is simultaneously involved in an internal H-bond as well as an external H-bond with a surrounding solvent molecule. The amphoteric nature of the alcohol solvents would allow us to explore this possibility through the magnitude of oxybenzone's calculated **A** solute descriptor. A very small calculated **A** value near zero would suggest the negligible involvement of the hydroxyl hydrogen atom in external H-bond formation with the H-bond acceptor sites on neighboring solvent molecules considered in the current study.

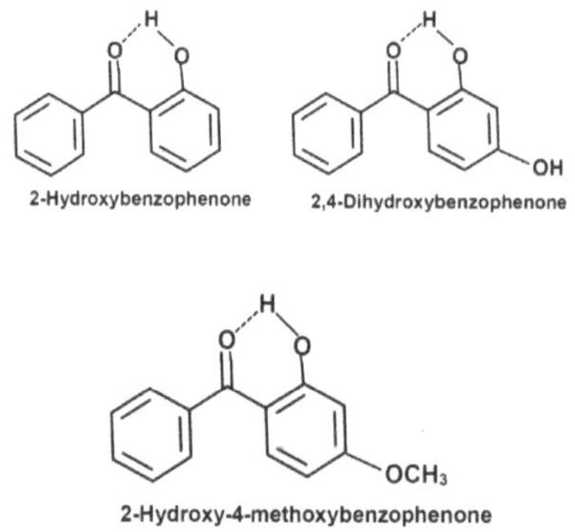

Figure 2. Intramolecular hydrogen bond formation in three 2-hydroxybenzophenone compounds (e.g., 2-hydroxybenzophenone, 2,4-dihydroxybenzophenone and 2-hydroxy-4-methoxybenzophenone).

We note that oxybenzone is widely used as a UV filter in a variety of personal sunscreen products to prevent sunburn, premature photoaging and skin cancer, or as a stabilizer in many commercial products (e.g., plastic surface coatings, outdoor building materials) to minimize photodegradation. It is reported that as many as 81% of the personal care products from both the U.S. and China contain oxybenzone [59,60]. The increased demand for UV filters has resulted in the substantial release of oxybenzone and other benzophenone-type filter materials into the environment, and a greater concern for the potential damage that these emerging contaminants might cause in aquatic and other living organisms [61]. To minimize adverse environmental effects, several countries now limit the amount of oxybenzone in sunscreen products [62], and Hawaii has even gone so far as to pass legislation banning the active chemical ingredients oxybenzone and octinoxate in sunscreens altogether so as to protect coral reefs from bleaching and disruptions of the ecosystem [63,64]. Laws banning the sale or use of sunscreen products containing oxybenzone have also been enacted in several Caribbean islands such as the U.S. Virgin Islands, Aruba and Bonaire [62].

The Abraham model experiment-based solute descriptors that are obtained from the solubility data measured in the current study, in combination with published Abraham model correlations, will allow researchers to evaluate potential biphasic aqueous–organic extraction systems and potential absorbent materials for the removal of oxybenzone from environmental waterways. At present, wastewater treatment facilities have problems removing UV filters [65].

2. Chemical Materials and Experimental Methodology

Oxybenzone was purchased from Sigma-Aldrich Chemical Company (Milwaukee, WI, USA, 0.98 mass fraction) in the highest purity available and was recrystallized three times from anhydrous methanol to remove any trace impurities that may have been present in the commercial sample. The recrystallized sample was further dried at 313 K for three days to remove any adsorbed methanol. Gas chromatographic analysis (thermal conductivity detector, carbowax stationary phase) indicated a purity of 0.995 mass fraction for the purified sample. The 21 different organic solvents were purchased from commercial chemical suppliers as follows: hexane (Aldrich Chemical Company, Milwaukee, WI, USA, 0.99+ mass fraction purity), heptane (Aldrich Chemical Company, anhydrous, 0.99 mass fraction), octane (Fluka Chemicals, Buchs, Switzerland, prem, 0.99+ mass fraction), dodecane (Aldrich Chemical Company, 0.99+ mass fraction), cyclohexane (Sigma-Aldrich Chemical Company, St. Louis, MO, USA, anhydrous, 0.995 mass fraction), methylcyclohexane (Aldrich Chemical Company, anhydrous, 0.99+ mass fraction), 2,2,4-trimethylpentane (Aldrich Chemical Company, anhydrous, 0.998 mass fraction), diisopropyl ether (Sigma-

Aldrich Chemical Company, anhydrous, 0.99 mass fraction), propan-1-ol (Alfa Aesar, Ward Hill, MA, USA, anhydrous, 0.999 mass fraction), butan-1-ol (Aldrich Chemical Company, HPLC grade, 0.998 mass fraction), pentan-1-ol (Sigma-Aldrich Chemical Company, ACS Reagent, 0.99+ mass fraction), heptan-1-ol (Alfa Aesar, 0.99 mass fraction), octan-1-ol (Sigma-Aldrich Chemical Company, anhydrous, 0.99+ mass fraction), 2-methylpropan-1-ol (Sigma-Aldrich Chemical Company, anhydrous, 0.995 mass fraction), butan-2-ol (Aldrich Chemical Company, anhydrous, 0.995 mass fraction), 2-methylbutan-1-ol (Sigma-Aldrich Chemical Company, 0.99+ mass fraction), 3-methylbutan-1-ol (Sigma-Aldrich Chemical Company, anhydrous, 0.99+ mass fraction), 4-methylpentan-2-ol (Acros Organics, Morris Plains, NJ, USA, 0.99+ mass fraction), 2-ethylhexan-1-ol (Acros Organics, 0.99 mass fraction), cyclopentanol (Sigma-Aldrich Chemical Company, anhydrous, 0.995 mass fraction), and ethylene glycol (Aldrich Chemical Company, spectrophotometric grade, 99+%). The organic solvents were stored over activated molecular sieves, and distilled prior to use. Gas chromatographic analysis showed the solvent purities to be 0.997 mass fraction or higher.

Mole fraction solubilities were determined using a spectrophotometric method of chemical analysis by placing excess oxybenzone and 20 mL of the respective organic solvent in sealed amber glass bottles. The sealed bottles were allowed to equilibrate in a constant-temperature water bath at 298.15 K for three days with periodic shaking to facilitate dissolution. Known aliquots of the clear saturated solutions were transferred into pre-weighed volumetric flasks. The flasks containing the transferred aliquot were then re-weighed on an analytical balance in order to determine the mass of the saturated solution removed for chemical analysis. By determining the amount of transferred saturated solution in this fashion we can calculate the solubility of oxybenzone as a mass fraction, which can be accurately converted to mole fraction solubilities, $X_{S,organic}$, without having to measure the density of each saturated solution. Many journals require that solubility data be reported in terms of mole fractions.

The transferred solutions were diluted quantitatively with propan-2-ol and their absorbances were recorded on a Milton Roy Spectronic 1000 Plus spectrophotometer (Milton Roy, Rochester, NY, USA) at an analysis wavelength of 284 nm. The 21 organic solvents were optically transparent at the analysis wavelength. The concentration of each diluted sample was obtained from a Beer–Lambert law graph prepared from the measured absorbances of nine standard solutions of known oxybenzone concentration. A new Beer–Lambert law graph was determined each day using nine freshly prepared standard solutions. The molar absorptivity coefficient, $\varepsilon \approx 14{,}000$ L/(mole cm), was found to be nearly constant over the concentration range from 3.29×10^{-5} Molar to 1.10×10^{-4} Molar.

The measured absorbances of the dilute solutions prepared from saturated solutions were converted to mass fraction solubilities and then mole fraction solubilities using the molar masses of oxybenzone and the respective organic solvent, the mass of the saturated solution taken for analysis, the volume of the volumetric flasks, calculated molar absorption coefficient, and any dilutions that were required to place the measured absorbance within the linear region of the Beer–Lambert law curve. Repetitive experimental measurements were performed on select samples after an additional two days of equilibration to ensure that saturation conditions had been attained. Melting point temperatures were determined on the equilibrated solid phases recovered from each saturated solution to verify that the solid phase did not change during the course of experimental measurements. For each of the 21 oxybenzone–organic solvent combinations studied in the current communication, the melting point temperature of the equilibrated oxybenzone solid phase was within ± 0.5 K of the melting point temperature of the recrystallized sample of oxybenzone prior to contact with the organic solvent. There was no evidence of solid-to-solid phase transitions or solvate formation.

3. Results and Discussion

The experimental mole fraction solubilities, $X_{S,organic}$, of the oxybenzone dissolved in seven different linear, cyclic and branched alkanes, in 12 primary and secondary mono-alcohols, in ethylene glycol, and in diisopropyl ether at 298.15 K, are reported in the second and fourth columns of Table 1. Each numerical value represents the average of 6 to 10 independent measurements, including the follow-up determinations to ensure that the samples had attained equilibrium after the initial three-day equilibration time. The follow up studies confirmed that in all cases, equilibrium had been reached. The tabulated $X_{S,organic}$ values were reproducible to within ±2.5% (relative error). Our search of the published chemical and engineering literature unfortunately failed to find any experimental solubility data that could be used to compare our measured $X_{S,organic}$ values against. In fact, we found no experimental $X_{S,organic}$ data for oxybenzone dissolved in organic solvents. We did find a practical logarithm of a water-to-octanol partition coefficient of log P = 3.79 [66]; however, that could be used in computing oxybenzone's Abraham model solute descriptors.

Table 1. Mole fraction solubilities, $X_{S,organic}$, of oxybenzone dissolved in select organic solvents at a temperature of 298.15 K and ambient atmospheric pressure of 101 kPa [a].

Organic Solvent	$X_{S,organic}$	Chemical Name	$X_{S,organic}$
Hexane	0.01298	Heptan-1-ol	0.03054
Heptane	0.01467	Octan-1-ol	0.04069
Octane	0.01554	Butan-2-ol	0.01698
Dodecane	0.02101	2-Methylpropan-1-ol	0.01161
Cyclohexane	0.01994	2-Methylbutan-1-ol	0.02262
Methylcyclohexane	0.02604	3-Methylbutan-1-ol	0.02313
2,2,4-Trimethylpentane	0.01264	4-Methylpentan-2-ol	0.02695
Diisopropyl ether	0.06958	Ethyl-1-hexan-1-ol	0.02726
Propan-1-ol	0.02016	Cyclopentanol	0.05380
Butan-1-ol	0.01819	Ethylene glycol	0.000994
Pentan-1-ol	0.01633		

[a] Standard uncertainties and relative uncertainties are $u(T)$ = 0.05 K; $u(p)$ = 5 kPa; and $u_r(x)$ = 0.025.

A major reason for performing experimental solubility measurements is to aid in selecting an organic mono-solvent or solvent mixture to serve as a reaction media for the synthesis of chemical compounds, for the purification of chemical products through either recrystallization or biphasic liquid–liquid extraction, or for dissolving medicinal drug compounds in controlled time-release drug formulations. The solubility data given in Table 1 provide useful oxybenzone solubility data for a very limited number of the solvents used in industrial manufacturing processes. The measurements become more useful if one discovers a way to utilize the $X_{S,organic}$ values given in Table 1 to predict oxybenzone solubilities in additional organic solvents. Linear free energy relationships and quantitative structure–property relationships provide a convenient means for researchers to extend their experimental studies by developing mathematical expressions to estimate additional physical and thermodynamic properties from a limited number of measured quantities. In the case of the current study, the measured $X_{S,organic}$ values in Table 1 can be used to calculate the Abraham model descriptor values of the dissolved oxybenzone solute molecule. The solute descriptors (**E**, **S**, **A**, **B**, **V**, and **L**) once calculated can be inserted into Equations (1) and (2) [67,68],

$$\log P \text{ or } \log (C_{S,organic}/C_{S,water}) = c_p + e_p \times \mathbf{E} + s_p \times \mathbf{S} + a_p \times \mathbf{A} + b_p \times \mathbf{B} + v_p \times \mathbf{V} \quad (1)$$

$$\log K \text{ or } \log (C_{S,organic}/C_{S,gas}) = c_k + e_k \times \mathbf{E} + s_k \times \mathbf{S} + a_k \times \mathbf{A} + b_k \times \mathbf{B} + l_k \times \mathbf{L} \quad (2)$$

thus allowing researchers to estimate the logarithm of the solute's water-to-octanol partition coefficient, log P, the logarithm of the solute's gas-to-organic solvent partition coefficient, log K, or the logarithm of the solute's molar solubility ratios, log ($C_{S,organic}/C_{S,water}$) and log ($C_{S,organic}/C_{S,gas}$), in the more than 130 different organic solvents for which the solvent's equation coefficients (c_p, e_p, s_p, a_p, b_p, v_p, c_k, e_k, s_k, a_k, b_k, and l_k) are known [67,68]. The phase to which each molar solute concentration refers is indicated by the subscripts "organic", "water" and "gas".

Each term on the right-hand side of Equations (1) and (2) represents a different type of solute–solvent interaction. The magnitude of the different molecular interactions is quantified by a solute property times the complementary solvent property. The solute descriptors are valuable regarding how the solute interacts with surrounding molecules. The solute descriptor **E** measures the molar refraction of the given solute in excess of that of a linear alkane with a comparable molecular size, which when multiplied by the complementary solvent property, e_p and e_k, describes the additional dispersion interactions possible for solute and solvent molecules with polarizable π-electrons and lone electron pairs. Intermolecular solute–solvent hydrogen-bonding interactions are quantified by the $a \times \mathbf{A} + b \times \mathbf{B}$ terms on the right-hand side of Equations (1) and (2). In the first of the two respective terms, the oxybenzone solute molecule acts as the H-bond donor, while the solubilizing solvent medium acts as the H-bond acceptor. The roles of dissolved oxybenzone and the neighboring solvent molecules are reversed in the case of the two $b \times \mathbf{B}$ terms.

The three remaining terms on the right-hand side of Equations (1) and (2) refer to contributions resulting from both orientation and induction dipole-type interactions (e.g., the $s \times \mathbf{S}$ terms), and the breaking of solvent–solvent interactions needed in the creation of the solvent cavity wherein the dissolved oxybenzone solute molecule will reside (e.g., the $v_p \times \mathbf{V}$ and $l_k \times \mathbf{L}$ terms). The size of the created solvent cavity depends on the volume of the solute molecule (e.g., the McGowan V solute descriptor). The various solute descriptors and complementary solvent equation coefficients are described in greater detail in several very informative review and research articles [69–73].

The first step in determining oxybenzone's solute descriptor values is to convert the measured $X_{S,organic}$ values to molar solubilities, $C_{S,organic}$. This conversion is easily accomplished through Equation (3).

$$C_{S,organic} \approx X_{S,organic}/[X_{S,organic}\, V_{Solute} + (1 - X_{S,organic})\, V_{Solvent}] \quad (3)$$

The numerical values of $X_{S,organic}$ given in Table 1 are divided by the respective ideal molar volume of the respective saturated solution. A value of V_{Solute} = 0.1948 L mol^{-1} was used for the molar volume of the hypothetical subcooled liquid hippuric acid. The mole fraction solubility of oxybenzone is sufficiently small in the studied organic solvents that any errors resulting from our estimation of oxybenzone's hypothetical subcooled liquid molar volume, V_{Solute}, or the ideal molar volume approximation should result in a negligible effect on each calculated $C_{S,organic}$ value.

The calculated molar solubilities (listed in Table 2), combined with the solvent equation coefficients (listed in Table 3), are substituted into Equations (1) and (2). They give 42 Abraham model expressions to use in our solute descriptor determination. Two additional expressions are obtained from the practical log P value contained in the published paper by Rodil and Moeder [66]. The equation coefficients that describe solute partitioning into a water-saturated octanol phase are identified by the word "wet", which follows the name of the organic solvent. The remaining equation coefficients in Table 3 that are not identified as "wet" mean that the organic solvent does not contain enough water to significantly alter its solubilizing character. In other words, it is these equation coefficients that are used in conjunction with measured molar solubility data. The practical log P value for "wet" octan-1-ol is converted into the corresponding water-to-wet octan-1-ol partition coefficient, log K, as follows:

$$\log K = \log P + \log (C_{S,water}/C_{S,gas}) \quad (4)$$

This uses oxybenzone's aqueous molar solubility, $C_{S,water}$, and molar gas phase concentration, $C_{S,gas}$. The low solubility of 1-octanol in water allows "wet" log P values to be converted to log K in this fashion.

Table 2. Comparison between inputted logarithms of the experimental molar solubility of oxybenzone dissolved in the 21 different organic mono-solvents, log $C_{S,organic}$, and back-calculated values based for Equations (1) and (2).

Solvent	Log $C_{S,organic}^{exp}$	Log $C_{S,organic}^{calc}$; Equation (1) [a]	Log $C_{S,organic}^{calc}$; Equation (2) [a]
Hexane	−1.008	−1.102	−1.077
Heptane	−1.004	−1.039	−1.070
Octane	−1.023	−1.031	−1.048
Dodecane	−1.035	−1.101	−0.999
Cyclohexane	−0.744	−0.678	−0.810
Methylcyclohexane	−0.698	−0.716	−0.822
Isooctane	−1.120	−1.184	−1.219
Propan-1-ol	−0.585	−0.659	−0.700
Butan-1-ol	−0.713	−0.631	−0.677
Pentan-1-ol	−0.822	−0.636	−0.682
Heptan-1-ol	−0.675	−0.601	−0.603
Octan-1-ol	−0.594	−0.613	−0.627
2-Methylpropan-1-ol	−0.908	−0.703	−0.732
Butan-2-ol	−0.744	−0.673	−0.670
2-Methylbutan-1-ol	−0.687	−0.667	−0.702
3-Methylbutan-1-ol	−0.684	−0.669	−0.687
4-Methylpentan-2-ol	−0.680	−0.707	−0.733
2-Ethylhexan-1-ol	−0.764	−0.677	−0.653
Cyclopentanol	−0.236	−0.455	−0.380
Ethylene glycol	−1.751	−1.554	−1.581
Diisopropyl ether	−0.324	−0.533	−0.491

[a] Calculated using numerical values of the solute descriptors of **E** = 1.500; **S** = 1.413; **A** = 0.000; **B** = 0.617; **V** = 1.7391; **L** = 8.660; and log $C_{S,water}$ = −4.558 and log $C_{S,gas}$ = −9.581.

Table 3. Equation coefficients for water solvent partition coefficients and air solvent partition coefficients, as log *P* and log *K*, and for molar solubility ratios, as log ($C_{S,org}/C_{S,water}$) and log ($C_{S,org}/C_{S,gas}$), at 298 K for the organic solvents considered in the current study.

Log (*P* or $C_{S,org}/C_{S,water}$)	c_p	e_p	s_p	a_p	b_p	v_p
Octan-1-ol (wet)	0.088	0.562	−1.054	0.034	−3.460	3.814
Hexane	0.333	0.560	−1.710	−3.578	−4.939	4.463
Heptane	0.297	0.634	−1.755	−3.571	−4.946	4.488
Octane	0.241	0.690	−1.769	−3.545	−5.011	4.511
Dodecane	0.114	0.668	−1.644	−3.545	−5.006	4.459
Cyclohexane	0.159	0.784	−1.678	−3.740	−4.929	4.577
Methylcyclohexane	0.246	0.782	−1.982	−3.517	−4.293	4.528
Isooctane	0.318	0.555	−1.737	−3.677	−4.864	4.417
Air-to-water	−0.994	0.577	2.549	3.813	4.841	−0.869
Propan-1-ol	0.139	0.405	−1.029	0.247	−3.767	3.986
Butan-1-ol	0.165	0.401	−1.011	0.056	−3.958	4.044
Pentan-1-ol	0.150	0.536	−1.229	0.141	−3.864	4.077
Heptan-1-ol	0.035	0.398	−1.063	0.002	−4.342	4.317
Octan-1-ol	−0.034	0.489	−1.044	−0.024	−4.235	4.218
2-Methylpropan-1-ol	0.188	0.354	−1.127	0.016	−3.568	3.986
Butan-2-ol	0.127	0.253	−0.976	0.158	−3.882	4.114
2-Methylbutan-1-ol	0.143	0.388	−1.173	−0.024	−3.817	4.129
3-Methylbutan-1-ol	0.111	0.337	−1.180	0.063	−3.880	4.218
4-Methylpentan-2-ol	0.096	0.301	−1.100	0.039	−4.081	4.242
2-Ethylhexan-1-ol	−0.033	0.566	−1.233	−0.068	−3.912	4.153
Cyclopentanol	0.332	0.522	−1.034	−0.106	−3.756	3.892
Ethylene glycol	−0.270	0.578	−0.511	0.715	−2.619	2.729
Diisopropyl ether	0.181	0.285	−0.954	−0.956	−5.077	4.542
Log (*K* or $C_{S,org}/C_{S,gas}$)	c_k	e_k	s_k	a_k	b_k	l_k
Octan-1-ol (wet)	−0.198	0.002	0.709	3.519	1.429	0.858
Hexane	0.320	0.000	0.000	0.000	0.000	0.945
Heptane	0.284	0.000	0.000	0.000	0.000	0.950

Table 3. *Cont.*

Log (P or $C_{S,org}/C_{S,water}$)	c_p	e_p	s_p	a_p	b_p	v_p
Octane	0.219	0.000	0.000	0.000	0.000	0.960
Dodecane	0.017	0.000	0.000	0.000	0.000	0.989
Cyclohexane	0.163	−0.110	0.000	0.000	0.000	1.013
Methylcyclohexane	0.318	−0.215	0.000	0.000	0.000	1.012
Isooctane	0.264	−0.230	0.000	0.000	0.000	0.975
Air-to-water	−1.271	0.822	2.743	3.904	4.814	−0.213
Propan-1-ol	−0.042	−0.246	0.749	3.888	1.076	0.874
Butan-1-ol	−0.004	−0.285	0.768	3.705	0.879	0.890
Pentan-1-ol	−0.002	−0.161	0.535	3.778	0.960	0.900
Heptan-1-ol	−0.056	−0.216	0.554	3.596	0.803	0.933
Octan-1-ol	−0.147	−0.214	0.561	3.507	0.749	0.943
2-Methylpropan-1-ol	−0.003	−0.357	0.699	3.595	1.247	0.881
Butan-2-ol	−0.034	−0.387	0.719	3.736	1.088	0.905
2-Methylbutan-1-ol	−0.055	−0.348	0.601	3.565	0.996	0.925
3-Methylbutan-1-ol	−0.040	−0.408	0.648	3.599	0.905	0.932
4-Methylpentan-2-ol	−0.013	−0.606	0.687	3.622	0.436	0.985
2-Ethylhexan-1-ol	−0.127	−0.339	0.551	3.397	0.722	0.963
Cyclopentanol	−0.151	−0.314	0.693	3.549	0.914	0.956
Ethylene glycol	−0.887	0.132	1.657	4.457	2.355	0.565
Diisopropyl ether	0.139	−0.473	0.610	2.568	0.000	1.016

The Abraham model correlations that we have constructed contain the six solute descriptors (**E**, **S**, **A**, **B**, **V** and **L**) that we wish to calculate, as well as the values of $C_{S,water}$ and $C_{S,gas}$. The latter two quantities will be floated and determined during the course of the solute description computation. Our search of the published literature did not find experimental solubility for oxybenzene in water. There are more than enough Abraham model expressions to calculate the eight desired quantities. The number of solute descriptors can be reduced to four by remembering that the values of both **V** and **E** can be obtained entirely from oxybenzone's molecular structure. For example, the characteristic McGowan volume, **V** = 1.7391, is calculated from the number of chemical bonds, as well as the number and sizes of the individual hydrogen, carbon and oxygen atoms in oxybenzone [74]. An estimated value for the **E** solute descriptor, **E** = 1.500, is obtained by inputting the oxybenzone's Canonical SMILES code, O=C(C=1C=CC=CC1)C2=CC=C(OC)C=C2O, into the estimation software program available on the UFZ-LSER website [45]. Our past experience with this software program is that this group contribution method provides a very good estimate of the **E** solute descriptor for organic molecules having few functional groups and relatively simple molecular structures. Estimated values of the **S**, **A**, **B**, and **L** solute descriptors for large, complex solute molecules containing multiple functional groups, however, can differ significantly from experiment-based solute descriptors deduced from measured molar solubilities and measured partition coefficients.

Preliminary regression analysis yielded a negative numerical value of **A** = −0.068 for oxybenzone's overall hydrogen bond acidity solute descriptor. A negative numerical value of **A** was also obtained when we analyzed only the solubility in the 12 alcohol mono-solvents having a single hydroxyl functional group (**A** = −0.126). A negative value of the **A** solute descriptor is unrealistic, so we set the value to zero, and performed the regression analysis one more time to give the set of experiment-based descriptor values of **E** = 1.500, **S** = 1.413, **A** = 0.000, **B** = 0.617, **V** = 1.7391 and **L** = 8.660, as well as two logarithms of molar concentrations of oxybenzone—log $C_{S,water}$ = −4.558 and log $C_{S,gas}$ = −9.581. Only a very slight increase was noted in the overall standard deviation for the solute descriptor regression analysis, from SD = 0.094 log units to SD = 0.115 log units, by our setting **A** = 0.000. Small individual standard deviations of SD = 0.114 log units and SD = 0.118 log units were obtained for the 21 calculated and observed log ($C_{S,organic}/C_{S,water}$) values and the 21 calculated and observed log ($C_{S,organic}/C_{S,gas}$) values, respectively, indicating that either Abraham model could be used in predicting the molar solubilities of oxybenzene in additional organic solvents (See the last two columns of Table 2 for the back-calculated values of log $C_{S,organic}{}^{calc}$ based on Equations (1) and (2)).

Our private database of experiment-based Abraham model solute descriptors contains numerical values for relatively few hydroxybenzophenone derivatives: 2-hydroxybenzophenone (**E** = 1.54, **S** = 1.46, **A** = 0.00, **B** = 0.46, **V** = 1.5395, **L** = 7.7950); 4-hydroxybenzophenone (**E** = 1.59, **S** = 1.89, **A** = 0.81, **B** = 0.59, **V** = 1.5395, **L** = 8.802); and 2,4-dihydroxybenzophenone (**E** = 1.73, **S** = 2.03, **A** = 0.49, **B** = 0.70, **V** = 1.5982, **L** = 9.062). Oxybenzone and 2-hydroxybenzophenone have a single -OH functional group located in the 2-position on the aromatic phenyl ring, and the calculated **A** solute descriptor of both compounds is equal to zero, as would be expected for strong intramolecular H-bond formation. 4-Hydroxybenzophenone also has a single -OH function group; however, its **A** solute descriptor value is much larger, as the hydroxyl functional group is too far away from the carbonyl group to permit intramolecular H-bond formation. The hydrogen atom on the hydroxyl functional group on 4-hydroxybenzophenone is thus available to form intermolecular H-bonds with surrounding solvent molecules. 2,4-Dihydroxybenzophenone has two hydroxyl functional groups, one that engages in intramolecular H-bond formation (-OH group at the 2-ring position) and one that is capable of forming intermolecular hydrogen bonds (-OH group at the 4-ring position) with lone electron pairs on surrounding solvent molecules.

An Abraham model hydrogen bond acidity descriptor value of zero indicates that the hydrogen atom of the hydroxyl functional group is engaged in strong intramolecular hydrogen bond formation with one of the lone electron pairs on the oxygen atom of the >C=O. This observation would be consistent with published spectroscopic [48–52] and computation studies [53,54] that suggest the formation of intramolecular H-bonds in 2-hydroxybenzophenone derivatives. Moreover, a value of **A** = 0.000 rules out the possibility of a bifurcated H-bond between the intramolecular H-bonded hydrogen atom and a neighboring diisopropyl ether or alcohol molecule. A bifurcated H-bond should result in a small, nonzero **A** value.

As a reminder, one of the reasons for selecting oxybenzone was to determine if the compound engaged in intramolecular hydrogen bond formation, and to ascertain if the existing group contribution and machine learning models used for estimating the Abraham model solute descriptors would be able to recognize the possibility that the molecule could form intramolecular H-bonds. In the case of 1,4-dihydroxyanthraquinone, 1,8-dihydroxyanthraquinone, and 4,5-dihydroxyanthraquinone-2-carboxylic acids, the two-group contribution methods [45,46] significantly overestimated the **A** solute descriptor value [47,75]. The MIT machine learning method [46] was found to provide better estimates of the **A** solute descriptor value in these three compounds.

Oxybenzone has a very different molecular structure (2-hydroxybenzophenone versus hydroxyanthraquinone motif), and it will informative to compare the experiment-based solute descriptors to estimated values based on the UFZ-LSER and MIT group contribution methods, as well as the MIT and newly proposed AbraLlama machine learning models. Inputting the Canonical SMILES of oxybenzone, O=C(C=1C=CC=CC1)C2=CC=C(OC)C=C2O, into the respective computational software programs yielded the following four sets of Abraham model solute descriptor values:

- Group contribution UFZ-LSER estimation [45]—**E** = 1.50, **S** = 1.66, **A** = 0.41, **B** = 0.71, **V** = 1.7391, and **L** = 9.214;
- Group contribution MIT estimation [46]—**E** = 1.668, **S** = 1.898, **A** = 0.431, **B** = 0.832, **V** = 1.7391, and **L** = 9.354;
- Machine learning MIT estimation [46]—**E** = 1.644, **S** = 1.714, **A** = 0.153, **B** = 0.697, **V** = 1.7391, and **L** = 8.915;
- Machine learning AbraLlama estimation [76]—**E** = 1.467, **S** = 1.599, **A** = 0.175, **B** = 1.049, and **V** = 1.767.

Both group contribution methods significantly overestimated oxybenzone's **A** solute descriptor. The two machine learning models performed much better in this regard. One must remember, however, that in comparing the different methods to estimate Abraham model solute descriptors, the goal is not simply to obtain a set of estimated values, but rather to obtain values that are of sufficient "quality" to be used in conjunction with existing

Abraham model correlations to predict a solute's physical and thermodynamic properties. To further assess the capability of the four estimation approaches for solute descriptors, we used the estimated values to predict the solubility of oxybenzone in the 21 organic solvents considered in the current study. As part of the comparison, we redetermined the numerical values of log $C_{S,water}$ and log $C_{S,gas}$ that were floated in our initial solute descriptor determination. The values of the two floated quantities need to be the ones that minimize the overall standard deviation for the predicted solute descriptors. The results of our calculations, which are summarized in Table 4, reveal that the four sets of estimated solute descriptors provide poor predictions for the solubility of oxybenzone when substituted into existing Abraham model log ($C_{S,organic}/C_{S,water}$) and log ($C_{S,organic}/C_{S,gas}$) correlations, as reflected in the much larger overall standard deviations. Slight differences between the experiment-based and estimated solute descriptor values can result in sizeable errors in the predicted physical and thermodynamic properties of the given solute molecule, particularly when the descriptor value is multiplied by a large complementary solvent equation coefficient.

Table 4. Comparison of the four different sets of estimated descriptor values used in predicting the molar solubility of oxybenzone in 21 organic mono-solvents.

Solute Descriptor Estimation Method	SD (Log Units)	Log $C_{S,water}$	Log $C_{S,gas}$
Experiment-Based Values	0.115	−4.558	−9.581
UFZ-LSER Group Contribution Model	0.896	−3.100	−11.485
MIT Group Contribution Model	1.025	−2.642	−11.551
MIT Machine Learning Model	0.439	−3.729	−10.347
AbraLlama Machine Learning Model	0.707 [a]	−2.397 [a]	

[a] Values for the AbraLlama machine learning model based only on Equation (1) because the predictive software does not calculate the Abraham model **L** solute descriptor.

The Canonical SMILES unfortunately does not incorporate all of the salient structural features that govern a molecule's physical and thermodynamic properties. The SMILES code provides information regarding the numbers and various types of atoms, and how the individual atoms are arranged within the molecule. There are arrangements that give rise to special intramolecular interactions and to steric hinderance, which can inhibit a functional group from efficiently interacting with surrounding molecules. We believe that a better set of solute descriptors could be obtained if such information were used in the initial training of the group contribution or machine learning method. For example, if one knows from independent spectroscopic or computational studies that a specific hydroxyl hydrogen atom forms a strong intramolecular H-bond, would it not be better to reduce the contribution of this particular hydrogen atom from the overall H-bond acidity descriptor?

In the case of group contribution methods, it should be fairly easy to introduce a new functional group and associated group value to account for hydrogen atoms engaged in forming an intramolecular H-bond. The Canonical SMILES code of oxybenzone could be modified to include "hb" after the oxygen atom that is covalently bonded to the hydrogen atom involved in intramolecular hydrogen bonding. Our suggested "Informed User-Modified Canonical SMILES code" for oxybenzone would take the form of O=C(C=1C=CC=CC1)C2=CC=C(OC)C=C2Ohb. Modified SMILES codes for other select molecules and their experiment-based Abraham model solute descriptors are provided in Tables 5 and 6, respectively. The tabulated information would provide a good starting point for anyone interested in incorporating intramolecular H-bond formation in the Abraham model solute descriptor predictions. The methodology can be extended to other structural features, such as intramolecular H-bond formation involving the hydrogen atom of a primary and/or secondary amine (Nhb) with lone electron pairs on neighboring oxygen or nitrogen atoms within the molecule [77,78]. Steric hindrance around an H-bond donor site (Osh and Nsh) could be introduced into the SMILES code as well. Each modification would permit the user of the predictive software to encode their knowledge of any special structural features that the solute molecule might possess that would affect the estimated Abraham model's solute descriptors. We note that Gani and coworkers [79] accounted for intramolecular H-bond formation and other steric factors by introducing second-order and third-order groups into their group contribution approach. The challenge with implementing higher-order groups into existing software programs for estimating solute descriptors is that much of the computational program will need to be completely rewritten. On the other hand, our suggested approach would require the addition of only a few select functional groups and the retraining of the method with a sufficient number of compounds and their modified Canonical SMILES codes in order to obtain group values with good predictive accuracy.

Table 5. Solute molecules that exhibit intramolecular hydrogen bond formation and their proposed Informed User-Modified Canonical SMILES Codes.

Solute	Informed User-Modified Canonical SMILES Code [a]
Oxybenzone	O=C(C=1C=CC=CC1)C2=CC=C(OC)C=C2Ohb
2-Hydroxybenzophenone	O=C(C=1C=CC=CC1)C=2C=CC=CC2Ohb
2,4-Dihydroxybenzophenone	O=C(C=1C=CC=CC1)C2=CC=C(O)C=C2Ohb
5-Hydroxyflavone	O=C1C=C(OC=2C=CC=C(Ohb)C12)C=3C=CC=CC3
3,5-Dihydroxyflavone	O=C1C(O)=C(OC=2C=CC=C(Ohb)C12)C=3C=CC=CC3
5,6-Dihydroxyflavone	O=C1C=C(OC2=CC=C(O)C(Ohb)=C12)C=3C=CC=CC3
5,7-Dihydroxyflavone	O=C1C=C(OC=2C=C(O)C=C(Ohb)C12)C=3C=CC=CC3
5,4′-Dihydroxyflavone	O=C1C=C(OC=2C=CC=C(Ohb)C12)C=3C=CC(O)=CC3
5,2′-Dihydroxyflavone	O=C1C=C(OC=2C=CC=C(Ohb)C12)C=3C=CC=CC3O
5,3′-Dihydroxyflavone	O=C1C=C(OC=2C=CC=C(Ohb)C12)C=3C=CC=C(O)C3
5,4′-Dihydroxy-7-methoxyisoflavone	O=C1C(=COC=2C=C(OhbC)C=C(O)C12)C=3C=CC(O)=CC3
1-Hydroxyanthraquinone	O=C1C=2C=CC=CC2C(=O)C=3C(Ohb)=CC=CC13
1,4-Dihydroxyanthraquinone	O=C1C=2C=CC=CC2C(=O)C=3C(Ohb)=CC=C(Ohb)C13
1,8-Dihydroxyanthraquinone	O=C1C=2C=CC=C(Ohb)C2C(=O)C=3C(Ohb)=CC=CC13
4,5-Dihydroxyanthraquinone-2-carboxylic acid	O=C(O)C1=CC(Ohb)=C2C(=O)C=3C(Ohb)=CC=CC3C(=O)C2=C1
2′-Hydroxyacetophenone	O=C(C=1C=CC=CC1Ohb)C
2-Nitrophenol	O=N(=O)C=1C=CC=CC1Ohb
2,4-Dinitrophenol	O=N(=O)C1=CC=C(Ohb)C(=C1)N(=O)=O
2,5-Dinitrophenol	O=N(=O)C1=CC=C(C(Ohb)=C1)N(=O)=O
2,6-Dinitrophenol	O=N(=O)C1=CC=CC(=C1Ohb)N(=O)=O
2-Methoxyphenol	OhbC=1C=CC=CC1OC
2-Methyl-4,6-dinitrophenol	O=N(=O)C=1C=C(C(Ohb)=C(C1)C)N(=O)=O

[a] The Informed User Canonical SMILES codes were obtained by modifying the compound's SMILES codes taken from SciFinder Scholar [80].

Steric factors have been used in the past to explain the dissimilar solubility behaviors of 2-acetyl-1-naphthol and 1-acetyl-2-naphthol in saturated hydrocarbons versus in alcohol solvents [81,82]. The measured solubilities of 2-acetyl-1-naphthol are significantly larger in alcohol solvents than in saturated hydrocarbons. In contract, there is very little difference in the solubility of 1-acetyl-2-naphthol in these two solvent types. Both acetylnaphthols exhibit intramolecular H-bond formation, as indicated by UV-Vis absorption [83] and dielectric studies [84]. It is believed that the formation of an intermolecular solute–alcohol H-bond in the case of 1-acetyl-2-naphthol is capable of breaking the weak intramolecular H-bond [81]. Our suggested Informed User-Modified Canonical SMILES, if incorporated into group contribution methods, will enable researchers to calculate different sets of solute descriptors for 1-acetyl-2-naphthol depending upon whether they wish to include the intramolecular H-bond. In other words, one could use one set of Abraham model solute descriptors to predict the solubility of 1-acetyl-2-naphthol in saturated hydrocarbons and a second set of values to predict solubilities in alcohol solvents.

Table 6. Experiment-based Abraham model solute descriptors for compounds that engage in intramolecular hydrogen bond formation.

Solute	E	S	A	B	V	L
Oxybenzone	1.500	1.413	0.000	0.617	1.7391	8.660
2-Hydroxybenzophenone	1.54	1.46	0.00	0.46	1.5395	7.95
2,4-Dihydroxybenzophenone	1.73	2.03	0.49	0.70	1.5982	9.062
5-Hydroxyflavone	1.91	1.91	0.02	0.52	1.7284	9.41
5,3-Dihydroxyflavone	2.09	2.07	0.37	0.73	1.7871	10.13
5,6-Dihydroxyflavone	2.06	1.96	0.63	0.73	1.7871	10.08
5,7-Dihydroxyflavone	2.13	1.98	0.69	0.73	1.7871	10.22
5,4′-Dihydroxyflavone	2.14	1.94	0.71	0.98	1.7871	10.18
5,2′-Dihydroxyflavone	2.14	1.90	0.42	0.84	1.7871	10.08
5,3′-Dihydroxyflavone	2.14	1.90	0.67	0.85	1.7871	10.14
1-Hydroxyanthraquinone	1.504	1.491	0.050	0.539	1.5874	9.075
1,4-Dihydroxyanthraquinone	2.455	1.743	0.000	0.669	1.6462	9.919
1,8-Dihydroxyanthraquinone	2.455	1.786	0.000	0.558	1.6462	9.907
4,5-Dihydroxyanthraquinone-2-carboxylic acid	2.340	2.195	0.755	0.596	1.8615	11.073
2′-Hydroxyacetophenone	0.948	1.32	0.00	0.37	1.0726	5.341
2-Nitrophenol	1.015	1.05	0.06	0.35	0.9493	4.778
2,4-Dinitrophenol	1.200	1.57	0.09	0.54	1.1235	6.078
2,5-Dinitrophenol	1.200	1.46	0.14	0.51	1.1235	5.990
2,6-Dinitrophenol	1.220	2.11	0.16	0.45	1.1235	6.565
2-Methoxyphenol	0.837	0.92	0.16	0.54	0.9747	4.494
2-Methyl-4,6-dinitrophenol	1.200	1.59	0.04	0.52	1.2644	5.34

4. Summary

The Abraham general solvation parameter model has been shown to provide a reasonably accurate mathematical description of the observed solubility behavior of oxybenzone dissolved in hexane, heptane, octane, dodecane, cyclohexane, methylcyclohexane, 2,2,4-trimethylpentane, diisopropyl ether, propan-1-ol, butan-1-ol, pentan-1-ol, heptan-1-ol, octan-1-ol, 2-methylpropan-1-ol, butan-2-ol, 2-methylbutan-1-ol, 3-methylbutan-1-ol, 4-methylpentan-2-ol, 2-ethylhexan-1-ol, cyclopentanol, and ethylene glycol at 298.15 K. The solute descriptors calculated from experimental solubility data for oxybenzone dissolved in 21 different organic solvents indicate that the hydrogen atom on the hydroxyl functional group forms an intramolecular hydrogen bond with the lone electron pair on the oxygen atom of the neighboring >C=O functional group. This observation is consistent with published spectroscopic [48–52] and computation studies [53,54] that suggest the formation of intramolecular H-bonds in 2-hydroxybenzophenone derivatives. Moreover, the calculated value of **A** = 0.000 rules out the possibility that oxybenzone forms a bifurcated H-bond between the intramolecular H-bonded hydrogen atom and a neighboring diisopropyl ether or alcohol molecule. A bifurcated H-bond should result in a small, nonzero **A** value.

Two existing group contribution methods and two machine learning methods were used to estimate the Abraham model's solute descriptors. Both group contribution methods were found to significantly overestimate oxybenzone's **A** solute descriptor because the methods did not properly account for the formation of the intramolecular H-bond. The two machine learning models performed much better in this regard. An "Informed User-Modified Canonical SMILES code" was proposed as an input structural parameter to identify which hydrogen atoms are involved in intramolecular H-bond formation in the hopes of improving the group contribution methods that have already been developed to estimate Abraham model solute descriptors.

Author Contributions: Conceptualization and writing—original draft preparation, W.E.A.; data curation, J.C., A.C. and Y.Y.; formal analysis, J.C., A.C. and Y.Y.; writing—review and editing, J.C., A.C. and Y.Y. All authors have read and agreed to the published version of the manuscript.

Funding: This research received no external funding.

Data Availability Statement: All experimental data are contained within the manuscript.

Acknowledgments: J.C. and A.C. thank the University of North Texas's Texas Academy of Mathematics & Science (TAMS) program for providing a summer research scholarship to each student.

Conflicts of Interest: The authors declare no conflicts of interest.

References

1. Katritzky, A.R.; Slavov, S.H.; Dobchev, D.A.; Karelson, M. Rapid QSPR model development technique for prediction of vapor pressure of organic compounds. *Comp. Chem. Eng.* **2007**, *31*, 1123–1130. [CrossRef]
2. Zushi, Y. Direct prediction of physicochemical properties and toxicities of chemicals from analytical descriptors by GC-MS. *Anal. Chem.* **2022**, *94*, 9149–9157. [CrossRef] [PubMed]
3. Carande, W.H.; Kazakov, A.; Muzny, C.; Frenkel, M. Quantitative structure-property relationship predictions of critical properties and acentric factors for pure compounds. *J. Chem. Eng. Data* **2015**, *60*, 1377–1387. [CrossRef]
4. Katritzky, A.R.; Petrukhin, R.; Jain, R.; Karelson, M. QSPR analysis of flash points. *J. Chem. Inf. Comp. Sci.* **2001**, *41*, 1521–1530. [CrossRef] [PubMed]
5. Katritzky, A.R.; Stoyanova-Slavova, I.B.; Dobchev, D.A.; Karelson, M. QSPR modeling of flash points: An update. *J. Mol. Graph. Model.* **2007**, *26*, 529–536. [CrossRef]
6. Liu, Y.; Yu, X.; Chen, J. Quantitative structure-property relationship of distribution coefficients of organic compounds. *SAR QSAR Environ. Res.* **2020**, *31*, 585–596. [CrossRef]
7. Fu, L.; Liu, L.; Yang, Z.-J.; Li, P.; Ding, J.-J.; Yun, Y.-H.; Lu, A.-P.; Hou, T.-J.; Cao, D.-S. Systematic modeling of log D7.4 based on ensemble machine learning, group contribution, and matched molecular pair analysis. *J. Chem. Inf. Model.* **2020**, *60*, 63–76. [CrossRef]
8. Kang, X.; Hu, B.; Perdana, M.C.; Zhao, Y.; Chen, Z. Extreme learning machine models for predicting the n-octanol/water partition coefficient (Kow) data of organic compounds. *J. Environ. Chem. Eng.* **2022**, *10*, 108552. [CrossRef]
9. Katritzky, A.R.; Oliferenko, A.A.; Oliferenko, P.V.; Petrukhin, R.; Tatham, D.B.; Maran, U.; Lomaka, A.; Acree, W.E., Jr. A general treatment of solubility. 2. QSPR prediction of free energies of solvation of specified solutes in ranges of solvents. *J. Chem. Inf. Comp. Sci.* **2003**, *43*, 1806–1814. [CrossRef]
10. Katritzky, A.R.; Oliferenko, A.A.; Oliferenko, P.V.; Petrukhin, R.; Tatham, D.B.; Maran, U.; Lomaka, A.; Acree, W.E., Jr. A general treatment of solubility. 1. The QSPR correlation of solvation free energies of single solutes in series of solvents. *J. Chem. Inf. Comp. Sci.* **2003**, *43*, 1794–1805. [CrossRef]
11. Brown, T.N. QSPRs for predicting equilibrium partitioning in solvent-air systems from the chemical structures of solutes and solvents. *J. Solut. Chem.* **2022**, *51*, 1101–1132. [CrossRef]
12. Fu, Z.; Chen, J.; Li, X.; Wang, Y.; Yu, H. Comparison of prediction methods for octanol-air partition coefficients of diverse organic compounds. *Chemosphere* **2016**, *148*, 118–125. [CrossRef] [PubMed]
13. Wang, Y.; Tang, W.; Xiao, Z.; Yang, W.; Peng, Y.; Chen, J.; Li, J. Novel quantitative structure activity relationship models for predicting hexadecane/air partition coefficients of organic compounds. *J. Environ. Sci.* **2023**, *124*, 98–104. [CrossRef] [PubMed]
14. Chung, Y.; Vermeire, F.H.; Wu, H.; Walker, P.J.; Abraham, M.H.; Green, W.H. Group contribution and machine learning approaches to predict Abraham solute parameters, solvation free energy, and solvation enthalpy. *J. Chem. Inf. Model.* **2022**, *62*, 433–446. [CrossRef]
15. Ebert, R.-U.; Kühne, R.; Schüürmann, G. Henry's law constant—A general-purpose fragment model to predict Log Kaw from molecular structure. *Environ. Sci. Technol.* **2023**, *57*, 160–167. [CrossRef]
16. Katritzky, A.R.; Chen, K.; Wang, Y.; Karelson, M.; Lucic, B.; Trinajstic, N.; Suzuki, T.; Schuurmann, G. Prediction of liquid viscosity for organic compounds by a quantitative structure-property relationship. *J. Phys. Org. Chem.* **2000**, *13*, 80–86. [CrossRef]
17. Mirkhani, S.A.; Gharagheizi, F. Predictive quantitative structure-property relationship model for the estimation of ionic liquid viscosity. *Ind. Eng. Chem. Res.* **2012**, *51*, 2470–2477. [CrossRef]
18. Paduszyński, K.; Domańska, U. Viscosity of ionic liquids: An extensive database and a new group contribution model based on a feed-forward artificial neural network. *J. Chem. Inf. Model.* **2014**, *54*, 1311–1324. [CrossRef] [PubMed]
19. Das, S.; Ojha, P.K.; Roy, K. Development of a temperature dependent 2D-QSPR model for viscosity of diverse functional ionic liquids. *J. Mol. Liq.* **2017**, *240*, 454–467. [CrossRef]
20. Lazzús, J.A.; Pulgar-Villarroel, G.; Cuturrufo, F.; Vega, P. Development of a group contribution method for estimating surface tension of ionic liquids over a wide range of temperatures. *J. Mol. Liq.* **2017**, *240*, 522–531. [CrossRef]
21. Naef, R. A generally applicable computer algorithm based on the group additivity method for the calculation of seven molecular descriptors: Heat of combustion, log Po/w, log S, refractivity, polarizability, toxicity and log BB of organic compounds; scope and limits of applicability. *Molecules* **2015**, *20*, 18279–18351. [CrossRef] [PubMed]
22. Gharagheizi, F.; Mirkhani, S.A.; Tofangchi Mahyari, A.R. Prediction of standard enthalpy of combustion of pure compounds using a very accurate group-contribution-based method. *Energy Fuels* **2011**, *25*, 2651–2654. [CrossRef]
23. Cardozo, R.L. Prediction of the enthalpy of combustion of organic compounds. *AIChE J.* **1986**, *32*, 844–848. [CrossRef]
24. Hukkerikar, A.S.; Meier, R.J.; Sina, G.; Gani, R. Method to estimate the enthalpy of formation of organic compounds with chemical accuracy. *Fluid Phase Equilibria* **2013**, *348*, 23–32. [CrossRef]

25. Golmohammadi, H.; Dashtbozorgi, Z.; Gholam Samani, M.; Acree, W.E., Jr. QSPR prediction of gas-to-methanol solvation enthalpy of organic compounds using replacement method and support vector machine. *Phys. Chem. Liq.* **2015**, *53*, 46–66. [CrossRef]
26. Toubaei, A.; Golmohammadi, H.; Dashtbozorgi, Z.; Acree, W.E., Jr. QSPR studies for predicting gas to acetone and gas to acetonitrile solvation enthalpies using support vector machine. *J. Mol. Liq.* **2012**, *175*, 24–32. [CrossRef]
27. Khajeh, A.; Modarress, H. Quantitative structure–property relationship prediction of liquid heat capacity at 298.15 K for organic compounds. *Ind. Eng. Chem. Res.* **2012**, *51*, 6251–6255. [CrossRef]
28. Khajeh, A.; Modarress, H. Quantitative structure-property relationship prediction of gas heat capacity for organic compounds. *Ind. Eng. Chem. Res.* **2012**, *51*, 13490–13495. [CrossRef]
29. Naef, R. Calculation of the isobaric heat capacities of the liquid and solid phase of organic compounds at 298.15 K by means of the group-additivity method. *Molecules* **2020**, *25*, 1147. [CrossRef]
30. Chickos, J.S.; Hesse, D.G.; Liebman, J.F. A group additivity approach for the estimation of heat capacities of organic liquids and solids at 298 K. *Struct. Chem.* **1993**, *4*, 261–269. [CrossRef]
31. Acree, W.; Chickos, J.S. Phase transition enthalpy measurements of organic and organometallic compounds and ionic liquids. sublimation, vaporization, and fusion enthalpies from 1880 to 2015. Part 2. C11-C192. *J. Phys. Chem. Ref. Data* **2017**, *46*, 013104. [CrossRef]
32. Kolska, Z.; Kukal, J.; Zabransky, M.; Ruazicka, V. Estimation of the heat capacity of organic liquids as a function of temperature by a three-level group contribution method. *Ind. Eng. Chem. Res.* **2008**, *47*, 2075–2085. [CrossRef]
33. Naef, R.; Acree, W.E., Jr. Calculation of five thermodynamic molecular descriptors by means of a general computer algorithm based on the group-additivity method: Standard enthalpies of vaporization, sublimation and solvation, and entropy of fusion of ordinary organic molecules and total phase-change entropy of liquid crystals. *Molecules* **2017**, *22*, 1059. [CrossRef]
34. Gharagheizi, F.; Ilani-Kashkouli, P.; Acree, W.E., Jr.; Mohammadi, A.H.; Ramjugernath, D. A group contribution model for determining the vaporization enthalpy of organic compounds at the standard reference temperature of 298 K. *Fluid Phase Equilibria* **2013**, *360*, 279–292. [CrossRef]
35. Gharagheizi, F.; Ilani-Kashkouli, P.; Acree, W.E., Jr.; Mohammadi, A.H.; Ramjugernath, D. A group contribution model for determining the sublimation enthalpy of organic compounds at the standard reference temperature. *Fluid Phase Equilibria* **2013**, *354*, 265–285. [CrossRef]
36. Dohnal, V. New QSPR molecular descriptors based on low-cost quantum chemistry computations using DFT/COSMO approach. *J. Mol. Liq.* **2024**, *407*, 125256. [CrossRef]
37. Yan, X.; Lan, T.; Jia, Q.; Yan, F.; Wang, Q. A norm indexes-based QSPR model for predicting the standard vaporization enthalpy and formation enthalpy of organic compounds. *Fluid Phase Equilibria* **2020**, *507*, 112437. [CrossRef]
38. Chickos, J.S.; Acree, W.E., Jr.; Liebman, J.F. Estimating solid-liquid phase change enthalpies and entropies. *J. Phys. Chem. Ref. Data* **1999**, *28*, 1535–1673. [CrossRef]
39. Chickos, J.S.; Acree, W.E., Jr. Total phase change entropies and enthalpies. An update on fusion enthalpies and their estimation. *Thermochim. Acta* **2009**, *495*, 5–13. [CrossRef]
40. Fredenslund, A.; Gmehling, J.; Michelsen, M.L.; Rasmussen, P.; Prausnitz, J.M. Computerized design of multicomponent distillation columns using the UNIFAC group contribution method for calculation of activity coefficients. *Ind. Eng. Chem. Proc. Des. Dev.* **1977**, *16*, 450–462. [CrossRef]
41. Gmehling, J.; Tiegs, D.; Knipp, U. A comparison of the predictive capability of different group contribution methods. *Fluid Phase Equilibria* **1990**, *54*, 147–165. [CrossRef]
42. Gmehling, J.; Lohmann, J.; Jakob, A.; Li, J.; Joh, R. A modified UNIFAC (Dortmund) model. 3. Revision and extension. *Ind. Eng. Chem. Res.* **1998**, *37*, 4876–4882. [CrossRef]
43. Currier, R.P.; O'Connell, J.P. An analysis of the solution of groups method for component activity coefficients. *Fluid Phase Equilibria* **1987**, *33*, 245–265. [CrossRef]
44. Tochigi, K.; Gmehling, J. Determination of ASOG parameters-extension and revision. *J. Chem. Eng. Jpn.* **2011**, *44*, 304–306. [CrossRef]
45. Ulrich, N.; Endo, S.; Brown, T.N.; Watanabe, N.; Bronner, G.; Abraham, M.H.; Goss, K.-U. *UFZ-LSER Database*; v 3.2.1; Helmholtz Centre for Environmental Research-UFZ: Leipzig, Germany, 2017; Available online: http://www.ufz.de/lserd (accessed on 27 June 2024).
46. Solvation Tools. Available online: https://rmg.mit.edu/database/solvation/search/ (accessed on 27 June 2024).
47. Acree, W.E., Jr.; Smart, K.; Abraham, M.H. Abraham model solute descriptors reveal strong intramolecular hydrogen bonding in 1,4-dihydroxyanthraquinone and 1,8-dihydroxyanthraquinone. *Phys. Chem. Liq.* **2018**, *56*, 416–420. [CrossRef]
48. Merrill, J.R. Measurements of intramolecular hydrogen bonding by nuclear magnetic resonance and infrared spectroscopy. *J. Phys. Chem.* **1961**, *65*, 2023–2026. [CrossRef]
49. Strat, M.; Umreiko, D.S.; Khovratovich, N.N. Spectroscopic manifestation of an intramolecular hydrogen bond in 2-hydroxybenzophenone derivatives. *Zhurnal Prikl. Spektrosk.* **1973**, *19*, 103–108.
50. Liebich, B.W.; Parthe, E. 2-Hydroxy-4-methoxybenzophenone. *Acta Crystallogr. B* **1974**, *30B*, 2522–2524. [CrossRef]
51. Liebich, B.W. The crystal and molecular structure of 2,4-dihydroxybenzophenone (HHB). *Acta Crystallogr. B* **1979**, *B35*, 1186–1190. [CrossRef]

52. Baumstark, A.L.; Boykin, D.W. Oxygen-17 NMR spectroscopy: Multiple intramolecular hydrogen-bonded phenolic carbonyl systems. *Nouv. J. Chim.* **1992**, *16*, 357–360.
53. Zahedi-Tabrizi, M.; Tayyari, S.F.; Badalkhani-Khamseh, F.; Ghomi, R.; Afshar-Qahremani, F. Molecular structure and intramolecular hydrogen bonding in 2-hydroxybenzophenones: A theoretical study. *J. Chem. Sci.* **2014**, *126*, 919–929. [CrossRef]
54. Fang, Z.; Zhang, X.; Wu, F.; Huang, B.; Au, C.; Yi, B. Effect of substituent groups on the strength of intramolecular hydrogen bonds in 2,4-dihydroxybenzophenone UV absorbers. *Molecules* **2023**, *28*, 5017. [CrossRef] [PubMed]
55. Baughman, B.M.; Stennett, E.; Lipner, R.E.; Rudawsky, A.C.; Schmidtke, S.J. Structural and spectroscopic studies of the photophysical properties of benzophenone derivatives. *J. Phys. Chem. A* **2009**, *113*, 8011–8019. [CrossRef]
56. Litwinienko, G.; Di Labio, G.A.; Mulder, P.; Korth, H.-G.; Ingold, K.U. Intramolecular and intermolecular hydrogen bond formation by some ortho-substituted phenols: Some surprising results from an experimental and theoretical investigation. *J. Phys. Chem. A* **2009**, *113*, 6275–6288. [CrossRef]
57. Bureiko, S.F.; Golubev, N.S.; Pihlaja, K. Spectroscopic studies of bifurcated hydrogen bonds in solution. *J. Mol. Struct.* **1999**, *480–481*, 297–301. [CrossRef]
58. Bureiko, S.F.; Golubev, N.S.; Mattinen, J.; Pihlaja, K. On the existence of bifurcated hydrogen bond in liquids: Interaction of 2,6-disubstituted phenols with proton acceptors. *J. Mol. Liq.* **1990**, *45*, 139–145. [CrossRef]
59. Liao, C.; Kannan, K. Widespread occurrence of benzophenone-type UV light filters in personal care products from China and the United States: An assessment of human exposure. *Environ. Sci. Technol.* **2014**, *48*, 4103–4109. [CrossRef] [PubMed]
60. Scheele, A.; Sutter, K.; Karatum, O.; Danley-Thomson, A.A.; Redfern, L.K. Environmental impacts of the ultraviolet filter oxybenzone. *Sci. Total Environ.* **2023**, *863*, 160966. [CrossRef] [PubMed]
61. Huang, Y.; Law, J.C.F.; Lam, T.K.; Leung, K.S.Y. Risks of organic UV filters: A review of environmental and human health concern studies. *Sci. Total Environ.* **2021**, *755*, 142486. [CrossRef]
62. Downs, C.A.; Bishop, E.; Diaz-Cruz, M.S.; Haghshenas, S.A.; Stien, D.; Rodrigues, A.M.S.; Woodley, C.M.; Sunyer-Caldu, A.; Doust, S.N.; Espero, W.; et al. Oxybenzone contamination from sunscreen pollution and its ecological threat to Hanauma Bay, Oahu, Hawaii, U.S.A. *Chemosphere* **2022**, *291*, 132880. [CrossRef]
63. Narla, S.; Lim, H.W. Sunscreen: FDA regulation, and environmental and health impact. *Photochem. Photobiol. Sci.* **2020**, *19*, 66–70. [CrossRef] [PubMed]
64. Raffa, R.B.; Pergolizzi, J.V., Jr.; Taylor, R., Jr.; Kitzen, J.M. Sunscreen bans: Coral reefs and skin cancer. *J. Clin. Pharm. Ther.* **2019**, *44*, 134–139. [CrossRef] [PubMed]
65. Schneider, S.L.; Lim, H.W. Review of environmental effects of oxybenzone and other sunscreen active ingredients. *J. Am. Acad. Dermatol.* **2019**, *80*, 266–271. [CrossRef] [PubMed]
66. Rodil, R.; Moeder, M. Development of a method for the determination of UV filters in water samples using stir bar sorptive extraction and thermal desorption–gas chromatography–mass spectrometry. *J. Chromatogr. A* **2008**, *1179*, 81–88. [CrossRef] [PubMed]
67. Abraham, M.H.; Smith, R.E.; Luchtefeld, R.; Boorem, A.J.; Luo, R.; Acree, W.E., Jr. Prediction of solubility of drugs and other compounds in organic solvents. *J. Pharm. Sci.* **2010**, *99*, 1500–1515. [CrossRef]
68. Acree, W.E., Jr.; Horton, M.Y.; Higgins, E.; Abraham, M.H. Abraham model linear free energy relationships as a means of extending solubility studies to include the estimation of solute solubilities in additional organic solvents. *J. Chem. Thermodyn.* **2016**, *102*, 392–397. [CrossRef]
69. Jalan, A.; Ashcraft, R.W.; West, R.H.; Green, W.H. Predicting solvation energies for kinetic modeling. *Annu. Rep. Prog. Chem Sec. C Phys. Chem.* **2010**, *106*, 211–258. [CrossRef]
70. Clarke, E.D. Beyond physical properties–application of Abraham descriptors and LFER analysis in agrochemical research. *Bioorg. Med. Chem.* **2009**, *17*, 4153–4159. [CrossRef]
71. Poole, C.F.; Ariyasena, T.C.; Lenca, N. Estimation of the environmental properties of compounds from chromatographic properties and the solvation parameter method. *J. Chromatogr. A* **2013**, *1317*, 85–104. [CrossRef]
72. Poole, C.F.; Atapattu, S.N. Recent advances for estimating environmental properties for small molecules from chromatographic measurements and the solvation parameter model. *J. Chromatogr. A* **2022**, *1687*, 463682. [CrossRef]
73. Endo, S.; Goss, K.-U. Applications of polyparameter linear free energy relationships in environmental chemistry. *Environ. Sci. Technol.* **2014**, *48*, 12477–12491. [CrossRef] [PubMed]
74. Abraham, M.H.; McGowan, J.C. The use of characteristic volumes to measure cavity terms in reversed phase liquid chromatography. *Chromatographia* **1987**, *23*, 243–246. [CrossRef]
75. Sinha, S.; Yang, C.; Wu, E.; Acree, W.E., Jr. Abraham solvation parameter model: Examination of possible intramolecular hydrogen-bonding using calculated solute descriptors. *Liquids* **2022**, *2*, 131–146. [CrossRef]
76. Lang, A.S.I.D.; Lee, Y. AbraLlama Hugging Face App: Predicting Abraham Model Solute Descriptors and Modified Solvent Parameters Using Llama. Available online: https://huggingface.co/spaces/ttmn/AbraLlama (accessed on 21 July 2024).
77. Hansen, P.K. A Spectroscopic overview of intramolecular hydrogen bonds of NH...O,S,N type. *Molecules* **2021**, *26*, 2409. [CrossRef]
78. Bertolasi, V.; Ferretti, V.; Gilli, P.; Gilli, G.; lssal, Y.M.; Sherif, O.E. Intramolecular N-H O hydrogen bonding assisted by resonance. Part 2. Intercorrelation between structural and spectroscopic parameters for five 1.3-diketone arylhydrazones derived from dibenzoylmethane. *J. Chem. Soc. Perkin Trans. 2* **1993**, 2223–2228. [CrossRef]

79. Marrero, J.; Gani, R. Group-contribution based estimation of pure component properties. *Fluid Phase Equilibria* **2001**, *183–184*, 183–208. [CrossRef]
80. SciFinder. Chemical Abstracts Service: Columbus, OH. Carbon-13 NMR Spectrum; Spectrum ID CC-03-C_SPC-3734; RN 50-52-2. Available online: https://scifinder.cas.org (accessed on 21 July 2024).
81. Domanska, U. Solubility of acetyl-substituted naphthols in binary solvent mixtures. *Fluid Phase Equilibria* **1990**, *55*, 125–145. [CrossRef]
82. Buchowski, H.; Domanska, U. Solubility and hydrogen bonding. Part VI. Evidence for intramolecular hydrogen bonds. *Pol. J. Chem.* **1980**, *54*, 97–102.
83. Skulski, L. Studies on intramolecular hydrogen bonding by ultraviolet spectroscopy. VI. Ultraviolet absorption spectra of acetyl- and benzoyl-substituted naphthols. *Bull. Acad. Polon. Sci. Ser. Sci. Chim.* **1969**, *17*, 253–258.
84. Skulski, L.; Waclawek, W. Studies on intramolecular hydrogen bonding by ultraviolet-visible absorption spectroscopy. VIII. Correlations between the spectral solvent effects and dioxane effect in the dipolometric measurements. *Bull. Acad. Polon. Sci. Ser. Sci. Chim.* **1971**, *19*, 277–286.

liquids

MDPI

Article

Quantum Chemical (QC) Calculations and Linear Solvation Energy Relationships (LSER): Hydrogen-Bonding Calculations with New QC-LSER Molecular Descriptors

Costas Panayiotou

Department of Chemical Engineering, Aristotle University of Thessaloniki, 54624 Thessaloniki, Greece; cpanayio@auth.gr

Abstract: A new method, based on quantum chemical calculations, is proposed for the thermodynamically consistent reformulation of QSPR-type Linear Free-Energy Relationship (LFER) models. This reformulation permits the extraction of valuable information on intermolecular interactions and its transfer in other LFER-type models, in acidity/basicity scales, or even in equation-of-state models. New molecular descriptors of electrostatic interactions are derived from the distribution of molecular surface charges obtained from COSMO-type quantum chemical calculations. The widely used and very successful Abraham's Linear Solvation Energy Relationship (LSER) model is selected as the reference LSER model for the calculations in solute–solvent systems as well as in solute self-solvation. Hydrogen-bonding free energies, enthalpies, and entropies are now derived for a variety of common solutes. The capacity of the method to address the role of conformational changes in solvation quantities is discussed. The perspectives of the LSER model with the implementation of the new descriptors are also discussed.

Keywords: solvation thermodynamics; hydrogen bonding; COSMO-RS; sigma profiles; intramolecular association

1. Introduction

The new technological advances and their ever-changing landscape bring up new challenging problems and open new research horizons for efficient and insightful solutions in new materials or new process designs. Current developments in modern Molecular Thermodynamics exploit the widely available high computational speed, even at the laptop level, and derive valuable information from quantum chemical calculations and molecular simulations. Quantum chemical suites at various levels are now widely and freely available. Atomistically detailed force fields developed with information from quantum chemical calculations are extensively used, allowing the computer-based reliable prediction of thermodynamic/thermophysical properties, phase equilibria, or solvation quantities [1–6].

On the other hand, in numerous sectors, including the pharmaceutical and biomedical sector, the need often arises for efficient fast solvent screening or reactant selection with specific properties. Quantitative structure–property or structure–activity relationships (QSPR/QSAR) are the preferred approach in most practical applications. One of the most successful QSPR-type approaches is Abraham's LSER (Linear Solvation Energy Relationship) model [7–11] and has led, ultimately, to the valuable and freely available comprehensive and rich in thermodynamic information LSER Database [12]. Abraham's legacy is continued by his collaborators and other developers worldwide. Bill Acree, having been one of Abraham's closest collaborators, is continuing the development and expansion of the LSER model with particular focus on solvation enthalpies [10,13,14]. The LSER model is currently one of the simplest, robust, and widely used tools for thermodynamic calculations in a remarkably broad range of applications and is often more successful than other more involved models [15].

The LSER model uses simple linearity equations that quantify solute transfer between two phases. These important linear relationships, in Abraham's LSER approach, take the form of Equations (1) and (2) for the equilibrium constant, K_G, of solute partitioning between gas and liquid phases and for the corresponding solvation energy (enthalpy) constant, K_E, respectively:

$$LogK_G = \frac{-\Delta G_{12}}{2.303RT} = c_{g2} + e_{g2}E_1 + s_{g2}S_1 + a_{g2}A_1 + b_{g2}B_1 + l_{g2}L_1 \quad \text{(LSER)} \quad (1)$$

and

$$LogK_E = -\frac{\Delta H_{12}}{2.303RT} = c_{e2} + e_{e2}E_1 + s_{e2}S_1 + a_{e2}A_1 + b_{e2}B_1 + l_{e2}L_1 \quad \text{(LSER)} \quad (2)$$

Analogous equations apply for the solute transfer/partitioning between two condensed phases. The upper-case letters in these equations indicate solute molecular LSER descriptors, while the lower-case letters indicate the corresponding complementary but solvent-phase-specific LFER coefficients. The solute molecular LSER descriptors V_x, L, E, S, A, and B correspond to McGowan's characteristic volume, the equilibrium constant for the gas–liquid partition in n-hexadecane at 298 K, the excess molar refraction, the dipolarity/polarizability, the hydrogen bonding (HB) acidity A, and hydrogen bonding basicity B, respectively [7–11].

Besides the traditional solute partitioning/transfer studies and the limited Henry's law, the solvation free energy, ΔG_{12}, and its components enthalpy, ΔH_{12}, and entropy, ΔS_{12}, are directly connected with phase equilibrium studies involving activity or fugacity coefficients through the following most useful classical working equation:

$$\frac{\Delta G_{12}}{RT} = \frac{\Delta H_{12} - T\Delta S_{12}}{RT} = \ln \frac{\varphi_1^0 P_1^0 V_{m2} \gamma_{1/2}^{\infty}}{RT} \quad (3)$$

V_{m2} in Equation (3) is the molar volume of the solvent (component 2) and $\gamma_{1/2}^{\infty}$ is the activity coefficient of solute 1 at infinite dilution in solvent 2. P_1^0 is the vapor pressure of pure solute at temperature, T, and φ_1^0 its fugacity coefficient (typically, set equal to 1 at ambient conditions). As seen, by knowing ΔG_{12} and ΔH_{12} from the LSER Equations (1) and (2), one has direct information on $\gamma_{1/2}^{\infty}$ over a range of external conditions, which is crucial, among others, for phase equilibrium calculations [16–18]. This explains the interest of thermodynamicists, especially chemical engineers, in LSER-type models and their solvation equations.

There are, however, two main drawbacks of this otherwise very useful LSER model [7–12]. The LSER descriptors and the corresponding LFER coefficients are typically determined by multilinear regression of experimental data. The model expansion is, thus, restricted by the availability of experimental data. The second drawback is the thermodynamic inconsistency of the way the linear Equations (1)–(2) are currently used. Although there is a sound thermodynamic basis for the LSER equations, their current use leads to peculiar results when applied to the self-solvation of hydrogen-bonded solutes and is very far from the expected equality of the complementary hydrogen-bonding interaction energies since solute and solvent become identical on self-solvation. This problem has been extensively discussed in recent literature, along with the significant advantages that an eventual, consistent reformulation would have [16–18].

New, reliable experimental data on solvation or hydrogen-bonding (HB) quantities are becoming more and more scarce following the related scarcity of research groups on experimental thermodynamics worldwide. There is, in fact, an enormous amount of experimental information and calculations on these quantities already available [4,12,19–34], but, unfortunately, the scatter of data often reached very high levels of several thermal energy

(RT) units, even for well-studied systems such as water or alkanols and their mixtures. It seems also that little care for thermodynamic consistency was applied to the proposed equations and scale correlations, which seriously restricts or rather forbids the reliable exchange of information between the various compilations and scales [16–18]. In this respect, the decisions to use data exclusively from ab initio or high-level DFT quantum chemical calculations for the development of predictive tools of solvation quantities [6] or the correlation of exclusively their own experimental solvation data for testing new descriptors [35] appear well plausible.

Strong specific interactions, such as hydrogen-bonding or Lewis acid/base interactions, are ubiquitous and very often of primordial importance in chemical and biochemical processes and systems of practical or industrial interest [20–25,36,37]. Developments of modern Molecular Thermodynamics do focus on these specific interactions of directional character. As an example, the core of one of the most advanced equation-of-state models today, the family of SAFT (statistical associating fluid thermodynamics) models [38–40], consists of Wertheim's perturbation approximation of these association interactions [41]. However, the above-mentioned uncertainty regarding hydrogen bonding quantities also pervades SAFT models, with often very large discrepancies in reported hydrogen bonding strengths [42]. An alternative equation-of-state model [43–45], which handles hydrogen bonding with the more versatile Veytsman statistics [46], was also developed and applied to numerous hydrogen-bonded systems exhibiting intermolecular, intramolecular or co-operative association, forming oligomeric associates up to three–dimensional network structures [47]. However, this NRHB (non-randomness with hydrogen-bonding) model, the SAFT models, and, essentially, all similar models of current Molecular Thermodynamics require external information on hydrogen bonding interactions as well as information on conformational changes upon solvation or mixing, typically obtained from quantum chemical calculations.

Nearly thirty years ago, Klamt published his pioneering work on the Conductor-like Screening Model for Real Solvents (COSMO-RS) [48], which became one of the central directions in the development of modern Molecular Thermodynamics. Sandler, Lin, and collaborators [49,50] have further contributed to familiarizing thermodynamicists, especially chemical engineers, with quantum–chemical/COSMO calculations. A key feature of COSMO-RS is the adoption of the simple nearest-neighbor pairwise additive interaction approach for the calculation of potential and solvation energies [48–54]. The combination of this approach with detailed quantum–chemical information on the distribution of molecular charge densities has turned COSMO-RS into a powerful a-priori predictive and popular tool in Molecular Thermodynamics, especially in solute solvation/partitioning calculations [55,56]. Due to the structure of the model, the calculation of the separate hydrogen-bonding contribution to the solvation-free energy cannot be made, but the corresponding contribution to the solvation enthalpy can be made. Thus, COSMO-RS may be used as a predictive tool for this contribution and compared with the corresponding LSER contribution. Above all, however, COSMO-RS may be used for a well-based development and expansion of a consistent LSER-type model, not only for solvent screening applications but also as a reliable source of information on intermolecular interactions and conformational changes needed in activity-coefficient or equation-of-state models [18].

Thus, the main task of the present work is to explore the potential of using COSMO-RS for the development of simple enough but thermodynamically consistent linear solvation energy relationships. These simple equations will be used to correlate experimental solvation data from recent compilations and extract their hydrogen bonding information. For this purpose, new molecular descriptors will be derived in the next section based on the molecular charge-density distributions or sigma profiles of the COSMO-RS model [48–54]. The new linear relationships will be tested against experimental solvation data, including self-solvation data, and compared with LSER calculations [12–14] and, of course, COSMO-RS estimations [57]. The significance and perspectives of these calculations will be critically discussed.

2. The New Quantum Chemistry-Based LSER (QC-LSER) Molecular Descriptors

In Abraham's LSER approach, only the molecular descriptors V_1 and E_1 are well defined by the molecular structure and the refractivity of the solute [7–12]. The descriptors S_1, A_1, and B_1, although attributed a more or less clear physical meaning, are determined, in essence, via global optimization of data correlations or error minimization in the prediction of various solute/solvent properties in a variety of solvents. Thus, our first task in this work is to derive new molecular descriptors obtained solely from the molecular structure of the solute and the COSMO-based molecular surface charge distributions or sigma-profiles [48–54,57], which could replace the LSER descriptors *S*, *A*, *B*. This is not a novel idea. Abraham, Klamt et al. [58] have already used the moments of the σ-profiles (cosmoments) in order to derive expressions for the LSER descriptors. These cosmoments are used extensively in the literature for the calculation of numerous quantities in a variety of applications [51]. More recently, Dohnal [35] has also used σ-profiles and derived new molecular descriptors. However, in none of these applications, the concern was the consistent reformulation of the linear relationships that would permit the extraction of reliable information on intermolecular interactions. This is what will be attempted in this and the next sections.

From its very definition, a sigma-profile (σ-profile) is the distribution of probabilities $P_i = A_i/A$ to find a surface charge density σ_i on the (COSMO solvation/screening) cavity of the studied molecule [51]. The sum of the partial surface areas A_i (not to be confused with the LSER descriptor A_i) gives the total surface area, *A*, of the molecular cavity, while the sum of the partial charges $\sigma_i A_i$ gives the total charge of the molecule or ion. Figure 1 is an example of a sigma profile.

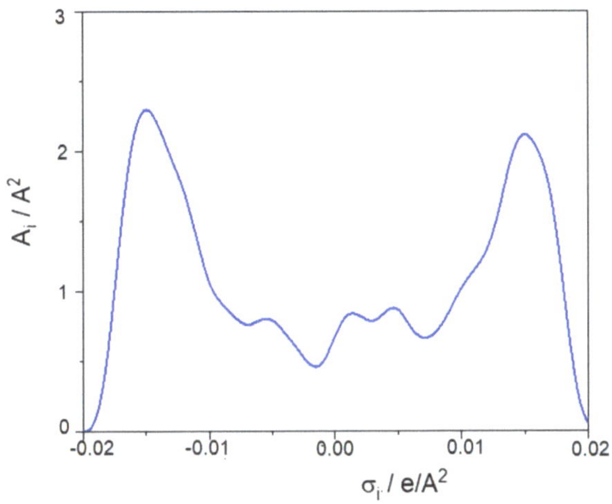

Figure 1. The σ-profile of water [data from [57]].

Klamt has derived from the σ-profile $A^X(\sigma)$ of a solute X a number of molecular descriptors or cosmoments M_i^X of type i, defined by the integral, $\int A^X(\sigma) f_i(\sigma) d\sigma$ for various functional forms $f_i(\sigma)$ [51]. Dohnal's hydrogen-bonding descriptors are derived by setting $f_i(\sigma) = \sigma_i$, integrating into the hydrogen-bonding ranges of the charge density, σ, and using the "centroid" concept [35].

Of interest to us is the derivation of descriptors reflecting the strength of intermolecular forces or interactions. A proper handling of the σ-profiles could give us information on these forces of electrostatic character. In this respect, the product of the surface charge density σ_i by the charge $\sigma_i A_i$ or the product $\sigma_i^2 A_i$ (being charged by a charge over the unit area with units $C^2 m^{-2}$) appears more appropriate for reflecting an electrostatic force. Such a distribution function turns Figure 1 into Figure 2. A key difference between the two distributions is in the central dome above zero charge density, which becomes a valley in Figure 2, touching the ground zero of electrostatic interactions.

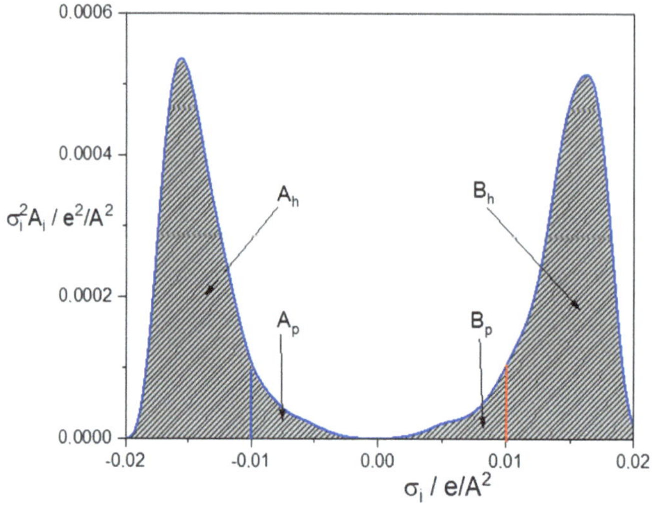

Figure 2. The distribution function of the product ${\sigma_i}^2 A_i$ of charge densities σ_i (units: e/Å^2) with the molecular charges $\sigma_i A_i$ of water and the definition areas of the QC-LSER molecular descriptors. The four surface areas under the distribution curve multiplied by 10^6 give the four QC-LSER descriptors of the molecule, as indicated.

According to Klamt's suggestion [51], charge densities to the left of the blue line in Figure 2 correspond to acidic or donor sites able to participate in strong specific or hydrogen-bonding interactions (HBD sites). Similarly, charge densities to the right of the red line in Figure 2 correspond to basic or acceptor (HBA) sites. The molecular sites with charge densities between the two lines interact with much weaker forces, primarily dispersion/induction forces. Of course, they may also interact with sites beyond the two cutoff lines when the latter are not involved in hydrogen bonding interactions. Thus, interactions due to polarizability do not exclusively involve sites from the two central sub-sections of Figure 2. Representative distributions for other molecules are reported in Figures S1 and S2 in the Supplementary Materials file.

The two cutoff lines and the zero charge density divide the total area under the distribution curve of Figure 2 into four sub-areas, two to the left towards the acidic sites and two to the right towards the basic sites. The areas pertaining to hydrogen bonding sites are marked by the subscript h (A_h, the HBD sites, and B_h, the HBA sites). The two central "polarizability" sub-areas are marked by the subscript p (A_p, the sites to the left with screening charge sign negative (as acidic ones) and B_p, the sites towards the basic sites of positive screening charge sign). The four new QC-LSER descriptors are just the areas of these four sub-sections in Figure 2, and they may be obtained from simple integration as follows (factor 10^6 accounts for the units of σ and A in σ profiles):

$$A_h = 10^6 \int_{-\infty}^{-0.01} A_i \sigma_i^2 d\sigma_i \tag{4}$$

$$A_p = 10^6 \int_{-0.01}^{0} A_i \sigma_i^2 d\sigma_i \tag{5}$$

$$B_p = 10^6 \int_0^{0.01} A_i \sigma_i^2 d\sigma_i \quad (6)$$

$$B_h = 10^6 \int_{0.01}^{\infty} A_i \sigma_i^2 d\sigma_i \quad (7)$$

σ-Profiles are available, free of charge, for thousands of molecules in the open literature, as an example in ref. [50]. They may, of course, be obtained also by using appropriate quantum–chemical calculation suites, such as the TURBOMOLE, DMol^3 of BIOVIA's MATERIALS STUDIO suite, or the SCM suite [59–61]. In this work, the σ-profiles from COSMObase [57] at the DFT/TZPVD-Fine level of quantum chemical calculations are used, and the corresponding QC-LSER descriptors for common solutes are reported in Table 1.

Table 1. The QC-LSER Molecular Descriptors of Common Solutes. Conformational energies are given in kJ/mol.

SOLUTE	A_h	A_p	B_p	B_h
ETHANE	0.00	0.28	0.23	0.00
PROPANE	0.00	0.32	0.29	0.00
n-BUTANE	0.00	0.37	0.37	0.00
n-PENTANE	0.00	0.42	0.40	0.00
n-HEXANE	0.00	0.47	0.46	0.00
n-HEPTANE	0.00	0.52	0.52	0.00
n-OCTANE	0.00	0.57	0.57	0.00
n-NONANE	0.00	0.62	0.63	0.00
n-DECANE	0.00	0.67	0.69	0.00
n-UNDECANE	0.00	0.71	0.73	0.00
n-DODECANE	0.00	0.77	0.78	0.00
n-TRIDECANE	0.00	0.80	0.84	0.00
n-TETRADECANE	0.00	0.85	0.89	0.00
n-PENTADECANE	0.00	0.90	0.94	0.00
n-HEXADECANE	0.00	0.96	1.02	0.00
n-OCTADECANE	0.00	1.05	1.11	0.00
ISOBUTANE	0.00	0.39	0.37	0.00
ISOPENTANE	0.00	0.44	0.41	0.00
2-METHYLPENTANE	0.00	0.49	0.46	0.00
3-METHYLPENTANE	0.00	0.49	0.45	0.00
3-METHYLHEXANE	0.00	0.54	0.51	0.00
2,2-DIMETHYLHEXANE	0.00	0.61	0.60	0.00
2,5-DIMETHYLHEXANE	0.00	0.59	0.58	0.00
CYCLOPENTANE	0.00	0.35	0.33	0.00
CYCLOHEXANE	0.00	0.35	0.33	0.00
CYCLOHEPTANE	0.00	0.38	0.37	0.00
CYCLOOCTANE	0.00	0.42	0.42	0.00
ETHYLENE	0.00	0.67	0.73	0.02
PROPYLENE	0.00	0.67	0.80	0.06

Table 1. *Cont.*

SOLUTE	A_h	A_p	B_p	B_h
1-BUTENE	0.00	0.70	0.84	0.06
1-PENTENE	0.00	0.74	0.90	0.07
1-HEXENE	0.00	0.79	0.96	0.06
1-HEPTENE	0.00	0.83	1.02	0.06
1-OCTENE	0.00	0.86	1.07	0.05
BENZENE	0.00	1.24	1.14	0.00
TOLUENE	0.00	1.22	1.23	0.00
ETHYLBENZENE	0.00	1.25	1.29	0.00
n-PROPYLBENZENE	0.00	1.29	1.36	0.00
n-BUTYLBENZENE	0.00	1.34	1.42	0.00
o-XYLENE	0.00	1.25	1.34	0.00
m-XYLENE	0.00	1.22	1.31	0.00
p-XYLENE	0.00	1.21	1.29	0.00
1-PENTYNE	0.49	1.01	1.51	0.16
1-HEXYNE	0.48	1.05	1.57	0.17
3-HEXYNE	0.00	1.10	1.39	0.34
DICHLOROMETHANE	0.94	0.94	1.08	0.00
CHLOROFORM	1.14	0.59	0.71	0.00
CARBON TETRACHLORIDE	0.00	0.71	0.37	0.00
2,2-DICHLOROPROPANE	0.00	1.48	1.43	0.00
CARBON TETRAFLUORIDE	0.00	0.33	0.14	0.00
DIETHYL ETHER	0.00	0.79	0.38	1.81
DI-n-PROPYL ETHER	0.00	0.85	0.47	1.79
DI-n-BUTYL ETHER	0.00	0.94	0.60	1.79
FURAN	0.24	1.40	1.24	0.13
TETRAHYDROFURAN	0.00	0.79	0.31	2.15
1,3-DIOXANE	0.00	1.50	0.54	3.12
1,4-DIOXANE	0.00	1.50	0.54	3.12
TETRAHYDROPYRAN	0.00	0.77	0.34	1.97
METHYL FORMATE	0.19	1.59	0.74	2.21
ETHYL FORMATE	0.15	1.56	0.73	2.36
ETHYL ACETATE	0.00	1.61	0.68	2.81
n-PROPYL ACETATE	0.00	1.60	0.72	2.80
ISOPROPYL ACETATE	0.00	1.65	0.68	2.68
n-BUTYL ACETATE	0.00	1.64	0.76	2.81
ETHYL PROPIONATE	0.00	1.53	0.73	2.66
ETHYL n-BUTYRATE	0.00	1.56	0.78	2.65
n-PROPYL PROPIONATE	0.00	1.54	0.74	2.69
ACETONE	0.02	1.47	0.38	2.95
METHYL ETHYL KETONE	0.00	1.41	0.43	2.78
2-PENTANONE	0.00	1.43	0.47	2.81
3-PENTANONE	0.00	1.35	0.51	2.58

Table 1. *Cont.*

SOLUTE	A_h	A_p	B_p	B_h
3-HEXANONE	0.00	1.37	0.53	2.62
FORMALDEHYDE	0.00	1.34	0.62	1.55
ACETALDEHYDE	0.00	1.39	0.47	2.36
PROPANAL	0.00	1.32	0.49	2.33
BUTANAL	0.00	1.34	0.54	2.32
PENTANAL	0.00	1.38	0.59	2.33
HEXANAL	0.00	1.43	0.66	2.32
OCTANAL	0.00	1.52	0.76	2.33
METHANOL	1.39	0.55	0.29	2.39
ETHANOL	1.23	0.66	0.35	2.46
1-PROPANOL	1.23	0.70	0.41	2.44
1-BUTANOL	1.22	0.74	0.44	2.47
1-PENTANOL	1.22	0.79	0.51	2.46
1-HEXANOL	1.22	0.83	0.55	2.48
1-HEPTANOL	1.21	0.90	0.61	2.48
1-OCTANOL	1.19	0.95	0.68	2.46
1-NONANOL	1.21	0.99	0.73	2.47
1-DECANOL	1.23	1.02	0.79	2.47
ISOPROPANOL	1.18	0.73	0.38	2.52
2-BUTANOL	0.91	0.81	0.39	2.46
2-PENTANOL	1.14	0.79	0.52	2.22
GLYCEROL	2.98	1.68	0.75	5.26
ETHYLENE GLYCOL, c0	2.13	1.10	0.57	3.87
ETHYLENE GLYCOL, c6 E_{conf} = 8.98	2.97	0.86	0.86	4.11
1,3-PROPYLENE GLYCOL, c0	1.58	1.17	0.53	3.92
1,3-PROPYLENE GLYCOL, c3 E_{conf} = 5.54	2.63	1.01	0.56	4.38
1,2-BUTANEDIOL	1.73	1.20	0.64	3.66
1,3-BUTANEDIOL	1.60	1.21	0.58	3.96
1,4-BUTANEDIOL	1.63	1.08	0.55	3.98
2-METHOXYETHANOL c0	0.77	1.22	0.60	3.13
2-METHOXYETHANOL c7 E_{conf} = 10.97	1.39	1.08	0.62	3.19
2-ETHOXYETHANOL c0	0.63	1.24	0.60	3.13
2-ETHOXYETHANOL c1 E_{conf} = 2.59	1.35	1.20	0.47	4.23
2-ETHOXYETHANOL c9 E_{conf} = 10.5	1.39	1.11	0.64	3.28
PHENOL	2.03	1.26	1.45	0.88
o-CRESOL	1.99	1.21	1.57	0.77
m-CRESOL	1.98	1.25	1.52	0.93
p-CRESOL	1.96	1.26	1.46	0.96
BENZYL ALCOHOL	1.23	1.59	1.18	2.07

Table 1. *Cont.*

SOLUTE	A_h	A_p	B_p	B_h
ACETIC ACID	2.04	1.11	0.76	2.70
PROPIONIC ACID	1.99	1.06	0.78	2.62
n-BUTYRIC ACID	1.98	1.07	0.83	2.62
n-PENTANOIC ACID	1.97	1.12	0.89	2.62
ETHYLAMINE	0.43	0.95	0.32	2.73
n-BUTYLAMINE	0.43	1.04	0.45	2.73
DIETHYLAMINE	0.20	0.84	0.40	2.02
DI-n-PROPYLAMINE	0.19	0.93	0.52	2.02
TRIMETHYLAMINE	0.00	0.65	0.35	1.49
ANILINE	1.62	1.41	1.82	0.78
PYRIDINE	0.00	1.60	0.56	2.11
FORMAMIDE	3.03	0.84	0.36	4.23
DIMETHYL SULFOXIDE	0.10	2.30	0.29	5.24
ACETONITRILE	0.31	1.67	0.91	2.00
ACRYLONITRILE	0.43	1.48	1.00	1.52
PROPIONITRILE	0.08	1.64	0.91	2.02
ACETAMIDE	2.61	1.03	0.34	4.74
N,N-DIMETHYLFORMAMIDE	0.00	1.66	0.25	4.17
N-METHYL FORMAMIDE	1.58	1.21	0.28	4.20
WATER	2.89	0.28	0.27	2.99

The descriptors A_h and B_h are considered to reflect the HBD and the HBA capacity of the molecule, respectively. They are the analogs of the A and B descriptors of the Abraham LSER model [7–12]. The descriptors A_p and B_p do not have direct analogs in Abraham's LSER model. Abraham's polarity/polarizability descriptor, S, is much different from descriptors A_p and B_p. Dipole or quadrupole moments and dipolar interactions cannot be extracted from σ-profiles since there is no information in them on the spatial (3D) charge distribution. Conformer distribution provides valuable information but not direct polarity information. Yet, σ-profiles may provide valuable information for the electrostatic interactions between sites of the A_h or B_h subareas with the B_p or A_p subareas of Figure 2. In fact, the COSMO-RS model has very successfully used this information via the "misfit" energy contribution to the potential energy [48–54]. Following this idea, the sum $A_hB_p + B_hA_p$ could be considered a substitute for the LSER S descriptor. This sum is considered to reflect the intermediate between hydrogen-bonding and dispersion interaction contributions to solvation energy, that is, to play the role the polarity/polarizability interactions are playing in Abraham's LSER model. It is in this sense that the descriptors A_p and B_p are used in this series of works, not as measures of the actual polarizability of the molecule.

An important advantage of the new QC-LSER descriptors of Table 1 over Abraham's LSER descriptors is their capacity to account for conformational changes in the solute upon transfer to the solvent. As seen in Table 1, two conformers are reported for ethylene glycol, 2-methoxyethanol, and 1,3-propylene glycol, and 3 conformers for 2-ethoxyethanol (many more conformers are reported in ref. [57]). The σ-profiles and sigma-surfaces of these conformers are shown in Figure 3 and in the Supplementary Materials (SI) file. As shown in Figure 3, the conformer c0 of 1,3-propylene glycol has a significantly reduced HBD capacity compared to conformer c3 (acidity peaks on the left-hand side of σ-profiles).

Conformer c0 is hydrogen-bonded intra-molecularly, and, thus, the proton donor is not much available for intermolecular association.

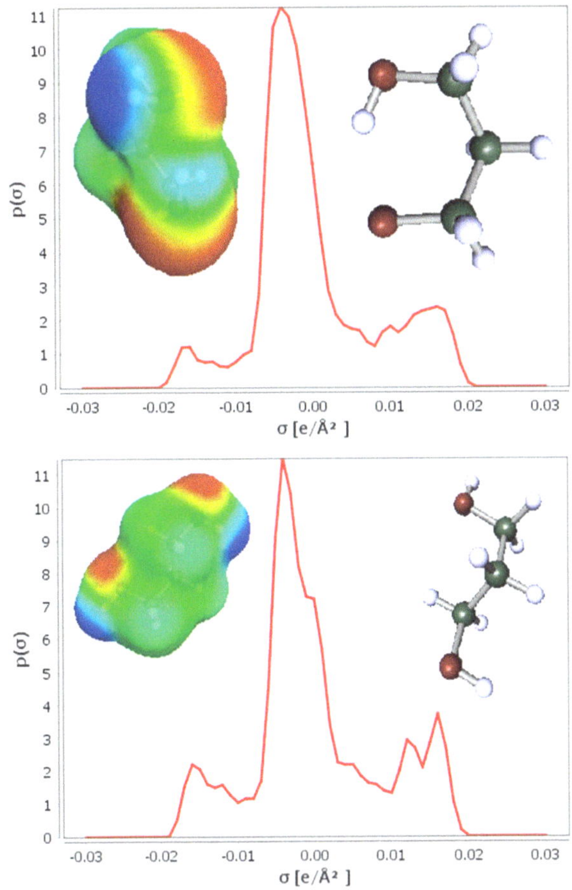

Figure 3. The σ-profiles and σ-surfaces of conformer c0 (**upper**) and c3 (**lower**) of 1,3-propylene glycol. Data from [57].

Table 1 also reports the conformational energies of the less stable conformers, namely their excess energy over the stable conformer c0. As shown also in the Supplementary Materials file, the c0 conformers favor intramolecular association, while the less stable ones expose their hydrogen-bonding sites much more and favor intermolecular association. In the case of 2-ethoxyethanol, both conformers c1 and c9 favor intermolecular hydrogen bonding. However, the conformational energy of conformer c9 is much higher than that of conformer c1, and this will diminish the gain from hydrogen bonding. Most likely, then, conformer c1 will be a more favorable structure of 2-ethoxyethanol for intermolecular hydrogen bonding. We will come back to these conformers in a later section.

Before using these new QC-LSER descriptors for calculations in real systems, it is worth comparing them with corresponding descriptors in the literature. As mentioned already, Klamt has proposed various cosmoments of the σ-profiles as molecular descriptors [51]. Figure 4 compares the cosmoment sig2 [51,57] with the square root of the product of the two QC-LSER polarizability descriptors of alkanes. As shown, the ratio of the two is nearly constant, as both seem to be nearly proportional to the molecular size.

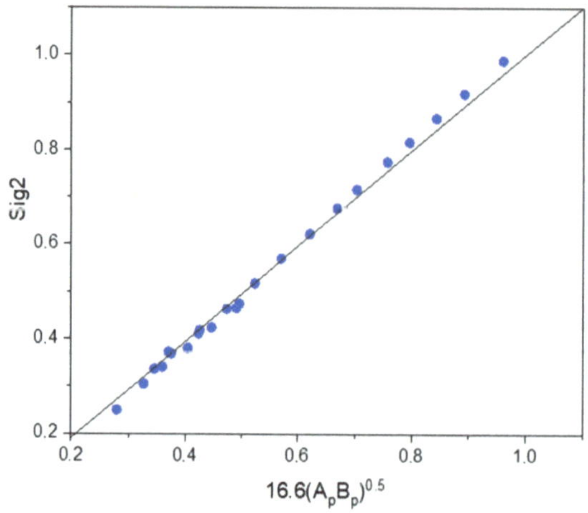

Figure 4. Comparison of Cosmoment sig2 [57] and the square root of the product of the descriptors A_p and B_p of saturated hydrocarbons. Symbols correspond to sig2 vs. $16.6\sqrt{A_pB_p}$. The straight line is the diagonal.

The third-order hydrogen-bonding cosmoments [51] are often considered in the literature as adequate hydrogen-bonding descriptors [14,15]. In Figure 5, the new hydrogen-bonding descriptors B_h are compared with the corresponding B LSER descriptor [12] and the cosmoment *HB_acc3* [57] for a number of common hydrogen-bonded hydroxyl (OH-) containing solutes. As shown, although the descriptors are not equal, their trends are rather similar. A corresponding graph with the HBD descriptors shows even more scatter, and thus, each set of descriptors reflects the hydrogen-bonding interaction capacity of the molecules in its own way. More information on the new QC-LSER descriptors is reported in the Supplementary Materials file.

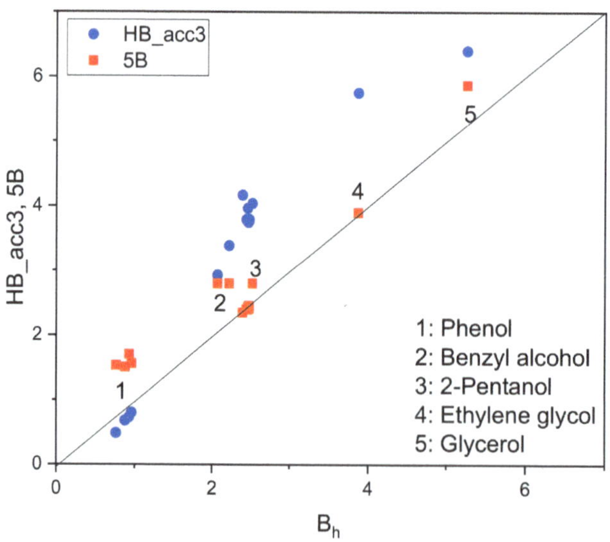

Figure 5. The cosmoment *HB_acc3* [57] and the basicity LSER descriptor *Bx5* [12] versus the corresponding QC-LSER basicity descriptor B_h of OH-containing compounds.

A noticeable discrepancy between QC-LSER and Abraham's LSER descriptors is in the case of aromatic hydrocarbons. In Abraham's LSER model and in much of the relevant literature [20–25], arenes are attributed to a non-zero basicity (proton-acceptor) character and descriptor. In contrast, COSMO-RS, due to the selection of the above cutoff limit of 0.01 e/Å^2, attributes no basic character (B_h = 0) to these compounds.

3. Linear Solvation Energy Relationships with the New QC-LSER Descriptors

Hydrogen bonding and strong polar interactions belong to the so-called "complementary" interactions in the sense that a donor site interacts with an acceptor site (not with a "similar" donor site), a positively charged site with a negatively charged one, or a basic with an acidic one. When two different molecules, 1 and 2, having both donor and acceptor sites interact, two types of intermolecular hydrogen bonds form, namely, one $HBD_1 \times HBA_2$ and one $HBA_1 \times HBD_2$. The strength of these two bonds is considered here to be proportional to the complementary products $A_{h1}B_{h2}$ and $B_{h1}A_{h2}$, respectively. This is in conformity with the thermodynamic basis of linearity of all these LSER-type models [16–18]. The form, then, of thermodynamically consistent linear relationships for the enthalpy of solvation of solute 1 by solvent 2 cannot be arbitrary, and, in the limit of self-solvation, the sum of the above two products should be equal to $2A_{h1}B_{h1} = 2A_{h2}B_{h2}$ since solute and solvent become identical on self-solvation (subscript 2 becomes subscript 1). This is guaranteed by the following simple expression for the hydrogen-bonding term of the linearity equation:

$$HB - term = c_{h12}(A_{h1}B_{h2} + B_{h1}A_{h2}) \tag{8}$$

where c_{h12} is a characteristic constant of the 1–2 pair.

In a similar manner, the contribution of the intermediate "polarity/polarizability" interactions is considered to be given by the expression:

$$Polar - term = c_{p12}(A_{h1}B_{p2} + B_{h1}A_{p2}) + c_{p12}(A_{h2}B_{p1} + B_{h2}A_{p1}) \tag{9}$$

where c_{p12} is a characteristic constant of the 1–2 pair. In conformity, however, with the mainstream of relevant linear relationships in the literature, the polar contribution is not activated when the solute is non-polar. In this way, the solute must possess at least one hydrogen bonding site, and only the first of the two terms on the right-hand side of Equation (9) must be kept. We will further expand on these two alternative views in a forthcoming publication.

As mentioned above, one of the deficiencies of the sigma profiles is the fact that they do not quite reflect the shape or the three-dimensional character of the molecule. The contribution to solvation energy of the dispersion interactions is affected by this 3D character, and thus, in this series of works, the polarizability descriptors are not used for the estimation of dispersion interactions except for a minor correction. Instead, the well-defined and easily obtained Abraham's descriptors V and E for the McGowan volume and the excess refractivity, respectively, will be used for the dispersion–induction interactions with the following simple expression:

$$Dispersion - term = c_v V_1^{\alpha} + c_e E_1 + c_2 \tag{10}$$

where c_v, c_e, and α are (expected to be) universal constants and c_2 is a characteristic constant of solvent 2. The sum of the above three terms gives the solvation enthalpy, ΔH_{12}, of the 1–2 pair of molecules. Thus, the logarithm of the solvation enthalpy (energy) constant, K_E, is given by the following working equation:

$$LogK_E = c_2 + c_v V_1^{\alpha} + c_e E_1 + c_{p12}(A_{h1}B_{p2} + B_{h1}A_{p2}) + c_{h12}(A_{h1}B_{h2} + B_{h1}A_{h2}) \tag{11}$$

or a similar equation containing both terms on the right-hand side of Equation (9). It is this form of Equation (9) that will facilitate the direct use of LSER descriptors V and E and also facilitate the direct comparison of the new molecular descriptors with the corresponding Abraham's LSER descriptors.

In addition to the above sum, the solvation-free energy, ΔG_{12}, will require solvation–entropy contribution, ΔS_{12}, so that at any temperature, T, the left-hand side of Equation (3) will hold true. In order to facilitate comparison and handle solvation entropy data, the final working equation of the present linear solvation energy relationship is cast in direct analogy with Abraham's LSER model in the following form:

$$LogK_G = \frac{-\Delta G_{12}}{2.303RT} = \frac{-\Delta H_{12}}{2.303RT} + \frac{\Delta S_{12}}{2.303R} = LogK_H - LogK_S \quad (12)$$

$LogK_G$ may also be cast in a form similar to Equation (11) but the coefficients cannot be arbitrary since always Equation (12) should be valid. When adopting the LSER Equations (1)–(2), thermodynamic consistency implies that the following equation should hold true:

$$LogK_S = (c_{e2} - c_{g2}) + (e_{e2} - e_{g2})E_1 + (s_{e2} - s_{g2})S_1 + (a_{e2} - a_{g2})A_1 + (b_{e2} - b_{g2})B_1 + (l_{e2} - l_{g2})L_1(\text{LSER}) \quad (13)$$

4. Applications

As mentioned already, one of the main tasks of this work is the extraction of information on hydrogen-bonding quantities from the corresponding solvation quantities. COSMO-RS [57] gives information on hydrogen bonding enthalpy but no information on hydrogen bonding free energy or entropy. Thus, this section is divided into two parts. The focus in the first part is on self-solvation calculations in a variety of common solutes and the extraction of hydrogen-bonding quantities of self-associated solutes. The focus of the second part is on the extraction of consistent hydrogen-bonding energies in representative mixtures, including aqueous systems. The calculations make extensive use of relevant experimental data on solvation quantities from the CompSol compilation [19] and COSMO-RS estimations [57].

4.1. Self-Solvation Calculations

The CompSol database [19] contains information on self-solvation free energy, enthalpy, and entropy for a number of common solutes and has systematically used the thermodynamic Equation (3). This is very useful information for model development and verification of calculations. In order to directly apply the equations of the previous section, the experimental data are, first, transformed into their logarithmic equivalents (Log_{10}), as is the practice of Abraham's LSER [7–12] model (cf. Equations (1)–(2)). In order to recover solvation free energies and enthalpies in kJ/mol units, one has to multiply $LogK_G$ and $LogK_E$ by the transformation factor −2.303RT/1000 = −5.71 kJ/mol. Similarly, the solvation entropy in $JK^{-1}mol^{-1}$ is recovered by multiplying LogK_S by the transformation factor −2.303R = −19.15 $JK^{-1}mol^{-1}$.

For the calculation of these self-solvation quantities, we must first determine the constants and coefficients in Equations (10) and (11). On the basis of Equation (12), if we are able to calculate the logarithms for energy (enthalpy) and entropy, the logarithm of self-solvation free energy will be obtained by simple subtraction. On the other hand, since there are plenty of experimental data on heats of vaporization (the negative of self-solvation energy), it would be useful to focus also on one of the other two quantities (their logarithms) and, by subtraction, obtain also the third quantity.

In order to simplify things, we will start with the solvation quantities of normal alkanes. For alkanes, only the dispersion–induction term will contribute to the solvation energy since there are no polarity or hydrogen-bonding interactions present. As mentioned above, the general dispersion–induction term is considered here to be given by Equation (10).

Since, by definition, the excess refractivity E is equal to zero for alkanes, their working equation will be:

$$LogK_E = c_v V_1^{\alpha} + c_1 \qquad \text{(n-alkanes)} \tag{14}$$

In fact, as shown in Figure 6, the experimental self-solvation enthalpies of n-alkanes are described well by the following simple equation:

$$LogK_E = 5.55V_1^{1.07} \qquad \text{(n-alkanes)} \tag{15}$$

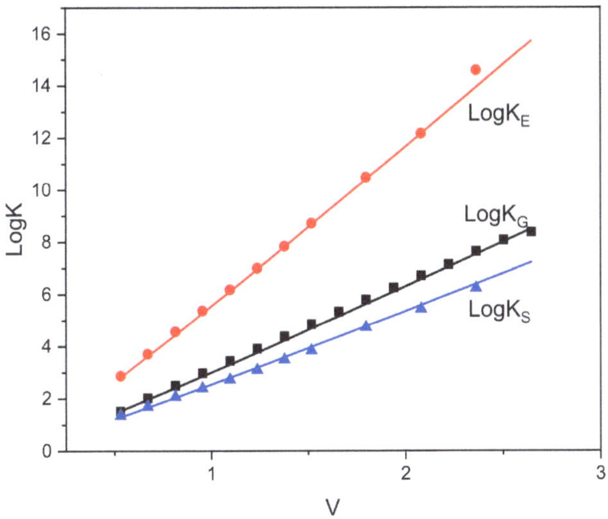

Figure 6. Experimental (symbols) [19] and calculated (lines) logarithms of solvation equilibrium constants of n-alkanes as a function of the McGowan volume [12]. Calculations were conducted with Equations (15)–(17).

In a similar manner, the self-solvation entropy of n-alkanes is described well by the following simple equation:

$$LogK_S = 2.5V_1^{1.07} \qquad \text{(n-alkanes)} \tag{16}$$

On the basis of the above Equations (15)–(16), the self-solvation free energy of n-alkanes is given by:

$$LogK_G = 3.05V_1^{1.07} \qquad \text{(n-alkanes)} \tag{17}$$

Calculations with Equations (15), (16), and (17) are compared with experimental data in Figure 6.

In view of the simplicity of the equation for $LogK_S$, it is interesting to examine the corresponding expression for non-hydrogen-bonded solutes. Obviously, some correction must be applied to Equation (16), which would take into account at least the size difference from the corresponding "homomorph" alkane with carbon atoms equal to the number of non-hydrogen atoms of the solute. In Table 2 are reported the experimental $LogK_s$ and the calculated ones for some representative solutes. The calculations were made with the following simple equation:

$$\begin{aligned} &LogK_S = 2.5V_1^{1.07} + 2\ln\frac{V_{hom}}{V_1} \\ &or \\ &LogK_S = 2.5V_1^{1.07} + 4.6Log\frac{V_{hom}}{V_1} \end{aligned} \tag{18}$$

Table 2. Experimental [19] equilibrium constants of solvation entropy and calculated ones for non-hydrogen-bonded solutes at 298.15 K.

Solute	$LogK_S$ exp [19]	$LogK_S$ Equation (18)
ISOBUTANE	1.57	1.63
ISOPENTANE	1.99	2.00
2-METHYLPENTANE	2.25	2.38
3-METHYLPENTANE	2.25	2.38
1-BUTENE	1.62	1.65
1-PENTENE	1.99	2.00
1-HEXENE	2.38	2.35
3-HEXYNE	2.40	2.40
CYCLOPENTANE	2.03	2.00
CYCLOHEXANE	2.25	2.33
CYCLOHEPTANE	2.64	2.67
CYCLOOCTANE	2.95	3.03
BENZENE	2.36	2.32
TOLUENE	2.58	2.61
ETHYLBENZENE	2.93	2.92
n-PROPYLBENZENE	3.19	3.25
n-BUTYLBENZENE	3.58	3.60
o-XYLENE	2.99	2.92
m-XYLENE	2.92	2.92
p-XYLENE	2.93	2.92
CARBON TETRACHLORIDE	2.23	2.23*
DIETHYL ETHER	2.21	2.00
DI-n-PROPYL ETHER	2.67	2.69
DI-n-BUTYL ETHER	3.39	3.42
ETHYL ACETATE	2.71	2.59
n-BUTYL ACETATE	3.41	3.16
ACETONE	2.27	2.10
METHYL ETHYL KETONE	2.49	2.33
2-PENTANONE	2.80	2.60
3-PENTANONE	2.60	2.60
ACETALDEHYDE	2.01	1.96
PROPANAL	2.21	2.10
BUTANAL	2.34	2.33
PENTANAL	2.51	2.60
HEXANAL	2.88	2.90

V_{hom} is the McGowan volume of the homomorph alkane of the solute with volume V_1 [7–12].

This definition of the homomorph alkane is operational for solvents with more than two non-hydrogen atoms. Our focus here is on the self-solvation of liquid solvents and their homomorph counterparts, which should also be in a condensed/liquid state. Thus, instead of methane, the homomorph volume for water is taken equal to 1/4 of the McGowan volume of butane, and instead of ethane, the homomorph volume for methanol is taken equal to 2/4 of the volume of butane. It is worth mentioning that this choice of homomorph volume for water turns its size correction term in Equation (18) negligible and, thus, the simple Equation (16) also applies to water for the non-hydrogen-bonding contribution to $LogK_S$. It is also worth mentioning that the general validity of Equation (18) may be used to define the homomorph in the case of halogenated compounds. Thus, for carbon tetrachloride, Equation (18) reproduces its known $LogK_s$ by setting its homomorph volume equal to 0.835 of the volume of n-heptane, or $V_{hom} = 0.835\ V_{n\text{-}C7}$.

Equation (18), though empirical, is an important result. The observed discrepancies in Figure 6 and Table 2 between experimental and calculated $LogK_S$ are rather tolerable, in view of the scatter of the experimental data [19]. In fact, a deviation of the order of 0.2 in $LogK_S$ implies a deviation of $0.2 \times 19.15 = 3.8\ JK^{-1}mol^{-1}$ in solvation entropy, and this, in turn, implies an error of $3.8 \times 0.298 = 1.13$ kJ/mol in solvation free energy or enthalpy, which is well within the experimental uncertainty of these quantities. Thus, Equation (18) will be adopted in the present work for the estimation of the non-hydrogen-bonding contribution, $LogK_{S,nhb}$, to the solvation entropy. If the overall solvation entropy is known, the corresponding hydrogen-bonding contribution is obtained by subtraction or:

$$LogK_{S,hb} = LogK_S - LogK_{S,nhb} \tag{19}$$

Equations (18) and (19) may now be coupled with information on hydrogen bonding enthalpy from COSMO-RS [57] in order to obtain the hydrogen-bonding free energy of the studied solute through the classical equation:

$$\Delta G_{HB} = \Delta H_{HB} - T\Delta S_{HB} \tag{20}$$

This comprehensive hydrogen-bonding information for a number of common self-associated solutes is reported in Table 3. The corresponding LSER estimations of ΔG_{HB} are available for some of the solutes [13] and are reported in the last column of Table 3 for comparison.

The results reported in Table 3 are particularly interesting. The hydrogen-bonding enthalpies are, essentially, very close to the values more often used in the current equation of state or activity coefficient models for the strength of hydrogen-bonding interactions [42,47]. Quite often, in practice, these self-association strengths are used as a basis, and the corresponding cross-association strengths are obtained by simple combining/mixing rules [42]. However, as we will see in the next sub-section, there is an alternative route for direct information on these cross-association strengths. On the other hand, the calculated hydrogen-bonding free energies are also close to the estimations by Abraham's LSER model [13], as shown by a comparison of the data in the last two columns of Table 3. The observed relatively large discrepancy in the case of acetic acid is, most likely, due to the fact that lower carboxylic acids are dimerized in their pure state. It is not certain that all reported estimations have properly taken into account this dimerization for acetic acid. As shown in the table, the impact of this dimerization is significantly reduced after pentanoic acid, and the entropy contribution assumes the expected larger absolute values of strongly hydrogen-bonded compounds. This is one example of the importance of accounting for self-solvation, especially for self-associated solutes, by the various linear relationships.

Table 3. The equilibrium constant of solvation entropy, $LogK_S$, with its hydrogen-bonding, $LogK_{S,hb}$, and non-hydrogen-bonding, $LogK_{S,nhb}$, components and the hydrogen-bonding contributions to self-solvation entropy, ΔS_{HB}, enthalpy, ΔH_{HB}, and free energy, ΔG_{HB}, of common solutes at 298.15 K.

Solute	$LogK_S$ exp [19]	$LogK_{S,nhb}$ Equation (18)	$LogK_{S,hb}$	$-\Delta S_{HB}$/ JK^{-1} mol^{-1}	$-\Delta H_{HB}$ [57] $kJmol^{-1}$	$-\Delta G_{HB}$/ $kJmol^{-1}$ Equation (20)	$-\Delta G_{HB}$/ $kJmol^{-1}$ LSER [13]
METHANOL	2.95	0.88	2.07	39.58	23.08	11.28	13.14
ETHANOL	3.47	1.40	2.08	39.78	24.93	13.07	11.49
1-PROPANOL	4.02	1.68	2.34	44.78	24.61	11.26	11.16
2-PROPANOL	4.06	1.68	2.38	45.62	23.74	10.14	10.43
1-BUTANOL	4.55	2.01	2.54	48.72	24.16	9.63	10.21
1-PENTANOL	4.79	2.34	2.45	46.85	24.01	10.04	10.59
2-PENTANOL	4.76	2.34	2.42	46.43	23.77	9.93	10.34
1-HEXANOL	5.44	2.69	2.75	52.67	24.24	8.54	10.06
1-HEPTANOL	6.01	3.05	2.96	56.64	23.64	6.75	9.77
1-OCTANOL	6.14	3.42	2.72	52.10	23.43	7.90	9.40
1-NONANOL	6.75	3.79	2.96	56.68	23.68	6.78	
1-DECANOL	6.84	4.17	2.67	51.11	23.26	8.02	9.40
ETHYLENE GLYCOL, c0	4.68	1.77	2.91	54.51	43.29	26.70	25.24
1,3-PROPYLENE GLYCOL, c0	4.74	2.02	2.72	51.10	50.21	34.70	
GLYCEROL	6.18	2.32	3.86	73.90	51.61	29.58	
BENZYL ALCOHOL	4.87	2.88	2.00	38.25	22.47	11.06	10.32
PHENOL	4.26	2.59	1.67	31.91	17.58	8.07	
o-CRESOL	4.55	2.88	1.67	31.98	13.22	3.69	
m-CRESOL	4.63	2.88	1.76	33.65	17.98	7.95	
p-CRESOL	4.96	2.88	2.09	39.92	18.59	6.69	
2-METHOXY-ETHANOL c0	3.28	2.03	1.25	23.99	19.53	12.38	9.06
2-ETHOXY-ETHANOL c0	3.93	2.32	1.61	30.89	19.21	10.00	9.45
ACETIC ACID	2.69	1.84	0.85	16.27	25.31	20.46	14.95
PROPIONIC ACID	3.43	2.05	1.38	26.43	25.39	17.51	
n-BUTYRIC ACID	4.09	2.32	1.77	33.84	25.37	15.28	
n-PENTANOIC ACID	4.85	2.62	2.23	42.71	24.89	12.16	
n-BUTYLAMINE	2.69	2.00	0.69	13.19	2.84	−1.10	
DIETHYLAMINE	2.38	2.00	0.38	7.33	2.69	0.50	
WATER	2.72	0.38	2.35	44.77	41.15	27.80	27.37

4.2. Hydrogen-Bonding Calculations in Solute (1)–Solvent (2) Systems

More often needed is the strength of cross-association interaction. If alcohol is interacting with an ester or an ether, the required information for alcohol is its donor capacity. If it is interacting with chloroform, the required information is its acceptor capacity. If it interacts with phenol or alkylamine, both its donor capacity and its acceptor capacity must be known. In view of this, it is crucial to extract information on hydrogen bonding interactions from solvation quantities in a consistent manner, caring for the separate donor and acceptor contributions.

Let us consider again a binary mixture of cross-associated solute (1) with solvent (2) and let their hydrogen-bonding donor and acceptor capacities be HBD_1 and HBA_1 for the solute and HBD_2 and HBA_2 for the solvent. The crucial information for Molecular Thermodynamic models is the strength of the two cross-association interactions $HBD_1 \times HBA_2$ and $HBA_1 \times HBD_2$ and, of course, the two self-association interactions $HBD_1 \times HBA_1$ and $HBD_2 \times HBA_2$. Obviously, when the solute becomes identical to the solvent, the above cross-association strengths should become identical to the corresponding self-association strengths. This is precisely what is guaranteed with Equation (8) or (11). Unfortunately, the existing LSER models and acidity/basicity scales in the literature have not paid much attention to this fact, and their interaction strengths cannot be used for thermodynamic calculations. Luckily, the overall hydrogen-bonding enthalpies calculated by Abraham's LSER model [7–12] are, very often, close to the corresponding COSMO-RS and equation-of-state calculations [18] and, thus, still are useful information.

In Table 4 are reported the HB interaction strengths for representative solute–solvent systems. The LSER calculations [13,14] for the $HBD_1 \times HBA_2$ and $HBA_1 \times HBD_2$ interactions are reported in the second and third columns, respectively. The fifth and sixth columns of the table report the corresponding calculations from the COSMO-RS estimations [57]. COSMO-RS estimates the overall HB interaction $HBD_1 \times HBA_2 + HBA_1 \times HBD_2$. This overall strength is split into its separate contributions through Equation (8). The equality of the two results specifies the binary coefficient c_{h12}, which is reported in the fourth column of Table 4. A number of comments regarding these calculations are in order.

Table 4. The two components of the hydrogen-bonding solvation energies calculated by Abraham's LSER model [13,14] and the corresponding components calculated by the new descriptors of Table 1 from COSMO-RS estimations of the overall hydrogen-bonding contribution to solvation energy of common solutes [57] at 298.15 K.

	LSER [13,14]		This Work / COSMO-RS [57]		
SOLUTE	$-A_1a_2$/ kJ/mol	$-B_1b_2$/ kJ/mol	c_{h12}	$-c_{h12}A_{h1}B_{h2}$ / kJ/mol	$-c_{h12B}B_{h1}A_{h2}$ / kJ/mol
			Solvent: **WATER**		
WATER	24.41	15.79	0.42	20.56	20.56
TETRAHYDROFURAN	0.00	20.03	0.69	0.00	24.52
TETRAHYDROPYRAN	0.00	21.43	0.72	0.00	23.26
ETHYL ACETATE	0.00	18.82	0.38	0.00	17.59
n-PROPYL ACETATE	0.00	18.82	0.38	0.00	17.61
n-BUTYL ACETATE	0.00	18.79	0.39	0.00	18.03
ACETONE	1.28	20.62	0.40	0.14	19.52
METHYL ETHYL KETONE	0.00	21.31	0.41	0.00	18.64
2-PENTANONE	0.00	21.30	0.42	0.00	19.64
ACETALDEHYDE	0.00	18.82	0.35	0.00	13.60
PROPANAL	0.00	18.82	0.35	0.00	13.48
BUTANAL	0.00	18.82	0.35	0.00	13.51
DMSO	0.00	38.05	0.32	0.57	27.81
METHANOL	13.76	19.65	0.49	11.53	19.17
ETHANOL	11.84	20.07	0.55	11.63	22.48

Table 4. *Cont.*

SOLUTE	LSER [13,14]		This Work / COSMO-RS [57]		
	$-A_1a_2$/ kJ/mol	$-B_1b_2$/ kJ/mol	c_{h12}	$-c_{h12}A_{h1}B_{h2}$ / kJ/mol	$-c_{h12B}B_{h1}A_{h2}$ / kJ/mol
1-PROPANOL	11.84	20.07	0.56	11.61	22.35
1-BUTANOL	11.80	20.16	0.55	11.51	22.52
1-PENTANOL	11.80	20.16	0.56	11.56	22.53
1-HEXANOL	11.76	20.29	0.55	11.48	22.55
1-HEPTANOL	11.78	20.19	0.55	11.40	22.59
1-OCTANOL	11.71	20.31	0.56	11.33	22.64
1-NONANOL	11.76	20.53	0.56	11.51	22.71
1-DECANOL	11.66	20.37	0.56	11.70	22.71
ISOPROPANOL	9.81	23.42	0.60	11.99	24.74
2-PENTANOL	10.44	23.42	0.64	12.38	23.31
PHENOL	19.50	12.63	0.50	17.17	7.18
ETHYLENE GLYCOL, c0	18.57	32.62	0.63	23.08	40.54
ETHYLENE GLYCOL, c6	18.57	32.62	0.54	27.23	36.39
2-ETHOXYETHANOL c0	9.72	34.10	0.89	9.56	45.97
2-ETHOXYETHANOL c1	9.72	34.10	0.60	13.74	41.79
2-ETHOXYETHANOL c9	9.72	34.10	0.71	16.90	38.63
1,2-BUTANEDIOL	18.57	35.13	0.65	19.28	39.46
1,3-BUTANEDIOL	20.65	35.13	0.71	19.52	46.63
ACETIC ACID	19.71	18.46	0.42	14.65	18.74
PROPIONIC ACID	19.42	18.54	0.44	15.12	19.23
n-BUTYLAMINE	5.12	25.51	0.41	2.98	18.35
DIETHYLAMINE	2.56	28.73	0.65	2.22	21.53
PYRIDINE	0.00	21.42	0.47	0.00	16.23
			Solvent: **ETHANOL**		
ETHANOL	18.08	5.53	0.73	12.58	12.58
1-PROPANOL	18.08	5.53	0.72	12.28	12.36
2-PROPANOL	14.97	6.45	0.77	12.56	13.55
1-OCTANOL	17.87	5.59	0.74	12.17	12.71
ETHYLENE GLYCOL, c0	28.34	8.98	0.70	20.75	19.05
ETHYLENE GLYCOL, c6	28.34	8.98	0.57	23.43	16.37
2-ETHOXYETHANOL c0	14.83	9.39	0.86	7.53	18.93
2-ETHOXYETHANOL c1	14.83	9.39	0.55	10.22	16.24
2-ETHOXYETHANOL c9	14.83	9.39	0.63	12.05	14.41
ACETIC ACID	30.08	5.08	0.47	13.39	8.96
n-BYTYL AMINE	7.82	7.02	0.82	4.87	15.69
DMSO	0.00	10.48	0.54	0.77	19.88
WATER	37.25	4.35	0.60	24.22	12.66
			Solvent: 1-**OCTANOL**		
1-OCTANOL	19.74	4.37	0.70	11.69	11.69
ETHANOL	19.98	4.32	0.69	11.91	11.52

Table 4. *Cont.*

SOLUTE	LSER [13,14]		This Work / COSMO-RS [57]		
	$-A_1a_2$/ kJ/mol	$-B_1b_2$/ kJ/mol	c_{h12}	$-c_{h12}A_{h1}B_{h2}$ / kJ/mol	$-c_{h12B}B_{h1}A_{h2}$ / kJ/mol
1-PROPANOL	19.98	4.32	0.73	12.55	12.09
2-PROPANOL	16.54	5.03	0.74	12.19	12.59
ETHYLENE GLYCOL, c0	31.31	7.01	0.65	19.52	17.15
ETHYLENE GLYCOL, c6	31.31	7.01	0.53	21.97	14.70
2-ETHOXYETHANOL c0	16.39	7.33	0.77	6.80	16.37
2-ETHOXYETHANOL c1	16.39	7.33	0.49	9.19	13.98
2-ETHOXYETHANOL c9	16.39	7.33	0.56	10.81	12.36
ACETIC ACID	33.24	3.97	0.47	13.37	8.56
n-BYTYL AMINE	8.64	5.48	0.83	4.98	15.35
DMSO	0.00	8.18	0.54	0.77	19.06
WATER	41.17	3.39	0.59	23.80	11.91
			Solvent: DMSO		
DMSO (Dimethyl Sulfoxide)	0.00	5.33	0.004	0.01	0.01
ETHANOL	17.55	2.81	0.36	13.17	0.52
1-PROPANOL	17.55	2.81	0.36	13.08	0.51
1-OCTANOL	17.34	2.85	0.36	12.94	0.53
2-PROPANOL	14.53	3.28	0.36	12.54	0.53
ETHYLENE GLYCOL, c0	27.50	4.57	0.41	25.79	0.92
ETHYLENE GLYCOL, c6	27.50	4.57	0.29	26.00	0.71
2-ETHOXYETHANOL c0	14.39	4.78	0.62	11.62	1.14
2-ETHOXYETHANOL c1	14.39	4.78	0.30	12.02	0.74
2-ETHOXYETHANOL c9	14.39	4.78	0.29	12.19	0.57
ACETIC ACID	29.20	2.59	0.31	18.81	0.49
n-BYTYL AMINE	7.59	3.58	0.04	0.50	0.06
WATER	36.16	2.21	0.33	28.28	0.58

As observed in Table 4, the overall LSER estimations of the cross-association interactions are often not very different from COSMO-RS estimations. It is their splitting in the separate interactions that matters for thermodynamic calculations since these values, extracted from infinitely dilute solutions, are typically used throughout the full composition range of the mixture. As seen, in all self-solvation cases, the LSER calculations give two significantly different estimations for the same quantity, $HBD_1 \times HBA_1$. When the solvents are ethanol or 1-octanol and the solute is an alkanol, the hydrogen-bonding interactions are of the OH---OH type, and one would expect some symmetry in the $HBD_1 \times HBA_2$ and $HBA_1 \times HBD_2$ interaction values. As seen in Table 4, the two LSER estimations are significantly different, and this is rather difficult to rationalize.

In contrast to LSER calculations, the splitting of the COSMO-RS estimations gives two identical values on self-solvation and symmetric values for the OH---OH interactions. In fact, the reported values in the last two columns of Table 4 are quite close to the typically adopted values in equation-of-state calculations [42,47]. This is important in practice and in model developments since there are no limitations in COSMO-RS calculations for both the hydrogen-bonding contribution to the solvation enthalpy as well as to the QC-LSER descriptors even for non-synthesized compounds. If not already available, the required quantum chemical calculations can be conducted with available software suites [59–61].

It is worth observing in Table 4 that the binary coefficient c_{h12} takes nearly a common value for each class of solute molecules or, rather, for each type of donor–acceptor pair. Thus, in the case of aqueous systems, this coefficient is nearly equal to 0.555 for all alkanols except for methanol. In the case of esters, it is nearly equal to 0.385, and in the case of alkanals, it is equal to 0.35. This indicates that the new QC-LSER descriptors discriminate properly the type of interacting HB sites, and the above coefficients may be used for the prediction of HB interaction strengths with other members of the homologous series. Of course, many more diverse systems have to be studied in order to derive safe rules for these predictive calculations.

Of much interest is the availability of QC-LSER descriptors for conformers when conformer-specific calculations are needed. Conformational changes may occur upon transfer of the solute from the gaseous to the liquid state. Of particular interest are conformers with intramolecular association. Missopolinou et al. [62,63] have studied the thermodynamics of alcoxyethanol mixtures with inert solvents and have shown that their skewed heat-of-mixing diagrams can be explained by accounting for the conformational changes in alcoxyethanols upon change in mixture composition. At an infinite dilution of ethoxyethanol in solvent n-octane, the prevailing conformer is the c0 shown in Figure 7 and Figure S5 of the Supplementary Materials file. Intramolecular hydrogen bonding is possible with this conformation. The practically isolated ethoxyethanol molecules in the dilute concentration region prefer this conformation since the intramolecular association reduces the system's potential energy and free energy. However, as the concentration increases, the probability of intermolecular association increases, and it is the preferred association since it is stronger than the intramolecular one. The more open conformers c1 or c9 of Figure 7 favor intermolecular interaction much more than the conformer c0. There is, of course, a penalty of conformation energy in going from conformer c0 to these more extended conformers, but this is offset by the gain from the stronger intermolecular interactions. These considerations explain the dispersive, intermolecular, and intramolecular contributions to the experimental heat of mixing, especially the last one, which is negative [62].

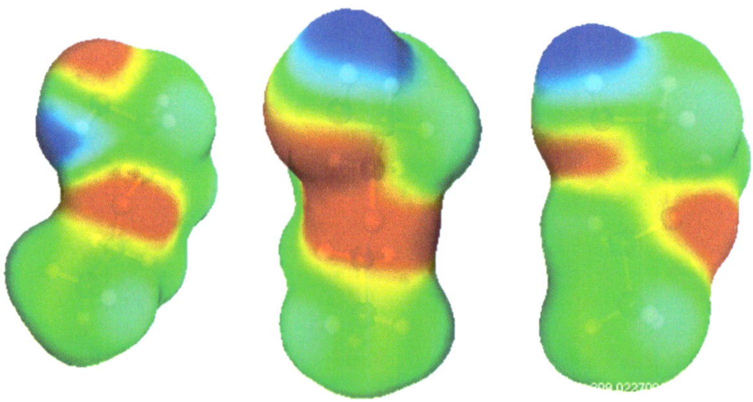

2-ethoxyethanol, c0

2-ethoxyethanol, c1
ΔE_{conf} = 2.6 kJ/mol

2-ethoxyethanol, c9
ΔE_{conf} = 10.5 kJ/mol

Figure 7. COSMO calculations of the surface profiles of three conformers of 2-ethoxyethanol [57]. Deep blue indicates donor sites and deep red acceptor sites.

As mentioned already, information on hydrogen-bonding strengths and conformational changes is essential but extraneous to activity-coefficient or equation-of-state thermodynamic models. It is on account of this that the above COSMO-RS information, incorporated in the new QC-LSER descriptors, is particularly useful. As seen in Table 4, the solute–solvent HB interactions change significantly with conformation change, and this must be accounted for by the molecular thermodynamic models, like the above-mentioned SAFT or NRHB models.

COSMO-RS also provides values for the total solvation energy [57]. In principle, then, one could combine it with the more easily available [19] total solvation free energy and obtain the corresponding total solvation entropy. A scheme analogous to the one discussed above for self-solvation (cf. Equations (16) and (18)) could be devised for the mixtures. Again, the starting point could be n-alkane solvents. In this case, a correction should be made to Equation (18), accounting at least for the size difference of solute and solvent. For all other solvents, the additional correction should again account at least for the size difference of the solvent from its homomorph n-alkane. In this way, both the hydrogen bonding and the non-hydrogen bonding contributions to the solvation entropy may be obtained. In a forthcoming publication, we will see how this scheme may be applied, not only to systems of inert solvents like n-hexane or n-hexadecane but also to self-associated solvents such as ethanol or water and aqueous systems.

5. Discussion and Conclusions

As mentioned in the Introduction, hydrogen-bonding (HB) interactions are ubiquitous, playing a vital role in biological systems and processes but also in systems of practical or industrial interest. As such, they must be attributed a concrete value as any other properties and parameters in thermodynamic calculations. Unfortunately, in spite of the enormous experimental and theoretical work conducted for nearly one century now, there is not even one single system whose strength of HB interaction is unanimously accepted as a precisely known quantity. Various research schools define and measure differently in various environments the presumably very same interaction. The various acidity/basicity scales have their own essentially arbitrary scale–origin or zero interaction strength, which prevents the reliable exchange of information. Often, no due attention is paid to the thermodynamic consistency of the proposed correlations and predictive equations. One example is the inconsistency of the current linear relationships of solvation quantities when applied to self-solvation. This is really unfortunate because self-solvation is probably more appropriate for devising a rational, widely acceptable scheme for the precise determination of HB quantities.

The present work is not solving this problem, but it indicates a potential route towards this end. Quantum chemical calculations or quantum mechanics-based molecular simulations may be used to determine the strength of HB interaction in a pure self-associated fluid under well-defined external conditions and express it in understandable and replicable terms. Ethanol or propanol are examples of molecules that might be carefully studied. They are small enough with no extensive conformer populations and large enough to sterically hinder the simultaneous attack of both acceptor sites (lone pairs) of oxygen. Well-founded averaging schemes could be adopted for the determination of the OH--O interaction. Various attempts from various research schools would, sooner or later, coincide on the value of it within a tolerance less than the thermal energy (RT) unit. This is the information needed by statistical thermodynamic equation-of-state models empowered with well-founded hydrogen-bonding statistics.

The solvation quantities are clearly defined from the thermodynamics standpoint and are measurable precisely. Consistent self-solvation entropy values appear highly informative, and the scheme leading to equations analogous to Equations (16) and (18) deserves further exploration as a potential route for the sufficiently precise determination of the non-hydrogen-bonding contribution to the self-solvation entropy and, thus, the hydrogen-bonding contribution as well. A comprehensive table analogous to the above Table 3 could be produced by a well-concerted international effort which could report unanimously accepted values for the hydrogen-bonding contributions to self-solvation entropy, ΔS_{HB}, enthalpy, ΔH_{HB}, and free energy, ΔG_{HB}, of well-studied solutes. Such a table could act as a reference or standard measure of hydrogen-bonding interactions.

In a self-associated pure solvent at equilibrium, at given external conditions, hydrogen bonds are continuously formed and disrupted, but, on average, there is one degree of association and, thus, one average number of hydrogen bonds that share the overall hydrogen bonding contribution to the solvation or potential energy of this well-defined pure system. It is this share of hydrogen bonding energy that may be determined with good and eventually widely accepted precision. Statistical thermodynamic HB modeling might, then, be used to decouple multi-site interactions in solutes like water, glycols, alkoxyalkanols, etc. The fact that ethylene glycol has two hydroxyls does not mean that its overall averaged HB interaction strength should be twice the strength of each separate OH--O interaction. The latter strength, when intermolecular, would be close to the one determined in self-associated propanol or butanol, but it would be different when it is an intramolecular interaction. Statistical thermodynamics may account for this with the proper information from quantum chemical calculations. This is the first key step for turning the vast available relevant information into a thermodynamically usable quantity with very important implications.

The second crucial step is the use of thermodynamically consistent solvation energy relationships, such as the above Equation (11). There is a key prerequisite for this step: the existence of consistent and coherent molecular descriptors, especially the descriptors of the HBD and HBA scales or of the acidity and basicity scales. Quantum mechanics seems very useful in this respect. Schemes analogous to the above Equations (4) to (7) could be implemented as scripts in the existing relevant databases and produce consistent descriptors for thousands of compounds. At the same time, quantum chemical calculations of the strength of hydrogen-bonding energies in solute–solvent systems could be split into their $HBD_1 \times HBA_2$ and $HBA_1 \times HBD_2$ interactions, as shown in Table 4. This may become very useful information for testing hydrogen-bonding theories and for the development of advanced molecular thermodynamic models of mixtures. The above two steps are essential for the systematic thermodynamic study of hydrogen bonding. Besides the descriptors used in the present work, analogous descriptors like the above-mentioned Klamt cosmoments [57], Dohnal descriptors [35], or any other analogous descriptors may be explored for analogous consistent HB calculations.

The HB enthalpies from COSMO-RS [57] or from the LSER model [13,14] are close to the widely used values by activity–coefficient and equation-of-state models [42,47]. In contrast, the corresponding HB entropies differ significantly. As an example, the HB entropies used in the NRHB equation-of-state model [47] for alkanols, are in the range from -25 to -30 $JK^{-1}mol^{-1}$, which differs by more than 50% from the reported ones in Table 3. Care must be exercised before judging this difference. Entropy changes are sensitive to density changes and depend on molecular sizes/shapes. Density changes with external conditions, as well as molecular sizes/shapes, are taken explicitly into account by equation-of-state models. The above type of linear relationships do not account explicitly for the impact of density changes on HB interactions, although their cavitation terms account indirectly for solvent densities. Thus, the transfer of HB entropies in an equation-of-state model must be adapted to the specifics of the system and its treatment of hydrogen bonding. As discussed before [16–18], partial solvation parameters or partial solubility parameters are a useful bridge for this transfer as they do take explicitly into account the molar volumes or densities of solutes and solvents. However, a systematic and conclusive study on this HB entropy transfer has not been conducted so far.

The focus in the present work was on hydrogen-bonding interactions. If these interactions were sufficiently precisely determined, the determination of the intermediate electrostatic or polarity/polarizability interactions could be very much facilitated. A forthcoming article is focusing on this and on calculations with Equations (9) and (11).

Summing up, this work has shown how quantum mechanics-based calculations may be combined with experimental information on solvation quantities in order to extract information on hydrogen-bonding interactions. A simple scheme was used for the derivation of two pairs of molecular descriptors, one for the hydrogen-bonding interactions and one for the electrostatic interactions of the intermediate strength of polarity/polarizability interactions from the σ-profiles of COSMO calculations [57]. These descriptors were used to split in a consistent manner the COSMO-RS estimations of the HB interaction energies into their $HBD_1 \times HBA_2$ and $HBA_1 \times HBD_2$ component interactions. These COSMO-RS calculations were combined with experimental solvation entropies, which permitted the extraction of information on hydrogen-bonding free energy, enthalpy, and entropy for a number of common solutes. The potential for extension and thermodynamic use of these calculations is significant for practical applications and molecular model developments.

Supplementary Materials: Additional information on the new QC-LSER descriptors is reported in terms of graphs and tables. Sigma profiles and sigma surfaces are reported for additional conformers. The following supporting information can be downloaded at: https://www.mdpi.com/article/10.3390/liquids4040037/s1, Figure S1: The distribution function of the product $\sigma_i^2 A_i$ of charge densities σ_i with the molecular charges σiAi of carbon tetrachloride and the definition areas of the QC-LSER molecular descriptors. CCl4 has only polarizability non-zero QC-LSER descriptors. Figure S2: The distribution function of the product $\sigma_i^2 A_i$ of **propanal** and the definition areas of the QC-LSER molecular descriptors. Propanal has one basic (hydrogen-bonding acceptor) and two polarizability non-zero QC-LSER descriptors. Figure S3: The distribution function of the molecular charges $\sigma_i A_i$ of water and the areas) for alternative definition of analogous to QC-LSER molecular descriptors. A_h and B_h would correspond, approximately, to the regions of definition of Dohnal's [35] α_{cosm} and β_{cosm} descriptors, respectively. Figure S4: The σ-profiles and σ-surfaces of conformer c0 (upper) and c6 (lower) of ethylene glycol. Data from [51]. Figure S5: The σ-profiles and σ-surfaces of conformer c0 (upper), c1 (middle) and c9 (lower) of 2-ethoxyethanol Data from [51]. Table S1: Comparison of hydrogen-bonding QC-LSER molecular descriptors, A_h and B_h, with the corresponding α_{cosmo} and β_{cosmo} Dohnal's [35] descriptors for some common solutes.

Funding: There was no funding for this work.

Data Availability Statement: Data supporting reported results can be found in Supplementary Information file and in cited references.

Conflicts of Interest: The authors declare no conflict of interest.

References

1. Moore, J.D.; Mountain, R.D.; Ross, R.B.; Shen, V.K.; Siderius, D.W.; Smith, K.D. The 9th Industrial Fluid Properties Simulation Challenge. *Fluid Phase Equilib.* **2018**, *476*, 1–5. [CrossRef] [PubMed]
2. Tillotson, M.J.; Diamantonis, N.I.; Buda, C.; Bolton, L.W.; Müller, E.A. Molecular modelling of the thermophysical properties of fluids: Expectations, limitations, gaps and opportunities. *Phys. Chem. Chem. Phys.* **2023**, *25*, 12607–12628. [CrossRef]
3. Sapir, L.; Harries, D. Revisiting Hydrogen Bond Thermodynamics in Molecular Simulations. *J. Chem. Theory Comput.* **2017**, *13*, 2851–2857. [CrossRef]
4. Matos, G.D.R.; Kyu, D.Y.; Loeffler, H.H.; Chodera, J.D.; Shirts, M.R.; Mobley, D.L. Approaches for Calculating Solvation Free Energies and Enthalpies Demonstrated with an Update of the FreeSolv Database. *J. Chem. Eng. Data* **2017**, *62*, 1559–1569. [CrossRef]
5. Hirata, F. *Exploring Life Phenomena with Statistical Mechanics of Molecular Liquids*; CRC Press: Boca Raton, FL, USA, 2020.
6. Pereyaslavets, L.; Kamath, G.; Butin, O.; Illarionov, A.; Olevanov, M.; Kurnikov, I.; Sakipov, S.; Leontyev, I.; Voronina, E.; Gannon, T.; et al. Accurate determination of solvation free energies of neutral organic compounds from first principles. *Nat. Commun.* **2022**, *13*, 414. [CrossRef]
7. Abraham, M.H.; McGowan, J.C. The use of characteristic volumes to measure cavity terms in reversed phase liquid chromatography. *Chromatographia* **1987**, *23*, 243–246. [CrossRef]
8. Abraham, M.H. Scales of solute hydrogen-bonding: Their construction and application to physicochemical and biochemical processes. *Chem. Soc. Rev.* **1993**, *22*, 73–83. [CrossRef]
9. Abraham, M.H.; Ibrahim, A.; Zissimos, A.M. Determination of sets of solute descriptors from chromatographic measurements. *J. Chromatogr. A* **2004**, *1037*, 29–47. [CrossRef]
10. Abraham, M.H.; Smith, R.E.; Luchtefeld, R.; Boorem, A.J.; Luo, R.; Acree, W.E., Jr. Prediction of solubility of drugs and other compounds in organic solvents. *J. Pharm. Sci.* **2010**, *99*, 1500–1515. [CrossRef]
11. Goss, K.-U. Predicting the equilibrium partitioning of organic compounds using just one linear solvation energy relationship (LSER). *Fluid Phase Equilibria* **2005**, *233*, 19–22. [CrossRef]

12. Endo, S.; Watanabe, N.; Ulrich, N.; Bronner, G.K.-U. Goss UFZ-LSER Database v 2.1. Helmholtz Centre for Environmental Research-UFZ, Leipzig, Germany, 2015. Available online: https://www.ufz.de/index.php?en=31698&contentonly=1&m=0&lserd_data[mvc]=Public/start (accessed on 12 June 2024).
13. Sinha, S.; Yang, C.; Wu, E.; Acree, W.E. Abraham Solvation Parameter Model: Examination of Possible Intramolecular Hydrogen-Bonding Using Calculated Solute Descriptors. *Liquids* **2022**, *2*, 131–146. [CrossRef]
14. Mintz, C.; Ladlie, T.; Burton, K.; Clark, M.; Acree, W.E., Jr.; Abraham, M.H. Enthalpy of Solvation Correlations for Gaseous Solutes Dissolved in Alcohol Solvents based on the Abraham Model. *QSAR Comb. Sci.* **2008**, *27*, 627–635. [CrossRef]
15. Egert, T.; Langowski, H.C. Linear Solvation Energy Relationships (LSERs) for Accurate Prediction of Partition Coefficients between Low Density Polyethylene and Water-Part II: Model Evaluation and Benchmarking. *Eur. J. Pharm. Sci.* **2022**, *172*, 106137. [CrossRef] [PubMed]
16. Panayiotou, C.; Zuburtikudis, I.; Abu Khalifeh, H. Linear Free-Energy Relationships and Solvation Thermodynamics: The Thermodynamic Basis of LFER Linearity. *Ind. Eng. Chem. Res.* **2023**, *62*, 2989–3000. [CrossRef]
17. Panayiotou, C.; Zuburtikudis, I.; Abu Khalifeh, H.; Hatzimanikatis, V. Linear Solvation–Energy Relationships (LSER) and Equation-of-State Thermodynamics: On the Extraction of Thermodynamic Information from the LSER Database. *Liquids* **2023**, *3*, 66–89. [CrossRef]
18. Panayiotou, C.; Acree, W.E., Jr.; Zuburtikudis, I. COSMO-RS and LSER Models of Solution Thermodynamics: Towards a COSMOLSER Equation of State Model of Fluids. *J. Molec. Liq.* **2023**, *390*, 122992. [CrossRef]
19. Moine, E.; Privat, R.; Sirjean, B.; Jaubert, J.-N. Estimation of Solvation Quantities from Experimental Thermodynamic Data: Development of the Comprehensive CompSol Databank for Pure and Mixed Solutes. *J. Phys. Chem. Ref. Data* **2017**, *46*, 033102. [CrossRef]
20. Joesten, M.D.; Schaad, L. *Hydrogen Bonding*; Marcel Dekker: New York, NY, USA, 1974.
21. Raevsky, O.A.; Grigor'Ev, V.Y.; Kireev, D.B.; Zefirov, N.S. Complete Thermodynamic Description of H-Bonding in the Framework of Multiplicative Approach. *Quant. Struct. Relationships* **1992**, *11*, 49–63. [CrossRef]
22. Gilli, G.; Gilli, P. *The Nature of the Hydrogen Bond*; Oxford University Press: Oxford, UK, 2009.
23. Baev, A.K. *Specific Intermolecular Interactions of Nitrogenated and Bioorganic Compounds;* Springer: Berlin/Heidelberg, Germany, 2014.
24. Katritzky, A.R.; Fara, D.C.; Yang, H.; Tämm, K.; Tamm, T.; Karelson, M. Quantitative Measures of Solvent Polarity. *Chem. Rev.* **2004**, *104*, 175–198. [CrossRef]
25. Laurence, C.; Gal, J.-F. *Lewis Basicity and Affinity Scales: Data and Measurements*; Wiley: New York, NY, USA, 2010.
26. Epley, T.D.; Drago, R.S. Calorimetric studies on some hydrogen-bonded adducts. *J. Am. Chem. Soc.* **1967**, *89*, 5770–5773. [CrossRef]
27. Murthy, A.S.; Rao, C. Spectroscopic Studies of the Hydrogen Bond. *Appl. Spectrosc. Rev.* **1968**, *2*, 69–191. [CrossRef]
28. Arnett, E.M.; Joris, L.; Mitchell, E.; Murty, T.; Gorrie, T.; Schleyer, P.V.R. Hydrogen-bonded complex formation. III. Thermodynamics of complexing by infrared spectroscopy and calorimetry. *J. Am. Chem. Soc.* **1970**, *92*, 2365–2377. [CrossRef]
29. Cabani, S.; Gianni, P.; Mollica, V.; Lepori, L. Group Contributions to the Thermodynamic Properties of Non-Ionic Organic Solutes in Dilute Aqueous Solution. *J. Solut. Chem.* **1981**, *10*, 563–595. [CrossRef]
30. Abbott, S.; Yamamoto, H.; Hansen, C.M. *Hansen Solubility Parameters in Practice, Complete with Software, Data and Examples*, 3rd ed.; version 3.1.20; Hansen-Solubility: Hørsholm, Denmark, 2010.
31. Dortmund Data Bank. Available online: https://www.ddbst.com/ddb.html (accessed on 28 June 2024).
32. Sedov, L.A.; Solomonov, B.N. Hydrogen bonding in neat aliphatic alcohols: The Gibbs free energy of self-association and molar fraction of monomer. *J. Mol. Liq.* **2012**, *167*, 47–51. [CrossRef]
33. Solomonov, B.N.; Yagofarov, M.I. Compensation relationship in thermodynamics of solvation and vaporization: Features and applications. I. Non-hydrogen-bonded systems. *J. Mol. Liq.* **2022**, *368*, 120762. [CrossRef]
34. Solomonov, B.N.; Yagofarov, M.I. Compensation relationship in thermodynamics of solvation and vaporization: Features and applications. II. Hydrogen-bonded systems. *J. Mol. Liq.* **2023**, *372*, 121205. [CrossRef]
35. Dohnal, V. New QSPR molecular descriptors based on low-cost quantum chemistry computations using DFT/COSMO approach. *J. Mol. Liq.* **2024**, *407*, 125256. [CrossRef]
36. Goodsell, D.S. *Bionanotechnology: Lessons from Nature*; Wiley & Sons: New York, NY, USA, 2003.
37. Qiu, X.; Li, H.; Ver Steeg, G.; Godzik, A. Advances in AI for Protein Structure Prediction: Implications for Cancer Drug Discovery and Development. *Biomolecules* **2024**, *14*, 339. [CrossRef]
38. Chapman, W.G.; Gubbins, K.E.; Jackson, G.; Radosz, M. New reference equation of state for associating liquids. *Ind. Eng.Chem. Res.* **1990**, *29*, 1709–1721. [CrossRef]
39. Vega, L.; Llovell, F. Review and new insights into the application of molecular-based equations of state to water and aqueous solutions. *Fluid Phase Equilibria* **2016**, *416*, 150–173. [CrossRef]
40. Torshizi, M.F.; Muller, E.A. Coarse-grained molecular dynamics study of the self-assembly of polyphilic bolaamphiphiles using the SAFT-γ Mie force field. *Mol. Syst. Des. Eng.* **2021**, *6*, 594–608. [CrossRef]
41. Wertheim, M. Fluids with highly directional attractive forces. III. Multiple attraction sites. *J. Statist. Phys.* **1986**, *42*, 459–476. [CrossRef]
42. Kontogeorgis, G.M.; Folas, G.K. *Thermodynamic Models for Industrial Applications. From Classical and Advanced Mixing Rules to Association Theories;* John Wiley and Sons, Ltd.: Chichester, UK, 2010.

43. Panayiotou, C.; Sanchez, I.C. Hydrogen Bonding in Fluids: An Equation of-State Approach. *J. Phys. Chem.* **1991**, *95*, 10090–10097. [CrossRef]
44. Panayiotou, C.; Pantoula, M.; Stefanis, E.; Tsivintzelis, I.; Economou, I.G. Nonrandom Hydrogen-Bonding Model of Fluids and Their Mixtures. 1. Pure Fluids. *Ind. Eng. Chem. Res.* **2004**, *43*, 6592–6606. [CrossRef]
45. Panayiotou, C.; Tsivintzelis, I.; Economou, I.G. Nonrandom Hydrogen-Bonding Model of Fluids and Their Mixtures. 2. Multicomponent Mixtures. *Ind. Eng. Chem. Res.* **2007**, *46*, 2628–2636. [CrossRef]
46. Veytsman, B.A. Are lattice models valid for fluids with hydrogen bonds? *J. Phys. Chem.* **1990**, *94*, 8499–8500. [CrossRef]
47. Mensitieri, G.; Scherillo, G.; Panayiotou, C.; Musto, P. Towards a predictive thermodynamic description of sorption processes in polymers: The synergy between theoretical EoS models and vibrational spectroscopy. *Mater. Sci. Eng. R Rep.* **2020**, *140*, 100525. [CrossRef]
48. Klamt, A. Conductor-like Screening Model for Real Solvents: A New Approach to the Quantitative Calculation of Solvation Phenomena. *J. Phys. Chem.* **1995**, *99*, 2224–2235. [CrossRef]
49. Lin, S.-T.; Sandler, S.I. A Priori Phase Equilibrium Prediction from a Segment Contribution Solvation Model. *Ind. Eng. Chem. Res.* **2001**, *41*, 899–913. [CrossRef]
50. Bell, I.H.; Mickoleit, E.; Hsieh, C.-M.; Lin, S.-T.; Vrabec, J.; Breitkopf, C.; Jäger, A. A Benchmark Open-Source Implementation of COSMO-SAC. *J. Chem. Theory Comput.* **2020**, *16*, 2635–2646. [CrossRef]
51. Klamt, A. *COSMO-RS from Quantum Chemistry to Fluid Phase Thermodynamics and Drug Design*; Elsevier: Amsterdam, The Netherlands, 2005.
52. Grensemann, H.; Gmehling, J. Performance of a Conductor-Like Screening Model for Real Solvents Model in Comparison to Classical Group Contribution Methods. *Ind. Eng. Chem. Res.* **2005**, *44*, 1610–1624. [CrossRef]
53. Pye, C.C.; Ziegler, T.; van Lenthe, E.; Louwen, J.N. An implementation of the conductor-like screening model of solvation within the Amsterdam density functional package—Part II. COSMO for real solvents. *Can. J. Chem.* **2009**, *87*, 790–797. [CrossRef]
54. Klamt, A.; Eckert, F.; Arlt, W. COSMO-RS: An Alternative to Simulation for Calculating Thermodynamic Properties of Liquid Mixtures. *Annu. Rev. Chem. Biomol. Eng.* **2010**, *1*, 101–122. [CrossRef] [PubMed]
55. Klamt, A.; Eckert, F.; Reinisch, J.; Wichmann, K. Prediction of cyclohexane-water distribution coefficients with COSMO-RS on the SAMPL5 data set. *J. Comput. Aided Mol. Des.* **2016**, *30*, 959–967. [CrossRef]
56. Reinisch, J.; Klamt, A. Prediction of free energies of hydration with COSMO-RS on the SAMPL4 data set. *J. Comput. Aided Mol. Des.* **2014**, *28*, 169–173. [CrossRef]
57. North Data. *COSMObase, Ver. 2019*; COSMOlogic GmbH & CoKG: Leverkusen, Germany; BIOVIA Dassault Systemes: Waltham, MA, USA, 2019.
58. Zissimos, A.M.; Abraham, M.H.; Klamt, A.; Eckert, F.; Wood, J. A Comparison between the Two General Sets of Linear Free Energy Descriptors of Abraham and Klamt. *J. Chem. Inf. Comput. Sci.* **2002**, *42*, 1320–1331. [CrossRef]
59. Program Package for Electronic Structure Calculations. Available online: https://www.turbomole.org (accessed on 28 September 2024).
60. Modeling & Simulation for Next-Generation Materials. Available online: https://www.3ds.com/products/biovia/materials-studio (accessed on 28 September 2024).
61. What is COSMO-RS/SAC. Available online: https://www.scm.com/product/cosmo-rs/ (accessed on 28 September 2024).
62. Missopolinou, D.; Ioannou, K.; Prinos, I.; Panayiotou, C. Thermodynamics of Alkoxyethanol + Alkane Mixtures. *Z. Phys. Chem.* **2002**, *216*, 905. [CrossRef]
63. Missopolinou, D.; Tsivintzelis, I.; Panayiotou, C. Excess enthalpies of binary mixtures of 2-ethoxyethanol with four hydrocarbons at 298.15, 308.15, and 318.15K: An experimental and theoretical study. *Fluid Phase Equilibria* **2006**, *245*, 89–101. [CrossRef]

liquids

MDPI

Article

Thermodynamic Properties of Two Cinnamate Derivatives with Flavor and Fragrance Features

Vera L. S. Freitas *, Carlos A. O. Silva and Maria D. M. C. Ribeiro da Silva

Research Center in Chemistry of University of Porto (CIQUP), Institute of Molecular Sciences (IMS), Department of Chemistry and Biochemistry (DQB), Faculty of Sciences of University of Porto (FCUP), Rua do Campo Alegre, P-4169-007 Porto, Portugal; carlosoliveirasilva@outlook.pt (C.A.O.S.); mdsilva@fc.up.pt (M.D.M.C.R.d.S.)

* Correspondence: vera.freitas@fc.up.pt

Abstract: The standard molar enthalpies of formation in the liquid phase for ethyl (*E*)-cinnamate and ethyl hydrocinnamate, two cinnamate derivatives with notable flavor and fragrance characteristics, were determined experimentally using combustion calorimetry in an oxygen atmosphere. To derive the gas-phase enthalpies of formation for these derivatives, their enthalpies of vaporization were measured using a high-temperature Calvet microcalorimeter and the vacuum drop microcalorimetric technique. Additionally, a computational analysis employing the G3(MP2)//B3LYP composite method was conducted to calculate the gas-phase standard enthalpies of formation at T = 298.15 K for both compounds. These findings enabled a detailed assessment and analysis of the structural and energetic effects of the vinyl and ethane moieties between the phenyl and carboxylic groups in the studied compounds. Considering the structural features of ethyl (*E*)-cinnamate and ethyl hydrocinnamate, a gas-phase enthalpy of hydrogenation analysis was conducted to explore their energetic profiles more thoroughly.

Keywords: fragrance compounds; enthalpy of combustion; enthalpy of vaporization; enthalpy of formation; hydrogenation enthalpy; ethyl (2*E*)-3-phenylacrylate; ethyl 3-phenylpropanoate

1. Introduction

Understanding the thermodynamic properties of ethyl (*E*)-cinnamate and ethyl hydrocinnamate in the liquid phase at room temperature, namely, their enthalpies of vaporization and formation, is crucial since this knowledge facilitates the prediction of their behavior under various conditions, ensuring safety to handle and to process, optimizing industrial procedures, and offering valuable insights for forecasting related compounds [1]. Moreover, in the fine chemicals industry, where most chemical reactions are exothermal, and compounds often exist in a metastable state, the proper management of the thermal energy involved is very relevant. The release of energy during these reactions, if not controlled, can lead to severe consequences, highlighting the importance of reliable data for the thermodynamic properties of chemicals required in industrial processes.

This study integrates a broad research project whose main goal is to determine the thermodynamic properties of essential fragrance classes, addressing current knowledge gaps and enhancing estimation methods for the environmental risk assessment of similar compounds [2–4]. Herein, a comprehensive comparative analysis of the thermodynamic properties of ethyl (*E*)-cinnamate and ethyl hydrocinnamate is provided, depicted by structural formulas in Figure 1.

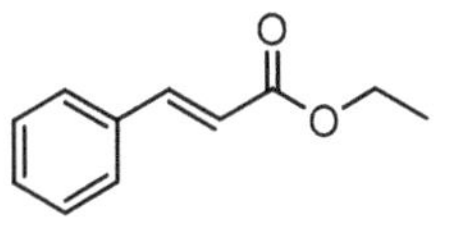

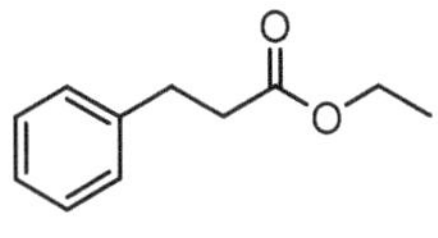

(A) (B)

Figure 1. Structural formulae of ethyl (*E*)-cinnamate (**A**) and ethyl hydrocinnamate (**B**).

Regarding the liquid compounds ethyl (*E*)-cinnamate and ethyl hydrocinnamate, they are widely found across several industries, such as flavor and fragrance, biomedical, pharmaceutical, agricultural, and polymer sectors. Despite their similar basic structures, characterized by an ethyl ester linked to a phenyl group, the structural differences exhibited by these compounds lead to varied chemical properties and applications. Ethyl (*E*)-cinnamate (IUPAC name: ethyl (2*E*)-3-phenylacrylate) features a phenyl ring attached to an acrylic acid ester with a *trans* double bond, giving it unique chemical reactivity and sensory properties, such as a sweet and fruity odor [5]. This makes its use common in perfumes, cosmetics, food products, household cleaners, and detergents [6–8]. In biomedical research, ethyl (*E*)-cinnamate is used in a new tissue-clearing method (2Eci) for imaging large 3D structures while preserving fluorescence and being non-toxic [9]. Concerning the polymer industry, this compound serves as a "green plasticizer" for polylactide (PLA), enhancing its ductility and impact strength, improving processability, and offering a sustainable alternative to petroleum-derived plasticizers [10]. Its molecular structure enhances chain mobility, cohesion, and compatibility, leading to better performance in biodegradable materials. Moreover, it is an intermediate in pharmaceutical synthesis [11] and agricultural chemical production [12]. On the other hand, ethyl hydrocinnamate (IUPAC name: ethyl 3-phenylpropanoate) presents a phenyl ring connected to a saturated propanoic acid ester. The absence of a double bond in the side chain makes its structure more stable and less reactive compared to ethyl (*E*)-cinnamate. This stability, combined with its sweet and floral odor, makes this cinnamate derivative a desirable component in the flavor and fragrance industry [13,14].

This work involved the use of calorimetric techniques to study the thermodynamic properties of the two cinnamate derivatives. The investigation encompassed the following steps: (a) the determination of the corresponding standard molar enthalpy of formation in the liquid phase, $\Delta_f H_m^\circ(l)$, by static-bomb combustion calorimetry experiments; (b) the determination of both enthalpies of vaporization, $\Delta_l^g H_m^\circ$, via high-temperature microcalorimetry. These two parameters enabled the derivation of standard molar enthalpies of formation for both molecules in the gaseous phase, $\Delta_f H_m^\circ(g)$, at $T = 298.15$ K. Additionally, computational studies were conducted using the G3(MP2)//B3LYP composite method to estimate the $\Delta_f H_m^\circ(g)$ of the two compounds.

The gas phase serves as a reference state in thermodynamics for establishing and discussing the relationships related to the structure, energy, and reactivity of molecules. In the gaseous phase, molecules are widely spaced as predominant intramolecular interactions, while intermolecular forces among molecules are often negligible. Focusing on the gaseous phase allows us to "isolate" the contributions of intramolecular interactions, thereby simplifying the analysis of the relationship between the molecular structure and thermodynamic properties. The standard molar enthalpy of formation in the gaseous state, $\Delta_f H_m^\circ(g)$, is a critical thermodynamic parameter used in this context.

The hydrogenation enthalpy of ethyl (*E*)-cinnamate to ethyl hydrocinnamate in the gas phase, $\Delta_{hyd} H_m^\circ(g)$, is calculated using the experimental enthalpy of formation values obtained in this study and compared with values from the literature, together with a detailed analysis of the hydrogenation enthalpy of isomers (*E*/*Z*), which offer a valuable foundation for future research in this area. Currently, hydrogenation reactions are a central focus of scientific research due to the quest for sustainable, efficient, and carbon-neutral energy storage solutions [15], since hydrogen, with its high energy output and environmentally friendly characteristics, is emerging as a promising candidate for future energy systems. However, storing and transporting hydrogen poses significant challenges due to its low volumetric energy density and the complexities of handling gaseous substances. Therefore, reactions that enable the chemical storage and release of hydrogen, particularly those involving Liquid Organic Hydrogen Carriers (LOHCs), are essential for efficient and secure hydrogen storage [16].

In relation to the current state of research, a study conducted by Kozlovskiy et al. [17] investigated a series of esters, including ethyl (*E*)-cinnamate, phenethyl acetate, phenethyl propionate, phenethyl butyrate, and phenethyl hexanoate. The vapor pressures of these compounds were determined within the range of 298.15 K up to their boiling temperatures using gas chromatography. The purity of the compounds does not seem to pose an obstacle in this study, as gas chromatography effectively separates impurities. This capability underscores a key advantage of the technique, enabling the evaluation of pure component properties within mixed substances. Previous thermodynamic investigations on three liquid pyridine acetyl derivatives, conducted using identical experimental methodologies [18], yielded good results.

2. Materials and Methods

2.1. Materials, Treatment, and Purity Control

Ethyl (*E*)-cinnamate (CAS Registry number 4192-77-2) and ethyl hydrocinnamate (CAS Registry number 2021-28-5) were obtained as commercial products from TCI Europe. According to the certificates of analysis for each batch, the sample of ethyl (*E*)-cinnamate (lot number HRRDM) had a mass fraction of 0.999, while ethyl hydrocinnamate (lot number 2PWUE) had a purity of 0.998. Prior to the thermodynamic studies, ethyl hydrocinnamate was further purified through fractional distillation under reduced pressure, achieving a purity greater than 0.999. In contrast, ethyl (*E*)-cinnamate was used directly in its as-received form, as its purity was already deemed sufficient for calorimetric analysis. To assess purity, liquid samples of ethyl (*E*)-cinnamate and ethyl hydrocinnamate were subjected to gas–liquid chromatography using a baseline subtraction or baseline correction technique. In this process, dimethylformamide (DMF) was initially injected as a solvent to establish a baseline and identify any background signals or residual compounds in the system before analyzing the sample. Following this, a mixture of the compound and DMF was injected. In the resulting chromatograms, peaks present in the DMF baseline were excluded, allowing for the isolation and integration of peaks corresponding to the compound and any potential impurities. This analysis determined the purity of ethyl (*E*)-cinnamate to be 0.9999 and that of ethyl hydrocinnamate to be 0.9990. Purity control analyses were conducted with gas–liquid chromatography on an Agilent 4890 apparatus equipped with an HP-5 column (15 m long and 0.530 mm in diameter, with 5% diphenyl and 95% dimethylpolysiloxane in the composition) with a flame ionization detector (FID) powered by hydrogen. The gas used during the experiments was a mixture of nitrogen and compressed air.

The density values used for ethyl (*E*)-cinnamate and ethyl hydrocinnamate were, respectively, 1.05 $g{\cdot}cm^{-3}$ [19] and 1.02 $g.cm^{-3}$ [20].

Other materials were used in this work to calibrate the calorimetric systems, including benzoic acid (CAS Registry number 65-85-0) with a purity of 0.999996, which served as the calorimetric standard for combustion calorimetry calibrations. This standard reference material (SRM) 39j [21] was obtained from the National Institute of Standards & Technology (NIST). Additionally, undecane (CAS Registry number 1120-21-4), with a purity of $\geq$0.990, was used as the recommended reference material for the calibration of the high-temperature Calvet microcalorimeter, sourced from Sigma-Aldrich®.

The standard atomic weights of the elements used throughout this paper were those suggested by the IUPAC Commission in 2021 [22].

2.2. Combustion Calorimetry

The combustion energies of the liquid samples of ethyl (*E*)-cinnamate and ethyl hydrocinnamate were obtained using static-bomb combustion calorimetry in oxygen equipped with a twin-valve 1108 Parr bomb (Parr Instrument Company, Moline, IL, USA), with an internal volume of 0.342 dm^3. The equipment and experimental procedure used have been previously described in detail [23], with the most relevant aspects summarized below.

The energy equivalent of the static-bomb combustion calorimeter, with the empty bomb, ε(cal), was determined using benzoic acid standard reference material (SRM) 39j from the NIST [21]. This certified material has a massic energy of combustion of (-26434 ± 3) $J{\cdot}g^{-1}$ when burned under standard bomb conditions. The energy equivalent of the calorimeter, ε(cal), was found to be (16002.6 ± 1.7) $J{\cdot}K^{-1}$, based on an average water mass added to the calorimeter of 3119.6 g; the uncertainty reported represents the standard deviation of the mean from six experiments.

In the sample preparation process, to prevent vaporization and the subsequent loss of mass, the liquid samples were placed inside polyester bags made of Melinex® (DuPont Teijin Films, Hopewell, VA, USA), following the procedure described by Skinner and Snelson [24]. During bomb assembly, the Melinex bag containing the sample was placed into a platinum crucible with a cotton thread fuse attached to the 0.05 mm platinum ignition wire (Goodfellow, Huntingdon, United Kingdom). Additionally, 1.00 cm^3 of deionized water was added to the bottom of the bomb to maintain a vapor phase saturated with water throughout the experiment. The bomb was sealed, purged twice to remove air, and then filled with oxygen ($x_{O2} \geq 0.99995$) at a pressure of 3.04 MPa. Subsequently, the bomb was placed inside the calorimeter along with 3119.6 g of distilled water, which served as the calorimetric liquid for this system.

The temperature of the calorimetric liquid was monitored using a 2804A quartz crystal thermometer (Hewlett-Packard, Palo Alto, CA, USA), which has an uncertainty of $\pm(1 \times 10^{-4})$ K and was interfaced with a computer running the LABTERMO program to acquire the temperature and calculate the adiabatic temperature [25]. The sample was ignited at $T = (298.1500 \pm 0.0001)$ K by discharging a 1400 μF capacitor through the platinum ignition wire. Prior to ignition, 100 temperature readings were recorded, followed by 200 readings after ignition. The surrounding thermostatic bath was maintained at approximately 301.15 K using a temperature controller. During the combustion experiments, gravimetric analysis was used to collect carbon dioxide, and acid–base titration was employed to quantify the nitric acid. The mass of carbon dioxide produced directly corresponded to the amount of compound burned in each experiment. The formation of nitric acid was attributed to possible contamination from nitrogen residues inside the bomb, potentially originating from oxygen used to fill the bomb and incomplete ventilation.

The compounds and combustion aids were weighed using a Mettler AE240 balance (Mettler Toledo, Greifensee, Switzerland) with a precision of $\pm(2 \times 10^{-5})$ g. The mass of water added to the calorimetric vessel was measured using a Mettler PC 8000 balance with a sensitivity of $\pm(1 \times 10^{-1})$ g. Additionally, a Mettler AT201 balance with a precision of $\pm(3 \times 10^{-5})$ g (for the range of 100 to 200 g) was used in the gravimetric determination of carbon dioxide.

2.3. High-Temperature Calvet Microcalorimetry

The standard molar enthalpies of vaporization of the two compounds were determined using a high-temperature Calvet microcalorimeter (Setaram HT1000D, Lyon, France) using the vacuum drop microcalorimetric technique, which has been described by Skinner et al. [26] for the study of solids and adapted and tested for liquid vaporizations by Ribeiro da Silva et al. [27]. The equipment characteristics, together with the procedure, have been described in detail in the literature [28].

The temperature, *T*, of the hot reaction vessel of the calorimeter, was set to 345 K for the vaporization measurements of ethyl (*E*)-cinnamate and 376 K for ethyl hydrocinnamate. The mass of the sample used in each run was 4 to 6 mg. The thermal corrections for the glass capillary tubes were determined from separate experiments and were minimized by dropping tubes of nearly equal mass, within $\pm(1 \times 10^{-4})$ g, into each of the twin calorimetric cells. The Calvet microcalorimeter was calibrated in situ for the predefined temperatures by determining the enthalpy of vaporization of undecane as a reference substance [29]. The procedure used in the calibration experiments was the same used as that for the compound experiments. The values used for the standard molar enthalpy of vaporization, at $T = 298.15$ K, for undecane was (56.6 ± 0.6) kJ·mol^{-1} [29]. The derived calibration constants for the different temperatures using undecane were k_{cal} ($T \approx 345$ K) = (1.0321 ± 0.0081) and k_{cal} ($T \approx 376$ K) = (1.0234 ± 0.0027); these values were obtained as the average of six independent experiments, where the uncertainty corresponds to the standard deviation of the mean.

2.4. Computational Methodology

Theoretical calculations of the molecular structures were performed using Gaussian-09 software [30] (Gaussian, Inc., Wallingford, CT, USA). These calculations were based on a variation in the Gaussian-3 (G3) theory [31], specifically the composite method G3(MP2)//B3LYP [32]. A conformational analysis was conducted to identify and quantify the various conformations for each cinnamate derivative. The conformer compositions, calculated according to the Boltzmann distribution, were subsequently used in the calculation of the final value of the gas-phase enthalpy of formation. Hypothetical gas-phase group substitution reactions were used to estimate the gas-phase enthalpy of formation of ethyl (*E*)-cinnamate and ethyl hydrocinnamate. The good agreement found between the experimental and the estimated values for the gas-phase enthalpy of formation of compounds studied in previous works [2–34] gives us confidence for the use of the G3(MP2)//B3LYP approach and the same type of reactions.

3. Results and Discussion

3.1. Experimental Studies

3.1.1. Enthalpies of Combustion and Formation in the Liquid Phase

The general combustion reactions for ethyl (*E*)-cinnamate and ethyl hydrocinnamate are illustrated in Equations (1) and (2), respectively:

$$C_{11}H_{12}O_2\ (l) + 13\ O_2\ (g) \rightarrow 11\ CO_2(g) + 6\ H_2O\ (l) \quad (1)$$

$$C_{11}H_{14}O_2\ (l) + 13.5\ O_2\ (g) \rightarrow 11\ CO_2(g) + 7\ H_2O\ (l) \quad (2)$$

The internal energy change associated with the isothermal bomb process, $\Delta U(\text{IBP})$, was calculated using Equation (3), where ΔT_{ad} refers to the temperature change in the calorimeter, which was adjusted for heat exchange and stirring work, $\Delta m(H_2O)$ is the difference between the mass of water added to the calorimeter and the baseline mass of 3119.6 g assigned to $\varepsilon(\text{cal})$, ε_f is the energy equivalent of the bomb contents in the final state, $\Delta U(\text{ign})$ is the electric energy necessary for ignition, and $c_p(H_2O, l)$ is the specific heat capacity of liquid water at constant pressure.

$$\Delta U(\text{IBP}) = -\{\varepsilon(\text{cal}) + \Delta m(H_2O)\cdot c_p(H_2O, l) + \varepsilon_f\}\Delta T_{ad} + \Delta U(\text{ign}) \quad (3)$$

The method outlined by Hubbard et al. [34] was employed to determine the standard massic energy of combustion, $\Delta_c u^\circ$, for combustions taking place in a bomb calorimeter at constant volume. Additionally, several factors must be considered to accurately account for the energy associated with side reactions. These factors include the standard formation energy of nitric acid solution, the energy from the combustion of the cotton fuse, the energy from the combustion of carbon resulting from incomplete combustion, the combustion energy of Melinex (used as a combustion aid), and the ignition energy. Detailed combustion results for the two cinnamate derivatives can be found in Tables S1 and S2, which are included in the Supplementary Materials as supplementary data. The average $\Delta_c u^\circ$, for ethyl

(*E*)-cinnamate and ethyl hydrocinnamate in the liquid phase obtained from six combustion experiments were (−32496.1± 4.3) J·g^{-1} and (−33110.5 ± 5.9) J·g^{-1}, respectively; the stated uncertainty represents the standard deviation of the mean. From these values, the standard molar internal energy, $\Delta_c U_m^\circ(l)$, and enthalpy, $\Delta_c H_m^\circ(l)$, for the combustion reactions in Equations (1) and (2) were derived, as presented in Table 1. The standard molar enthalpies of formation in the liquid phase, $\Delta_f H_m^\circ(l)$, for each cinnamate derivative, at T = 298.15 K, were calculated from the $\Delta_c H_m^\circ(l)$ values and the following $\Delta_f H_m^\circ$ values for the species involved in the combustion reaction at 298.15 K: (−285.830 ± 0.042) kJ·mol^{-1} [35] for H_2O (l) and (−393.51 ± 0.13) kJ·mol^{-1} [35] for CO_2 (g).

Table 1. Standard (p° = 0.1 MPa) molar values for ethyl (*E*)-cinnamate (EEC) and ethyl hydrocinnamate (EHC) at T = 298.15 K in the liquid phase obtained from combustion calorimetry experiments.

Compound	$\Delta_c U_m^\circ$(l)/kJ·mol^{-1}	$\Delta_c H_m^\circ$(l)/kJ·mol^{-1}	$\Delta_f H_m^\circ$(l)/kJ·mol^{-1}
EEC	−5726.2 ± 2.4 [1]	−5731.2 ± 2.4 [1]	−312.4 ± 2.8 [2]
EHC	−5901.2± 2.8 [1]	−5907.4 ± 2.8 [1]	−422.0 ± 3.2 [2]

[1] The stated uncertainty was determined by accounting for combined standard uncertainty, which includes the contributions from the calibration with benzoic acid and Melinex®, along with the coverage factor k = 2, corresponding to a 0.95 confidence level [36]; [2] The stated uncertainty was calculated by considering the combined standard uncertainty, which factors in the contribution of the species involved in the respective combustion reactions (Equations (1) and (2)), and the coverage factor k = 2, corresponding to a 0.95 confidence level [36].

3.1.2. Enthalpy of Vaporization

The main vaporization results for the two cinnamate derivatives are summarized in Table 2, with detailed results available in Tables S3 and S4 included in the Supplementary Materials as supplementary data. The observed enthalpy of vaporization at the experimental temperature T for each compound, represented by $\Delta_{l,298.15K}^{g,T} H_m^\circ$, is the average of six experiments. These values were adjusted to T = 298.15 K using Equations (4) and (5) and the molar heat capacities in the gaseous phase for each compound. These capacities, within the temperature range of 200 K to 600 K (Table S5 of Supplementary Materials), were calculated through statistical thermodynamics using vibrational frequencies from B3LYP calculations with the 6-31G(d) basis set, scaled by a factor of 0.960 ± 0.022 [37].

$$\Delta_{298.15K}^{T} H_m^o(g) = \int_{298.15K}^{T} C_p^\circ(g)dT \quad (4)$$

$$\Delta_l^g H_m^\circ = \Delta_{l,\,298.15K}^{g,\,T} H_m^\circ - \Delta_{298.15\,K}^{T} H_m^\circ(g) \quad (5)$$

Table 2. Standard (p° = 0.1 MPa) molar enthalpies of vaporization for ethyl (*E*)-cinnamate (EEC) and ethyl hydrocinnamate (EHC) obtained from high-temperature microcalorimetry experiments.

Compound	T/K	$\Delta_{l,298.15K}^{g,T} H_m^\circ$/kJ·mol^{-1}	$\Delta_{298.15K}^{T} H_m^\circ$(g)/kJ·mol^{-1}	$\Delta_l^g H_m^\circ$ /kJ·mol^{-1}
EEC (l)	345.6	82.8 ± 0.8 [1]	10.4 ± 0.2 [2]	72.4 ± 2.5 [3]
EHC (l)	376.1	86.4 ± 0.3 [1]	18.6 ± 0.4 [2]	67.9 ± 1.7 [4]

[1] The quoted uncertainty corresponds to the estimated standard deviation of the mean for six experiments; [2] the quoted uncertainty includes the root-mean-square deviation of the third-degree polynomial used and the uncertainty of the vibrational frequency scaling factor for the B3LYP/6-31G(d) method [37]; [3] the combined standard uncertainty (which includes the uncertainty associated with calibration experiments with undecane) and the coverage factor k = 2.10 (for a 0.95 level of confidence with 18 effective degrees of freedom, calculated by the Welch–Satterthwaite formula) were used to determine the quoted uncertainty [36]; and [4] The combined standard uncertainty (which includes the uncertainty associated with calibration experiments with undecane) and the coverage factor k = 2.18 (for a 0.95 level of confidence with 12 effective degrees of freedom, calculated by the Welch–Satterthwaite formula) were used to determine the quoted uncertainty [36].

The standard molar enthalpies of vaporization at T = 298.15 K were found to be (72.4 ± 2.5) kJ·mol^{-1} for ethyl (E)-cinnamate and (67.9 ± 1.7) kJ·mol^{-1} for ethyl hydrocinnamate.

As mentioned in the literature review presented in the introduction, Kozlovskiy et al. [17] studied the vapor pressures of ethyl (E)-cinnamate over a temperature range from 298.15 K to its boiling point using gas chromatography. They reported a value of (70.4 ± 1.4) kJ·mol^{-1}, which is consistent with the value obtained through our calorimetric technique, considering the associated uncertainties.

3.2. Computational Studies

The computational study of ethyl (E)-cinnamate and ethyl hydrocinnamate, using the G3(MP2)//B3LYP approach, revealed several minima on the potential energy surface, as anticipated, due to the various possible rearrangements of each substituent in the gaseous phase. Figure 2 reports the conformational composition, χ_i, for the most stable predominant molecular geometries, calculated assuming a Boltzmann distribution of the possible equilibrium structures. Conformations with residual probability and negative frequencies were excluded from the final conformational composition. Detailed results of the conformational analysis are provided in Tables S6 and S7 of Supplementary Materials.

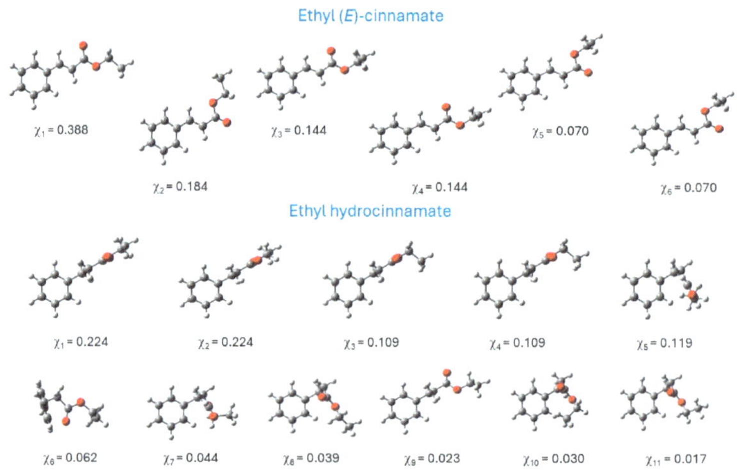

Figure 2. Conformational composition, χi, for the most stable predominant molecular geometries, corresponding to minima on the potential energy surface, obtained using the G3(MP2)//B3LYP composite method for cinnamate derivatives. Atom color code: grey, C; red, O; and white, H.

The only structural differences between the compounds are that ethyl (E)-cinnamate features a vinyl group (–CH=CH–) with a trans double bond relative to the ester group, whereas ethyl hydrocinnamate has an ethyl group (CH_2-CH_2-) instead, making it the ethyl ester of hydrocinnamic acid. In ethyl (E)-cinnamate, the vinyl group is conjugated with the benzene ring, creating a significant energy barrier to rotation around the single bond connecting them. This conjugation restricts rotation, keeping the vinyl group coplanar with the ring and limiting the number of conformers to six. In contrast, ethyl hydrocinnamate lacks this conjugation, as the benzene ring connects to the side chain via an ethyl group, allowing more flexible rotation and resulting in eleven conformers.

The estimation of the gas-phase enthalpy of formation for both compounds under investigation, using hypothetical substitution reactions (isodesmic or homodesmotic), relies on the similarity of bonding environments between reactants and products, which helps to cancel out systematic errors. However, a limitation of these reactions is that they require experimental gas-phase enthalpies of formation for all auxiliary molecules involved. To reduce the risk of depending on potentially inaccurate experimental data for any given species, multiple working reactions are used, replicating a procedure that has proven to be quite effective in previous studies [2–34].

In this approach, the reaction enthalpy, $\Delta_r H_m^\circ(g)$, is calculated as outlined in Equation (6) for each of the hypothetical gas-phase reactions, using the absolute standard enthalpies, $H_{298.15K}^\circ$, of the molecules involved. Subsequently, Equation (7), along with the experimentally determined standard molar enthalpies of formation for all the chemical species in the hypothetical reactions, was used to estimate $\Delta_f H_m^\circ(g)$ for the species of interest, specifically for the following two molecules under study: ethyl (*E*)-cinnamate (see Table 3) and ethyl hydrocinnamate (see Table 4). The relevant data for these chemical species are provided in Table S8 of the Supplementary Materials.

Table 3. Hypothetical gas-phase reactions proposed for calculating the gas-phase enthalpy of formation, $\Delta_f H_m^\circ(g)$, for ethyl hydrocinnamate, along with the corresponding calculated values for the enthalpies of reaction $\Delta_r H_m^\circ(g)$, at T = 298.15 K. All values account for the conformer distribution of the six minimum conformers.

Hypothetical Gas-Phase Reactions	Equation	$\Delta_r H_m^\circ(g)$ /kJ·mol^{-1}	$\Delta_f H_m^\circ(g)$ /kJ·mol^{-1}
	I	24.46	−245.66
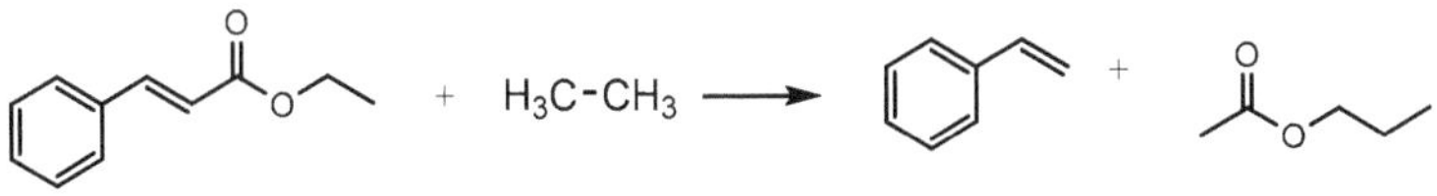	II	13.55	−246.54
	III	29.83	−249.33
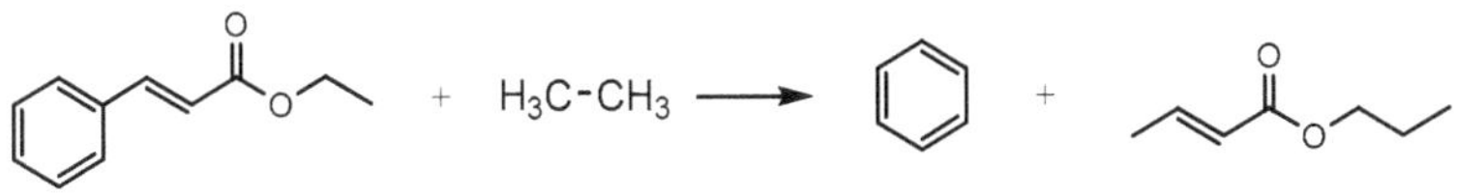	IV	18.95	−247.75
Mean value			−247.3 ± 2.5 [1]

[1] The quoted uncertainty defines an interval with a 0.95 level of confidence (coverage factor used k = 3.18 for three degrees of freedom.

$$\Delta_r H_m^\circ(g) = \Sigma H_{298.15K}^\circ\ (g,\ \text{products}) - \Sigma H_{298.15K}^\circ\ (g,\ \text{reagents}) \tag{6}$$

$$\Delta_r H_m^\circ(g) = \Sigma \Delta_f H_m^\circ\ (g,\ \text{products}) - \Sigma \Delta_f H_m^\circ\ (g,\ \text{reagents}) \tag{7}$$

The estimated values of the gas-phase standard molar enthalpy of formation, $\Delta_f H_m^\circ(g)$, obtained for ethyl (*E*)-cinnamate was (−247.3 ± 2.5) kJ·mol^{-1} based on four reactions, while for ethyl hydrocinnamate, it was (−353.7 ± 1.7) kJ·mol^{-1} from five reactions.

Table 4. Hypothetical gas-phase reactions proposed for calculating the gas-phase enthalpy of formation, $\Delta_f H_m^\circ(g)$, for ethyl hydrocinnamate, along with the corresponding calculated values for the enthalpies of reaction $\Delta_r H_m^\circ(g)$, at T = 298.15 K. All values account for the conformer distribution of the eleven minimum conformers.

Hypothetical Gas-Phase Reactions	Equation	$\Delta_r H_m^\circ(g)$ /kJ mol^{-1}	$\Delta_f H_m^\circ(g)$ /kJ mol^{-1}
	I	25.27	−353.27
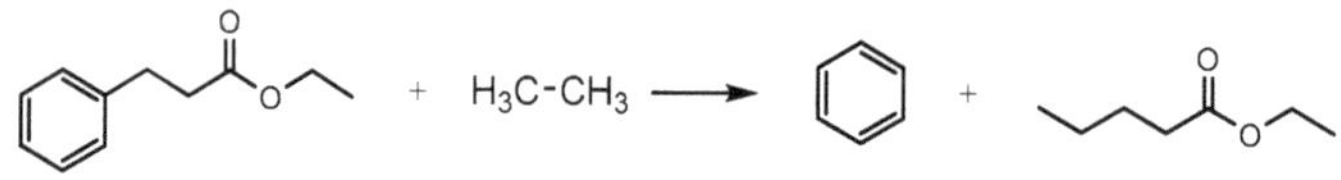	II	13.34	−352.84
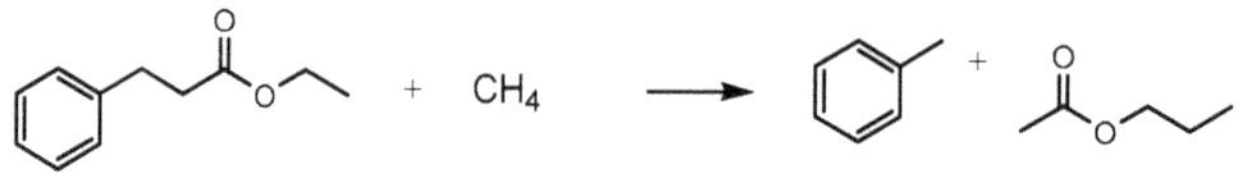	III	15.42	−355.31
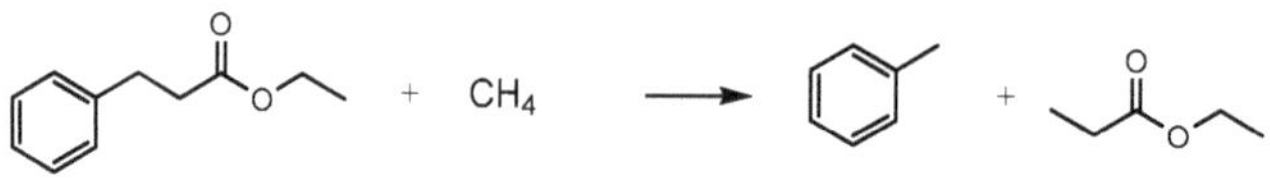	IV	13.67	−355.31
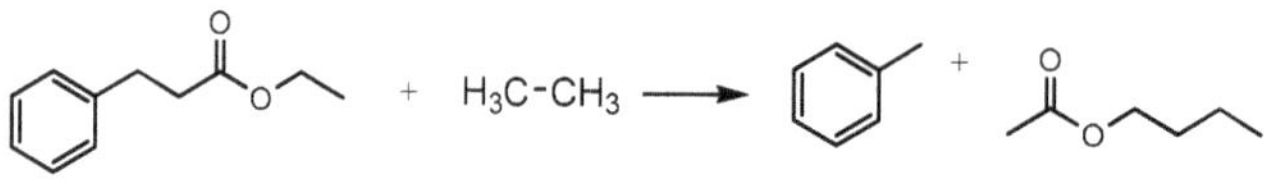	V	3.86	−354.86
Mean value			−353.7 ± 1.7 [1]

[1] The quoted uncertainty defines an interval with a 0.95 level of confidence (coverage factor used k = 2.78) for four degrees of freedom.

4. Discussion

The experimental determination of the standard (p° = 0.1 MPa) molar gas-phase enthalpy of formation, $\Delta_f H_m^\circ(g)$, often involves starting with the enthalpy of formation in the condensed state, where both intermolecular and intramolecular interactions are considered. By combining this with the enthalpy of phase transition, the intermolecular interactions can be effectively canceled out, isolating the contributions of intramolecular interactions for analysis. This approach fulfills the initial objective of the work by allowing for a clearer understanding of the effects of molecular structure on thermodynamic properties and chemical behavior.

In Table 5, the standard (p° = 0.1 MPa) molar enthalpies of formation (liquid and gas) and vaporization for ethyl (*E*)-cinnamate and ethyl hydrocinnamate obtained in this study are compiled. Comparing the experimental values with the computational values obtained for $\Delta_f H_m^\circ(g)$, the difference for ethyl (*E*)-cinnamate is 7.3 kJ·mol^{-1}, while for ethyl hydrocinnamate, it is 0.4 kJ·mol^{-1}. The difference for ethyl (*E*)-cinnamate falls within the acceptable range established for experimental and computational values [38].

Table 5. Compilation of the standard (p° = 0.1 MPa) molar enthalpies of formation (liquid and gaseous) and vaporization for ethyl (*E*)-cinnamate (EEC) and ethyl hydrocinnamate (EHC).

	$\Delta_l^g H_m^\circ$/kJ·mol^{-1}	$\Delta_f H_m^\circ(l)$/kJ·mol^{-1}	$\Delta_f H_m^\circ(g)$/kJ·mol^{-1} Experimental	$\Delta_f H_m^\circ(g)$/kJ·mol^{-1} Computational
EEC	72.4 ± 2.5 [1]	−312.4 ± 2.8 [1]	−240.0 ± 3.8 [2]	−247.3 ± 2.5 [1]
EHC	67.9 ± 1.7 [1]	−422.0 ± 3.2 [1]	−354.1 ± 3.6 [2]	−353.7 ± 1.7 [1]

[1] Standard uncertainties with 0.95 confidence level. [2] Standard uncertainties calculated using the RSS method.

Considering the structural differences between the compounds studied, understanding the enthalpic difference associated with the hydrogenation of ethyl (*E*)-cinnamate to ethyl hydrocinnamate requires knowledge of the experimental $\Delta_f H_m^\circ(g)$ values. This information is relevant for a comprehensive understanding of the hydrogenation process. Figure 3 illustrates the hydrogenation reaction of ethyl (*E*)-cinnamate to ethyl hydrocinnamate (Equation (8)), with the gas-phase hydrogenation enthalpy at 298.15 K being $\Delta_{hyd} H_m^\circ(g) = -114.1 \pm 5.2$ kJ·mol^{-1}, calculated using Hess Law. The hydrogenation enthalpy for the reaction of ethyl (2*E*)-butenoate to ethyl butyrate (Equation (9)) was also calculated, which involved only the substituents without the benzene ring, yielding a value of -109.5 ± 2.6 kJ·mol^{-1}; this result is consistent with the previous value when accounting for the associated uncertainty.

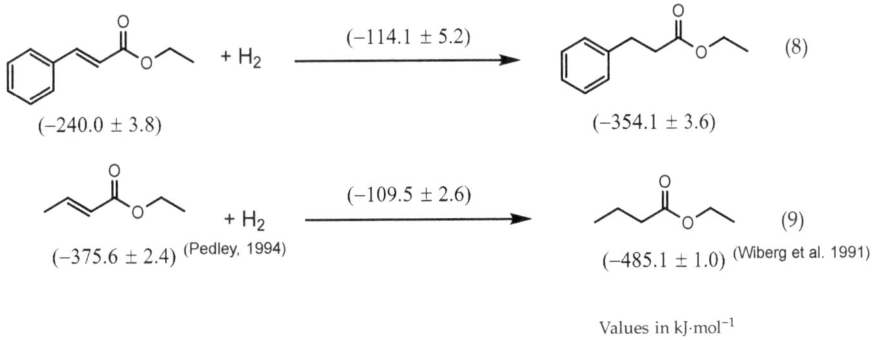

Figure 3. Hydrogenation reactions with corresponding enthalpy of hydrogenation values for the conversion of ethyl (*E*)-cinnamate to ethyl hydrocinnamate and ethyl (2*E*)-2-butenoate to ethyl butyrate. Refs. [39,40].

Williams [41] pioneered the design of the first hydrogen calorimeter specifically for measuring the enthalpy of hydrogenation in solution. The setup used a conventional reaction calorimeter with a Dewar flask but required hydrogen-tight conditions. The sample was in a glass ampoule, broken by an externally controlled mechanical device to start the reaction. Williams investigated the hydrogenation of methyl (Z)-cinnamate and methyl (*E*)-cinnamate to methyl hydrocinnamate at 302 K (Figure 4), using glacial acetic acid as the solvent for both reactions. The enthalpy value at this temperature is close to 298 K, given the minimal difference in heat capacities (at constant pressure) between the methyl cinnamate derivatives and hydrogen, making any corrections unnecessary. The enthalpy of hydrogenation was -117.9 ± 1.0 kJ·mol^{-1} for methyl (Z)-cinnamate and -101.2 ± 1.0 kJ·mol^{-1} for methyl (*E*)-cinnamate. It is important to note that for methyl (Z)-cinnamate, the fusion enthalpy value was used to calculate gas-phase hydrogenation enthalpy due to the compound being solid, which may explain the observed lower enthalpy value.

Table 6 presents the hydrogenation enthalpy values at 298.15 K compiled by Rogers [42] (NIST–JANAF Thermochemical Tables) for some linear alkenes, specifically for the *E* and *Z* isomers. A brief analysis of these values shows that the hydrogenation enthalpy for the *Z* (cis) isomers tends to be more negative compared to the *E* (trans) isomers. This can be explained by the fact that the Z isomer is generally less stable than the E isomer, which causes the Z isomer to release more energy upon hydrogenation due to the greater stability difference between the isomer and the resulting saturated product (alkane). Therefore, the hydrogenation enthalpy of the *Z* isomer tends to be more negative than that of the *E* isomer.

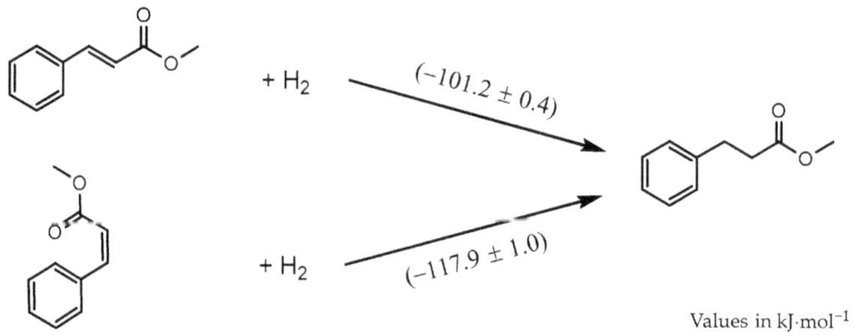

Figure 4. Hydrogenation reactions with corresponding experimental enthalpy of hydrogenation values for the conversion of methyl (*E*)-cinnamate and methyl (Z)-cinnamate to methyl hydrocinnamate at a temperature of 302 K.

Table 6. Compilation of experimental values for the gas-phase hydrogenation enthalpy of some unbranched (*E/Z*) isomeric alkenes at T = 298.15 K [42].

Reactions	$\Delta_{hyd}H^{\circ}_{m}(g)$/kJ·mol^{-1}	
	Isomer E	**Isomer Z**
But-2-ene + H_2 → Butane	−114.5 ± 0.4	−118.5 ± 0.4
Pent-2-ene + H_2 → Pentane	−114.6 ± 0.4	−118.5 ± 0.4
Hex-2-ene + H_2 → Hexane	−113.8 ± 1.3	−115.8 ± 0.8
Hex-3-ene + H_2 → Hexane	−112.3 ± 1.7	−122.6 ± 1.3
Hept-2-ene + H_2 → Heptane	−114.1 ± 0.5	−117.9 ± 0.4
Hept-3-ene + H_2 → Heptane	−114.7 ± 0.3	−120.0 ± 2.9
Oct-2-ene + H_2 → Octane	−115.5 ± 0.7	−119.4 ± 1.1
Oct-3-ene + H_2 → Octane	−115.8 ± 0.4	−117.8 ± 0.4
Oct-4-ene + H_2 → Octane	−115.0 ± 0.7	−114.6 ± 0.4

5. Conclusions

With the present experimental and computational work on ethyl (*E*)-cinnamate and ethyl hydrocinnamate, two cinnamate derivatives with flavor and fragrance features, the main conclusions that can be taken are as follows: (a) from a structural perspective in the gas phase, the vinyl group in ethyl (*E*)-cinnamate restricts conformational flexibility due to conjugation with the benzene ring, whereas the ethyl group in ethyl hydrocinnamate allows greater flexibility and more conformers; (b) minor discrepancies between experimental and computational values for the gas-phase enthalpy of formation were observed for both compounds, validating the computational method used; and (c) the enthalpy of hydrogenation in the gas-phase from ethyl (*E*)-cinnamate to ethyl hydrocinnamate is −114.1 ± 5.2 kJ·mol^{-1}, which is consistent with values for similar reactions involving ethyl (2*E*)-butenoate.

Supplementary Materials: The following supporting information can be downloaded at https://www.mdpi.com/article/10.3390/liquids4040038/s1. Supplementary Materials: (ethyl (*E*)-cinnamate is abbreviated as EEC and ethyl hydrocinnamate as EHC): Tables S1 and S2. Combustion results and standard (p° = 0.1 MPa) massic energy of combustion at T = 298.15 K for EEC and EHC, respectively. Tables S3 and S4. Calvet microcalorimetry results for the processes of vaporization of EEC and EHC, respectively. Table S5. Standard (p° = 0.1 MPa) molar heat capacities in the gaseous phase for EEC and EHC, obtained from statistical thermodynamics using the vibrational frequencies calculated at the B3LYP/6-31G(d) level of theory (scaled by a factor of 0.960 ± 0.022). Tables S6 and S7. Absolute standard enthalpies, $H^{\circ}_{298.15\ K}$, and entropies, $S^{\circ}_{298.15\ K}$, obtained by the G3(MP2)//B3LYP composite method and the corresponding derived gas-phase standard molar enthalpies, $\Delta_f H^{\circ}_{m}(g)$, entropies, $\Delta_f S^{\circ}_{m}(g)$, and Gibbs energy of formation, $\Delta_f G^{\circ}_{m}(g)$ at T = 298.15 K as well as the conformational composition, χ_I, obtained for EEC and EHC, respectively. Data from [43,44] Table S8.

G3(MP2)//B3LYP enthalpies with corresponding gas-phase standard (p^o = 0.1 MPa) molar enthalpies of formation at T = 298.15 K for the molecular species. Data from [45,46]

Author Contributions: Experimental and computational investigation, C.A.O.S. and V.L.S.F.; Supervision, V.L.S.F. and M.D.M.C.R.d.S.; Writing—review and editing, V.L.S.F. and M.D.M.C.R.d.S. All authors have read and agreed to the published version of the manuscript.

Funding: The authors acknowledge Fundação para a Ciência e Tecnologia (FCT) for the financial support through the projects UIDB/QUI/00081/2020 (https://doi.org/10.54499/UIDB/00081/2020 accessed on 8 October 2024) and IMS-(LA/P/0056/2020). C.A.O.S. thanks FCT for the doctoral grant SFRH/BD/137672/2018. VLSF thanks FCT for founding with the reference https://doi.org/10.54499/DL57/2016/CP1454/CT0019 (accessed on 8 October 2024).

Data Availability Statement: The data supporting the results reported in this study are not publicly available due to privacy and ethical restrictions. No new datasets were created beyond those already available in the literature and the data provided in this paper. For further information about the study, please contact the corresponding author.

Conflicts of Interest: The authors declare no conflicts of interest. The funders had no role in the design of the study; in the collection, analyses, or interpretation of data; in the writing of the manuscript; or in the decision to publish the results.

References

1. Stoessel, F. *Thermal Safety of Chemical Processes: Risk Assessment and Process Design*, 1st ed.; Wiley-VCH Verlag GmbH & Co. KGaA: Weinheim, Germany, 2020.
2. Silva, C.A.O.; Freitas, V.L.S.; Ribeiro da Silva, M.D.M.C. Determination and Analysis of Thermodynamic Properties of Methyl Methylanthranilate Isomers. *Molecules* **2023**, *28*, 6686. [CrossRef] [PubMed]
3. Ledo, J.M.; Flores, H.; Freitas, V.L.S.; Solano-Altamirano, J.M.; Hernández-Pérez, J.M.; Ribeiro da Silva, M.D.M.C.; Camarillo, E.A. Thermal and structural properties of ethyl 2- and 3-aminobenzoates: Experimental and computational approaches. *J. Chem. Thermodyn.* **2019**, *133*, 93–99. [CrossRef]
4. Ximello-Hernández, A.; Freitas, V.L.S.; Ribeiro da Silva, M.D.M.C. Assessment of thermochemical data of γ butyrolactone from experimental and computational studies. *J. Chem. Eng. Data* **2020**, *65*, 1968–1975. [CrossRef]
5. Fahlbusch, K.-G.; Hammerschmidt, F.-J.; Panten, J.; Pickenhagen, W.; Schatkowski, D.; Bauer, K.; Garbe, D.; Surburg, H. Flavors and Fragrances. In *Ullmann's Encyclopedia of Industrial Chemistry*; Wiley-VCH: Weinheim, Germany, 2003; Volume 11.
6. Xiao, Q.; Zhou, X.; Xiao, Z.; Niu, Y. Characterization of the differences in the aroma of cherry wines from different price segments using gas chromatography–mass spectrometry, odor activity values, sensory analysis, and aroma reconstitution. *Food Sci. Biotechnol.* **2017**, *26*, 331–338. [CrossRef]
7. Xu, Y.; Wang, D.; Li, H.; Hao, J.; Jiang, W.; Liu, Z.; Qin, Q. Flavor Contribution of Esters in Lager Beers and an Analysis of Their Flavor Thresholds. *J. Am. Soc. Brew. Chem.* **2017**, *75*, 201–206. [CrossRef]
8. Bhatia, S.P.; Wellington, G.A.; Cocchiara, J.; Lalko, J.; Letizia, C.S.; Api, A.M. Fragrance material review on ethyl cinnamate. *Food Chem. Toxicol.* **2007**, *45*, S90–S94. [CrossRef]
9. Masselink, W.; Reumann, D.; Murawala, P.; Pasierbek, P.; Taniguchi, Y.; Bonnay, F.; Meixner, K.; Knoblich, J.A.; Tanaka, E.M. Broad applicability of a streamlined ethyl cinnamate-based clearing procedure. *Development* **2019**, *146*, dev166884. [CrossRef]
10. Barandiaran, A.; Gomez-Caturla, J.; Ivorra-Martinez, J.; Lascano, D.; Selles, M.A.; Moreno, V.; Fenollar, O. Esters of Cinnamic Acid as Green Plasticizers for Polylactide Formulations with Improved Ductility. *Macromol. Mater. Eng.* **2023**, *308*, 2300022. [CrossRef]
11. Morais, M.C.; Oliveira Lima, E.; Perez-Castillo, Y.; Sousa, D.P. Synthetic Cinnamides and Cinnamates: Antimicrobial Activity, Mechanism of Action, and In Silico Study. *Molecules* **2023**, *28*, 1918. [CrossRef]
12. Araújo, M.O.; Pérez-Castillo, Y.; Oliveira, L.H.G.; Nunes, F.C.; de Sousa, D.P. Larvicidal Activity of Cinnamic Acid Derivatives: Investigating Alternative Products for Aedes aegypti L. Control. *Molecules* **2021**, *26*, 61. [CrossRef]
13. Ribeiro, S.G.; Martins, C.; Tavares, T.; Rudnitskaya, A.; Alves, F.; Rocha, S.M. Volatile Composition of Fortification Grape Spirit and Port Wine: Where Do We Stand? *Foods* **2023**, *12*, 2432. [CrossRef]
14. Zieniuk, B.; Fabiszewska, A.; Wołoszynowska, M.; Białecka-Florjańczyk, E. Synthesis of flavor compound ethyl hydrocinnamate by Yarrowia lipolytica lipases. *Biocatal. Biotransform.* **2021**, *39*, 455–464. [CrossRef]
15. Wei, D.; Shi, X.; Qu, R.; Junge, K.; Junge, H.; Beller, M. Toward a Hydrogen Economy: Development of Heterogeneous Catalysts for Chemical Hydrogen Storage and Release Reactions. *ACS Energy Lett.* **2022**, *7*, 3734–3752. [CrossRef]
16. Preuster, P.; Papp, C.; Wasserscheid, P. Liquid Organic Hydrogen Carriers (LOHCs): Toward a Hydrogen-Free Hydrogen Economy. *Acc. Chem. Res.* **2017**, *50*, 74–85. [CrossRef] [PubMed]
17. Kozlovskiy, M.; Gobble, C.; Chickos, J. Vapor pressures and vaporization enthalpies of a series of esters used in flavors by correlation gas chromatography. *J. Chem. Thermodyn.* **2015**, *86*, 65–74. [CrossRef]
18. Freitas, V.L.S.; Oliveira, L.I.P.; Ribeiro da Silva, M.D.M.C. Standard molar enthalpies of formation of the acetylpyridine isomers. *J. Chem. Thermodyn.* **2007**, *39*, 39–43. [CrossRef]

19. TCI Europe, N.V. "Ethyl (E)-Cinnamate" Safety Data Sheet. 2024. Available online: https://tcichemicals.com (accessed on 3 July 2024).
20. TCI Europe, N.V. "Ethyl 3-Phenylpropionate" Safety Data Sheet. 2024. Available online: https://tcichemicals.com (accessed on 3 July 2024).
21. *Certificate of Analysis, Standard Reference Material 39j*; Benzoic Acid Calorimetric Standard, N.B.S.: Washington, DC, USA, 1995.
22. Prohaska, T.; Irrgeher, J.; Benefield, J.; Böhlke, J.; Chesson, L.; Coplen, T.; Ding, T.; Dunn, P.; Gröning, M.; Holden, N.; et al. Standard atomic weights of the elements 2021 (IUPAC Technical Report). *Pure Appl. Chem.* **2022**, *94*, 573–600. [CrossRef]
23. Ribeiro da Silva, M.A.V.; Ribeiro da Silva, M.D.M.C.; Pilcher, G. The construction, calibration and use of a new high-precision static bomb calorimeter. *Rev. Port. Quim.* **1984**, *26*, 163–172.
24. Skinner, H.A.; Snelson, A. The heats of combustion of the four isomeric butyl alcohols. *Trans. Faraday Soc.* **1960**, *56*, 1776–1783. [CrossRef]
25. Santos, L.M.N.B.F.; Silva, M.T.; Schröder, B.; Gomes, L. LABTERMO: Methodologies for the calculation of the corrected temperature rise in isoperibol calorimetry. *J. Therm. Anal. Calorim.* **2007**, *89*, 175–180. [CrossRef]
26. Adedeji, F.A.; Brown, D.L.S.; Connor, J.A.; Leung, W.L.; Paz-Andrade, I.M.; Skinner, H.A. Thermochemistry of arene chromium tricarbonyls and the strengths of arene-chromium bonds. *J. Organomet. Chem.* **1975**, *97*, 221–228. [CrossRef]
27. Ribeiro da Silva, M.A.V.; Matos, M.A.R.; Amaral, L.M.P.F. Thermochemical study of 2-, 4-, 6-, and 8-methylquinoline. *J. Chem. Thermodyn.* **1995**, *27*, 565–574. [CrossRef]
28. Santos, L.M.N.B.F.; Schröder, B.; Fernandes, O.O.P.; Ribeiro da Silva, M.A.V. Measurement of enthalpies of sublimation by drop method in a Calvet type calorimeter: Design and test of a new system. *Thermochim. Acta* **2014**, *415*, 15–20. [CrossRef]
29. Sabbah, R.; Xu-Wu, A.; Chickos, J.S.; Planas Leitão, M.L.; Roux, M.V.; Torres, L.A. Reference materials for calorimetry and differential thermal analysis. *Thermochim. Acta* **1999**, *331*, 93–204. [CrossRef]
30. Frisch, M.J.; Trucks, G.W.; Schlegel, H.B.; Scuseria, G.E.; Robb, M.A.; Cheeseman, J.R.; Scalmani, G.; Barone, V.; Petersson, G.A.; Nakatsuji, H.; et al. *Gaussian 09, Revision A.02*; Gaussian, Inc.: Wallingford, CT, USA, 2016.
31. Curtiss, L.A.; Raghavachari, K.; Redfern, P.C.; Rassolov, V.; Pople, J.A. Gaussian-3 (G3) theory for molecules containing first and second-row atoms. *J. Chem. Phys.* **1998**, *109*, 7764–7776. [CrossRef]
32. Baboul, A.G.; Curtiss, L.A.; Redfern, P.C.; Raghavachari, K. Gaussian-3 theory using density functional geometries and zero-point energies. *J. Chem. Phys.* **1999**, *110*, 7650–7657. [CrossRef]
33. Freitas, V.L.S.; Lima, A.C.M.O.; Sapei, E.; Ribeiro da Silva, M.D.M.C. Comprehensive thermophysical and thermochemical studies of vanillyl alcohol. *J. Chem. Thermodyn.* **2016**, *102*, 287–292. [CrossRef]
34. Hubbard, W.N.; Scott, D.W. Standard states and corrections for combustions in a bomb at constant volume. In *Experimental Thermochemistry*; Rossini, F.D., Ed.; Interscience: New York, NY, USA, 1956; Volume 1.
35. Cox, J.D.; Wagman, D.D.; Medvedev, V.A. *CODATA Key Values for Thermodynamics*; Hemisphere: New York, NY, USA, 1989.
36. Taylor, B.N.; Kuyatt, C.E. *Guidelines for Evaluating and Expressing the Uncertainty of NIST Measurement Results*; NIST Technical Note 1297; National Institute of Standards and Technology: Gaithersburg, MD, USA, 1994.
37. NIST Computational Chemistry Comparison and Benchmark Database. *NIST Standard Reference Database Number 101*; Release 16a; Johnson, R.D., III, Ed.; National Institute of Standards and Technology (NIST): Gaithersburg, MD, USA, 2013. Available online: https://cccbdb.nist.gov/vibscalex.asp (accessed on 5 July 2023).
38. Freitas, V.L.S.; Ribeiro da Silva, M.D.M.C.; Gomes, J.R.B. A computational study on the thermochemistry of methylbenzo- and methyldibenzothiophenes. *J. Mol. Struct. THEOCHEM* **2010**, *946*, 20–25. [CrossRef]
39. Pedley, J.B. *Thermochemical Data and Structures of Organic Compounds Thermodynamics*; Research Center: College Station, TX, USA, 1994.
40. Wiberg, K.B.; Waldron, R.F. Lactones. 2. Enthalpies of hydrolysis, reduction, and formation of the C4-C13 monocyclic lactones. strain energies and conformations. *J. Am. Chem. Soc.* **1991**, *113*, 7697–7705. [CrossRef]
41. Williams, R.B. Heats of Catalytic Hydrogenation in Solution. I. Apparatus, Technique, and the Heats of Hydrogenation of Certain Pairs of Stereoisomers. *J. Am. Chem. Soc.* **1942**, *64*, 1395–1404. [CrossRef]
42. Rogers, D.W. *Heats of Hydrogenation: Experimental and Computational Hydrogen Thermo-Chemistry of Organic Compounds*; World Scientific: Singapore, 2006. Available online: https://webbook.nist.gov/chemistry/ (accessed on 4 August 2024).
43. Merrick, P.; Moran, D.; Radom, L. An evaluation of harmonic vibrational frequency scale factor. *J. Phys. Chem. A* **2007**, *111*, 11683–11700. [CrossRef]
44. Chase, M.W., Jr. NIST-Janaf Thermochemical Tables. Available online: https://janaf.nist.gov/ (accessed on 1 May 2021).
45. Pedley, J.B. *Thermochemical Data and Structures of Organic Compounds Thermodynamics*; CRC Press: College Station, TX, USA, 1994.
46. Albahri, T.A.; Aljasmi, A.F. SGC method for predicting the standard enthalpy of formation of pure compounds from their molecular structures. *Thermochim. Acta* **2013**, *415*, 46–60. [CrossRef]

liquids

MDPI

Article

On the Diffusion of Anti-Tuberculosis Drugs in Cyclodextrin-Containing Aqueous Solutions

M. Melia Rodrigo [1], Ana M. T. D. P. V. Cabral [2], Sónia I. G. Fangaia [3], Afonso C. Nogueira [4], Artur J. M. Valente [4], Ana C. F. Ribeiro [4,*] and Miguel A. Esteso [1,5]

1 U.D. Química Física, Universidad de Alcalá, 28805 Alcalá de Henares, Spain; mmelia.rodrigo@uah.es (M.M.R.); mangel.esteso@ucavila.es (M.A.E.)
2 Faculty of Pharmacy, University of Coimbra, 3000-295 Coimbra, Portugal; acabral@ff.uc.pt
3 Institute of Implantology and Prosthodontics, Faculty of Medicine, University of Coimbra, 3000-075 Coimbra, Portugal; sfangaia@fmed.uc.pt
4 Institute of Molecular Sciences, Department of Chemistry, CQC, University of Coimbra, Rua Larga, 3004-535 Coimbra, Portugal; afonsocardoso127af@gmail.com (A.C.N.); avalente@ci.uc.pt (A.J.M.V.)
5 Universidad Católica de Ávila, Calle Los Canteros s/n, 05005 Ávila, Spain
* Correspondence: anacfrib@ci.uc.pt

Abstract: In this work, we propose a comprehensive experimental study of the diffusion of isoniazid, one of the first-line anti-tuberculosis drugs, in combination with another drug (ethambutol dihydrochloride) and with different cyclodextrins as carrier molecules, for facilitated transport and enhanced solubility. For that, ternary mutual diffusion coefficients measured by the Taylor dispersion method (D_{11}, D_{22}, D_{12}, and D_{21}) are determined for aqueous solutions containing isoniazid and different cyclodextrins (that is, α–CD, β–CD, and γ–CD) at 298.15 K. From the significant effect of the presence of these carbohydrates on the diffusion of this drug, interactions between these components are suggested. Support for this arose from models, which shows that these effects may be due to the formation of 1:1 (CDs:isoniazid) complexes.

Keywords: diffusion coefficient; cyclodextrins; isoniazid; solutions; transport properties

1. Introduction

Although there are many studies on the structure of free drugs or drugs complexed with different carrier vehicles, such as cyclodextrins (CDs), in different media involving techniques such as voltammetry, Raman, and NMR [1–3] there are still little data on the transport behavior of drugs in different aqueous media, such as those used in tuberculosis (TB).

Tuberculosis (TB) is one of the most life-threatening chronic bacterial infections that affects people around the world, which has led the World Health Organization (WHO) to declare this disease a global emergency; it causes around 3 million deaths annually [4,5]. Currently, to treat TB, medications are used, including first-line drugs with a high specificity for *Mycobacterium tuberculosis*, also known as Koch's bacillus, as currently recommended by international guidelines. Some examples of these drugs include isoniazid, ethambutol dihydrochloride, rifampicin, and pyrazinamide. To get rid of any persistent tuberculosis bacillus, TB patients will receive an initial extensive treatment with isoniazid, pyrazinamide, and rifampicin for the first two months, followed by another one with isoniazid and rifampicin for the next four months [6]. Since all these drugs are administered and absorbed orally, the extent of their absorption and their clinical effects depend on several factors, including the stomach and intestine's pH values. Several studies have showed that despite their therapeutic effectiveness, many variables are still difficult to control, which has been heightened by the increasing incidence of multidrug-resistant TB, caused by resistant strains of bacteria [4,5,7].

The mitigation of these effects can be achieved by increasing bioavailability, with a consequent reduction in the dose of the drug needed, as well as its stabilization [8,9]. Thus, other approaches might be developed to reach these objectives. A few decades ago, the use of cyclodextrins (Figure 1) in pharmaceutical formulations became quite common due to the possibility of them forming supramolecular compounds with drugs, contributing to an increase in their solubility and stability [9,10]. In brief, cyclodextrins are cyclic oligosaccharides with a hollow truncated cone shape and amphiphilic characteristics, i.e., while their cavity has hydrophobicity, their hydrophilic exterior gives cyclodextrins water solubility [11].

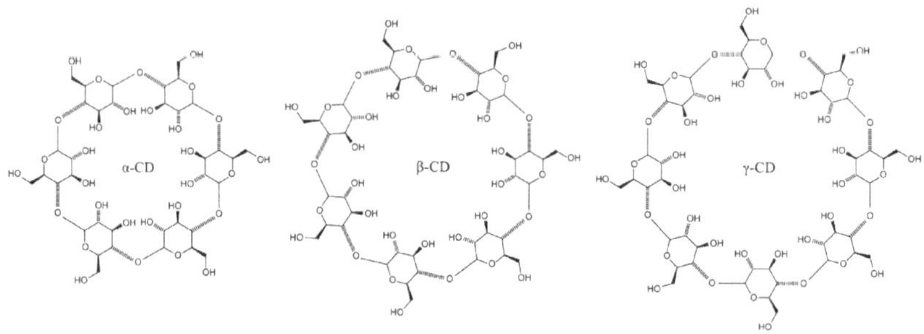

Figure 1. Representation of the structures of the most common natural cyclodextrins.

The measurement of the intra- and intermolecular diffusion of multicomponent systems is, among the several techniques used, highly effective at evaluating the formation of supramolecular complexes. In the present work, the mass transport of isoniazid in different multicomponent systems (i.e., isoniazid + ethambutol dydrochloride and isoniazid + cyclodextrins (α–, β–, and γ–)) has been studied. For that, intermolecular diffusion coefficients were obtained using the well-known Taylor dispersion technique [9]. This technique uses a small pulse of a solution in another flowing solution and its evaluation results in the obtainment of the binary and ternary coefficients of isoniazid with different compounds in the solution.

As mentioned before, isoniazid, the hydrazide of isonicotinic acid, is an antibiotic used for standard anti-tuberculosis therapy.

The effect of molar fractions of different components of aqueous solutions on the intermolecular diffusion coefficients of ternary systems were measured and quantified. The obtained values were discussed in terms of the contributions of their molecular components to the formation of a supramolecular structure and how the main and cross-diffusion coefficients are affected by those interactions.

2. Materials and Methods

2.1. Materials

Table 1 indicates all reagents, which were used as received, in the present work: α–cyclodextrin, β–cyclodextrin, γ–cyclodextrin, ethambutol dihydrochloride, and isoniazid. All of these compounds were used without further purification. The solutions were prepared in calibrated volumetric glass flasks, using ultrapure water as solvent (Millipore, Darmstadt, Germany, Milli-Q Advantage A10, specific resistance = 1.82×10^5 Ω m, at 298.15 K). The weighing was carried out using a Radwag AS 220C2 balance with an accuracy of ± 0.0001 g.

Table 1. Sample description.

Chemical Name	Source	CAS Number	Mass Fraction Purity [a]
Isoniazid	Sigma-Aldrich(Sigma-Aldrich Quimica S.L., Lisboa, Portugal)	54–85–3	>0.99
Ethambutol dihydrochloride	Sigma-Aldrich (Sigma-Aldrich Quimica S.L., Lisboa, Portugal)	1070–11–7	>0.99
α–Cyclodextrin [b]	Sigma-Aldrich (Sigma-Aldrich Quimica S.L., Lisboa, Portugal)	10016–20–3	$\geq$0.98
β–Cyclodextrin [c]	Sigma-Aldrich (Sigma-Aldrich Quimica S.L., Lisboa, Portugal)	7585–39–9	> 0.97
γ–Cyclodextrin [d]	Sigma-Aldrich (Sigma-Aldrich Quimica S.L., Lisboa, Portugal)	17465–86–0	$\geq$0.98
H_2O	Millipore-Q water (ρ = 1.82 $\times$ 10^5 Ω m at 298.15 K)	7732–18–5	

[a] The mass fraction purity is on a water-free basis; these data are provided by the suppliers. [b] α–cyclodextrin with a water mass fraction of 0.14. [c] β–cyclodextrin with a water mass fraction of 0.13. [d] γ–cyclodextrin with a water mass fraction of 0.10.

2.2. pH Measurements

The pH measurements of the solutions were carried out with a radiometer pH meter PHM 240 with an Ingold U457-K7 pH conjugated electrode; pH was measured in fresh solutions, and the electrode was calibrated immediately before measuring each set of solutions using the IUPAC-recommended pH 4.0, 7.0, and 10.0 buffers. From the pH meter calibration, a zero pH of 6.10 $\pm$ 0.07 and sensitivity higher than 98.5% were obtained.

2.3. Taylor Dispersion Method

The fundamentals and general technical details of the Taylor dispersion technique for measuring mutual diffusion coefficients are well described in the literature [12,13]. However, the specific details of the apparatus used in this study are described below.

Dispersion profiles were generated by injecting, at the start of each run, via a 6-port Teflon injection valve (Rheodyne, model 5020), 0.063 mL of solution into a laminar carrier stream of a slightly different composition at the entrance to a Teflon capillary dispersion tube of length 3048.0 ($\pm$0.1) cm and an internal radius 0.03220 $\pm$ (0.00003) cm. This tube and the injection valve were kept at 298.15 ($\pm$0.01) K in an air thermostat. The broadened distribution of the disperse samples was monitored at the tube outlet by a differential refractometer (Waters model 2410). The refractometer output voltages, $V(t)$, were measured at 5 s intervals by a digital voltmeter (Agilent 34401 A; Agilent Technologies, Santa Clara, CA, USA).

$$V(t) = V_0 + V_1 + V_{max}(t_R/t)^{1/2}\left[W_1 exp\left(-\frac{12D_1(t-t_R)^2}{r^2 t}\right) + (1-W_1)exp\left(-\frac{12D_2(t-t_R)^2}{r^2}\right)\right] \quad (1)$$

Equation (1) was fitted to the experimental dispersion profiles of these ternary solutions {isoniazid plus cyclodextrins (α, β or γ–CDs) or isoniazid plus ethambutol dydrochloride}, to attain the values of the different D_{ik} coefficients.

In Equation (1), D_1 and D_2 represent the eigenvalues of the matrix of the ternary D_{ik} coefficients; V_0, V_1, and $V_{\max}$ are the baseline voltage, the baseline slope, and the peak high, respectively; and W_1 and $(1 - W_1)$ are normalized pre-exponential factors.

The mutual diffusion of isoniazid-containing multicomponent solutions can be described by the following ternary diffusion equations (Equations (2) and (3)):

$$J_1 = -D_{11}\nabla c_1 - D_{12}\nabla c_2 \quad (2)$$

$$J_2 = -D_{21}\nabla c_1 - D_{22}\nabla c_2 \quad (3)$$

where J_1 and J_2 are the molar fluxes of isoniazid (1) and cyclodextrins or ethambutol dydrochloride (2), driven by the concentration gradients ∇c_1 and ∇c_2 of solute 1 and 2, respectively. The cross-diffusion coefficients D_{12} and D_{21} represent the coupled flux of each solute, driven by the concentration gradient in the other solute. The main diffusion coefficients D_{11} and D_{22} give the flux of each solute, driven by its own concentration gradient. A positive D_{ik} cross-coefficient ($i \neq k$) indicates the co-current coupled transport of solute i from regions of higher to lower concentrations of solute k. On the other hand, a negative D_{ik} coefficient indicates the counter-current coupled transport of solute i from regions of lower to higher concentrations of solute k.

3. Results and Discussion

3.1. Aqueous Isoniazid (1) + Cyclodextrin (2) Solutions

Table 2 shows the average experimental diffusion coefficients for aqueous isoniazid (component 1) + CDs (component 2). These data were obtained from at least four independent measurements. The main diffusion coefficients D_{11} and D_{22} were generally reproducible within $\pm(0.010 \times 10^{-9}$ m^2 s$^{-1})$, and their cross-coefficients were, in general, reproducible within about $\pm(0.020 \times 10^{-9}$ m^2 s$^{-1})$. The pH of the aqueous isoniazid solutions measured was approximately 6.50, very close to the pH value of the water (6.52) used in the preparation of these solutions.

Table 2. Ternary diffusion coefficients, D_{11}, D_{12}, D_{21}, and D_{22} of the aqueous isoniazid (component 1) + CDs (component 2) + solutions and their respective standard deviations of the means, S_D, at 25.00 °C.

C_1 [a]	C_2 [a]	X [b]	$D_{11} \pm S_D$ [c]	$D_{12} \pm S_D$ [c]	$D_{21} \pm S_D$ [c]	$D_{22} \pm S_D$ [c]	D_{12}/D_{22} [d]	D_{21}/D_{11} [e]
			Isoniazid (C_1) + $\alpha-$CD (C_2)					
0.000	0.010	0.000	0.787 ± 0.007	0.018 ± 0.011	0.001 ± 0.001	0.379 ± 0.001	0.047	0.001
0.005	0.005	0.500	0.785 ± 0.009	−0.033 ± 0.012	0.009 ± 0.022	0.380 ± 0.006	−0.087	0.011
0.010	0.000	1.000	0.786 ± 0.010	−0.046 ± 0.017	0.003 ± 0.014	0.382 ± 0.008	−0.120	0.004
			Isoniazid (C_1) + $\beta-$CD (C_2)					
0.000	0.010	0.000	0.714 ± 0.011	0.020 ± 0.003	0.008 ± 0.001	0.349 ± 0.001	0.057	0.011
0.005	0.005	0.500	0.740 ± 0.011	−0.029 ± 0.006	0.005 ± 0.013	0.350 ± 0.004	−0.083	0.007
0.010	0.000	1.000	0.787 ± 0.010	−0.040 ± 0.015	0.002 ± 0.014	0.352 ± 0.003	−0.114	0.002
			Isoniazid (C_1) + $\gamma-$CD (C_2)					
0.000	0.010	0.000	0.789 ± 0.010	0.011 ± 0.008	0.010 ± 0.006	0.329 ± 0.003	0.033	0.013
0.005	0.005	0.500	0.787 ± 0.009	0.010 ± 0.005	0.014 ± 0.006	0.310 ± 0.008	0.032	0.018
0.010	0.000	1.000	0.791 ± 0.008	0.005 ± 0.009	0.002 ± 0.003	0.332 ± 0.005	0.015	0.002

[a] C_1 and C_2 in mol dm^{-3}. [b] $X = C_1/C_2$ for ternary systems. [c] $D \pm S_D/(10^9$ m^2 s$^{-1})$ represents the mean diffusion coefficients of 4 to 6 replicate measurements. [d] D_{12}/D_{22} gives the number of moles of isoniazid counter-transported per mole of cyclodextrin. [e] D_{21}/D_{11} gives the number of moles of cyclodextrins co-transported per mole of isoniazid.

The binary diffusion coefficients of α–, β–, and γ–cyclodextrins at 0.01 mol dm^{-3} are as follows: (0.368 ± 0.005) × 10^{-9} m^2 s^{-1}, (0.337 ± 0.004) × 10^{-9} m^2 s^{-1}, and (0.321 ± 0.003) × 10^{-9} m^2 s^{-1}, respectively [14]. Taking into account that the binary diffusion coefficients for isoniazid are 0.826 ± 0.003 and 0.823 ± 0.004 × 10^{-9} m^2 s^{-1} (for initial concentrations equal to 0.001 and 0.010 mol dm^{-3}, respectively), and from the analysis of Table 2, it can be concluded that the main diffusion coefficient D_{11} is slightly lower than those obtained for the respective binary diffusion coefficients (within 4% deviations) [15], whilst the D_{22} values are similar to their respective binary diffusion coefficients [14,16,17].

Regarding the experimental D_{12} and D_{21} values, the results also indicate a weak interaction between isoniazid and alpha– and beta–cyclodextrins ($D_{12} < 0$). On the other hand, no interaction was detected with gamma–cyclodextrin. In fact, for the γ–CD-containing system, the D_{12} values are approximately zero, given the uncertainty of this method (that is, 2–3%). The supramolecular interactions between γ–CD and small molecules are complicated by the size of CD's cavity: 427 Å^3 [18]; such a volume allows for the occurrence of stoichiometric interactions greater than 1:1, as well as the total inclusion of the guest molecule. Therefore, we can hypothesize that such behavior, in terms of the cross-diffusion coefficients, might be due to the whole encapsulation of the isoniazid; consequently, the guest molecule is hindered by the host. This is supported by the behavior of the host–guest association of surfactants with cyclodextrins seen in the self-diffusion coefficients [19].

In contrast, for all cyclodextrins, their D_{21} values should be zero, within the experimental error. In other words, it can be inferred that cyclodextrin diffusion is not affected by the isoniazid gradient.

Considering that D_{12}/D_{22} gives the number of moles of isoniazid counter-transported per mole of β–CD, we may predict that, at these concentrations, one mole of diffusing β–CD counter-transports at most 0.12 mol of isoniazid. Using D_{21}/D_{11} values, at the same concentrations, we can expect that a mole of diffusing isoniazid co-transports, at most, 0.092 mol of β–CD.

Support for the possible coupled diffusion resulting from the formation of inclusion complexes can be obtained from a more detailed treatment of the diffusion on multicomponent systems, developed and described in the literature (e.g., [19,20]). Assuming that there is an inclusion equilibrium between isoniazid and β–CD (Equations (4) and (5)) (or α–CD) and using the limiting diffusion coefficients of the free and complexed species indicated in Table 3, an estimation of the respective constants of equilibrium were computed and are equal to $K = 20\ \mathrm{M}^{-1}$ and $K = 22\ \mathrm{M}^{-1}$, respectively. It should also be stressed that the measurement of the interdiffusion coefficients does not allow us to conclude whether the interaction occurs through a preferential rim of cyclodextrins [21].

$$\mathrm{ISO\ (aq)} + \beta\text{–}\mathrm{CD\ (aq)} \leftrightarrows \mathrm{ISO}\text{–}\beta\text{–}\mathrm{CD\ (aq)} \tag{4}$$

$$K = \frac{[\mathrm{ISO} - \beta\text{–CD}]}{[\mathrm{ISO}]\,[\beta\text{–CD}]} \tag{5}$$

Table 3. Limiting diffusion coefficients, D_s, of free and complexed species at T = 298.15 K.

Species	$D_s/(10^{-9}\ \mathrm{m^2\ s^{-1}})$
Isoniazid	0.826 [a]
β–CD	0.326 [b]
α–CD	0.353 [c]
ISO–β–CD	0.320 [d]

[a] [15]; [b] [16]; [c] [14]; [d] considering that the volume of the complexed species (Isoniazid–β–CD) is the sum of the volumes of the isoniazid molecules and β–cyclodextrin molecules isolated, the diffusion coefficient of these complexes was estimated from the known values of D_{ISO} and $D_{\beta\text{-CD}}$ using $(D_{\mathrm{ISO}}{}^{-3} + D_{\beta\text{-CD}}{}^{-3})^{-1/3}$.

3.2. Aqueous Isoniazid (1) + Ethambutol Dihydrochloride (2) Solutions

Table 4 shows the ternary mutual diffusion coefficients for dilute aqueous isoniazid (1) + ethambutol dihydrochloride (2) solutions containing ($C_1 + C_2$) = 0.010 mol dm^{-3}. The reported mutual diffusion coefficients are the average of four to six replicate D_{ik} measurements of each composition. The main diffusion coefficients, D_{11} and D_{22}, were generally reproducible within $\pm 0.02 \times 10^{-9}$ m^2 s^{-1}. The cross-coefficients were reproducible within $\pm 0.10 \times 10^{-9}$ m^2 s^{-1}. For the sake of comparison, the binary diffusion coefficients of ethambutol dihydrochloride (0.001 and 0.010 mol dm^{-3}), in aqueous solutions, are $(1.055 \pm 0.003) \times 10^{-9}$ m^2 s^{-1} and $(0.907 \pm 0.002) \times 10^{-9}$ m^2 s^{-1}, respectively [15].

Table 4. Ternary diffusion coefficients, D_{11}, D_{12}, D_{21}, and D_{22} of aqueous isoniazid (component 1) + ethambutol dihydrochloride (component 2) + solutions and the respective standard deviations of their means, S_D, at 25.00 °C.

C_1 [a]	C_2 [a]	X [b]	$D_{11} \pm S_D$ [c]	$D_{12} \pm S_D$ [c]	$D_{21} \pm S_D$ [c]	$D_{22} \pm S_D$ [c]	D_{12}/D_{22} [d]	D_{21}/D_{11} [e]
Isoniazid (C_1) + Ethambutol dihydrochloride (C_2)								
0.000	0.010	0.000	0.775 ± 0.012	0.011 ± 0.003	0.008 ± 0.037	0.903 ± 0.001	0.012	0.010
0.005	0.005	0.500	0.780 ± 0.009	0.109 ± 0.012	0.009 ± 0.022	0.890 ± 0.005	0.122	0.011
0.010	0.000	1.000	0.785 ± 0.003	0.478 ± 0.099	0.014 ± 0.008	0.815 ± 0.021	0.590	0.018

[a] C_1 and C_2 in mol dm^{-3}. [b] $X = C_1/C_2$ for ternary systems. [c] $D \pm S_D/(10^9\ m^2\ s^{-1})$ represents the mean diffusion coefficients of 4 to 6 replicate measurements. [d] D_{12}/D_{22} gives the number of moles of isoniazid co-transported per mole of cyclodextrin. [e] D_{21}/D_{11} gives the number of moles of cyclodextrins co-transported per mole of isoniazid.

From the analysis of Table 4, it can be seen that the ratio D_{12}/D_{22}, at tracer concentrations of isoniazid and 0.010 mol dm^{-3} ethambutol dihydrochloride, is almost zero (i.e., 0.012). Additionally, all D_{21}/D_{22} values are almost zero. However, we can highlight the higher positive value obtained for D_{12}/D_{22} at tracer concentrations of ethambutol dihydrochloride and isoniazid at 0.010 mol dm^{-3}. We can say that in these circumstances, the coupled diffusion that occurs is not negligible. One mole of diffused ethambutol dihydrochloride co-transports up to 0.6 mol of isoniazid. These interactions can be harmful to living organisms, since both drugs are used together to treat tuberculosis and their effectiveness may be shortened, with a consequent need for the use of an overdosage. Since there is this unforeseen transport, the concentration gradient of isoniazid may be smaller in the target areas of the compound in the human body, which may render it ineffective in fulfilling the purpose of the treatment.

4. Conclusions

Based on the measured ternary diffusion coefficients of the aqueous ternary systems studied (that is, isoniazid/cyclodextrins and isoniazid/ethambutol dihydrochloride), it can be concluded that coupled diffusion does take place, affecting the diffusion behavior of this anti-tuberculosis drug in those media. The data obtained for the systems isoniazid/β–CD and isoniazid/α–CD were interpreted on the basis of the formation of complexes between isoniazid and β–cyclodextrin molecules, whose K values, obtained for these equilibria, were 20 mol dm^{-3} and 22 mol dm^{-3}.

In relation to isoniazid and ethambutol, through the analysis of secondary diffusion coefficients, a favorable transport of isoniazid in the direction of the ethambutol concentration gradient ("co-transport") was found. This interaction can be harmful, as the two medicines in question are used together to treat tuberculosis and their effectiveness can be altered. Once this unforeseen transport occurs, the concentration gradient of isoniazid may be lower in the target areas of this compound in the human body, which may make it ineffective in achieving its treatment purpose.

Author Contributions: Conceptualization, M.M.R., A.M.T.D.P.V.C., A.C.F.R., A.J.M.V. and M.A.E.; Methodology, M.M.R., A.C.N., S.I.G.F. and A.C.F.R.; Software M.M.R., S.I.G.F., A.C.F.R. and A.M.T.D.P.V.C.; Validation, S.I.G.F., A.C.F.R., A.J.M.V., A.M.T.D.P.V.C., A.C.N. and M.A.E.; Formal Analysis, S.I.G.F., A.C.F.R., A.J.M.V., A.C.N. and M.A.E.; Investigation, S.I.G.F., A.C.N., A.C.F.R., A.M.T.D.P.V.C., A.J.M.V., M.M.R. and M.A.E.; Resources, S.I.G.F., A.C.F.R., A.J.M.V., M.M.R. and M.A.E.; Data Curation, S.I.G.F., A.C.F.R., A.J.M.V., M.M.R. and M.A.E.; Writing—Original Draft Preparation, S.I.G.F., A.C.F.R. and M.M.R.; Writing—Review and Editing, S.I.G.F., A.M.T.D.P.V.C., A.C.F.R., A.J.M.V., M.M.R. and M.A.E.; Visualization, S.I.G.F., A.M.T.D.P.V.C., A.C.F.R., A.C.N., A.J.M.V., M.M.R. and M.A.E.; Supervision, S.I.G.F. and A.C.F.R.; Project Administration, A.M.T.D.P.V.C. and A.C.F.R.; Funding Acquisition, M.M.R., A.J.M.V., A.M.T.D.P.V.C., S.I.G.F., A.C.F.R. and M.A.E. All authors have read and agreed to the published version of the manuscript.

Funding: The authors are grateful for funding from the Coimbra Chemistry Centre, which is supported by the Fundação para a Ciência e a Tecnologia (FCT), Portuguese Agency for Scientific Research, through the projects UID/QUI/UI0313/2013 and the COMPETE Programme (Operational Programme for Competitiveness).

Data Availability Statement: The original contributions presented in this study are included in the article; further inquiries can be directed to the corresponding author.

Conflicts of Interest: The authors declare no conflicts of interest.

References

1. Mazurek, A.H.; Szeleszczuk, Ł. A Review of Applications of Solid-State Nuclear Magnetic Resonance (SsNMR) for the Analysis of Cyclodextrin-Including Systems. *Int. J. Mol. Sci.* **2023**, *24*, 3648. [CrossRef] [PubMed]
2. González-Méndez, I.; Aguayo-Ortiz, R.; Sorroza-Martínez, K.; Solano, J.D.; Porcu, P.; Rivera, E.; Dominguez, L. Conformational Analysis by NMR and Molecular Dynamics of Adamantane-Doxorubicin Prodrugs and Their Assemblies with β–Cyclodextrin: A Focus on the Design of Platforms for Controlled Drug Delivery. *Bioorganic Med. Chem.* **2020**, *28*, 115510. [CrossRef] [PubMed]
3. Pathania, V.; Garg, A.; Kaur, N.; Neha. Molecular Interaction Studies of Antituberculosis Drug Isoniazid in Aq. β–Cyclodextrin Solution: A Volumetric, Spectroscopic and Molecular Docking Approach. *J. Chem. Thermodyn.* **2024**, *189*, 107194. [CrossRef]
4. World Health Organization. *Global Tuberculosis Report 2022*; World Health Organization: Geneva, Switzerland, 2022. Available online: https://www.who.int/teams/global-tuberculosis-programme/tb-reports/global-tuberculosis-report-2022 (accessed on 3 July 2024).
5. World Health Organization. *Global Tuberculosis Report 2013*; World Health Organization: Geneva, Switzerland, 2013; ISBN 978 92 4 156465 6.
6. Almeida Da Silva, P.E.; Palomino, J.C. Molecular Basis and Mechanisms of Drug Resistance in Mycobacterium Tuberculosis: Classical and New Drugs. *J. Antimicrob. Chemother.* **2011**, *66*, 1417–1430. [CrossRef] [PubMed]
7. Saviola, B. Mycobacterium Tuberculosis Adaptation to Survival in a Human Host. In *Tuberculosis—Current Issues in Diagnosis and Management*; InTech: Chennai, India, 2013.
8. Dandawate, P.; Vemuri, K.; Venkateswara Swamy, K.; Khan, E.M.; Sritharan, M.; Padhye, S. Synthesis, Characterization, Molecular Docking and Anti-Tubercular Activity of Plumbagin—Isoniazid Analog and Its β—Cyclodextrin Conjugate. *Bioorganic Med. Chem. Lett.* **2014**, *24*, 5070–5075. [CrossRef] [PubMed]
9. Aiassa, V.; Garnero, C.; Zoppi, A.; Longhi, M.R. Cyclodextrins and Their Derivatives as Drug Stability Modifiers. *Pharmaceuticals* **2023**, *16*, 1074. [CrossRef]
10. Kali, G.; Haddadzadegan, S.; Bernkop-Schnürch, A. Cyclodextrins and Derivatives in Drug Delivery: New Developments, Relevant Clinical Trials, and Advanced Products. *Carbohydr. Polym.* **2024**, *324*, 121500. [CrossRef] [PubMed]
11. Jansook, P.; Ogawa, N.; Loftsson, T. Cyclodextrins: Structure, Physicochemical Properties and Pharmaceutical Applications. *Int. J. Pharm.* **2018**, *535*, 272–284. [CrossRef] [PubMed]
12. Tyrrell, H.J.V.; Harris, K.R. *Diffusion in Liquids: A Theoretical and Experimental Study*; Butterworths: London, UK, 1984.
13. Fangaia, S.I.G.; Nicolau, P.M.G.; Guerra, F.A.D.R.A.; Rodrigo, M.M.; Ribeiro, A.C.F.; Valente, A.J.M.; Santos, J.R.C.; Marques, J.M.C.; Esteso, M.A. The Behaviour of Aluminium Ions in Artificial Saliva and the Impact of the Chlorhexidine Digluconate on Its Removal—A Diffusion Model. *J. Mol. Liq.* **2022**, *353*, 118749. [CrossRef]
14. Ribeiro, A.C.F.; Valente, A.J.M.; Santos, C.I.A.V.; Prazeres, P.M.R.A.; Lobo, V.M.M.; Burrows, H.D.; Esteso, M.A.; Cabral, A.M.T.D.P.V.; Veiga, F.J.B. Binary Mutual Diffusion Coefficients of Aqueous Solutions of α–Cyclodextrin, 2-Hydroxypropyl-α–Cyclodextrin, and 2-Hydroxypropyl-β–Cyclodextrin at Temperatures from (298.15 to 312.15) K. *J. Chem. Eng. Data* **2007**, *52*, 586–590. [CrossRef]
15. Ribeiro, A.C.F.; Santos, A.C.G.; Lobo, V.M.M.; Veiga, F.J.B.; Cabral, A.M.T.D.P.V.; Esteso, M.A.; Ortona, O. Binary Mutual Diffusion Coefficients of Isoniazid Aqueous Solutions at (298.15 and 310.15) K. *J. Chem. Eng. Data* **2009**, *54*, 3235–3237. [CrossRef]
16. Ribeiro, A.C.F.; Leaist, D.G.; Esteso, M.A.; Lobo, V.M.M.; Valente, A.J.M.; Santos, C.I.A.V.; Cabral, A.M.T.D.P.V.; Veiga, F.J.B. Binary Mutual Diffusion Coefficients of Aqueous Solutions of β–Cyclodextrin at Temperatures from 298.15 to 312.15 K. *J. Chem. Eng. Data* **2006**, *51*, 1368–1371. [CrossRef]
17. Ribeiro, A.C.F.; Santos, C.I.A.V.; Valente, A.J.M.; Ascenso, O.S.; Lobo, V.M.M.; Burrows, H.D.; Cabral, A.M.T.D.P.V.; Veiga, F.J.B.; Teijeiro, C.; Esteso, M.A. Some Transport Properties of γ–Cyclodextrin Aqueous Solutions at (298.15 and 310.15) K. *J. Chem. Eng. Data* **2008**, *53*, 755–759. [CrossRef]
18. Szejtli, J. Introduction and General Overview of Cyclodextrin Chemistry. *Chem. Rev.* **1998**, *98*, 1743–1754. [CrossRef] [PubMed]
19. Paduano, L.; Sartorio, R.; Vitagliano, V.; Albright, J.G.; Miller, D.G. Measurement of the Mutual Diffusion Coefficients at One Composition of the Four-Component System α–Cyclodextrin-L-Phenylalanine-Monobutylurea-Water at 25 °C. *J. Phys. Chem.* **1992**, *96*, 7478–7483. [CrossRef]

20. Paduano, L.; Sartorio, R.; Vitagliano, V.; Albright, J.G.; Miller, D.G.; Mitchell, J. Diffusion Coefficients in Systems with Inclusion Compounds. 1. α–Cyclodextrin-L-Phenylalanine-Water at 25 °C. *J. Phys. Chem.* **1990**, *94*, 6885–6888. [CrossRef]
21. Sun, Z.; Zheng, L.; Wang, K.; Huai, Z.; Liu, Z. Primary vs. Secondary: Directionalized Guest Coordination in β–Cyclodextrin Derivatives. *Carbohydr. Polym.* **2022**, *297*, 120050. [CrossRef] [PubMed]

liquids

MDPI

Article

A Machine Learning Free Energy Functional for the 1D Reference Interaction Site Model: Towards Prediction of Solvation Free Energy for All Solvent Systems

Jonathan G. M. Conn, Abdullah Ahmad and David S. Palmer *

Department of Pure and Applied Chemistry, University of Strathclyde, Thomas Graham Building, 295 Cathedral Street, Glasgow G1 1XL, UK
* Correspondence: david.palmer@strath.ac.uk

Abstract: Understanding the interactions between solutes and solvents is vital in many areas of the chemical sciences. Solvation free energy (SFE) is an important thermodynamic property in characterising molecular solvation and so accurate prediction of this property is sought after. The One-Dimensional Reference Interaction Site Model (RISM) is a well-established method for modelling solvation, but it is known to yield large errors in the calculation of SFE. In this work, we show that a single machine learning free energy functional for RISM can accurately model solvation thermodynamics in multiple solvents. A convolutional neural network is trained on solvation free energy density functions calculated by RISM for small organic molecules in approximately 100 different solvent systems. We achieve an average RMSE of 1.41 kcal/mol and an R^2 of 0.89 across all solvent systems. We also compare the performance for the most and least commonly represented solvents and show that higher accuracy is generally seen with higher volumes of data, with RMSE values of 0.69–1.29 kcal/mol and R^2 values of 0.78–0.97 for solvents with more than 50 data points. We have shown that machine learning can greatly improve solvation free energy predictions in RISM, while demonstrating that the methodology is generalisable across solvent systems. This represents a significant step towards a universal machine learning SFE functional for RISM.

Keywords: solvation free energy; RISM; machine learning; solvent systems; convolutional neural network

1. Introduction

Solvation Free Energy

The process of solvation, as defined by Ben-Naim, is the transfer of one solute molecule from a fixed position in the ideal gas phase to a fixed position in the liquid phase at a given temperature and pressure [1]. The solvation free energy (SFE) is the reversible work associated with the transfer of the solute molecule from the gas phase to the liquid phase. SFE is important in many areas, such as drug discovery, as it may be used in the predictions of solubility and octanol-water partition coefficient (logP) [2–4], and environmental chemistry, as it is useful in understanding pollutant distributions in different aqueous environments [5,6].

Generally, the SFE of solutes with low vapour pressures must be obtained indirectly from separate measurements of the solubility and pure compound vapour pressure by a thermodynamic cycle via the gas-phase. Experimental SFE determination is often a lengthy and difficult process, so many computational approaches have been developed for its prediction.

SFE prediction methods can be separated into implicit and explicit methods. Implicit methods are characterised by a simplified representation of the solvent in which no solvent particles are explicitly modelled. Continuum models are the most common form of implicit methods, in which the solvent is represented by a homogeneous polarisable medium described by a dielectric constant. A commonly used implicit model was developed by Cramer et al. [7], a universal model named SMD, which uses full solute electron density in place of atomic partial charges. The SFE consists of two components: bulk electrostatics calculated using the nonhomogeneous Poisson equation, and a cavity-dispersion term, which calculates the free energy required for forming a cavity in the solvent in which to place the solute, and short-range interactions between solute and solvent within the region of the first solvation shell. Cramer et al. also proposed the SMx models, the latest of which is SM12 [8–10], which implements the Generalised-Born (GB) approximation in conjunction with SASA to compute solvent effects. This model was parameterised for many different thermodynamic properties, including SFE predictions for both neutral and ionised species in both aqueous and organic solvents. The PCM by Tomasi et al. [11] implements a non-uniform dielectric treatment by accounting for the polarisability of the solvent and local polarisation effects around the solute. The PCM has been further developed in many ways, including the conductor-like PCM [12], which implements a boundary condition for conductor-like solvation screening, and the integral equation formalism [13,14], which allows bespoke dielectric treatment depending on the nature of chemical species. Additionally, the COSMO [15] approach calculates interaction energies between the solute and the dielectric continuum by determining the response of the continuum to screening charge distributions formed at the solvent accessible surface of the cavity formed around the solute. This boundary is discretised and the charge distributions are calculated, with exact solutions achievable for electric conductors.

Whilst implicit methods do provide reasonable accuracy and are computationally efficient, many physical phenomena which affect the solvation process, such as the reorientation of particles, are crudely approximated. This results in failure to accurately capture many local effects. The dielectric constant treatment may exhibit shortcomings in capturing local interactions which would be observed between solute and solvent molecules, such as electrostatic interactions at the solute–solvent interface, charge distribution changes in solvent molecules, and spatially induced changes to the solvent dielectric constant due to solute behaviour.

Explicit methods offer a more rigorous approach to simulating chemical systems by explicitly modelling both the solute and solvent particles. This allows the simulation of specific solute–solvent interactions in which local effects may be thoroughly probed. This additional level of molecular representation allows for a more physically realistic approach to the simulation of chemical systems. Populating a virtual space with a large number of additional simulated particles, i.e., solvent molecules, increases the computational expense considerably when compared to implicit models. Explicit solvent simulations are not generally amenable to high-throughout computing because of their computational expense. Additionally, the values of structural and thermodynamic data obtained from simulation may depend on the extent to which the configurational space has been sampled, with the result that there is a trade-off between efficiency (simulation time) and repeatability that introduces statistical noise to property estimates.

Explicit solvent methods for the calculation of SFE are commonly based on molecular dynamics (MD) or Monte Carlo simulations. Alchemical free energy calculations can be used to compute SFE. Sherman et al. [16] employed MD/free energy perturbation simulations to calculate the SFE as a transfer energy of a solute molecule transitioning from the ideal gas phase to the solvent, reporting an error of 1.1 kcal/mol. Leung et al. [17] used ab initio MD alongside thermodynamic integration to calculate the SFE of various ions to within approximately 2.5% of experimental observations. Although methods based on explicit solvent simulation (MD/MC) can provide accurate estimates of solvation free energies for small organic solutes in favourable circumstances, they are less reliable for larger, polyfunctional organic molecules, where issues caused by inaccurate forcefields or incomplete sampling of conformational/configurational space become more pronounced. A good example of these issues is provided by the results of the SAMPL challenges, in which entrants were asked to make blinded predictions of the HFEs of pharmaceutically and environmentally relevant molecules [18–20]. The best predictions were in the range RMSE = 2.5–3.5 kcal/mol, which equates to an ∼2 log unit error in the related equilibrium physicochemical property (e.g., solubility, pKa, etc.). As the authors of these challenges observed, "Accurate calculations of more complex properties or events will remain over the horizon until these more basic values [HFEs] can be predicted with greater accuracy" [18]. Moreover, computing power limits the size of the system that can be reliably modelled using explicit solvent simulation, and even for smaller systems may limit the number of simulations that can be run. Given that every solute and solvent molecule within the system is modelled, many interactions must be computed at any given system state, particularly when compared to implicit methods. This computational expense increases with larger systems, requiring the use of more powerful processing units. For large systems, additional considerations are often required to improve the computational feasibility, and techniques such as coarse-graining have proven effective in this endeavour [21,22].

Comparisons between the use of implicit and explicit models for the prediction of SFE have been the focus of many studies. Shivakumar et al. [23] calculated the absolute hydration free energies of 239 ligands commonly used in drug-like compounds using both explicit and implicit water models. It was found that the explicit method generally outperformed the implicit method, but that the implicit method had good agreement with the explicit method when using the Poisson–Boltzman (PB) model. Additionally, the implicit method was much more sensitive to parameter changes, such as atomic charges and dielectric coefficients. The ESE-GB-DNN method proposed by Vyboishchikov employs terms calculated from a Generalized-Born implicit solvent model as input to a deep neural network for the calculation of SFE [24]. Steinmann et al. [25] focused on an organic adsorbate chemisorbed at a metal–liquid interface and compared the SFE calculations of an implicit and explicit approach. It was found that each method yielded the same general conclusions in terms of the stability profiles of different chemical species, but that the explicit method had generally better agreement with experimental findings. Interestingly, neither method accurately described the adsorption strength of water on the metal surface. VanderSpoel et al. [26] explored the use of implicit and explicit approaches to calculate the SFE of organic solutes in organic solvents, using various different methods. It was observed that the implicit methods were less accurate than explicit methods, but that the PB implicit method was in much closer agreement to both the explicit and experimental results than the GB variants. Errors of 15 kJ/mol and 6 kJ/mol were observed for the implicit GB and PB methods, respectively, while the explicit method yielded an error of 4.8 kJ/mol.

An alternative approach for modelling molecular solvation is the Reference Interaction Site Model (RISM), initially proposed by Chandler et al. [27], which captures specific solute–solvent interactions while circumventing the need for large-scale modelling of many solute and solvent molecules. The RISM models are a method derived from integral equation theory (IET), which captures how a solute and solvent interact in such a way that the solvent density distribution around a solute molecule can be calculated. RISM comes in the form of 1D-RISM and 3D-RISM, which differ in the dimensionality of the integral equations involved. The approach of 1D-RISM solves integral equations between solute and solvent sites (atoms) over a radial distance, while 3D-RISM solves integral equations over a set of 3D coordinates, which allows for modelling of the spatial distributions of solvent sites around a solute molecule. With 1D-RISM models, both the solute and the solvent molecules are represented as a set of sites (atoms), while 3D-RISM considers the full solute molecule as a single entity. In recent years, 3D-RISM has been more commonly used than 1D-RISM.

In 2010, Palmer et al. [28] showed that 3D-RISM lacked the ability to accurately predict SFE, resulting in errors of approximately 17 kcal/mol, and so proposed a universal correction. Based on earlier findings by Chuev et al. [29], Palmer et al. [30], and Ratkova et al. [31], where it was found that the partial molar volume correlated with the difference between experimental SFE and that predicted by 1D-RISM methods, the partial molar volume was used with a scaling coefficient and a bias correction term to improve the 3D-RISM SFE predictions. Truchon et al. [32] showed that the non-polar component of 3D-RISM is the cause for inaccurate SFE prediction, while the electrostatic component requires no correction. They implemented a correction factor such that the non-polar component better describes repulsions between the oxygen in water and the solute such that solvent cavity formation is better described. Sergiievsky et al. [33] developed a pressure correction (PC) and an expansion upon this correction (PC+), originally for molecular density functional theory, in order to compensate for the overpressure resulting from homogeneous reference fluid approximation. Misin et al. then applied these corrections to 3D-RISM calculations in SFE prediction, finding that the accuracy did increase with the corrections [34]. This correction was later applied alongside solvent LJ approximations for coarse-grained solvent models and non-polar or weakly polar solvents [35], followed by an additional study in the use of this pressure correction as a means to describe the effects that salt has on solutes [36].

The standard 1D-RISM methods commonly produce errors of up to 20 kcal/mol in the prediction of SFE and so are not routinely used for the calculation of solvation thermodynamics. Ratkova and Fedorov [37] developed a model in which hydration free energy (HFE) calculations using 1D-RISM are improved using a cheminformatics-based correction. A multilinear regression model is trained on molecular features, such as the partial molar volume and various structural descriptors, to offset the error between the experimental and 1D-RISM-calculated HFE values. Fowles et al. [38] developed a machine learning approach trained on 1D-RISM calculations to predict the SFE in water, chloroform, and carbon tetrachloride. RMSE values with the 1D-RISM calculations themselves ranged between approximately 16 and 44 kcal/mol for two of three investigated approaches, with the third approach showing much lower error of approximately 1.8 to 6 kcal/mol. After training a machine learning model on these data, the RMSE values then dropped to below 1 kcal/mol. This approach was later applied to a multi-output model for the simultaneous prediction of the free energy, enthalpy, and entropy of hydration [39].

In this work, we use a machine learning model trained on the output of 1D-RISM calculations to accurately predict the SFE of various solute/solvent combinations. Additionally, we expand the number of solvents with the aim to make it feasible to make accurate SFE predictions for any combination of solute and solvent using this approach.

2. Theory

2.1. 1D-RISM

RISM is a method derived from the Integral Equation Theory (IET) of liquids, which is an implicit solvent model which uniquely describes interactions between solute and solvent particles to yield radial distribution functions (RDFs), which describe solvent density around the solute. This approach is effective as it provides a sufficient molecular description of solute molecules while being less computationally intensive than explicit solvent simulation, allowing for fast calculations of SFE.

The IET of atomic liquids is described by the Ornstein–Zernike (OZ) equation, which defines the total correlation function between two spherically symmetric particles in a homogeneous solvent, shown in Equation (1):

$$h(r_{12}) = c(r_{12}) + \rho \int c(r_{13})h(r_{32})dr_3, \tag{1}$$

where $h(r_{ij})$ and $c(r_{ij})$ are the total and direct correlation functions, respectively, between particles i and j, and ρ is the solvent density. The OZ equation is only applicable to simple atomic liquids, which is insufficient for chemical applications where more complex liquids are routinely used. Hence, the OZ equation has been generalised to the Molecular Ornstein–Zernike (MOZ) equation, which considers non-spherical molecules and is shown in Equation (2):

$$h(r_{12}, \Theta_1, \Theta_2) = c(r_{12}, \Theta_1, \Theta_2) + \frac{\rho}{\mathbb{Z}} \int c(r_{13}, \Theta_1, \Theta_3)h(r_{32}, \Theta_3, \Theta_2)dr_3 d\Theta_3, \tag{2}$$

where r_{12} and Θ_1 and Θ_2 are the displacement and the orientation, respectively, of particles 1 and 2. In this case, r is defined by a set of coordinates, (x, y, z), and Θ is defined by three Euler angles, (ψ, θ, φ). $\mathbb{Z}$ is equal to 4π for linear molecules, where only two angles are required to describe the molecule, or $8\pi^2$ for non-linear molecules, where three angles are required.

The 1D-RISM approach uses a one-dimensional approximation of the MOZ equation, based on spherically symmetric site–site correlation functions. In this approach, a modelled molecule is considered as a set of sites, each of which is spherically symmetric, where every atom is a single site. There are three types of correlation function present in the RISM theory: intramolecular correlation functions, $\omega(r)$, total correlation functions, $h(r)$, and direct correlation functions, $c(r)$, each of which depends only on the radial distance, r, between sites and has no angular dependencies.

In RISM calculations, the derived integral equations are solved in combination with a closure relation. The closure relation imposes limitations on the derived equations and ensures that there is self-consistency between pair correlation functions and the direct correlation function. To obtain solutions which are accurate beyond the mean-field approximation, it may be necessary for the integration over an effectively infinite series of integrals, which is not computationally feasible. Hence, a bridge functional is introduced to the closure relation, which quantifies the spatial contributions in the solvent density distribution and accounts for the correlations between solvent molecules. This reduces the necessity for an infinitely expansive integral series by helping to describe the part of the direct correlation function which cannot be easily determined from the solvent structure.

The structure of a molecule is described by the intramolecular correlation functions, $\omega(r)$. For two sites in a given molecule, denoted u and u', the intramolecular correlation function is shown in Equation (3):

$$\omega_{uu'}(r) = \frac{\delta(r - r_{uu'})}{4\pi r_{uu'}^2}, \tag{3}$$

where $r_{uu'}$ is the distance between the two sites and $\delta(r - r_{uu'})$ is the Dirac delta function. A single molecule will be described by a set of intramolecular correlation functions as these functions are site-pairwise in their description.

Intermolecular correlations between solute and solvent molecules are described by both pairwise total correlation functions and direct correlation functions, and are shown in Equation (4) for the solute site u and the solvent site v:

$$h_{uv}(r) = g_{uv} - 1, \tag{4}$$

where $h_{uv}(r)$ is the total correlation function and $g_{uv}(r)$ is the RDF of the solvent site around the solute site. Additionally, a total correlation function between sites of different solvent molecules, $h_{vv'}(r)$, is used to describe the distribution of solvent sites, v', of one molecule around the solvent site, v, of another solvent molecule. Hence, both solute–solvent and solvent–solvent total correlation functions are calculated.

In 1D-RISM, the total and direct correlation functions are related to one another through the derived integral equations, shown in Equation (5):

$$h_{uv}(r) = \sum_{u'=1}^{M} \sum_{v'=1}^{N} \int_{R^3} \int_{R^3} \omega_{uu'}(|\mathbf{r}_1 - \mathbf{r}'|) \times c_{u'v'}(|\mathbf{r}' - \mathbf{r}''|) \chi_{vv'}(|\mathbf{r}'' - \mathbf{r}_2|) d\mathbf{r}' d\mathbf{r}'', \tag{5}$$

where $r = |\mathbf{r}_1 - \mathbf{r}_2|$ and $\chi_{vv'}(r)$ are the bulk solvent susceptibility functions and M and N are the number of sites of the solute and solvent, respectively. Bulk solvent sites have mutual correlations which are described by $\chi_{vv'}(r)$, which may be obtained via the solvent total correlation function, $h_{vv'}(r)$, and a 3D structure of the solvent molecule, and therefore, the intramolecular correlation function, $\omega_{vv'}(r)$. Equation (6) shows the mutual correlation function:

$$\chi_{vv'}(r) = \omega_{vv'}(r) + \rho h_{vv'}(r), \tag{6}$$

where ρ is the bulk number density of the solvent.

To allow the total and direct correlation functions to be obtained by numerical solution of the RISM equations, $M \times N$ closure relations are introduced, the general form of which is shown in Equation (7):

$$h_{uv}(r) = e^{(-\beta u_{uv}(r)+\gamma_{uv}(r)+B_{uv}(r))} - 1, \tag{7}$$

where $u = 1, ..., M$, $v = 1, ..., N$, $\beta = 1/k_B T$, $u_{uv}(r)$ is a pair interaction potential between solute and solvent sites, and $B_{uv}(r)$ is a site–site bridge functional.

Generally, the interaction potential consists of short-range and long-range components. It is common for the short-range interactions to be described by an LJ potential, while the long-range interactions are described by an electrostatic term. Equation (8) shows the components of the interaction potential:

$$\begin{aligned} u_{uv}(r) &= u_{uv}^{el}(r) + u_{uv}^{LJ}(r), \\ u_{uv}^{el}(r) &= \frac{q_u q_v}{r}, \\ u_{uv}^{LJ}(r) &= 4\varepsilon_{uv}^{LJ}\left[\left(\frac{\sigma_{uv}^{LJ}}{r}\right)^{12} - \left(\frac{\sigma_{uv}^{LJ}}{r}\right)^{6}\right], \end{aligned} \tag{8}$$

where q_u and q_v are the partial charges of the solute and solvent sites, respectively, of interest, and ε_{uv}^{LJ} and σ_{uv}^{LJ} are the solute–solvent LJ parameters.

A bridge functional can be selected to increase the accuracy of the closure relations by accounting for higher-order correlations that are not considered in the mean-field approximation. The simplest bridge functional is the Hypernetted-chain (HNC) approximation, which sets $B_{uv}(r) = 0$. While this may be the most computationally inexpensive approximation, this results in a lesser degree of control in the exponent of Equation (7), which often leads to convergence issues, in which case, no solution is found. An improvement to this approximation is the Kovalenko–Hirata (KH) closure, which linearises the exponent above a threshold constant, C. The linearisation is shown in Equation (9):

$$h_{uv}(r) = \begin{cases} e^{\Xi_{uv}(r)} - 1 & \Xi_{uv}(r) \leq C, \\ \Xi_{uv}(r) + e^C - C - 1 & \Xi_{uv}(r) > C, \end{cases} \tag{9}$$

where $\Xi_{uv}(r) = -\beta u_{uv}(r) + \gamma_{uv}(r)$. When C tends to infinity, the KH closure becomes equal to the HNC closure. This approach allows the capturing of short-range interactions, while still setting them to 0 via the HNC closure in cases where observations of short-range interactions are not likely.

2.2. Solvation Free Energy Functionals

In a given system, the free energy of solvation may be obtained analytically through single-point calculations after the total and direct correlations functions have been computed by RISM. The HNC and KH closures have been used to derive SFE functionals for 1D-RISM, shown by Equations (10) and (11), respectively:

$$\Delta G_{solv}^{HNC} = 2\pi\rho k_B T \sum_{uv} \int_0^\infty [-2c_{uv}(r) - c_{uv}(r)h_{uv}(r) + h_{uv}^2(r)]r^2 dr, \tag{10}$$

$$\Delta G_{solv}^{KH} = 2\pi\rho k_B T \sum_{s=1}^{N}\sum_{a=1}^{M} \int_0^\infty [-2c_{uv}(r) - c_{uv}(r)h_{uv}(r) + h_{uv}^2(r)\Theta(-h_{uv}(r))]r^2 dr, \tag{11}$$

where ρ is the solvent bulk number density, k_B is the Boltzmann constant, T is the temperature, and $h(r)$ and $c(r)$ are the total and direct correlation functions, respectively. It has been found that the error of these functionals is too high for common use, with observed errors of the order of 20 kcal/mol. An alternative method is the Gaussian Fluctuations (GF) functional, shown in Equation (12):

$$\Delta G^{GF}_{solv} = 2\pi\rho k_B T \sum_{s=1}^{N}\sum_{a=1}^{M}\int_0^{\infty}[-2c_{uv}(r) - c_{uv}(r)h_{uv}(r)]r^2 dr. \tag{12}$$

Additionally, an improvement to the HNC functional was proposed: the repulsive bridge extension, shown in Equation (13):

$$\Delta G^{HNCB}_{solv} = \Delta G^{HNC}_{solv} + 2\pi\rho k_B T \sum_{uv}\int_0^{\infty}(h_{uv}(r)+1)(e^{-B^R_{uv}(r)-1})r^2 dr, \tag{13}$$

where $B^R_{uv}(r)$ is the repulsive bridge correction function. The bridge functional for a given pair of solute, u, and solvent, s, sites is shown in Equation (14):

$$e^{-B^R_{uv}(r)} = \prod_{v' \neq v}\left\langle \omega_{vv'} \times e^{(\beta\varepsilon_{uv'}(\frac{\sigma_{uv'}}{r})^{12})}\right\rangle, \tag{14}$$

where $\omega_{uv'}$ are intramolecular correlation functions, and $\varepsilon_{uv'}$ and $\sigma_{uv'}$ are the LJ parameters for the site–site pairwise potential. Both the GF functional and the repulsive bridge extension to the HNC functional provide improved descriptions of the solvation thermodynamics compared to standard HNC and KH functionals [38,40].

2.3. Solvation Free Energy Densities

To define a machine learning SFE functional, we begin by noting that the standard RISM SFE functionals as described previously may be reduced into a general form, shown by Equation (15):

$$\Delta G_{RISM} = \int_0^{\infty} w(r)dr, \tag{15}$$

where the integrand, $w(r)$, consists of a function of the total and direct correlation functions of a single solute, as well as the prefactor, $2\pi\rho kT$. $w(r)$ may be used to discern a set of variables which quantify the solvation effects between solute and solvent molecules at a set distance, r, from the solute. This function, known as the solvation free energy density (SFED), is the input to the ML SFE functional. SFED functions may be derived from the previously mentioned functionals, HNC, KH, and GF, shown in Equations (16)–(18), respectively:

$$HNC_w(r) = 2\pi\rho kT \times \sum_{uv}[-2c_{uv}(r) - h_{uv}(r)(c_{uv}(r) - h_{uv}(r))], \tag{16}$$

$$KH_w(r) = 2\pi\rho kT \times \sum_{uv}[-2c_{uv}(r) - h_{uv}(r)c_{uv}(r) + h^2_{uv}\Theta(-h_{uv}(r))], \tag{17}$$

$$GF_w(r) = 2\pi\rho kT \times \sum_{uv}[-2c_{uv}(r) - h_{uv}(r)c_{uv}(r)]. \tag{18}$$

2.4. Model Performance

The predictive accuracy of a model is typically quantified by statistical analysis. Once a set of predicted values are output, they can be compared to the known target values and the agreement between these two sets can be measured in many ways.

Equations (19) and (20) show the formulae for the coefficient by determination (R^2) and the root mean squared error (RMSE), respectively.

$$R^2 = 1 - \frac{\sum_{i=1}^{N}(y^i - y^i_{exp})^2}{\sum_{i=1}^{N}(y^i - M(y^i_{exp}))^2}, \tag{19}$$

$$RMSE(y, y_{exp}) = \sqrt{\frac{1}{N}\sum_{i=1}^{N}(y^i - y^i_{exp})^2}, \tag{20}$$

where index i runs through the set of N selected samples, $M()$ indicates a mean, and y^i and y^i_{exp} are the predicted and experimental target values, respectively. R^2 provides a measure of how well the output predictions match the target values with respect to the $y = x$ line, as a perfect set of predictions will reside on this line. RMSE provides a measure of the average difference between the predicted and target values, and also describes the combination of both the systematic and random error. The total deviation can be split into two parts: bias (or mean displacement, M), and the standard deviation of the error of prediction (SDEP), shown in Equations (21) and (22), respectively.

$$bias = M(y - y_{exp}) = \frac{1}{N}\sum_{i=1}^{N}(y^i - y^i_{exp}), \tag{21}$$

$$SDEP = \sigma(y - y_{exp}) = \sqrt{\frac{1}{N}\sum_{i=1}^{N}(y^i - y^i_{exp} - M(y - y_{exp}))^2} \tag{22}$$

The bias gives the systematic error, which can be corrected by the addition of a simple constant term in the final model. The SDEP gives the random error which is not explained by the model and cannot be corrected as bias can. The bias and SDEP are connected through the RMSE, given by Equation (23).

$$RMSE(y, y_{exp})^2 = M(y - y_{exp})^2 + \sigma(y - y_{exp})^2. \tag{23}$$

Models which report an RMSE greater than the standard deviation of the experimental data offer less accurate predictions than the null model provided by the mean of the experimental data.

3. Materials and Methods

3.1. Overview

Previously, work was undertaken to predict the SFE of small organic molecules in four different solvents; water, methanol, chloroform, and carbon tetrachloride [39]. A convolutional neural network (CNN) was trained on SFED data that resulted from 1D-RISM calculations for the accurate prediction of SFE. The focus of the work detailed in this paper is to expand this approach to a larger number of solvents such that the SFE of any solvent/solute combination can be predicted. Hence, SFE predictions have been made using a reoptimised CNN for over 100 different solvent systems with a variety of different small organic solutes. A workflow was developed for this to produce the required input files for the RISM calculations, as well as performing the RISM calculations and CNN predictions.

3.2. Dataset Compilation

Vermeire et al. [41] compiled several datasets containing various experimental SFE and enthalpy values at different temperatures. The work discussed in this paper made use of the CombiSolv-Exp (CSE) dataset. After applying structural filters and completing RISM calculations, the resulting dataset consisted of 2698 samples, each consisting of a unique combination of solute and solvent.

In order to perform 1D-RISM calculations, the generation of several inputs was required. Forcefield parameters and structure files were required for both the solute and solvent molecule for a given pair, and the density of the pure solvent was also required. The general AMBER forcefield (GAFF) was used to model both the solute and solvent molecules and non-bonded parameters were obtained using Antechamber and tLEaP [42]. GAFF was used because it is a high-throughput method able to create a consistent set of parameters for all samples. The structure files were generated by converting SMILES strings to pdb files using OpenBabel [43] for all solutes and solvents. Solvent densities were obtained from various freely available libraries.

3.3. 1D RISM Calculations

PyRISM [44] was used to carry out 1D-RISM calculations using the KH closure within a system consisting of 16384 grid points over a 20.48 Å radius from the solute. The extended reference interaction site model (XRISM) was used for all calculations to allow for the same treatment of both organic and non-organic solvents.

The SFE calculations were performed using the KH, HNC, and GF free energy functionals, with the assumption of infinite dilution. The modified direct inversion of the iterative subspace (MDIIS) solver was used for these calculations as it achieves convergence quickly when compared to other solvers. The following parameters were used to solve the equations: lambda = 10, picard damping = 0.1, depth = 16, and tolerance = 10^{-5}. The lambda parameter dictates the extent of discretisation in the RISM calculation. With a lambda value of 1, a single calculation is performed before an output is given, while a lambda value of 10 means the calculation is performed a total of 10 times. A random guess, as described earlier, is used to perform the first calculation, while the output serves as the initial guess for the next calculation. This allows the calculation to iterate over several initial guesses to allow for a more robust methodology at the cost of increased calculation time. The Picard damping parameter applies an operator to the output of the RISM calculations to manipulate the extent to which the resulting $c(r)$ deviates from the initial $c(r)$. The higher the Picard damping value, the more the initial $c(r)$ is retained, resulting in slower convergence and decreased chance of divergence. With the MDIIS solver, a solution plane is explored and an optimal orthogonal plane is identified to increase the iterative subspace and increase chances of converging on a solution. The depth parameter sets the number of previous solutions that are saved, along with residuals, which are the difference between consecutive solutions. The residuals are then compared to the tolerance parameter. If the tolerance is met, then the calculation ends and produces an output, otherwise the calculation will continue until either the tolerance is met or it is determined that convergence is not possible.

3.4. SFED Processing

The SFED is a functional of the total and direct correlation functions between a solute and solvent molecule, dependent on the systems conditions (such as temperature) and the choice of RISM bridge and SFE functional. Here, SFED was calculated from Equation (18) based on the GF SFE functional. The choice of SFE functional has previously been shown to have little influence on the obtained form of the SFED function [38]. System conditions were kept constant throughout all calculations.

Post-processing of the SFED outputs involved grid reduction to remove highly correlated data. Every 40th grid point up to an 8Å radius was retained, which additionally removed noisy data corresponding to long-range separation between the solute and solvent. This resulted in a feature vector of length 160 that was used in the training of a machine learning model.

Figure 1 shows the total and direct correlation functions between solute and solvent sites. Figure 1a shows a selection of the solute–solvent site pairs between which total correlation functions, $h_{s\alpha}$, are calculated. In practice, a total correlation function is calculated between every possible solute–solvent site pair, though in practice some may be identical due to symmetry in the solute or solvent. Hence, in the example of methanol and water, there are 18 total correlation functions calculated. The figure also illustrates an intramolecular correlation function, $\omega_{ss'}(r)$, in the methanol molecule. Each molecular structure involved in the RISM calculation will be described by intramolecular correlation functions, where one of these functions will be calculated for each site-pair within the molecule, e.g., water will be described by three intramolecular correlation functions; $\omega_{OH1}(r)$, $\omega_{OH2}(r)$, and $\omega_{H1H2}(r)$. Figure 1b shows the direct correlation functions between different particle sites. Figure 1c shows the total correlation functions highlighted by Figure 1a, while Figure 1d shows the direct correlation functions for the same solute–solvent site pairs, all plotted against the distance, r, between solute–solvent sites in Angstroms. Figure 1e shows the SFED calculated from the data in Figure 1c,d with Equation (15).

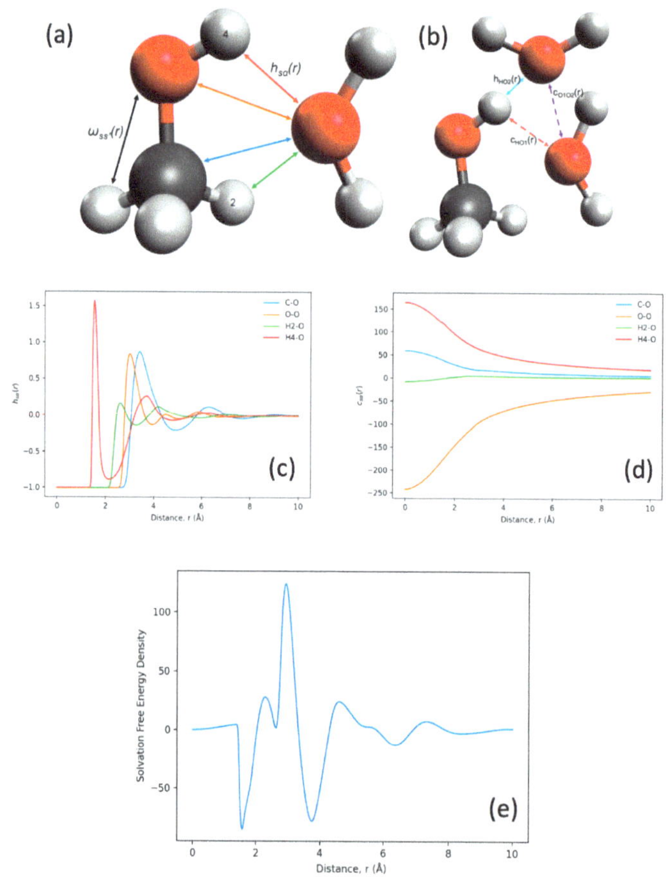

Figure 1. 1D-RISM correlation functions. (**a**) Intermolecular total correlation functions, $h_{s\alpha}(r)$, between a selection of solute sites and solvent sites. A total correlation function between every solute–solvent site-pair is calculated. Also shown is an intramolecular correlation function, $\omega_{ss'}(r)$, between two sites of the methanol. (**b**) Decomposition of $h_{OH4}(r)$ in terms of direct correlation functions $c_{s\alpha}(r)$ involving a third particle. (**c**) Total correlation functions corresponding to those shown in (**a**). (**d**) Direct correlation functions between the same solute–solvent site-pairs described in (**c**). (**e**) SFED calculated for methanol–water using $h_{s\alpha}(r)$ and $c_{s\alpha}(r)$, as well as a prefactor, $2\pi\rho kT$.

3.5. Final Dataset Analysis

The dataset contained 2689 data points comprising binary combinations of 108 unique solvents and 427 unique solutes. Figure 2a shows scores plots for principal component analysis (PCA) against the SFEDs calculated for all solute–solvent systems, while Figures 2b and 2c show the explained variance ratio (EVR) and cumulative explained variance (CEV), respectively.

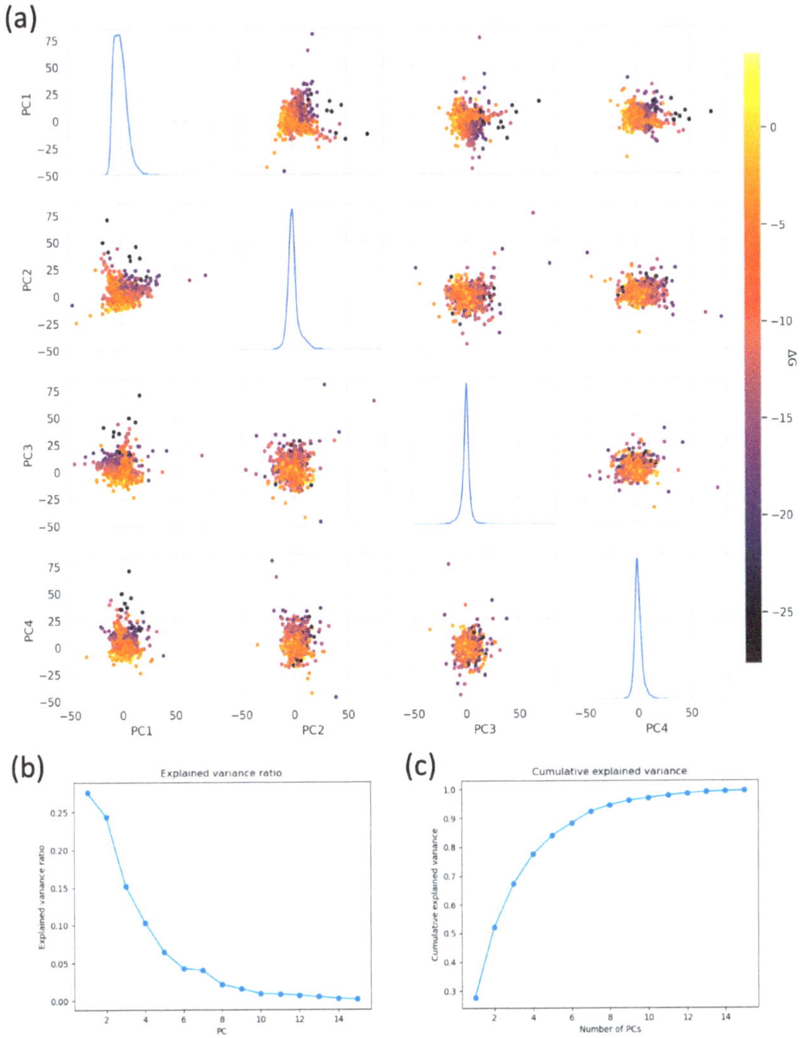

Figure 2. PCA plots showing the chemical space in which the final compiled dataset resides, with distribution plots of the first four PCs shown in the diagonal (**a**). Plots of the fraction of variance described by each principal component (**b**) and the cumulative variance explained by multiple principal components (**c**).

The EVR shows the variance captured by each PC, while the CEV shows the cumulative variance that is captured by all PCs. It is worth noting that because the PCA was carried out against SFED functions, it captures information about the diversity of solute–solvent systems. The first two PCs capture approximately 28% and 25% variance, respectively, with a decrease in variance captured by each consecutive PC. Approximately 80% of the variance is captured cumulatively by the first four PCs, and the plots of Fig. (a) show each of these PCs plotted against one another. In each plot, it can be seen that the majority of samples reside within a tight chemical space, with a few outliers, which include solute–solvent systems with a wide range of values of ΔG_{solv}, as indicated by the shading.

Figures 3 and 4 show the distributions of experimental SFE values for those solvents and solutes, respectively, that appear in the dataset most commonly.

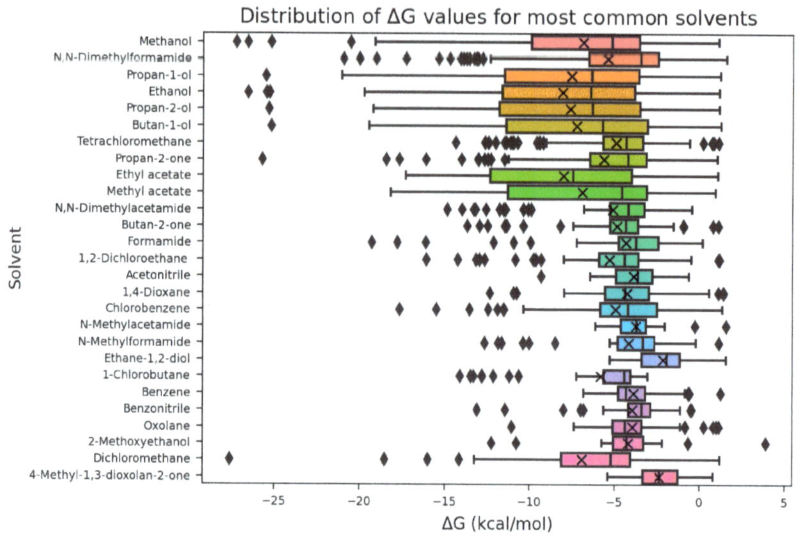

Figure 3. Boxplots showing the SFE distribution for the top 25% of solvents with respect to how frequently they appear in the dataset. The mean SFE value for each solvent is shown by a cross.

The SFE distributions for each solute (in different solvents) are much tighter than those for each solvent (with different solutes). This arises partly due to the sparsity of the dataset in which experimental SFEs were only available for some of the possible solute–solvent pairs. The most common solvent, methanol, appears 150 times, while the most common solute, benzene, appears only 52 times. There are many more solutes than solvents that appear only a small number of times, or even once. While there is a large degree of overlap between the SFE distributions observed between solvents, there is much more variation in the common solutes. Pyrene has measured ΔG values from −13.34 to −11.20 kcal/mol, while molecular nitrogen has measured ΔG values from 0.75 to 1.54 kcal/mol amongst 19 and 23 instances, respectively. The solvent distributions show a much higher number of outliers when compared to the solutes. This may be due to the fact that common solvents will often be paired with solutes for which there are very little data, while conversely, many solutes which appear a small number of times may be paired with a solvent for which a lot of data are present.

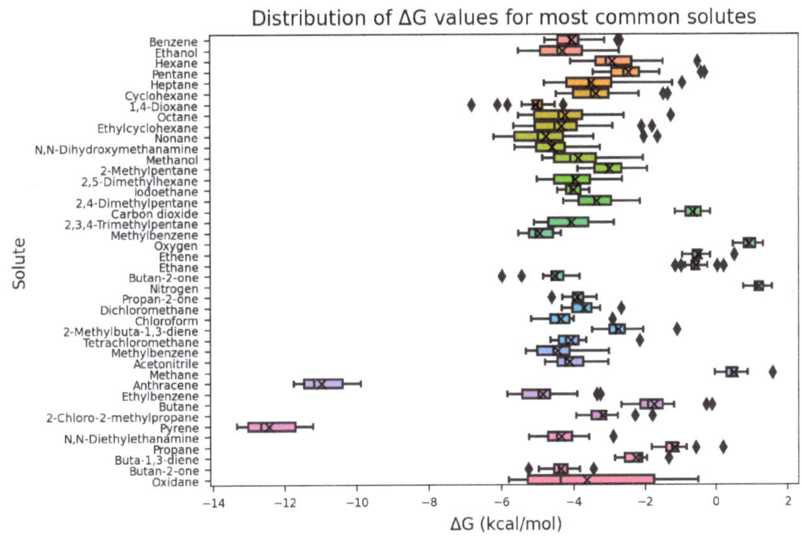

Figure 4. Boxplots showing the SFE distribution for the top 10% of solutes with respect to how frequently they appear in the dataset. The mean SFE value for each solute is shown by a cross.

3.6. CNN Training

A Convolutional Neural Network (CNN) was developed using TensorFlow V2.12.0 for the prediction of SFE based on 1D-RISM calculations. The model was trained and validated by nested cross-validation (CV). An outer cross-validation consisted of 50 resamples, each with a different random 70%/30% train/test split. Hyperparameter optimisation was performed to minimise the mean square error (MSE) estimated from 5-fold CV against the 70% training set, using the GridSearchCV function in SciKit-Learn and the SciKeras package [45]. Only the batch size was optimised. Validation metrics were computed for each of the (30%) test sets and then averaged over the 50 resamples.

The CNN was reoptimised from the initial work [39]. The reoptimised architecture was built using the Keras Sequential API and consists of three convolution blocks, a flatten layer, three densely connected hidden layers, and an output layer. The hidden layers consisted of 64, 32, and 16 nodes, based on the architecture used in previous work [46]. Each convolution block consisted of a Conv1D, MaxPooling1D, and BatchNormalisation layer, in that order. The Conv1D layer had 32 output filters, a kernal size of 3, a stride length of 2, and did not make use of padding, while the MaxPooling1D layers had a max pool size of 2. The ReLU activation function was used in each convolution block and hidden layer. The SciKit-Learn StandardScaler was used to autoscale the feature values such that the mean and standard deviation of the training data were used to scale the training, testing, and validation sets. The Adam optimiser was used throughout the network and had a set learning rate of 0.001. The mean signed error was used for the loss function. Each model was trained for up to 200 epochs, with the Keras callback EarlyStopping used to cease training when the validation loss stopped decreasing with a patience of 20.

4. Results

Model Validation

The CNN model was trained and validated by nested cross-validation in which the inner loop was used to optimise model hyperparameters and the outer loop was used to estimate the model performance. The R^2, RMSE, SDEP, and bias are reported for both the inner and outer loops in Table 1. The metrics from the inner loop are for 5-fold CV against the 70% training set reported as means (with associated standard deviations) over 50 resamples. The metrics from the outer loop are for prediction of the 30% testing set reported as means (and associated standard deviations) over 50 resamples. In all cases, the performance of the model on the outer loop is consistent with that on the inner loop, with each metric the same to within 1 standard deviation, which shows that the model is performing consistently without significant under- or overfitting. The bias is small compared to the RMSE, showing that the majority of the error is random rather than systematic.

The effect of reoptimising the model hyperparameters against the inner loop is evident in Figure 5, which shows that, as expected, the new model is more accurate and more consistent than the model taken from Ref. [38] without reoptimisation of the hyperparameters. This is not surprising given that the model from Ref. [38] was trained on data from only four solvents, whereas the dataset considered here has over 100 different solvents covering a wider array of solvation chemistries.

Table 1. Table showing mean statistics for the inner and outer loop of the nested cross-validation. The metrics from the inner loop are for 5-fold CV against the 70% training set reported as means (with associated standard deviations) over 50 resamples. The metrics from the outer loop are for prediction of the 30% testing set reported as means (and associated standard deviations) over 50 resamples.

	R^2	RMSE (kcal/mol)	Bias (kcal/mol)	SDEP (kcal/mol)
Inner Loop (5-fold CV)	0.87 (0.02)	1.50 (0.14)	0.04 (0.13)	1.49 (0.14)
Outer Loop (Testing Set)	0.89 (0.02)	1.41 (0.11)	−0.02 (0.14)	1.41 (0.11)

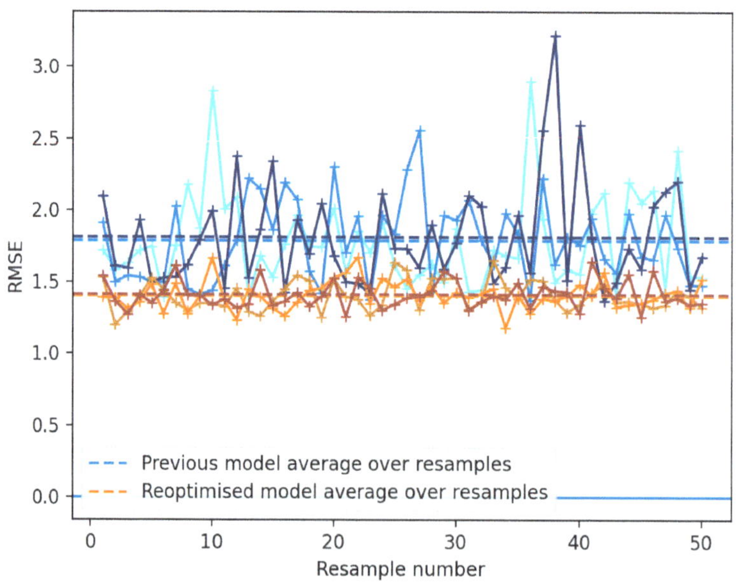

Figure 5. Plots showing the predictive performance of both the previous model (blue) and reoptimised model (orange). Each model was run in triplicate and the RMSE at each resample is plotted. Also plotted is the mean RMSE for each run (dotted lines).

The correlation between the experimental and predicted values for each sample throughout the full set of 50 resamples can be seen in Figure 6, which shows the predictive correlation coloured by (a) solvent and (b) solute, respectively. The overall agreement between the experimental and predicted values is high with few outliers, showing the model has performed well. The line of best fit is very close to the $x = y$ line, and when considered alongside the R^2 value of 0.89, illustrates that the model is performing satisfactorily. The largest outlier is the solute–solvent system 1,1-difluoroethane-1,4-dimethylbenzene, with an average error of 12 kcal/mol. The least consistently predicted solute–solvent system is niflumic acid-formamide, with a standard deviation of 5.02 kcal/mol.

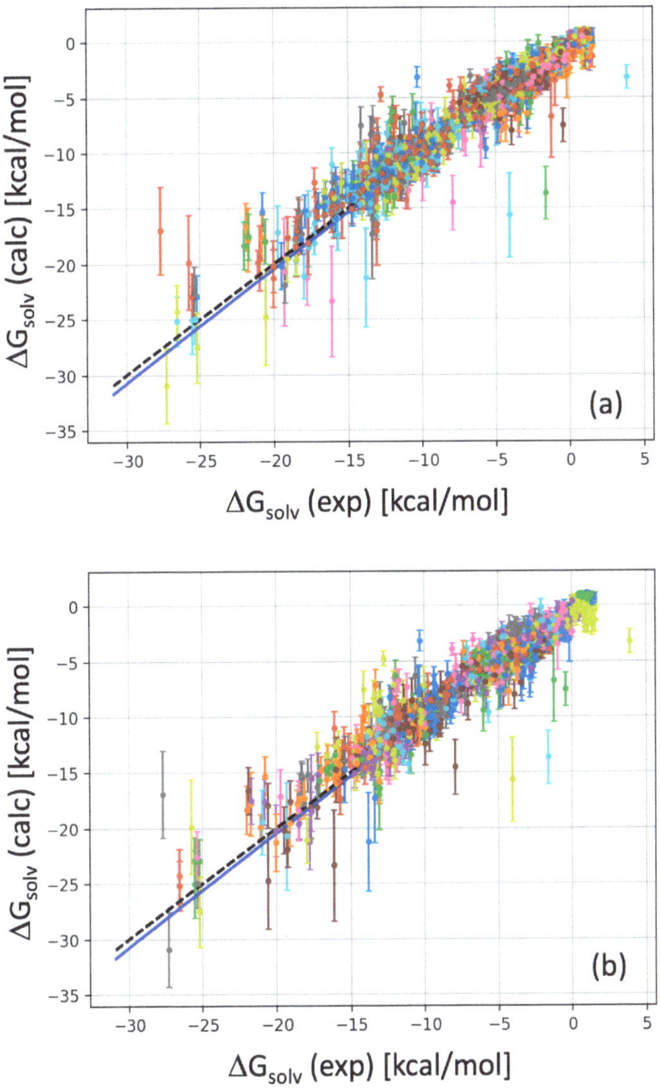

Figure 6. Plots showing the mean prediction vs. the experimentally measured SFE with standard deviations included on each data point, coloured according to (**a**) solvent used and (**b**) solute used. The $x = y$ line is included as a black dotted line, while the trendline is shown as a solid blue line.

Given the sparsity of the experimental data, in which not all solutes have been analysed in all solvents, it is interesting to consider how the availability of training data influences the prediction accuracy. Figure 7 shows that the ML SFE functional is more accurate for solvents that are well represented in the training data. The solvents with 50 or more data points have significantly higher correlations between experimental and predicted values ($0.76 < R^2 < 1$) than those solvents that are under-represented in the dataset. Indeed, all solvents with 80 or more data points are modelled accurately ($R^2 > 0.9$), as is evident in Table 2. These solvents are generally small and contain some polar heteroatoms (Figure 8), but these characteristics are not directly related to the predictive accuracy since several apolar solutes with fewer than 80 data points are also modelled accurately (e.g., 1,2-demethylbenzene, $R^2 = 0.78$). The performance of the CNN model validates the main hypothesis of this work—that a single ML SFE functional can accurately model different solvent systems—but it also suggests that expanding the dataset size to ensure an even representation of all solvents would improve the model. Unfortunately, the availability of experimental data in the published literature is a limitation. Despite the challenges associated with measuring SFE, there is a clear need for new experimental data.

Table 2. The most common solvents within the dataset with the number of solutes for which each solvent has a measured SFE. The R^2, RMSE, Bias, and SDEP are given for each solvent, calculated across all predictions made on samples containing these solvents.

Solvent	Solutes	R^2	RMSE	Bias	SDEP
Methanol	150	0.96	1.07	−0.19	1.05
N,N-Dimethylformamide	143	0.95	1.06	0.08	1.06
Propan-1-ol	138	0.96	1.00	0.08	1.00
Ethanol	133	0.96	1.10	−0.20	1.08
Propan-2-ol	124	0.96	1.00	0.07	1.00
Butan-1-ol	123	0.97	0.90	−0.15	0.88
Tetrachloromethane	118	0.95	0.70	0.05	0.70
Propan-2-one	96	0.93	1.18	−0.28	1.15
Ethyl acetate	83	0.96	0.97	−0.06	0.97
Methyl acetate	78	0.94	1.27	−0.41	1.20
N,N-Dimethylacetamide	77	0.93	0.88	−0.03	0.88
Butan-2-one	72	0.92	0.79	−0.23	0.76
Formamide	71	0.89	1.21	0.45	1.13
1,2-Dichloroethane	70	0.85	1.29	−0.34	1.24
Acetonitrile	64	0.78	0.69	−0.16	0.67
1,4-Dioxane	61	0.82	1.16	0.36	1.10

Given the overall high performance of the most common solvents, the model was retrained on a filtered variation of the dataset such that only samples corresponding to a solvent which appeared 50 or more times were retained. This reduced the dataset from 2698 samples to 1705. The predictive performance is shown in Table 3.

Table 3. Table showing the mean and standard deviation of 30% testing split prediction statistics calculated over 50 resamples.

	R^2	RMSE (kcal/mol)	Bias (kcal/mol)	SDEP (kcal/mol)
Mean	0.91	1.35	0.01	1.34
Standard Deviation	0.02	0.13	0.18	0.13

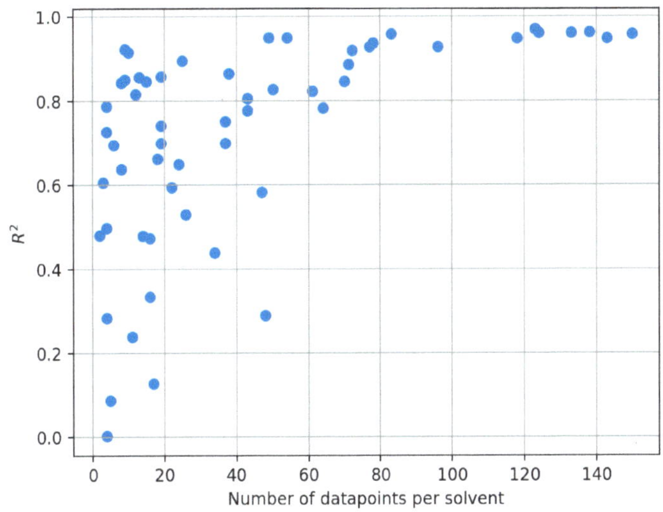

Figure 7. Plot showing the R^2 between the experimental and mean prediction ΔG values vs. the number of data points per solvent.

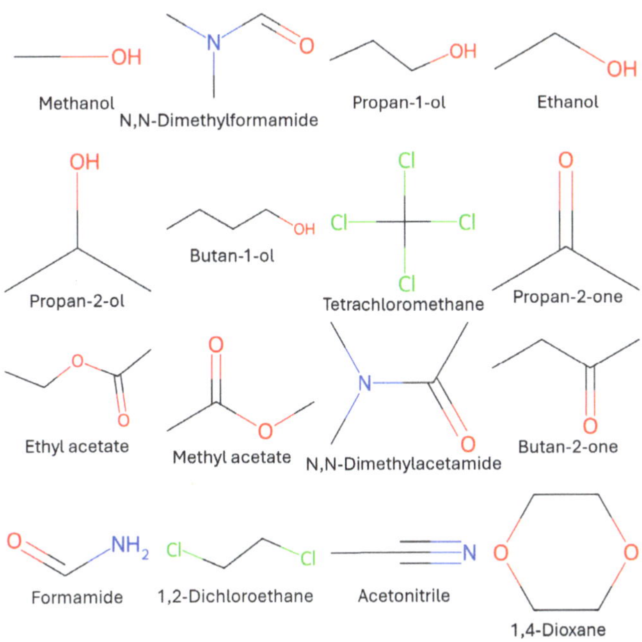

Figure 8. The most common solvents which appear in the dataset used for training.

All the mean values have improved slightly, but only the R^2 has had a significant increase, as the other statistics are within a standard deviation of the earlier results. Interestingly, the standard deviations for RMSE, bias, and SDEP have all slightly increased. Additionally, the sign on the bias has switched, meaning that this model is slightly overpredicting, when it was previously slightly underpredicting. Figure 9 shows the experimental vs. mean predicted SFE values for the common solvents dataset, coloured by (a) solvent and

(b) solute. The same general trends can be drawn from this plot as with the full dataset. The overall agreement between the experimental and mean predictions is satisfactory. There are fewer outliers with the common solvents, which is unsurprising as some scarce data have been removed. The region up to −15 kcal/mol experimental SFE and lower is still less accurately predicted, with larger standard deviations than the region above −15 kcal/mol. It is interesting to note that the highest standard deviation is once again observed on the niflumic acid-formamide data point.

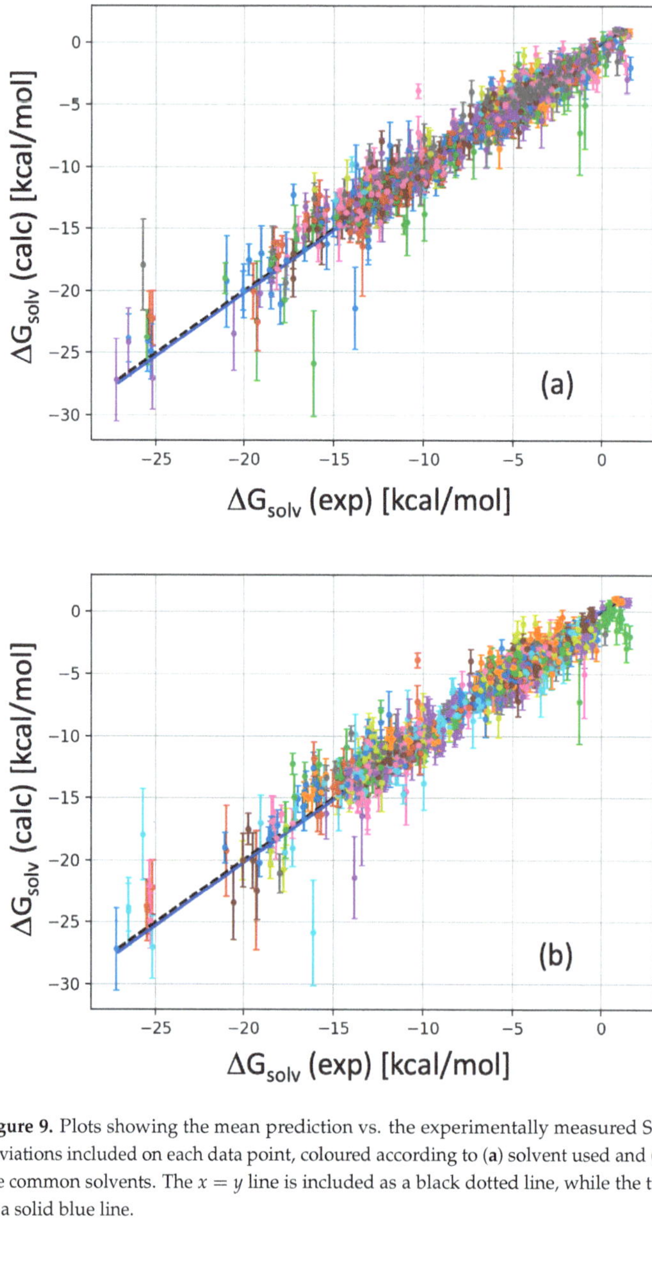

Figure 9. Plots showing the mean prediction vs. the experimentally measured SFE with standard deviations included on each data point, coloured according to (**a**) solvent used and (**b**) solute used for the common solvents. The $x = y$ line is included as a black dotted line, while the trendline is shown as a solid blue line.

Although only common solvents are retained, there has been no direct filtering of the solute data, meaning there are still solutes in the dataset which have a low amount of data. This is a limitation of the dataset and will likely also limit the predictive accuracy of the model. In order to improve the model performance, there would need to be a sufficiently large amount of data for each solute as well as for each solvent. Unfortunately, filtering the dataset by common solvents and solutes reduces the overall sample size to only 14 when setting a threshold of 50 data points on each, and 378 when setting a threshold of 50 for solvents and 20 for solutes. This is not an appropriately large enough volume of data to train a neural network and so more data would be required to verify this.

5. Conclusions

We have developed an ML SFE functional for 1D-RISM that gives accurate predictions for multiple different solvents. A CNN model was trained on SFED functions obtained from 1D-RISM calculations. We achieved accurate predictions, with an RMSE of 1.41 kcal/mol and R^2 of 0.89, using a dataset comprised of 2698 solute/solvent pairs. The CNN was retrained on a reduced version of this dataset, which included only solvents with a large volume of data present in the original dataset. A slight improvement to the model was observed, with an RMSE of 1.35 kcal/mol and R^2 of 0.91. Analysing these results has shown that the predictions are generally more accurate and less varied for those solvents which have larger volumes of data. We have also successfully treated a large number of solvent systems in this way, with a total of 108 solvents being considered in this dataset. The code and datasets used in this work are available at https://github.com/PalmerChem/Conn_Liquids_SI (accessed on 3 November 2024) [47]. The pyRISM software can be accessed at https://github.com/2AUK/pyRISM (accessed on 3 November 2024) [48].

Author Contributions: J.G.M.C. developed the workflow for performing the RISM calculations and developed and trained the CNN used for predictions, as well as curated the datasets used for training. A.A. developed the pyRISM software which was used to perform the RISM calculations. J.G.M.C. and D.S.P. conceptualised the work, methodological approach, and validation processes, as well as analysed the results produced in this work. D.S.P. supervised all work discussed in this manuscript. The manuscript was primarily written by J.G.M.C. with input from co-authors. All authors have read and agreed to the published version of the manuscript.

Funding: J.G.M.C. and D.S.P. would like to thank the EPSRC and IBM for funding via an i-Case Ph.D. studentship.

Data Availability Statement: The datasets used in this work, the CNN python script, and the conda environment yaml file are available at https://github.com/PalmerChem/Conn_Liquids_SI (accessed on 3 November 2024). The pyRISM software can be accessed via the GitHub page of A.A.: https://github.com/2AUK/pyRISM (accessed on 3 November 2024)

Acknowledgments: J.G.M.C., A.A., and D.S.P. would like to thank the ARCHIE-WeSt High-Performance Computing Centre (www.archie-west.ac.uk) for computational resources. J.G.M.C. would like to thank D.J.F. for support in automating RISM calculations.

Conflicts of Interest: The authors declare no conflicts of interest.

Abbreviations

The following abbreviations are used in this manuscript:

AMBER	Assisted Model Building with Energy Refinement
CEV	Cumulative Explained Variance
CSE	CombiSolv-Exp
CV	Cross-Validation
EVR	Explained Variance Ratio
GAFF	General AMBER Forcefield
GB	Generalised-Born
GF	Gaussian Fluctuations

HFE	Hydration Free Energy
HNC	Hypernetted-Chain
IET	Integral Equation Theory
KH	Kovalenko–Hirata
LJ	Lennard-Jones
MD	Molecular Dynamics
MDIIS	Modified Direct Inversion of the Iterative Subspace
MOZ	Molecular Ornstein–Zernike
OZ	Orstein–Zernike
PB	Poisson–Boltzmann
PC	Pure Pressure Correction
PCM	Polarisable Continuum Model
PC+	Expanded Pure Pressure Correction
RDF	Radial Distribution Function
ReLU	Rectified Linear Unit
RISM	Reference Interaction Site Model
RMSE	Root Mean Squared Error
SASA	Surface Accessible Surface Area
SDEP	Standard Deviation of the Error of Prediction
SFE	Solvation Free Energy
SFED	Solvation Free Energy Density
SMILES	Simplified Molecular-Input Line-Entry System
XRISM	Extended Reference Interaction Site Model

References

1. Ben-Naim, A. *A Molecular Theory of Solutions*; Oxford University Press: New York, NY, USA, 2006.
2. Abel, R.; Wang, L.; Harder, E.D.; Berne, B.J.; Friesner, R.A. Advancing Drug Discovery through Enhanced Free Energy Calculations. *Acc. Chem. Res.* **2017**, *50*, 1625–1632. [CrossRef] [PubMed]
3. Ganguly, A.; Tsai, H.; Fernández-Pendás, M.; Lee, T.; Giese, T.J.; York, D.M. AMBER Drug Discovery Boost Tools: Automated Workflow for Production Free-Energy Simulation Setup Analysis (ProFESSA). *J. Chem. Inf. Model.* **2022**, *62*, 6069–6083. [CrossRef] [PubMed]
4. Skyner, R.E.; McDonagh, J.L.; Groom, C.R.; Mourick, T.V.; Mitchell, J.B.O. A review of methods for the calculation of solution free energies and the modelling of systems in solution. *Phys. Chem. Chem. Phys.* **2015**, *17*, 6174–6191. [CrossRef] [PubMed]
5. Li, W.; Ding, G.; Gao, H.; Zhuang, Y.; Gu, X.; Peijnenburg, W.J.G.M. Prediction of octanol-air partition coefficients for PCBs at different ambient temperatures based on the solvation free energy and the dimer ratio. *Chemosphere* **2020**, *242*, 125246. [CrossRef] [PubMed]
6. Ding, W.; Chen, Y.; Ge, Z.; Cao, W.; Jin, H. A molecular simulation study on solvation free energy and structural properties of polycyclic aromatic hydrocarbons in supercritical water environment. *J. Mol. Liq.* **2020**, *318*, 114274. [CrossRef]
7. Marenich, A.V.; Cramer, C.J.; Truhlar, D.G. Universal solvation model based on solute electron density and on a continuum model of the solvent defined by the bulk dielectric constant and atomic surface tensions. *J. Phys. Chem. B.* **2009**, *113*, 6378–6396. [CrossRef]
8. Marenich, A.V.; Olson, R.M.; Kelly, C.P.; Cramer, C.J.; Truhlar, D.G. Self-Consistent Reaction Field Model for Aqueous and Nonaqueous Solutions Based on Accurate Polarized Partial Charges. *J. Chem. Theory Comput.* **2007**, *3*, 2011–2033. [CrossRef]
9. Cramer, C.J.; Truhlar, D.G. A Universal Approach to Solvation Modeling. *Acc. Chem. Res.* **2008**, *41*, 760–768. [CrossRef]
10. Marenich, A.V.; Cramer, C.J.; Truhlar, D.G. Generalized Born Solvation Model SM12. *J. Chem. Theory Comput.* **2013**, *9*, 609–620. [CrossRef]
11. Miertuš, S.; Scrocco, E.; Tomasi, J. Electrostatic interaction of a solute with a continuum. A direct utilization of AB initio molecular potentials for the prevision of solvent effects. *Chem. Phys.* **1981**, *55*, 117–129. [CrossRef]
12. Cossi, M.; Rega, N.; Scalmani, G.; Barone, V. Energies, structures, and electronic properties of molecules in solution with the C-PCM solvation model. *J. Comput. Chem.* **2003**, *24*, 669–681. [CrossRef] [PubMed]
13. Cancès, E.; Mennucci, B.; Tomasi, J. A new integral equation formalism for the polarizable continuum model: theoterical background and applications to isotropic and anisotropic dielectrics. *J. Chem. Phys.* **1997**, *107*, 3032–3041. [CrossRef]
14. Cancès, E.; Mennucci, B. New applications of integral equations methods for solvation continuum models: ionic solutions and liquid crystals. *J. Math. Chem.* **1998**, *23*, 309–326. [CrossRef]
15. Klamt, A.; Schüürmann, G. COSMO: a new approach to dielectric screening in solvents with explicit expressions for the screening energy and its gradient. *J. Chem. Soc.* **1993**, *2*, 799–805. [CrossRef]
16. Shivakumar, D.; Williams, J.; Wu, Y.; Damm, W.; Shelley, J.; Sherman, W. Prediction of Absolute Solvation Free Energies using Molecular Dynamics Free Energy Perturbation and the OPLS Force Field. *J. Chem. Theory Comput.* **2010**, *6*, 1509–1519. [CrossRef]

17. Leung, K.; Rempe, S.B.; von Lilienfeld, O.A. Ab Initio molecular dynamics calculations of ion hydration free energies. *J. Chem. Phys.* **2009**, *130*, 204507. [CrossRef]
18. Geballe, M.T.; Skillman, A.G.; Nicholls, A.; Guthrie, J.P.; Taylor, P.J. The SAMPL2 blind prediction challenge: introduction and overview. *J. Comput.-Aided Mol. Des.* **2010**, *24*, 259–279. [CrossRef]
19. Geballe, M.T.; Guthrie, J.P. The SAMPL3 blind prediction challenge: transfer energy overview. *J. Comput. Aided Mol. Des.* **2012**, *26*, 489–496. [CrossRef]
20. Mobley, D.L.; Wymer, K.L.; Lim, N.M.; Guthrie, J.P. Blind prediction of solvation free energies from the SAMPL4 challenge. *J. Comput. Aided Mol. Des.* **2014**, *28*, 135–150. [CrossRef]
21. Varilly, P.; Patel, A.J.; Chandler, D. An improved coarse-grained model of solvation and the hydrophobic effect. *J. Chem. Phys.* **2011**, *134*, 074109. [CrossRef]
22. Genheden, S. Solvation free energies and partition coefficients with the coarse-grained and hybrid all-atom/coarse-grained MARTINI models. *J. Comput. Aided Mol. Des.* **2017**, *31*, 867–876. [CrossRef] [PubMed]
23. Shivakumar, D.; Deng, Y.; Roux, B. Computations of Absolute Solvation Free Energies of Small Molecules Using Explicit and Implicit Solvent Model. *J. Chem. Theory Comput.* **2009**, *5*, 919–930. [CrossRef] [PubMed]
24. Vyboishchikov, S.F. Predicting Solvation Free Energies Using Electronegativity-Equalization Atomic Charges and a Dense Neural Network: A Generalized-Born Approach. *J. Chem. Theory Comput.* **2023**, *19*, 8340–8350. [CrossRef] [PubMed]
25. Steinmann, S.; Sautet, P.; Michel, C. Solvation free energies for periodic surfaces: comparison of implicit and explicit solvation models. *Phys. Chem. Chem. Phys.* **2016**, *18*, 31850. [CrossRef] [PubMed]
26. Zhang, J.; Zhang, H.; Wu, T.; Wang, Q.; van der Spoel, D. Comparison of Implicit and Explicit Solvent Models for the Calculation of Solvation Free Energy in Organic Solvents. *J. Chem. Theory Comput.* **2017**, *13*, 1034–1043. [CrossRef]
27. Chandler, D.; Anderson, H.C. Optimized cluster expansions for classical fluids. 2. Theory of molecular liquids. *J. Chem. Phys.* **1972**, *57*, 1930–1937. [CrossRef]
28. Palmer, D.S.; Frolov, A.I.; Ratkova, E.L.; Federov, M.V. Towards a universal method for calculating hydration free energies: 3D refernce interaction site model with partial molar volume correction. *J. Phys. Condens. Matter* **2010**, *22*, 492101. [CrossRef]
29. Chuev, G.N.; Fedorov, M.V.; Crain, J. Improved estimates for hydration free energy obtained by the reference interaction site model. *Chem. Phys. Lett.* **2007**, *448*, 198–202. [CrossRef]
30. Palmer, D.S.; Sergiievskyi, V.P.; Jensen, F.; Fedorov, M.V. Accurate calculations of the hydration free energies of druglike molecules using the reference interaction site model. *J. Chem. Phys.* **2010**, *133*, 044104. [CrossRef]
31. Ratkova, E.L.; Chuev, G.N.; Sergiievskyi, V.P.; Federov, M.V. An Accurate Prediction of Hydration Free Energies by Combination of Molecular Integral Equations Theory with Structural Descriptors. *J. Phys. Chem. B* **2010**, *114*, 12068–12079. [CrossRef]
32. Truchon, J.F.; Pettitt, B.M.; Labute, P. A Cavity Corrected 3D-RISM Functional for Accurate Solvation Free Energies. *J. Chem. Theory Comput.* **2014**, *10*, 934–941. [CrossRef] [PubMed]
33. Sergiievskyi, V.P.; Jeanmairet, G.; Levensque, M.; Borgis, D. Solvation free-energy pressure corrections in the three dimensional reference interaction site model. *J. Chem. Phys.* **2015**, *143*, 184116. [CrossRef] [PubMed]
34. Misin, M.; Federov, M.V.; Palmer, D.S. Hydration Free Energies of Molcular Ions from Theory and Simulation. *J. Phys. Chem. B* **2016**, *120*, 975–983. [CrossRef] [PubMed]
35. Misin, M.; Palmer, D.S.; Federov, M.V. Predicting Solvation Free Energies Using Parameter-Free Solvent Models. *J. Phys. Chem B* **2016**, *120*, 5724–5731. [CrossRef]
36. Misin, M.; Vainikka, P.A.; Federov, M.V.; Palmer, D.S. Salting-out effects by pressure-corrected 3D-RISM. *J. Chem. Phys.* **2016**, *145*, 194501. [CrossRef]
37. Ratkova, E.L.; Fedorov, M.V. Combination of RISM and Cheminformatics for Efficient Predictions of Hydration Free Energy of Polyfragment Molecules: Application to a Set of Organic Pollutants. *J. Chem. Theory Comput.* **2011**, *7*, 1450–1457. [CrossRef]
38. Fowles, D.J.; McHardy, R.G.; Ahmad, A.; Palmer, D.S. Accurately predicting solvation free energy in aqueous and organic solvents beyond 298 K by combining deep learning and the 1D reference interaction site model. *Digit. Discov.* **2022**, *2*, 177–188. [CrossRef]
39. Fowles, D.J.; Palmer, D.S. Solvation entropy, enthalpy, and free energy prediction using a multi-task deep learning functional in 1D-RISM. *Phys. Chem. Chem. Phys.* **2023**, *25*, 6944–6954. [CrossRef]
40. Kovalenko, A.; Hirata, F. Hydration free energy of hydrophobic solutes studied by a reference interaction site model with repulsive bridge correction and a thermodynamic perturbation method. *J. Chem. Phys.* **2000**, *113*, 2793–2805. [CrossRef]
41. Vermeire, F.H.; Chung, Y.; Green, W.H. Predictinf Solubility Limits of Organic Solutes for a Wide Range of Solvents and Temperatures. *J. Am. Chem. Soc.* **2022**, *144*, 10785–10797. [CrossRef]
42. Wang, J.; Wolf, R.M.; Caldwell, J.W.; Kollman, P.A.; Case, D.A. Development and Testing of General AMBER Force Field. *J. Comput. Chem.* **2004**, *25*, 1157–1174. [CrossRef] [PubMed]
43. The Open Babel Package, Version 3.1.1. Available online: http://openbabel.org (accessed on 1 November 2020)
44. Ahmad, A. *2AUK/pyRISM: pyRISM 0.3.0*; University of Strathclyde: Glasgow, UK, 2023. [CrossRef]
45. Badaracco, A.G. Adriangb/Scikeras: Scikeras 0.11.0. Available online: https://pypi.org/project/scikeras/ (accessed on 3 November 2024).

46. Conn, J.G.M.; Carter, J.W.; Conn, J.J.A.; Subramanian, V.; Baxter, A.; Engkvist, O.; Llinas, A.; Ratkova, E.L.; Pickett, S.D.; McDonagh, J.L.; et al. Blinded Predictions and Post Hoc Analysis of the Second Solubility Challenge Data: Exploring Training Data and Feature Set Selection for Machine and Deep Learning Models. *J. Chem. Inf. Model.* **2023**, *63*, 1099–1113. [CrossRef] [PubMed]
47. Conn, J.G.M. *PalmerChem/Conn_Liquids_SI, 2024*; University of Strathclyde: Glasgow, UK. Available online: https://github.com/PalmerChem/Conn_Liquids_SI (accessed on 3 November 2024)
48. Ahmad, A. *2AUK/pyRISM, 2024*; University of Strathclyde: Glasgow, UK. Available online: https://github.com/2AUK/pyRISM (accessed on 3 November 2024)

liquids MDPI

Article

A Basic Approach to Equations of States for Studying the Real Behavior of Noble Gases

Heinz Langhals

Department of Chemistry, LMU University of Munich, Butenandtstr. 13, D-81377 Munich, Germany; langhals@lrz.uni-muenchen.de

Abstract: The relation between the pressure and molar concentration (in mol/L) of real gases in a low- to medium-pressure range is precisely described by a logarithmic two-parameter equation. Increasing the concentration caused an increase in pressure and also the weakening of the effect due to intermolecular interactions, forming the basis for an equation with an adjusted parameter. Exceeding a critical concentration by a further increase caused a switch to another set of parameters in the same equation. At high pressure, a second switch to an exponential term was observed. This equation of state, defined segment by segment, was attributed to three different structures of the medium. The validity of the equations found was verified with experimental data reported in the literature for helium, neon, argon, krypton, and xenon and is discussed in more detail for argon. The temperature dependence of the parameters of the equations is reported and the formation of a liquid phase is discussed.

Keywords: equation of states; noble gases; condensed phases; pressure dependence on concentration (in mol/L)

1. Introduction

The dependence of the volume of a gas (v) on pressure (p) has been of general interest for a long time [1]. Reference is made to various applications, such as the detection of the influence of intermolecular interactions. In principle, equations of states allow quantitative descriptions of such interactions. A complete absence of interactions would lead to the equation for ideal gases being supported by the kinetic gas theory and realized by highly diluted permanent gases. Real gases exhibit more complex behavior caused by noticeable interactions where suitable equations of state are still under discussion. The van der Waals equation is one of the most established; however, it requires a Maxwell construction as a further assumption to cover the two-phase region of real systems and is still under discussion [2]. This creates a general problem, as there are no analytic functions consisting of curved and linear segments. Thus, even the introduction of more parameters in more complex functions maintains them as approximations, such as in the Clausius [3], Dieterici, Redlich–Kwong [4], Redlich–Kwong–Soave [5], Peng–Robinson [6], Benedict–Webb–Rubin [7], Bertholet [8], and Wohl [9] equations. More recent concepts have applied a perturbation theory of intermolecular interactions, such as the Statistical Association Fluid Theory (SAFT), considering various types of such interactions [10–13]. A global uniform treatment of these intermolecular interactions would bring about further progress.

2. General Intermolecular Interactions

The pressure p of a gas or liquid depends on the number of particles in a volume. This is more conveniently indicated by the applied molar concentration c (the inverse of volume/mol), widely applied in chemistry. Starting from highly diluted gases with essentially independent, non-interacting particles, the pressure p is expected to be proportional to the concentration c. This results in the differential Equation (1), where $const_1$ is a constant for the proportionality between the effect of concentration and pressure, considering the different dimensions of c and p.

$$dp = const_1 \cdot dc \tag{1}$$

The intermolecular interactions of particles become increasingly important at higher concentrations of c. Only a fraction of the effects of c on p remains, whereby this becomes smaller at higher concentrations of c so that an inverse proportionality becomes an approach [14]. Accordingly, the differential Equation (2), with a different constant $const_2$ for proportionality, was obtained for higher concentrations.

$$dp = const_2/c \cdot dc \tag{2}$$

A scaling factor c^* is defined as the shift from ideal behavior, according to Equation (1), to more real behavior in (2); for experimental verification and further details, see below. Equation (2) was divided by this scaling factor c^*, multiplied by c, and added to Equation (1). A separation of variables resulted in the differential Equation (3), in which the denominator is dimensionless.

$$dp = \frac{const_1 + const_2/c*}{\left(\frac{c}{c*} + 1\right)} dc \tag{3}$$

Equation (4) was obtained by integrating Equation (3) with the integration constant p_o.

$$p = (c^* \cdot const_1 + const_2) \cdot \ln\left(\frac{c}{c*} + 1\right) + p_o \tag{4}$$

All constants of the pre-factor were combined to form the constant E for simplification, which characterized the energetic effect of concentration c on pressure p [$E = (c^* \cdot const_1 + const_2)$]) to obtain Equation (5). Finally, p_o was expected to be close to zero because of infinite diluted gases.

$$p = E \cdot \ln\left(\frac{c}{c*} + 1\right) + p_o \tag{5}$$

Equation (5) had to be tested for practical applications. Firstly, Equation (5) was converted into the Ideal Gas Equation for diluted gases by a Taylor series expansion and truncation after the linear term, resulting in Equation (6), where $c << c^*$ and the concentration c is the number of moles n over the volume V.

$$p \approx \frac{E}{(c + c*)} \cdot c \approx \frac{E}{c*} \cdot c = \frac{E}{c*} \frac{n}{V} \qquad c << c^* \tag{6}$$

This resulted in the Ideal Gas Equation (7) with $R \cdot T = E/c^*$.

$$p \cdot V \approx n \cdot \frac{E}{c*} = n \cdot R \cdot T \qquad c << c^* \tag{7}$$

Secondly, the application of Equation (5) to the description of the behavior of real gases will be tested. The noble gas argon was used as a model based on the precise measurements of Michels, Levelt, and De Graaff [15] (compare also Refs. [16,17]) because its atoms form spherically symmetrical simple spheres with no tendency to form any chemical bonds. An isotherm at −120 °C, just above the critical point [18] of about −122 °C, was used for the test, since appreciable real behavior could be expected.

3. Argon

The linear correlation between the pressure p of Ar at -120 °C and the function of the concentration c represented by Equation (5) are shown in Figure 1 (performance of target to actual comparison, where r means the correlation number [19] and σ the standard deviation. r^2 means the measure of determination) and indicates a good accordance with the real system because of statistical scatter within the limits of experimental error. The standard deviation of 0.3 atm corresponds to a relative error of 1.6% at the medium pressure of 19.3 atm. The calculated function compared to the experimental data is shown in the inset of Figure 1. The limiting concentration c of 5.4 mol/L corresponds to a permanent contact of argon atoms (see below). The agreement of Equation (5) with the experimental data is not limited to the isotherm near the critical point, but also applies for all reported isotherms above and below the critical point: see Figure 2 where the data of all linear correlation are given in the caption.

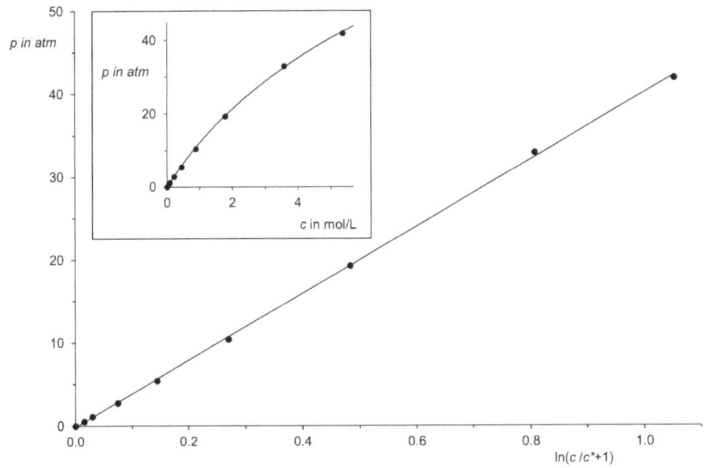

Figure 1. Linear correlation between measurements of the pressure p in atm (filled spheres) and ln $(c/c^* + 1)$ according to Equation (5): Ar at -120 °C, $c < 5.4$ mol·L^{-1}; $E = 40.3$ atm, $c^* = 2.88$ mol·L^{-1}, $r = 0.999\,8$, $n = 9$, standard deviation = 0.3 atm, intercept = -0.18 atm. Inset: measurements of the pressure p in atm (filled spheres) at -120 °C as the function of the molar concentration c of Ar (filled circles) and calculated pressure p in atm according to Equation (5) (solid curve).

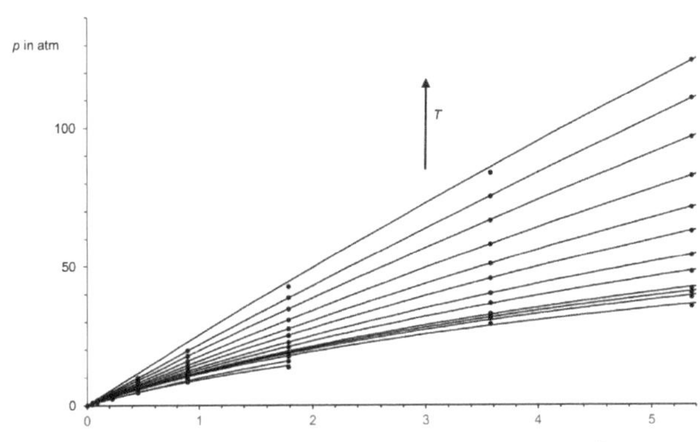

Figure 2. Diagram of state of Ar for concentrations $c < 5.4$ mol·L^{-1}. The discontinuous measurements were taken from [15] and are shown as filled circles. The curves were calculated using Equation (5)

and applying the least squares method to fit c^* to the experimental data. The corresponding increasing temperatures T from bottom to top and the data of linear correlations according to Equation (5) are: $T = -140$ °C, $E = 36.3$ atm, $c^* = 3.25$ mol/L, $r = 0.999996$, $n = 7$, $\sigma = 0.02$ atm, intercept = −0.02 atm; $T = -135$ °C, $E = 19.5$, $c^* = 1.67$, $r = 0.999996$, $n = 6$, $\sigma = 0.02$, intercept = −0.005; $T = -130$ °C, $E = 27.9$, $c^* = 2.00$, $r = 0.9995$, $n = 9$, $\sigma = 0.43$, intercept = −0.25; $T = -125$ °C, $E = 33.8$, $c^* = 4.23$, $r = 0.9997$, $n = 9$, $\sigma = 0.4$, intercept = −0.2; $T = -122.5$ °C, $E = 36.9$, $c^* = 2.65$, $r = 0.9998$, $n = 9$, $\sigma = 0.3$, intercept = −0.2; $T = -120$ °C, $E = 40.3$, $c^* = 2.88$, $r = 0.9998$, $n = 9$, $\sigma = 0.3$, intercept = −0.2; $T = -110$ °C, $E = 55.5$, $c^* = 3.86$, $r = 0.99994$, $n = 9$, $\sigma = 0.2$, intercept = −0.1; $T = -100$ °C, $E = 74.1$, $c^* = 4.98$, $r = 0.99998$, $n = 9$, $\sigma = 0.3$, intercept = −0.09; $T = -85$ °C, $E = 110.1$, $c^* = 6.97$, $r = 0.999996$, $n = 9$, $\sigma = 0.07$, intercept = −0.05; $T = -70$ °C, $E = 158.7$, $c^* = 9.45$, $r = 0.9999997$, $n = 9$, $\sigma = 0.02$, intercept = −0.02; $T = -50$ °C, $E = 250.7$, $c^* = 13.7$, $r = 0.9999997$, $n = 9$, $\sigma = 0.02$, intercept = 0.01; $T = -25$ °C, $E = 439.2$, $c^* = 21.8$, $r = 0.999999$, $n = 9$, $\sigma = 0.06$, intercept = 0.04; $T = 0$ °C, $E = 791.3$, $c^* = 35.8$, $r = 0.999998$, $n = 9$, $\sigma = 0.08$, intercept = 0.05; $T = 25$ °C, $E = 564.9$, $c^* = 21.8$, $r = 0.9998$, $n = 9$, $\sigma = 1.4$, intercept = −0.9.

4. Further Noble Gases

Equation (5) also applies to the other noble gases for similar concentration ranges as for the heavier noble gas krypton until 3.5 mol/L, as shown in Figure 3. The correlations according to Equation (5) are even better than for argon, where an almost ideal behavior is observed for temperatures of 200 °C and more. The even heavier noble gas xenon also fulfills Equation (5) and can be seen in Figure 4; the statistical data are given the caption.

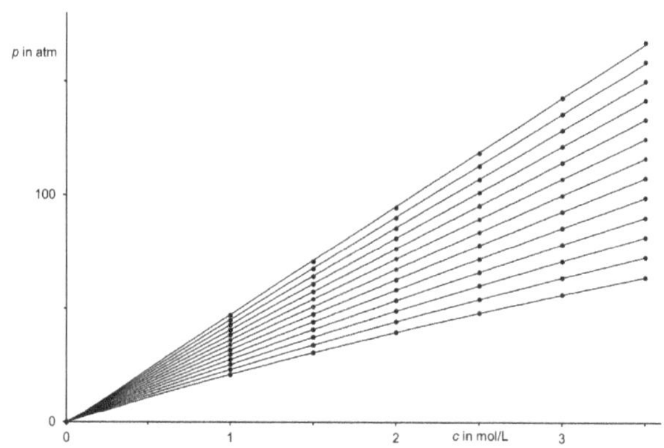

Figure 3. Diagram of state of Kr for concentrations $c < 3.5$ mol·L^{-1}. The discontinuous measurements were taken from [20,21] and shown as filled circles. Curves were calculated using Equation (5) and applying the least squares method to fit c^* to the experimental data. The corresponding increasing temperatures T from bottom to top and the data of the linear correlations according to Equation (5) are: $T = 0$ °C, $E = 154.0$ atm, $c^* = 6.81$ mol/L, $r = 0.9999993$, $n = 7$, $\sigma = 0.001$ atm, intercept = −0.01 atm; $T = 25$ °C, $E = 232.4$, $c^* = 9.53$, $r = 0.9999992$, $n = 7$, $\sigma = 0.04$, intercept = 0.03; $T = 50$ °C, $E = 334.7$, $c^* = 12.7$, $r = 0.9999996$, $n = 7$, $\sigma = 0.03$, intercept = 0.03; $T = 75$ °C, $E = 487.0$, $c^* = 17.2$, $r = 0.9999992$, $n = 7$, $\sigma = 0.04$, intercept = 0.05; $T = 100$ °C, $E = 699.0$, $c^* = 23.0$, $r = 0.9999991$, $n = 7$, $\sigma = 0.05$, intercept = 0.03; $T = 125$ °C, $E = 1050$, $c^* = 32.5$, $r = 0.9999989$, $n = 7$, $\sigma = 0.06$, intercept = 0.04; $T = 150$ °C, $E = 1668$, $c^* = 48.6$, $r = 0.9999987$, $n = 7$, $\sigma = 0.07$, intercept = −0.04; $T = 175$ °C, $E = 1791$, $c^* = 48.6$, $r = 0.9999895$, $n = 7$, $\sigma = 0.2$, intercept = −0.2; $T = 200$ °C, $E = 7302$, $c^* = 190.2$, $r = 0.9999986$, $n = 7$, $\sigma = 0.09$, intercept = 0.06; $T = 225$ °C, $E = 7769$, $c^* = 190$, $r = 0.999994$, $n = 7$, $\sigma = 0.2$, intercept = −0.2; $T = 250$ °C, $E = 8236$, $c^* = 190$, $r = 0.99998$, $n = 7$, $\sigma = 0.3$, intercept = −0.4; $T = 275$ °C, $E = 8701$, $c^* = 190$, $r = 0.99996$, $n = 7$, $\sigma = 0.5$, intercept = −0.5; $T = 300$ °C, $E = 9167$, $c^* = 190$, $r = 0.99994$, $n = 7$, $\sigma = 0.7$, intercept = −0.8.

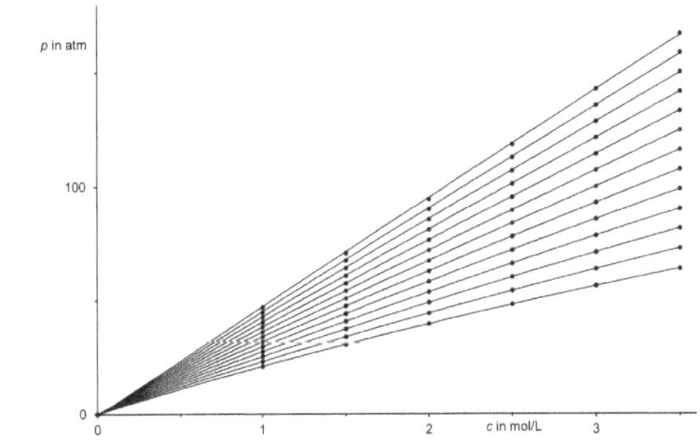

Figure 4. Diagram of state of Xe for concentrations $c < 3.5$ mol·L^{-1}. Discontinuous measurements were taken from [22] and shown as filled circles. The curves were calculated using Equation (5) and applying the least squares method to fit c^* to the experimental data. The corresponding increasing temperatures T from bottom to top and the data of linear correlations according to Equation (5) are: T = 16.65 °C, E = 41.16 atm, c^* = 1.47 mol/L, r = 0.9997, n = 7, σ = 0.5 atm, intercept = −0.12 atm; T = 25 °C, E = 48.24, c^* = 1.72, r = 0.9998, n = 7, σ = 0.4, intercept = −0.11; T = 50 °C, E = 74.04, c^* = 2.59, r = 0.99995, n = 7, σ = 0.2, intercept = −0.09; T = 75 °C, E = 107.4, c^* = 3.61, r = 0.999986, n = 7, σ = 0.2, intercept = −0.06; T = 100 °C, E = 150.9, c^* = 4.83, r = 0.999998, n = 7, σ = 0.06, intercept = −0.03; T = 125 °C, E = 207.4, c^* = 6.30, r = 0.99999994, n = 7, σ = 0.01, intercept = −0.004; T = 150 °C, E = 281.2, c^* = 8.12, r = 0.999995, n = 7, σ = 0.04, intercept = 0.02; T = 175 °C, E = 377.0, c^* = 10.3, r = 0.9999987, n = 7, σ = 0.07, intercept = 0.04; T = 200 °C, E = 511.0, c^* = 13.3, r = 0.999997, n = 7, σ = 0.1, intercept = 0.06; T = 225 °C, E = 695.8, c^* = 17.3, r = 0.999996, n = 7, σ = 0.1, intercept = 0.7; T = 250 °C, E = 958.6, c^* = 22.7, r = 0.999996, n = 7, σ = 0.1, intercept = 0.08; T = 275 °C, E = 1358, c^* = 30.7, r = 0.999996, n = 7, σ = 0.2, intercept = 0.096; T = 300 °C, E = 2038, c^* = 44.0, r = 0.999996, n = 7, σ = 0.2, intercept = 0.10.

The application of Equation (5) can be verified for the lighter neon and is shown in Figure 5; the smaller size of the atoms includes increased concentrations until 5.5 mol/L.

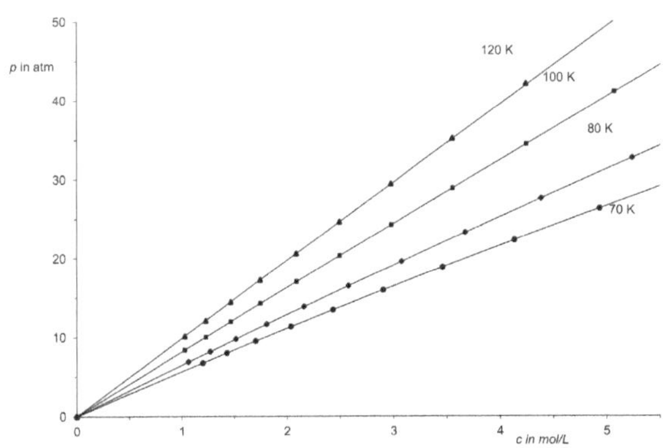

Figure 5. Diagram of state of Ne for concentrations $c < 5.5$ mol·L^{-1}. The discontinuous measurements were taken from [23] and shown as filled circles. Curves were calculated using Equation (5) and applying the least squares method to fit c^* to experimental data. The corresponding increasing temperatures T from bottom to top and the data of linear correlations according to Equation (5) are: T = 70 °K, E = 183.4 atm, c^* = 31.9 mol/L, $r \approx 1$, n = 9, σ = 0.003 atm, intercept = 0.0002 atm; T = 80 °K, E = 342.2, c^* = 52.2, r = 0.99999997, n = 9, σ = 0.002, intercept = 0.002; T = 100 °K, E = 1797, c^* = 219, r = 0.99999996, n = 9, σ = 0.002, intercept = 0.001; T = 120 °K, E = 870700000, c^* = 87080000, r = 0.999998, n = 9, σ = 0.02, intercept = −0.01.

The application of Equation (5) can even be extended to the very light helium and is verified by a linear correlation of the precise measurement of the isotherm at 4.245 K (Leyden isotherm), see Figure 6. The application of Equation (5) to the isotherm at 4.245 K is shown in Figure 7, where the given function closely follows the experimental data. The application for even lower temperatures is shown in the inset.

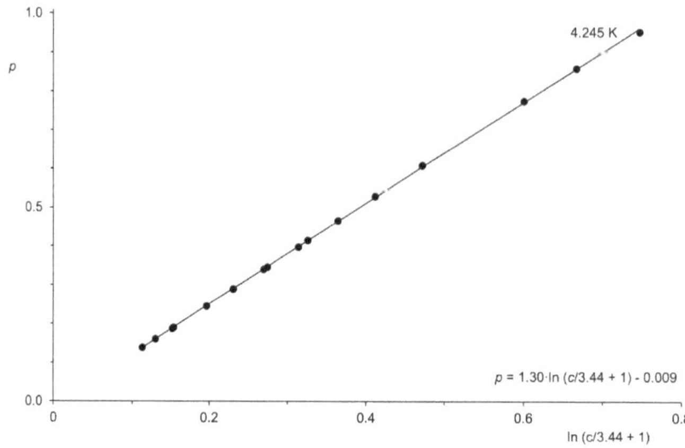

Figure 6. Linear correlation between the pressure p of helium in atm at 4.245 K and the concentration-derived Equation (5): $c \leq 1.51$ mol/L: E= 1.30 atm, c^* = 3.44 mol/L, r = 0.99997, r^2 = 0.999943, n = 11, σ = 0.0009, intercept = −0.009; $c \leq 3.82$ mol/L: r = 0.99993, r^2 = 0.9999, n =17, σ = 0.003.

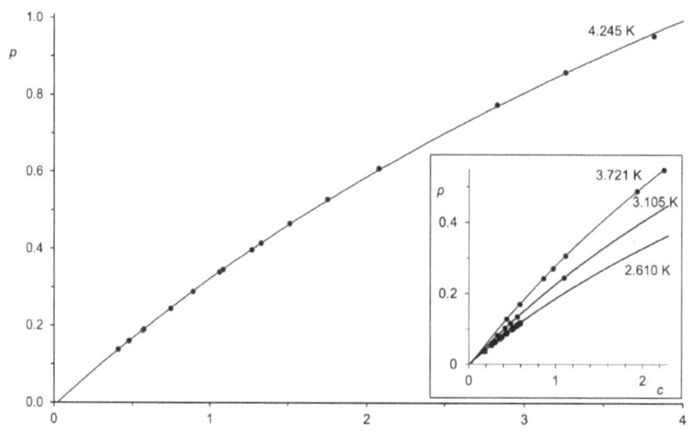

Figure 7. Diagram of state for Helium at 4.245 K for concentrations $c < 4$ mol·L^{-1}. The discontinuous measurements were taken from [24–26] and shown as filled circles. The curves were calculated using Equation (5) and applying the least squares method to fit c^* to the experimental data; $c \leq 1.51$ mol/L: E= 1.30 atm, c^* = 3.44 mol/L, r = 0.99997, r^2 = 0.999943, n = 11, σ = 0.0009, intercept = −0.009; $c \leq 3.82$ mol/L: r = 0.99993, r^2 = 0.9999, n =17, σ = 0.003. Inset: measurements and calculation according to Equation (5) for various temperatures. From top to bottom: T = 3.721 °K, E = 1.009 at, c^* = 3.059 mol/L, r = 0.999996, n = 7, σ = 0.0005 atm, intercept = −0.0069 atm; T = 3.105 °K, E = 0.788, c^* = 2.99, r = 0.99999997, n = 5, σ = 0.00006, intercept = 0.002; T = 2.610 °K, E = 0.585 atm, c^* = 2.67, r = 0.999992, n = 12, σ = 0.0009, intercept = −0.0005.

5. Exceeding Limiting Concentrations

The concentration dependence of the argon pressure below the limiting concentration c of 5.4 mol/L is shown in Figure 1 and essentially corresponds to the individual movement of atoms with occasional contact with other atoms. The dependence of p on the concentration c can be realized by means of Equation (5) up to about 8.6 mol/L, as is shown in Figure 8. Exceeding this concentration leads to an abrupt change in the parameters E and c^* of Equation (5) and a non-zero intercept p_o. This change is shown in Figure 8 by the solid curve up to 8.6 mol/L and the dotting continued in the extrapolation to higher concentrations. The function for higher concentration corresponds to the experimental data up to 18 mol/L and is indicated by the solid curve and the dashed extrapolation to higher concentrations. For even higher concentrations, the pressure p increases exponentially with the concentration c, according to Equation (8), with the parameters E and p_o and the intercept p^*.

$$p = p_o \cdot e^{E \cdot c} + p^* \tag{8}$$

Table 1. Parameters of Equation (5) (region i and ii) and Equation (8) (region iii) for argon. E, p_o, p^* and σ in atm, c^* in mol/L. Nine points for the correlation numbers r in region (i) except six for −135 °C and seven for −140 °C; seven points for the correlation numbers r in region (ii) and (iii). E and c^* were divided by 10 000 000 for better visualization.

	Region (i)					Region (ii)					Region (iii)				
T in °C	E	c^*	p_o	r	s	E	c^*	p_o	r	s	E	p^*	p_o	r	s
25	564.9	21.8	−0.931	0.9997645	1.42	907.9	28.4	−79.2	0.993	15.6	43.35	0.123	121.4	0.9999998	0.2
0	791.3	35.8	0.055	0.9999980	0.08	795.6	29.9	−57.5	0.993	12.9	29.75	0.131	126.4	1	0.1
−25	439.2	21.8	0.037	0.9999987	0.06	601.0	28.4	−35.6	0.993	10.3	18.32	0.142	127.0	0.9999999	0.1
−50	250.7	13.7	0.012	0.9999997	0.02	445.6	28.4	−13.5	0.994	7.5	9.79	0.156	119.0	0.9999999	0.1
−70	158.7	9.45	−0.017	0.9999997	0.02	321.0	28.4	4.1	0.994	5.3	5.02	0.172	105.3	0.9999993	0.2
−85	110.1	6.97	−0.049	0.9999958	0.07	240.0	29.9	17.1	0.994	3.7	2.45	0.190	92.0	0.9999974	0.4
−100	74.1	4.98	−0.092	0.9999773	0.14	143.1	29.9	29.4	0.995	2.1	0.88	0.217	73.6	0.9999904	0.6
−110	55.5	3.86	−0.129	0.9999399	0.21	80.7	29.9	36.9	0.994	1.2	0.41	0.237	54.4	0.9999529	1.3
−120	40.3	2.88	−0.178	0.9998387	0.30	22.5	30.8	43.0	0.970	0.8	0.11	0.275	35.6	0.9998646	1.8
−122.5	36.9	2.65	−0.193	0.9997909	0.33	9.4	30.8	43.5	0.915	0.9	0.07	0.287	30.4	0.9998158	2.0
−125	33.8	2.43	−0.209	0.9997282	0.36						0.04	0.302	25.1	0.9997335	2.2
−130	27.9	2.00	−0.246	0.9995254	0.44						0.05	0.296	−2.1	0.9999351	0.9
−135	19.5	1.67	−0.005	0.9999901	0.02						0.04	0.301	−27.0	0.9999923	0.3
−140	36.3	3.25	−0.015	0.9999958	0.02						0.03	0.300	−57.5	1	0.0

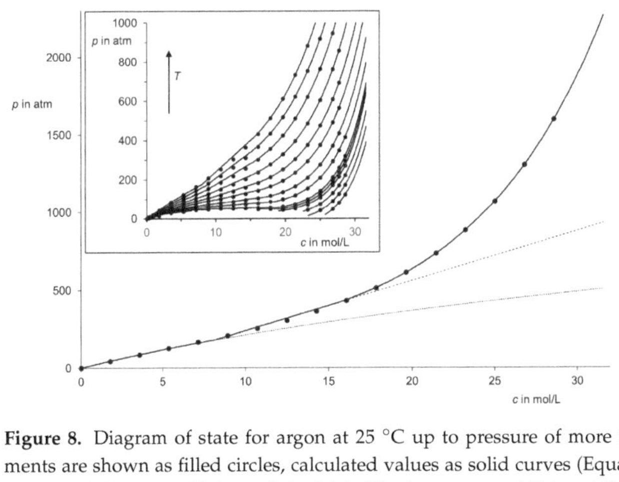

Figure 8. Diagram of state for argon at 25 °C up to pressure of more than 1500 atm; measurements are shown as filled circles, calculated values as solid curves (Equations (5) and (8), respectively). (i) Equation (5) is used for high dilution up to $c \leq 8.6$ mol/L in the most left branch: $E = 565$ atm, $c^* = 21.8$ mol/L, $p_o = 0.93$ atm (extrapolated to $c = 0$), $r = 0.9998$, $n = 9$, $\sigma = 1.4$ atm; dotted curve for extrapolation to higher concentration. (ii) Equation (5) for mean concentration 8.6 mol/L $\leq c \leq$ 18 mol/L: $E = 9.08 \cdot 10^9$ atm, $c^* = 2.84 \cdot 10^{10}$ mol/L, $p_o = -79.2$ atm, $r = 0.993$, $n = 7$, $\sigma = 16$ atm; dashed curve for extrapolation to higher concentration. (iii) Equation (8) is used for $c \geq 18$ mol/L: $p_o = 43.3$ atm, $E = 0.123$ L/mol, $p^* = 121.4$ atm, $r = 0.9999998$, $n = 5$, $\sigma = 0.19$ atm. Inset: overview with isotherms of argon at various temperatures T in °C. From top to bottom: 25, 0, −25, −50, −70, −85, −100, −110, −120, −122.5, −125, −130, −135, −140. For detailed information about the parameters of Equations (5) and (8) and statistical data see Table 1.

No further deviations were found. The isotherms of argon for lower temperatures were investigated to obtain more information, since the influence of the interactions on p is significantly stronger, as is shown in the inset of Figure 8. Equations (5) and (8) are valid for the entire range of measured temperatures from 25 until −140 °C and cover even the range below the critical point of −122.3 °C for both monophasic branches; detailed information on the parameters of Equations (5) and (8) as well as statistical data can be found in Table 1.

The application of Equation (5) and (8), respectively, was compared with the well-established van der Waals equation $p = n \cdot R \cdot T/(V\text{-}n \cdot b) - a \cdot n^2/V^2)$ where p means the pressure, V the volume, n the moles, T the absolute temperature, and R the universal gas constant. a and b are adjustable coefficients. The van der Waals equation was transformed to use the concentration $c = n/V$ to obtain $p = c \cdot R \cdot T/(1\text{-}c \cdot b) - c^2 \cdot a$. The parameters a and b were adapted to the experimental values of the isotherm at −120 °C by means of the least squares, as is shown in Figure 9. The segmental adaptation using Equations (5) and (8) (solid curves) gives appreciably better results than the van der Waals equation (dashed curve).

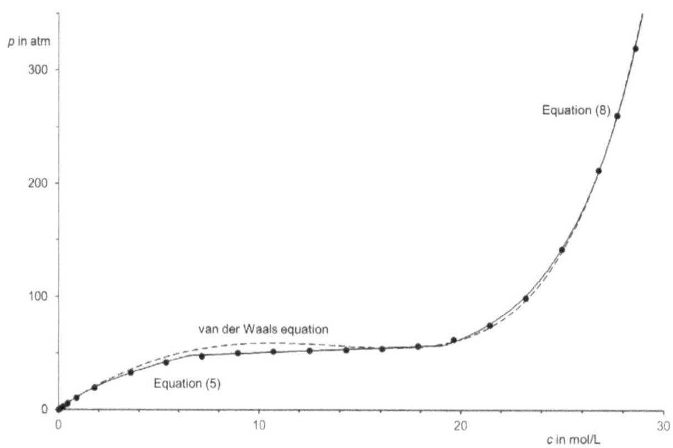

Figure 9. Isotherm of argon at −120 °C. Filled circles: experimental data. Solid curves calculated by Equation (5) (region i and ii) and (8) (region iii). Dashed curve calculated by the van der Waals equation; parameters obtained by the least squares method: a = 1.08 atm, b = 0.0246 L/mol; R = 0.0820 atm· L·mol^{-1}·K^{-1}; σ = 3.2 atm corresponds to 5.4% deviation at 8.9 mol/L.

The results shown in Figure 8 are interpreted in relation to the atomic contacts. At low concentrations, there are only random sporadic interactions between the particles, which can essentially still enable their free movement. As the concentration increases, the distance between the particles decreases until permanent contact is achieved and a comparably abrupt transition to a changed structure of the medium occurs. The concentration c_k(i–ii) of about 8.62 mol/L observed for this transition corresponds to a cube of 5.78 Å for the volume for one atom. The statistical average distance [27,28] is smaller by a factor of about 0.55 and is 3.20 Å; this indicates an effective diameter of one argon atom. A further increase in the concentration c reduces the distance between the particles more and more until a dense packing at c_k(ii–iii) of 18 mol/L is achieved; this corresponds to a cube of 4.52 Å for one atom. The most dense are hexagonal or cubic faced packings. Thus, a dense cubic-face-centered packing can be used for an approximation to c_k(ii–iii) where the lattice constant is larger by a factor of $\sqrt{2}$ than the interatomic distance so that 3.19 Å is obtained; this is in good agreement with the distance 3.20 Å found for c_k(i–ii).

For comparison, the reported [29] van der Waals radius for argon of 1.88 Å gives a diameter of 3.76 Å for argon atoms. This value is in acceptable agreement with the distance of 3.2 Å found, for which the application of a completely different model has to be considered.

A further increase in concentration c results in a strong repulsion of the electron shell of the atoms, verified by the exponential Equation (8) and the right branch in Figure 8, with a strong increase in pressure p.

6. The Two-Phasic Region

Figure 8 represents the isotherm of argon at 25 °C, which is appreciable above the critical point of 150.87 K and −122.28 °C, respectively. Similar isotherms are obtained for lower temperatures up to the critical point; see the inset of Figure 8 and Table 1. Even below the critical point, Equations (5) and (8) can be applied to the branches in region (i) with low concentration and region (iii) with high concentration, indicating a generally similar behavior of the medium. The formation of two phases is interpreted as limited miscibility of the two structures of the medium, the densely packed arrangement in region (iii) and the less densely arrangement in region (i) and (ii), respectively. As a consequence, a surface between these phases is formed which is characterized by the surface tension. The surface tension between these structures decreases with increasing temperatures and vanishes at the critical point.

7. Isotherms of Other Noble Gases

Isotherms similar to those interpreted for argon were found for the other noble gases. The critical concentration c_k(i–ii) of krypton was found at about 5.0 mol/L and corresponds to the larger dimension of krypton atoms, while xenon reaches the transition at 3.2 mol/L. The smaller homologue neon requires 15 mol/L. The relative concentrations of transition are approximately proportional to the relative volume of the elementary cells of crystalline noble gases because they present the volumes of the atoms and are given in Figure 10 (all noble gases form cubic face centered crystal lattices). The c_k(i–ii) value of helium can be estimated from the lattice constant of crystalline material by extrapolation and would require more than 30 mol/L and a very high pressure, respectively.

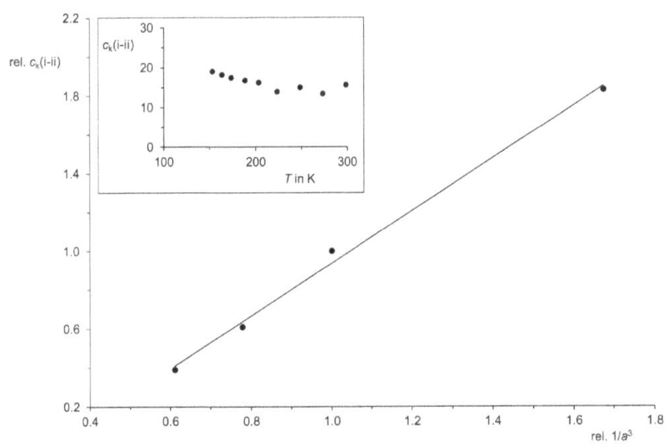

Figure 10. Linear correlation between the relative volume of the elementary cell of crystalline noble gases, indicated by the relative reciprocal of the cubic lattice constant (rel. $1/a^3$) and the relative concentrations (rel. c_k(i–ii)) of the transitions between region (i) and (ii), from left to right: filled circles for xenon, krypton, argon, and neon. Slope of the regression line 1.35, intercept 0.42, correlation number 0.998 for four points, coefficient of determination 0.996. Inset: the critical concentration c_k(i-ii) in mol/L of argon at various temperatures.

The c_k(i–ii) values of argon were only slightly temperature-dependent, as shown in the inset in Figure 10; the concentration may be essentially determined by the atomic dimensions of argon. There was a sufficiently large experimental basis for the determination at various temperatures, where there seems to be a slight increase at lower temperatures. However, considerable experimental uncertainties stand in the way of an exact determination.

8. The Temperature Dependence of Parameters

A linear temperature dependence of the quotient E/c^* is to be expected according to Equation (7) based on the comparison of the coefficients with the equation of ideal gases. To check this, the ln E/c^* of the experimental values were plotted against ln T in Figure 11. Slopes were found close to 1. The slope of 0.939 was found for argon for region (i) above the critical point (see Figure 11); the slope decreases slightly to 0.87 with stronger scattering of the points when lower temperatures are included. The slope for the heavier Kr was 1.022 and for Xe 0.922. Similar results were obtained for the less heavy Ne of 1.01 and for He of 1.007. The approximation to the slope to 1 is remarkable because this means an extrapolation to ideal gases from dense gases with strong real behavior.

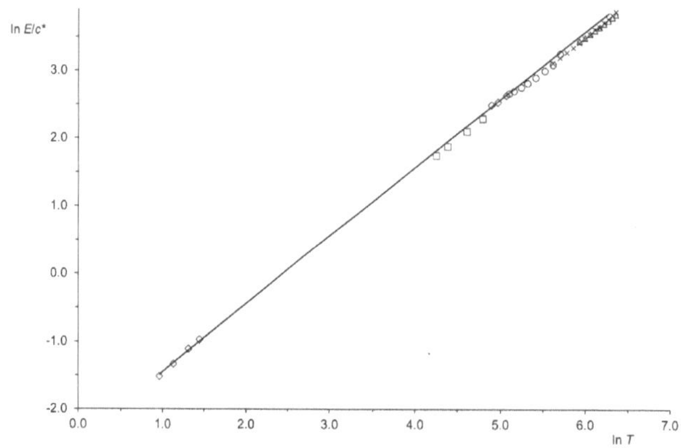

Figure 11. Temperature dependence of the quotient E/c^* in region (i) for the noble gases. Diamonds for He and line of linear regression: slope 1.007, intercept −2.46, correlation number 0.99994 for seven points, number of determinations R^2 = 0.999. Squares for Ne: slope 1.011, intercept −2.55, correlation number 0.99994 for four points, r^2 = 0.9994. Circles for Ar above the critical point: slope 0.939, intercept −2.15, correlation number 0.99 for eight points, coefficient of determination r^2 = 0.979. Crosses for Kr: slope 1.022, intercept −2.68, correlation number 0.9991 for 13 points, R^2 = 0.998. Triangles for Xe: slope 0.922, intercept −2.55, correlation number 0.9994 for nine points, R^2 = 0.999.

The temperature dependence of the parameter c^* is completely different from the quotient E/c^* as there is initially a strong increase with the temperature and finally a plateau is reached at very high temperatures. The primary increase can be described by an Arrhenius diagram, which is shown in Figure 12. Remarkably, the data for all noble gases cover the same linear correlation. Apparently, activation is involved in the atomic interactions where the activation energy (E_a) can be calculated from the Arrhenius diagram (b/R) and is 1.8 J/mol for He, 95 J/mol for Ne, 201 J/mol for Ar, 346 J/mol for Kr, and 447 J/mol for Xe. These small values of activation energy correspond approximately to the volume of the elementary cells of crystalline noble gases, as can be seen in the inset lower right in Figure 12. This could be due to the fact that the electron shell becomes softer as the diameter of the atoms increases, allowing more interactions at atomic contacts.

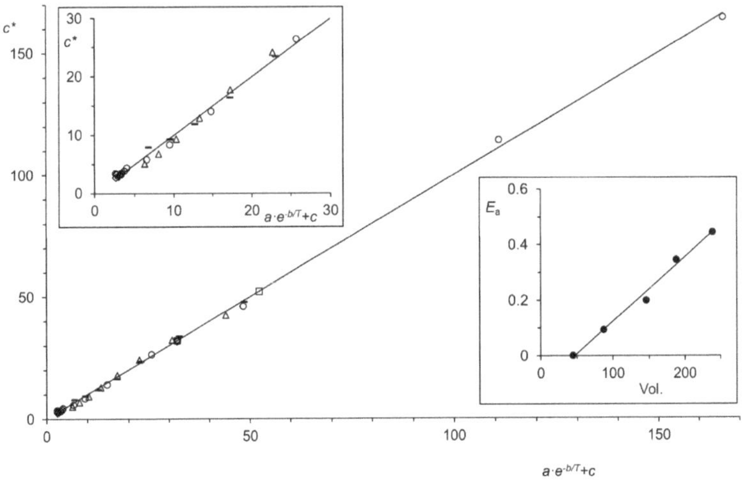

Figure 12. Temperature dependence (T) of the parameter c^* in mol/L obtained by an Arrhenius plot ($a \cdot e^{-b/T} + c$). Circles for Ar: slope 0.9853, intercept 1.34 mol/L, correlation number $r = 0.9997$ for 14 points, $r^2 = 0.9993$, $\sigma = 1.4$ mol/L, $a = 19{,}700$ mol/L, $b = 1674$ K, $c = 3.13$ mol/L. Triangles Xe: slope 1.02, intercept -0.64 mol/L, correlation number 0.996 for 13 points, $r^2 = 0.992$, $\sigma = 1.2$ mol/L, $a = 25{,}860$ mol/L, $b = 3715$ K, $c = 2.81$ mol/L. Dashes for Kr: slope 0.829, intercept 3.15 mol/L, correlation number 0.998 for seven points, $r^2 = 0.997$, $\sigma = 0.94$ mol/L, $a = 36{,}960$ mol/L, $b = 2879$ K, $c = 6.88$ mol/L. Squares for Ne: slope 0.9853, intercept 1.34 mol/L, three points, $a = 534{,}000$ mol/L, $b = 792$ K, $c = 25.4$ mol/L. Diamonds for He: slope 0.9999, intercept 0.002 mol/L, correlation number 0.99994 for seven points, $r^2 = 0.99987$, $\sigma = 0.2$ mol/L, $a = 32.63$ mol/L, $b = 15.36$ K, $c = 2.61$ mol/L. Inset upper left: extended region for small c^* values. Inset lower right: activation energies E_a in kJ/mol obtained from Figure 11 versus the volume (Vol. in Å^3) of the elementary cells of crystalline noble gases (as circles from bottom to top for He, Ne, Ar, Kr and Xe); line for linear correlation; slope 0.0023 kJ·mol^{-1}/Å^3, intercept -0.108 kJ/mol, correlation number 0.995 for five points, $r^2 = 0.990$, $\sigma = 0.002$.

9. Conclusions

The concept that the pressure of gases increases with the concentration, where the increase is gradually damped by intermolecular interactions, led to a logarithmic two-parameter equation for pressure and concentration. A continuous increase in the concentration led to a critical concentration bring exceeded with a switch to a second set of the parameters for the same equation. A further continuous increase in concentration causes a second critical value and switch to an exponential dependency of the pressure on concentration. This was verified by experimental data for the noble gases He, Ne, Ar, Kr, and Xe. The behavior was interpreted in terms of three different structures of media in which (i) at low concentrations only random contacts exist, which (ii) after exceeding a critical concentration form a structure with permanent contacts, and which (iii) after exceeding a second critical concentration form densely packed particles with increasing repulsion of the electron shell. No further change could be detected (it would be of interest whether the exponential Equation (8) remains suitable for extremely high pressure). The formation of liquid phases was attributed to a limited miscibility of structure with free particles and densely packed particles. Investigation of the temperature dependence of the parameter yielded the law for ideal gases at high dilution and a verification of the Arrhenius equation for the critical concentration c^* of the two-parameter equation. This was interpreted as a contribution of an activated process to the temperature dependence of parameters, whereby only small activation energies of a few to several hundred J/mol were found. The activation energies correlate linearly with the volume of the elementary cells of crystalline noble gases, indicating the importance of increasing size-dependent softening of the atoms for their interaction.

The developed equations for the state of gases and liquids can be useful for determining the concentration of gases by pressure measurements [30], whereby the high precision of the developed equations is of particular advantage.

Funding: This research received no external funding.

Data Availability Statement: No new primary data were created. All used primary data were taken from the cited literature available in numerical data. The generated secondary data are obtained by standard calculations and are obvious from the text (all necessary statistical data are also reported there).

Conflicts of Interest: The author declares no conflicts of interest.

References

1. Eucken, A.; Wicke, E. *Grundriss der Physikalischen Chemie*, 10th ed.; Akademische Verlagsgesellschaft Geest & Portig: Leipzig, Germany, 1959.
2. Woodcock, L.V. On Failures of van der Waals' Equation at the Gas–Liquid Critical Point. *Int. J. Thermophys.* **2018**, *39*, 120. [CrossRef]
3. Clausius, R. Über das Verhalten der Kohlensäure in Bezug auf Druck, Volumen und Temperatur. *Ann. Phys. Chem.* **1880**, *9*, 337–357. [CrossRef]
4. Redlich, O.; Kwong, J.N.S. On the Thermodynamics of Solutions. V. An Equation of State. Fugacities of Gaseous Solutions. *Chem. Rev.* **1949**, *44*, 233–244. [CrossRef]
5. Soave, G. Equilibrium Constants from a Modified Redlich-Kwong Equation of State. *Chem. Eng. Sci.* **1972**, *27*, 1197–1203. [CrossRef]
6. Peng, D.-Y.; Robinson, D.B. A New Two-Constant Equation of State. *Ind. Eng. Chem. Fundam.* **1976**, *15*, 59–64. [CrossRef]
7. Benedict, M.; Webb, G.B.; Rubin, L.C. An Empirical Equation for Thermodynamic Properties of Light Hydrocarbons and Their Mixtures: I. Methane, Ethane, Propane, and *n*-Butane. *J. Chem. Phys.* **1940**, *8*, 334–345. [CrossRef]
8. Ree, F.H. Thermodynamic Functions at Liquid–Vapor Transition Range of the van der Waals, the Berthelot, and the Dieterici Equations of State. *J. Chem. Phys.* **1962**, *36*, 3373–3378. [CrossRef]
9. Wohl, A. Untersuchungen über die Zustandsgleichung III. Die Hauptzustandsgleichung und die Zustandsgleichungen der Einzelstoffe. *Z. Phys. Chem.* **1921**, *99*, 226–233. [CrossRef]
10. Chapman, W.G.; Gubbins, K.E.; Jackson, G.; Radosz, M. Equation-of-state solution model for associating fluids. *Fluid Phase Equilib.* **1989**, *52*, 31–38. [CrossRef]
11. Nezbeda, I. On Molecular-Based Equations of State: Perturbation Theories, Simple Models, and SAFT Modeling. *Front. Phys.* **2020**, *8*, 287. [CrossRef]
12. Huang, S.H.; Radosz, M. Equation of State for Small, Large, Polydisperse, and Associating Molecules. *Ind. Eng. Chem. Res.* **1990**, *29*, 2284–2294. [CrossRef]
13. Dufal, S.; Lafitte, T.; Haslam, A.J.; Galindo, A.; Clark, G.N.I.; Vega, C.; Jackson, G. The A in SAFT: Developing the contribution of association to the Helmholtz free energy within a Wertheim TPT1 treatment of generic Mie fluids. *Mol. Phys.* **2015**, *113*, 948–984. [CrossRef]
14. Langhals, H. How the Concept of Solvent Polarity Investigated with Solvatochromic Probes Helps Studying Intermolecular Interactions. *Liquids* **2023**, *3*, 481–511. [CrossRef]
15. Michels, A.; Levelt, J.M.; De Graaff, W. Compressibility Isotherms of Argon at Temperatures Between −25 °C and −155 °C, and at Densities up to 640 Amagat (Pressures up to 1050 Atmospheres). *Physica* **1958**, *24*, 659–671. [CrossRef]
16. Michels, A.; Wijker, H.; Wijker, H.K. Isotherms of Argon Between 0 °C and 150° C and Pressures up to 2900 Atmospheres. *Physica* **1949**, *15*, 627–633. [CrossRef]
17. Michels, A.; Levelt, J.M.; Wolkers, G.J. Thermodynamic Properties of Argon at Temperatures Between 0 °C and −140 °C and at Densities up to 640 Amagat (Pressures up to 1050 atm). *Physica* **1958**, *24*, 769–794. [CrossRef]
18. McCain, W.D., Jr.; Ziegler, W.T. Critical Temperature, Critical Pressure, and Vapor Pressure of Argon. *J. Chem. Eng. Data* **1967**, *12*, 199–202. [CrossRef]
19. Taylor, J.R. *Error Analysis: The Study of Uncertainties in Physical Measurements*, 2nd ed.; University Science Books: Sausalito, CA, USA, 1997; ISBN 0-935702-42-3.

20. Beattie, J.A.; Brierley, J.S.; Barriault, R.J. The compressibility of krypton. I. An equation of State for krypton and the weight of a liter of krypton. *J. Chem. Phys.* **1952**, *20*, 1613–1615. [CrossRef]
21. Beattie, J.A.; Brierley, J.S.; Barriault, R.J. The Compressibility of Gaseous Krypton. II. The Virial Coefficients and Potential Parameters of Krypton. *J. Chem. Phys.* **1952**, *20*, 1615–1618. [CrossRef]
22. Beattie, J.A.; Barriault, R.J.; Brierley, J.S. The Compressibility of Gaseous Xenon. I. An Equation of State for Xenon and the Weight of a Liter of Xenon. *J. Chem. Phys.* **1951**, *19*, 1219–1221. [CrossRef]
23. Michels, A.; Wassenaar, T.; Louwerse, P. Isotherms of Neon at Temperatures Between 0 °C and 150 °C and at Densities up to 1100 Amagat (Pressure up to 2900 Atmospheres). *Physica* **1960**, *26*, 539–543. [CrossRef]
24. Keesom, W.H.; Kraak, H.H. The compressibility of helium gas between 2.6 °K and 4.2 °K. *Physica* **1935**, *2*, 37–44. [CrossRef]
25. Keesom, W.H.; Walstra, W.K. Isotherms of helium at liquid helium temperatures. *Physica* **1940**, *7*, 985–991. [CrossRef]
26. Canfield, F.B.; Leland, T.W.; Kobashy, R. Compressibility Factors for Helium-Nitrogen Mixtures. *J. Chem. Eng. Data* **1965**, *10*, 92–96. [CrossRef]
27. Hertz, P. Über den gegenseitigen durchschnittlichen Abstand von Punkten, die mit bekannter mittlerer Dichte im Raume angeordnet sind. *Math. Ann.* **1909**, *67*, 387–398. [CrossRef]
28. Chandrasekhar, S. Stochastic Problems in Physics and Astronomy. *Rev. Mod. Phys.* **1943**, *15*, 1–89. [CrossRef]
29. Bondi, A. van der Waals Volumes and Radii. *J. Phys. Chem.* **1964**, *68*, 441–451. [CrossRef]
30. Langhals, H. Determination of the Concentration of Gases by Measurement of Pressure. *Anal. Lett.* **1987**, *20*, 1595–1610. [CrossRef]

liquids

MDPI

Article

Green Plasticizers from Dimer Acids with Selected Esters Classified Through the Nile Red [E(NR)] Polarity Scale

Franco Cataldo

Actinium Chemical Research Institute, Via Casilina 1626A, 00133 Rome, Italy; franco.cataldo@fastwebnet.it

Abstract: Dimer and trimer acids are interesting viscous liquids produced from fatty acids derived from renewable sources. The chemical structures of dimer and trimer acids are known and quite complex and are presented here, discussed and further elucidated through electronic absorption spectroscopy, FT-IR and Raman spectroscopy. Dimer and trimer acids have a number of applications in their original form or in the form of derivatives. In the present study, a series of esters of dimer and trimer acids with alcohols from renewable sources were synthesized for use as plasticizers for rubber and plastics. The polarity of the dimer and trimer acids as well as their esters with alcohols from renewable sources (dimerates and trimerates) were systematically studied using a Nile red solvatochromic probe. The resulting E(NR) values were compared with the E(NR) values of the most common types of rubber and plastics. Compatibility and other physical properties expected from the E(NR) scale were studied and successfully confirmed in tire tread rubber compound formulations and in nitrile rubber and PVC matrices, confirming once again the sensitivity and the validity of the Nile red solvatochromic polarity scale for the development of new plasticizers. The validity of the liquids polarity measured with the Nile Red dye is supported by the correlation found between the E(NR) scale and the dielectric constants (ε) of carboxylic acids (including dimer and trimer acids, hydrogenated dimer acids and isostearic acid) and alcohols. A correlation was even found linking the E(NR) values the with the ε values of thin solid films of rubbers and plastics. In the case of the esters the correlation of their E(NR) values was found with the length of the aliphatic chains of the alcohols used in the esterification.

Keywords: dimer acid; trimer acids; dimerates; trimerates; plasticizers; renewable sources; Nile red dye; solvatochromism; polarity probe; compatibility

1. Introduction

As discussed in a previous paper [1], there is great attention on and a demand for new plasticizers for rubber and plastics derived from renewable sources. In fact, the traditional petroleum-derived plasticizers typically used in the tire and rubber technical products industry are already partially substituted by vegetable oils (e.g., sunflower oil, soybean oil [2–4]) or by esters of fatty acids (e.g., methyl esters of fatty acids [5]).

Much less known are dimer acids, which are obtained from naturally occurring unsaturated fatty acids by a thermal treatment under pressure with variable amounts of clay (typically, montmorillonite clay is the catalyst of choice) [6–16]. The clay catalyst causes the dehydrogenation, hydrogen shift reaction and double bond conjugation of the treated fatty acids, favoring the subsequent Diels–Alder addition reaction and leading to dimer acids that are characterized by a 1,2,3,4-substituted-cyclohexene dimer structure, as shown in Scheme 1, accompanied by a minor component of 1,2,5,6-tetrasubstituted-1,2,4a,5,6,8a-hexahydronaphthalene fatty acid bicyclic dimer, shown in Scheme 1 [6–12]. In addition to the dimer acids mentioned, which are generally the main products, trimer acids are produced that have, for example, the bicyclic structure 1,2,3,4,5,8- substituted 1,2,3,4,4a,5,8,8a--octahydronaphthalene (see Scheme 1) [6–12]. A linear dimer may also be present as a minor component (shown at the bottom of Scheme 1), which is mainly produced by a free radical mechanism [11]. Furthermore, a clay catalyzed hydrogen transfer reaction may lead to dehydrogenation of the 1,2,3,4-substituted cyclohexene ring, leading to the corresponding aromatic derivative, i.e., the 1,2,3,4-substituted benzene shown in Scheme 1. The earlier literature [6–12] claims that the key mechanism for the formation of dimers and trimers of fatty acids involves the Diels–Alder reaction and free radical mechanism for linear dimer formation. More recently, it has been suggested that an ionic mechanism, also promoted by clay, may contribute to dimer and trimer formation [13–16].

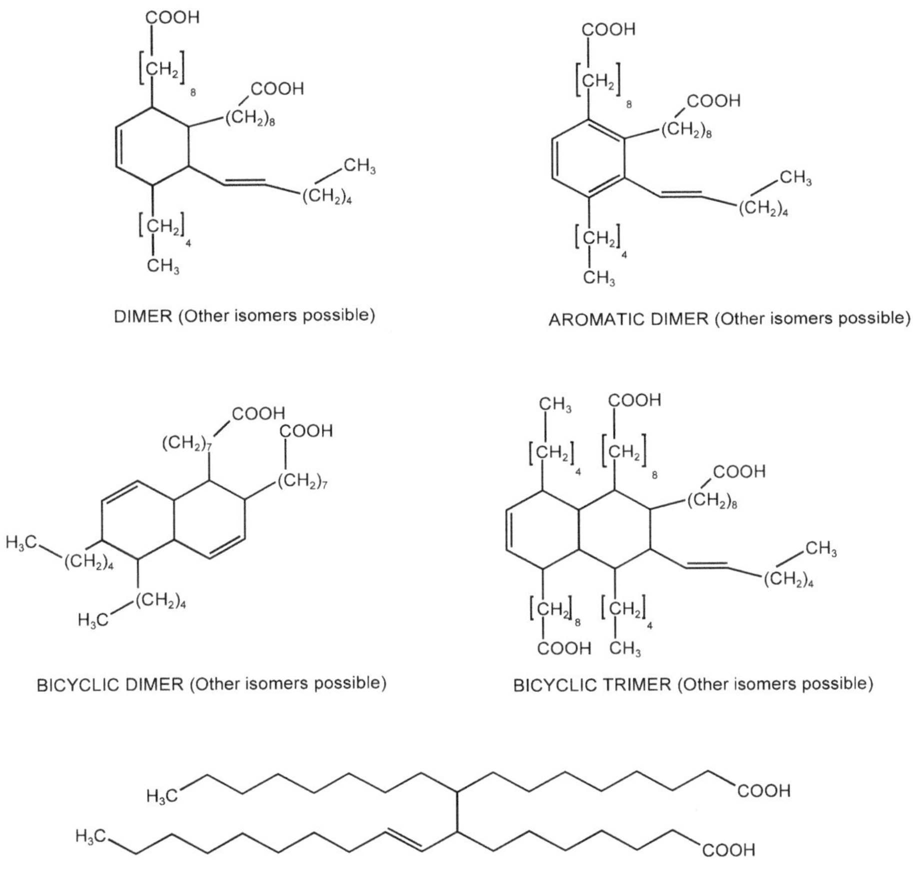

Scheme 1. Typical chemical structures of dimer and trimer acids; many other isomeric structures other than those shown can be conceived.

Dimer acids are isolated by distillation after the removal of the monomers and the isostearic acid (*vide infra*). The standard dimer acids available in the market are viscous liquids (8000–9000 cP at 25 °C), with a trimer content in the order of 20% [11].

However, by an additional distillation step, the dimer content can be increased to >96%, and high purity dimer is a clear, viscous liquid (7100 cP at 25 °C). The distillation residue is mainly composed of trimers in the range of 55–70%, with the rest still composed of dimers [11]. The a trimer residue is, in general, a dark-colored liquid, characterized by an extremely high viscosity (23,000–30,000 cP at 25 °C) [11].

Both the 80/20 dimer/trimer and the high purity >96% dimer can be subjected to mild, catalytic hydrogenation. All the ethylenic double bonds are removed, but the aromatic rings present in the dimer mixture are preserved (see the Results Section for further insights) [11]. In Scheme 2, some chemical structures of hydrogenated dimers are shown. The hydrogenation does not significantly change the viscosity of these liquids but improves the color and the thermo-oxidative stability.

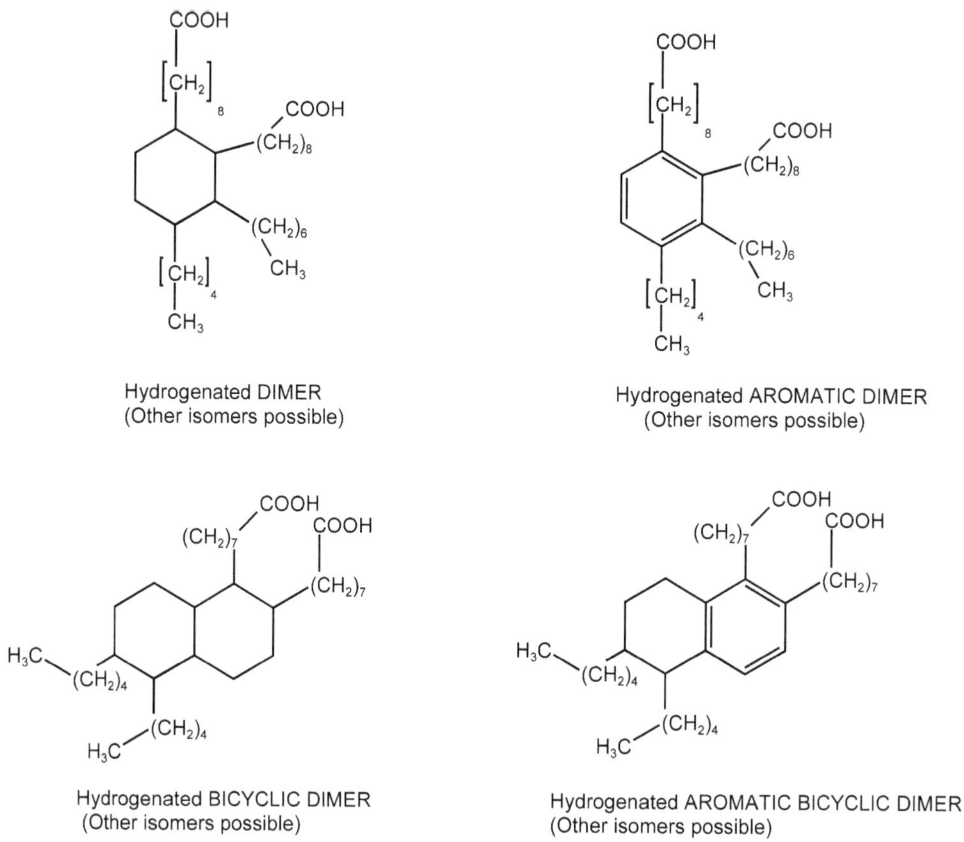

Scheme 2. Typical chemical structures of hydrogenated dimer acids; many other isomeric structures other than those shown can be conceived.

The hydrogenation can be pushed further to reduce the carboxylic functionalities into alcohol groups. In such a way, dimer alcohols are also produced, with the same structure as hydrogenated dimer acids but with –OH groups in place of –COOH groups. Examples of chemical structures present in dimer alcohols are shown in Scheme 3 [11].

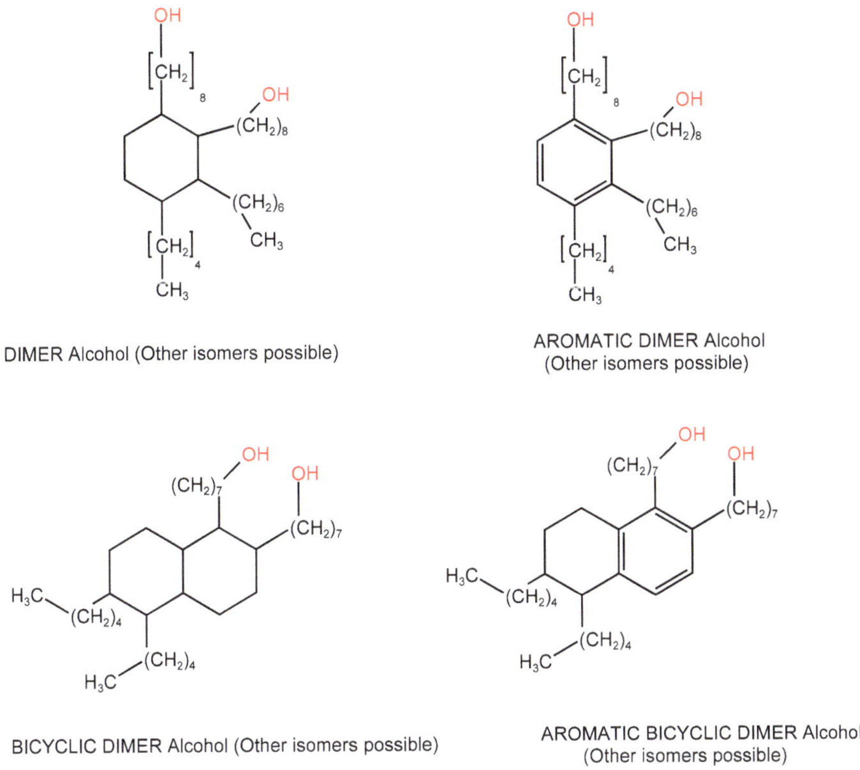

Scheme 3. Typical chemical structures of hydrogenated dimer alcohols; many other isomeric structures other than those shown can be conceived.

The fields of application of dimer acids are numerous; for instance, they can be used as corrosion inhibitors in petroleum processing equipment, as additives in jet fuels, as synthetic lubricants and as a hardener in epoxy resins and nylon fibers [11]. Indeed, the most comprehensive review on the applications of dimer acids and their derivatives is provided in ref. [11]. Large volumes of dimer and trimer acids are employed in the synthesis of fatty polyamides [17,18], which are used in coatings as additives in flexographic inks and in hot melts. However, the main use of liquid fatty polyamides is in epoxy resin curing [17,19]. Furthermore, polyesters based on dimer acids have been developed [20]. Dimer acids are also employed to produce polyester polyols, suitable for the production of high-quality polyurethanes [12].

Esters of dimer acids have found applications as plasticizers for plastics, especially for PVC [11]. Very recently it was found that certain esters of dimer and trimer acids are also excellent plasticizers for rubber compounds, especially for passenger tire tread rubber compounds [21]. Esters of dimer and trimer acids made with alcohols from renewable sources are 100% from renewable sources since dimer and trimer acids are also fully derived from renewable sources. Furthermore, dimer and trimer acids are neither toxic nor irritants [11,17] and are not considered to be flammable materials [11]. The advantages offered by a series of selected dimer and trimer acid esters with alcohols from renewable sources in rubber tread formulations pertain to the viscoelastic properties of the resulting compounds, offering better dynamic properties with respect to the traditional petroleum-based plasticizers used by the tire industry, such as T-DAE (treated distillate aromatic extract) or MES (mild extract solvate), and also to vegetable oils such as sunflower oil [21].

In this paper, dimer acids are first characterized with electronic absorption spectroscopy and then with FT-IR and Raman spectroscopy. The polarity of a selection of esters of dimer and trimer acids and other plasticizers is studied with a Nile red polarity probe [1].

2. Materials and Methods

2.1. Materials and Equipment

Dimer acid samples were kindly supplied by Croda B.V. from their Dutch production plant in Gouda (Gouda, The Netherlands). Croda is now part of the Cargill Corporation, Minnetonka, MN, USA. The commercial dimer acid samples received are as follows: Pripol 1017 (approx. 80% dimer and 20% trimer), Pripol 1025 (hydrogenated dimer acids; it is the hydrogenated version of Pripol 1017), Pripol 1040 (approx. 80% trimer and 20% dimer) and Pripol 2030 dimer alcohols (the hydrogenation was pushed to the conversion level of all -COOH groups into –OH groups).

The dimer and trimer acid esters were prepared as detailed in ref. [21].

Glycerol formal alcohol as well as the fatty acid esters of glycerol formal and its epoxidized derivatives were donated by Fluos S.A.S. company, Samarate plant, Italy.

The FT-IR spectra were collected with a Nicolet iS50 spectrometer from Thermo Fischer (Waltham, MA, USA).

The Raman spectra were collected on a BWTek Inc (Newark, DE, USA) dispersive spectrometer model BWS415i using a laser source at 785 nm.

The electronic absorption spectra were recorded on a Shimadzu (Kyoto, Japan) model UV-2450 spectrophotometer operating with a resolution of 0.1 nm.

2.2. Determination of the Maximum Absorbance with Solvatochromic Dyes in Liquid Samples

An evaluation of the polarity of dimer acids and their derivatives with Nile red was performed following the procedure detailed in the previous paper [1].

3. Results and Discussion

3.1. Electronic Absorption Spectra of Dimer Acids and Hydrogenated Dimer Acids

Dimer acids are made from a mixture of fatty acids primarily derived from tall oil fatty acids, oleic acids and linoleic acids [17]. As shown in Figure 1, the electronic absorption spectra of a commercial dimer acid are rather complex, with a series of absorption bands at 215, 237, 267, 280, 301 and 328 nm. These electronic transitions are essentially due to the conjugated diene and triene moieties present in fatty acid chains and, moreover, to a benzene ring conjugated with a double bond, as shown in Scheme 1.

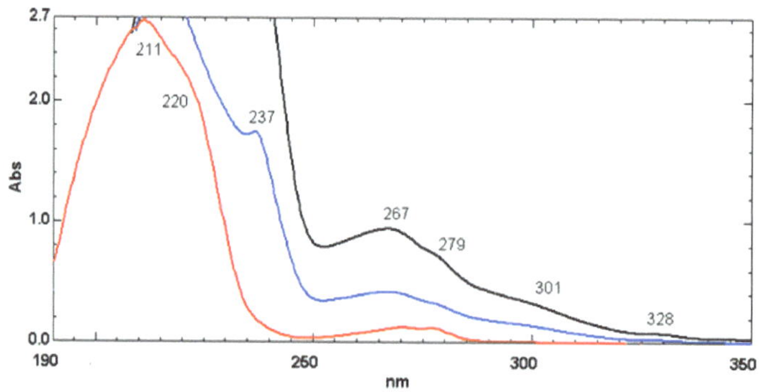

Figure 1. Electronic absorption spectra in n-hexane: hydrogenated dimer acids 1.17×10^{-3} M (red trace); dimer acids (blue trace 1.1×10^{-5} M and black trace 2.3×10^{-5} M).

As shown in Figure 1 and, especially, in Figure 2, the hydrogenation of dimer acids causes the removal of all the electronic transitions at longer wavelengths, with the exclusion of a benzenoid band characterized by two maxima at 270 and 277 nm (Figure 2). A similar electronic absorption spectrum is also displayed by the dimer alcohol, obtained when hydrogenation is pushed further, not only saturating the ethylenic double bonds but also converting the carboxylic functionalities into hydroxy groups. The benzenoid band in Figure 2 confirms that only the aromatic rings originally present in the dimer have survived the hydrogenation step, leading to the chemical structures shown in Schemes 2 and 3. The skeleton of the aromatic rings present in the hydrogenated dimer acids and in dimer alcohol is that of 1,2,3,4-tetraalkyl-substituted benzene. The simplest 1,2,3,4-tetraalkyl-substituted benzene is 1,2,3,4-tetramethylbenzene or prehnitene. For the latter hydrocarbon, the molar extinction coefficient of the strongest band (i.e., that at 270 nm) is reported at 301 M^{-1} cm^{-1} [22]. By using hydrogenated dimer acids and dimer alcohol solutions in n-hexane at the known concentration C_{total}, it is possible to make an estimation of the concentration of the aromatic fraction $\%C_{arom}$ in these commercial products using the Lambert–Beer law:

$$C_{arom} = A\varepsilon^{-1}b^{-1} \quad (1)$$

where A is the absorbance at 270 nm, ε is the molar extinction coefficient (301 M^{-1} cm^{-1}) and b is the pathlength of the cuvette (1 cm).

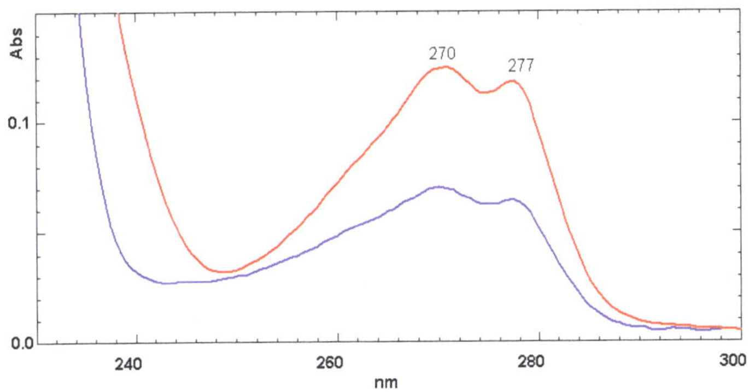

Figure 2. Electronic absorption spectra in n-hexane. Hydrogenated dimer acid concentration weighing 1.17×10^{-3} M (red trace); using ε = 301 $M^{-1}cm^{-1}$ at 270 nm, the aromatic concentration is 4.15×10^{-4} M; hence, the aromatic fraction is 35.5%. Dimer alcohol weighing 7.73×10^{-4} M (violet trace); using ε = 301 $M^{-1}cm^{-1}$ at 270 nm, the aromatic concentration is 2.87×10^{-4} M; hence, the aromatic fraction is 30.1%.

Thus, the aromatic fraction is then determined:

$$\%C_{arom} = (C_{arom}/C_{total}) \times 100 \quad (2)$$

where both C_{arom} and C_{total} are expressed in $mol \cdot L^{-1}$. From these determinations, it can be concluded that the aromatic fraction in commercial hydrogenated dimer acids (and hence in dimer acids) is 35.5%, and a little bit lower in the case of dimer alcohols, where the aromatic fraction is found at 30.1%.

It is interesting to make a comparison of our results with those reported by earlier investigators [23], who made a similar estimation of the aromatic fraction content of hydrogenated dimer acids produced on a lab scale (not a commercial product, as in our case). They concluded that the aromatic fraction is in the range of 20–25% [23]. The lower aromatic content reported is explainable by the different origins of the hydrogenated dimer acids and the fact that their samples were hydrogenated on a lab scale. Furthermore, in the current determination, we have correctly used the molar extinction coefficient of prehnitene, the correct hydrocarbon, while they have adopted the ε value of durene, which is more than double the value of the ε of prehnitene [24].

3.2. *Raman Spectroscopy on Dimer Acids and Hydrogenated Dimer Acids*

Dimer and trimer acids and hydrogenated dimer acids display essentially the same band pattern in the infrared spectra (see the FT IR spectra in Figure 3). Even the spectrum of oleic acid, taken as a reference, appears practically identical to the spectra of dimer and trimer acids (Figure 3). The unique distinctive feature in the oleic acid infrared spectrum is the weak band at 3005 cm^{-1} due to ethylenic =CH stretching. In all cases, the infrared spectra of the oleic, dimer, trimer and hydrogenated dimer acids are dominated by the C=O stretching of the COOH group at about 1710 cm^{-1}, while the C=C stretching expected at 1600 cm^{-1} is not detectable in the infrared spectra.

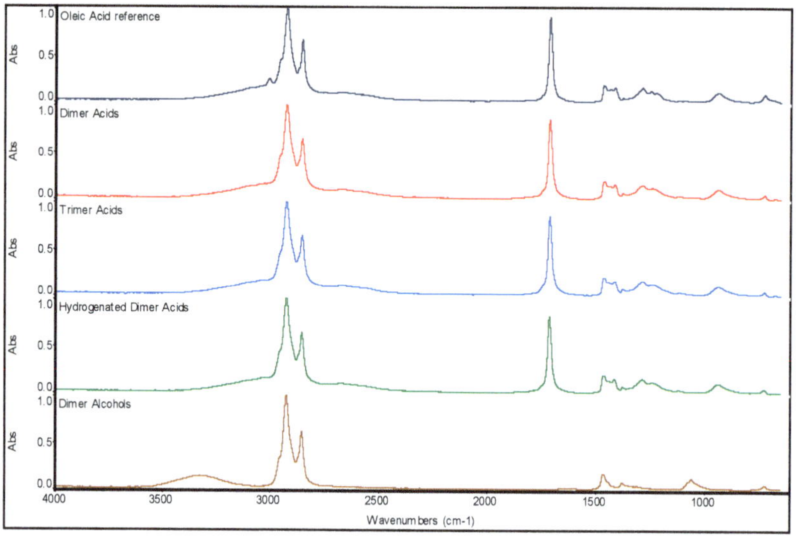

Figure 3. Normalized FT-IR spectra in ATR from top to bottom: oleic acid (reference); dimer acids; trimer acids; hydrogenated dimer acids; dimer alcohols.

FT-IR spectroscopy is very useful in the identification of dimer alcohol (see Figure 3 at the bottom). Hydrogenation has completely removed the C=O stretching band, and the presence of the OH groups is confirmed either from the ν_{OH} at 3328 cm^{-1} or from the C-O bending at 1057 cm^{-1}.

Raman spectroscopy is a very useful tool for the analysis of lipids thanks to its complementarity to infrared spectroscopy [25]. For the first time, the Raman spectra were successfully recorded in dimer acids, hydrogenated dimer acids and dimer alcohol, as shown in Figure 4. Dimer acid shows only the C=C stretching band at about 1655 cm^{-1}, while the carboxylic C=O band is very weak and not detectable with our instrumentation. The hydrogenated dimer acid shows a strong reduction in the intensity of the $\nu_{C=C}$ band, as shown in Figure 4 (compare a vs b), with respect to pristine dimer acids. As expected, in the Raman spectrum of dimer alcohol (bottom of Figure 4c), there is no more evidence of the $\nu_{C=C}$ band.

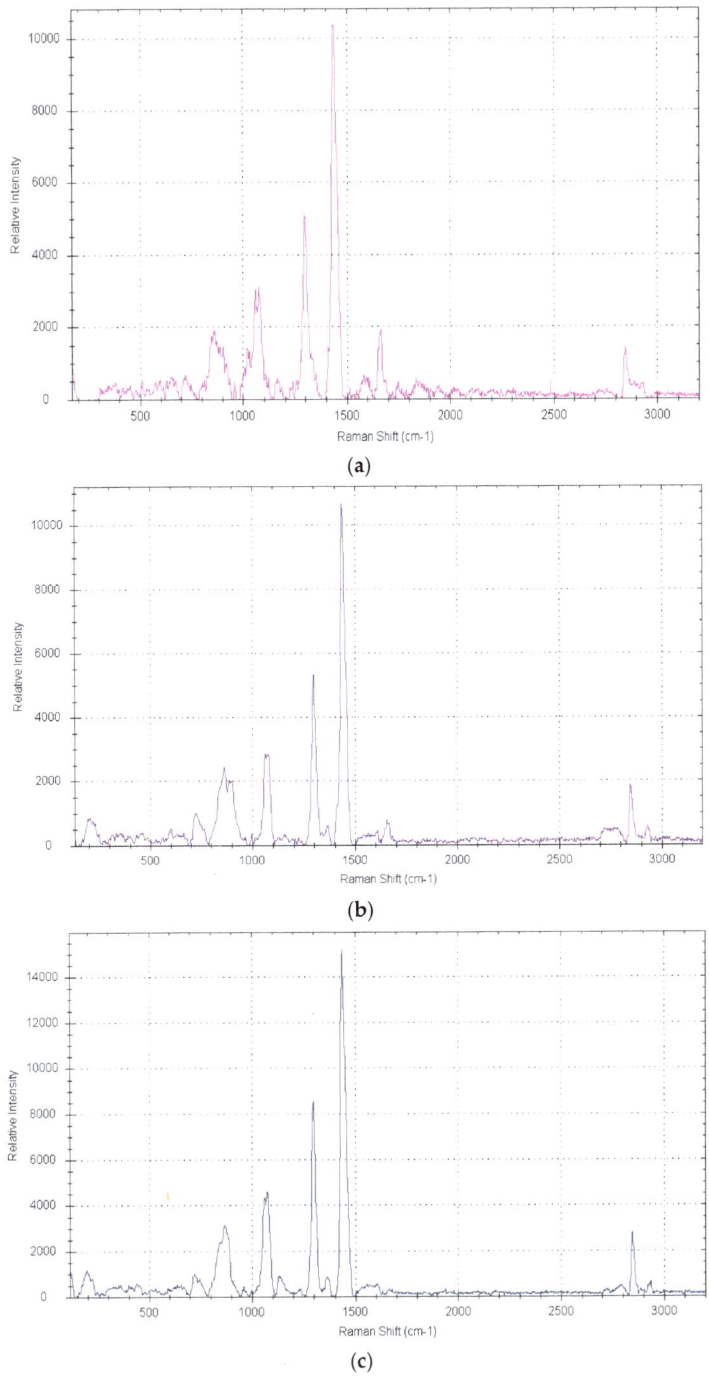

Figure 4. Raman spectra: (**a**) dimer acids; (**b**) hydrogenated dimer acids; (**c**)—dimer alcohols.

We have also tried to record the Raman spectrum of trimer acids. We were successful in obtaining a Raman spectrum of this substrate but at a relatively low quality due to fluorescence since the trimer is a dark viscous liquid. The Raman spectrum of trimer acids is completely comparable to that of the dimer acid in Figure 4a.

Regarding the other Raman lines observed in the spectrum of the dimer acids, it is interesting to notice that these bands are very similar to those observed, for instance, in the Raman spectrum of oleic acid (Figure 5a). In the case of oleic acid, the cis-C=C stretching band is very strong, even stronger than that observed in the spectrum of dimer acids. However, also in the Raman spectrum of oleic acid, the C=O stretching is not detectable (Figure 5a). The C−O stretching band appears in the Raman spectrum at about 1740 cm^{-1} only in ester ethyl oleate (Figure 5b) and is easily distinguished from the C=C stretching band. All esters, including the glycerides of oleic acid, show both the $\nu_{C=C}$ and $\nu_{C=O}$ bands in the Raman spectra.

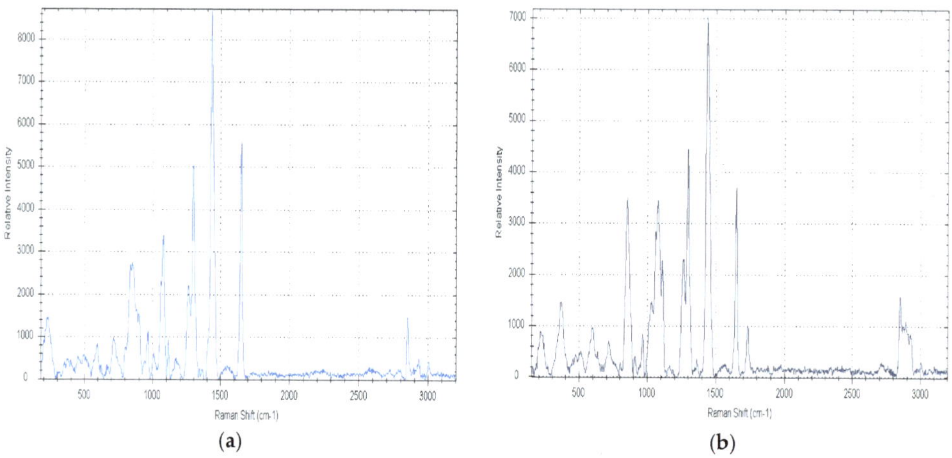

Figure 5. Raman spectra: oleic acid (**a**); ethyl oleate (**b**).

Following the assignment made by other authors in the case of the Raman spectra of fatty acids [25], Raman bands (Figures 4 and 5) in the range of 1500–1400 cm^{-1} are due to the scissoring and twisting vibrations of the CH_2 groups, while the CH_3 groups display the same types of vibration at about 1300 cm^{-1}. The =CH bending mode is reported at 1265 cm^{-1}, but it is evident only in the Raman spectra of oleic acid and ethyl oleate. In the range of 1200–1040 cm^{-1}, the skeletal optical modes due to $\nu_{C\text{-}C}$ appear [25].

3.3. Dimer Acids and Their Esters as Plasticizers Evaluated in the Nile Red Polarity Scale E(NR)

3.3.1. Dimer Acids and Other Carboxylic Acids Evaluated with the Nile Red Polarity Scale E(NR)

In the previous work [1], it was shown that one of the best ways to measure the polarity of plasticizers is to use Nile red dye (see Scheme 4 for the chemical structure) instead of Reichardt's $E_T(30)$ dye. The main problem with 2,6-diphenyl-4-(2,4,6-triphenyl-1-pyridinio) phenolate (i.e., $E_T(30)$ dye) is its sensitivity to acids. Even the trace residual acidity present in ester plasticizers (or other liquids) causes protonation of the phenolate oxygen anion, leading to the disappearance of the long-wavelength solvatochromic charge transfer (CT) band of the $E_T(30)$ dye, hindering any possible measurement. Other problems hindering the wider application of Reichardt's $E_T(30)$ dye are connected with its insolubility in certain liquids. The use of Nile red dye and the known correlation found between the E(NR) scale and $E_T(30)$ may represent a solution to evaluating the polarity of esters, plasticizers, carboxylic acids and other "difficult" liquids or certain polymers experimentally [1].

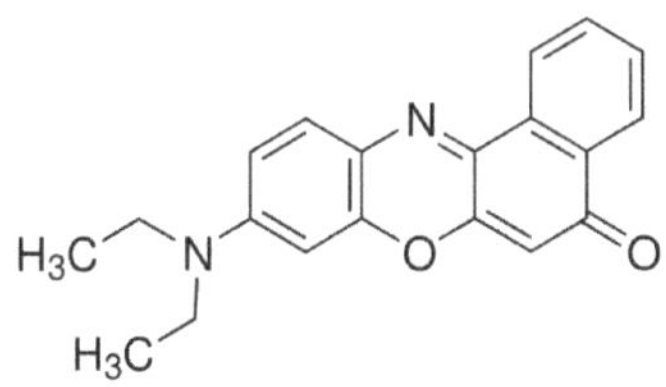

Scheme 4. Nile red dye, i.e., 9-diethylamino-5H-benzo[α]phenoxazinone.

Another powerful alternative to the above-mentioned experimental approaches is represented by a sophisticated semi-empirical approach proposed by Acree and Lang [26] that permits the estimation of the $E_T(30)$ values of certain "difficult" liquid and solid substrates where the experimental polarity measurement is inaccessible for a number of reasons.

Remaining with the purely experimental approach of polarity measurement, the use of Nile red dye represents the tool of choice for difficult substrates. Nile red dye is soluble in innumerable media, is completely insensitive to acidity and exhibits positive solvatochromism in different liquid media in contrast with the negative solvatochromic behavior of Reichardt's dye. A drawback of Nile red dye is that the absorption band of NR is less sensitive to the polarity changes of liquids than the absorption band of Reichardt's dye. This is manifested in smaller solvatochromic shifts as a change in the polarity of the given liquid, requiring higher sensitivity spectrophotometers.

Another advantage offered by Nile red solvatochromic dye compared with Reichardt's $E_T(30)$ dye is the possibility of measuring the polarity of liquid carboxylic acids. As shown in Table 1, a series of liquid carboxylic acids was studied with Nile red solvatochromic dye, including dimer and trimer acids, hydrogenated dimer acids and even isostearic acid (see Scheme 5 for the chemical structure). The latter compound is formed during the dimerization reaction when tall oil fatty acids (or other mixture of fatty acids) are heated with a montmorillonite catalyst. Isostearic acid is then separated from dimer acids, and it is characterized by a branched fatty acid chain instead of a linear one, as in the case of common stearic acid, as shown in Scheme 4 [6–12]. The unique property of isostearic acid is that it is liquid at ambient temperatures, while stearic acid is a solid with a melting point at about 70 °C. All the E(NR) data in Table 1 were measured in the present study, with the exclusion of formic, acetic and propionic acids, whose E(NR) values were taken from ref. [27].

Table 1. Polarity of liquid carboxylic acids measured with Nile red dye and the corresponding dielectric constant (or relative permittivity) ε.

Liquid Carboxylic Acid	λ (nm)	E(NR) kcal/mol	References or Notes on the E(NR) Values	Dielectric Constant ε (*)
Oleic acid	524.2	54.54	this work	2.34
Isostearic acid	525.2	54.44	this work	2.3 (**)
Dimer acids	528.7	54.08	this work	2.6 (**)
Hydrogenated dimer acids	528.7	54.08	this work	2.5 (**)
Trimer acids	531.5	53.79	this work	2.9 (**)
Propionic acid	542.4	52.71	Ref. [27]	3.3
Acetic acid	557.2	51.31	Ref. [27]	6.2
Levulinic acid (γ-ketovaleric acid)	595.5	48.01	this work	19
Lactic acid 90%	596.6	47.92	this work	22
Formic acid	634.0	45.10	Ref. [27]	54.5

(*) from refs. [28,29]; (**) estimated values, see text.

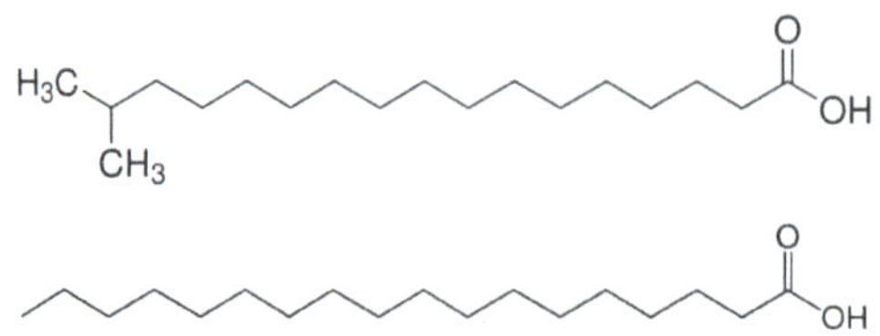

Scheme 5. Chemical structure of isostearic acid (**top**) versus the structure of stearic acid (**bottom**).

To show the validity of the E(NR) measurements in the case of these liquid carboxylic acids, an attempt was made to correlate the E(NR) values with the dielectric constants (ε) of these liquids. Fortunately, the dielectric constants (or relative permittivity) of the carboxylic acids studied in Table 1 are known [28,29], with the obvious exception of the dimer, trimer, hydrogenated dimer and isostearic acids. Since the dielectric constants of fatty acids are fully known [28,29], it was assumed that the dielectric constants of the dimer, trimer, hydrogenated dimer and isostearic acids were similar to the ε values of oleic, stearic and palmitic acids, with some minimal corrections derived from the E(NR) measurements. As shown in Figure 6, this approach has led to an interesting correlation between the E(NR) of liquid carboxylic acids and ε through a power law:

$$E(NR) = 57.09\ \varepsilon^{-0.058} \quad (3)$$

with an excellent R^2 value of 0.996.

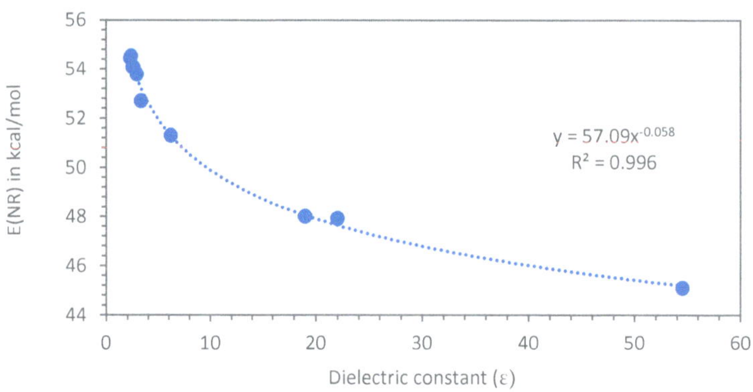

Figure 6. Correlation between the E(NR) values and the dielectric constant (or relative permittivity) of the liquid carboxylic acids.

3.3.2. Dimer Alcohols and Other Alcohols Evaluated with the Nile Red Polarity Scale E(NR)

Because of the interesting results obtained with liquid carboxylic acids in terms of E(NR) values and their correlation with the dielectric constants of the corresponding acids, a selection of alcohols typically used to produce ester plasticizers were subjected to the same methodological approach.

Table 2 shows the E(NR) values of selected alcohols measured in this work or derived from ref. [27]. The dielectric constant values were taken from refs. [28,29]. The dielectric constant of dimer alcohols was not found in the literature and was estimated as similar to that of 1-octanol (i.e., 11.3) since its structure recalls that of dimer alcohols (see Scheme 3) or t-butanol because dimer alcohols share a similar E(NR) value to t-butanol (see Table 2). Glycerol formal (see Scheme 6) is another interesting alcohol, similar to dimer alcohols, since it is >95% derived from glycerol from renewable sources, which is then reacted with one molecule of formaldehyde to produce glycerol formal (see structure in Scheme 6). Thus, it has been assumed that glycerol formal maintains the same dielectric constant value as glycerol, i.e., 46.5 [28]. This assumption is also justified from the experimental value of the E(NR) transition, which places glycerol formal just between ethylene glycol and methanol (see Table 2).

Scheme 6. Chemical structure of glycerol formal.

Table 2. Polarity of selected alcohols measured with Nile red dye and the corresponding dielectric constant (or relative permittivity) ε.

Alcohol	λ (nm)	E(NR) kcal/mol	References or Notes on the E(NR) Values	Dielectric Constant ε (*)
1-Dodecanol	534.7	53.47	this work	8.1
2-Methyl-2-propanol (t-butanol) (***)	536.5	53.29	this work	10.5
Dimer alcohol	536.5	53.29	this work	10.4 (**)
2-Propanol (isopropanol) (***)	539.2	53.03	Ref. [27]	18.3
1-Nonanol	541.0	52.85	this work	9.2
1-Octanol	544.0	52.56	Ref. [27]	10.3
2-Ethylhexanol (***)	544.8	52.48	this work	7.6
1-Propanol	545.6	52.40	Ref. [27]	20.1
3-Methyl-1-butanol (***)	545.7	52.39	this work	15.6
1-Pentanol	547.2	52.25	this work	16.2
1-Butanol	547.6	52.21	Ref. [27]	17.8
Tetrahydrofurfuryl alcohol (***)	548.2	52.16	this work	13.6
Ethanol	548.3	52.15	Ref. [27]	24.3
2-Methyl-1-butanol (***)	548.3	52.15	this work	15.6
Methanol	549.6	52.02	Ref. [27]	32.6
Glycerol formal	559.5	51.10	this work	46.5 (**)
Ethylene glycol	565.2	50.59	Ref. [27]	41.4
Water (***)	593.2	48.20	Ref. [27]	80.0

(*) from refs. [28,29]; (**) estimated values, see text; (***) included in Figure 7 but excluded from Figure 8.

Figure 7 shows the attempt to correlate the E(NR) values of all the selected alcohols in Table 2 with the relative dielectric constants of each alcohol. This time, the correlation can be considered reasonably good, although not as good as in the case of the carboxylic acids studied in Figure 6. In fact, this time, $R^2 = 0.890$, and the linear equation linking E(NR) with ε is the following:

$$E(NR) = -0.063\varepsilon + 53.54 \quad (4)$$

It is reasonable to think that the dispersion of values in Figure 7 is due to the relatively large variety of chemical structures of the alcohols considered. As reported in Table 2, if all the alcohols with non-linear structures and water (all marked with three asterisks) are excluded from the E(NR) vs. ε correlation, then the best fit of the data is offered by a power law equation (similar to that found for carboxylic acids) with $R^2 = 0.831$, and Equation (4) is turned into the following Equation (5):

$$E(NR) = 56.06\ \varepsilon^{-0.024} \tag{5}$$

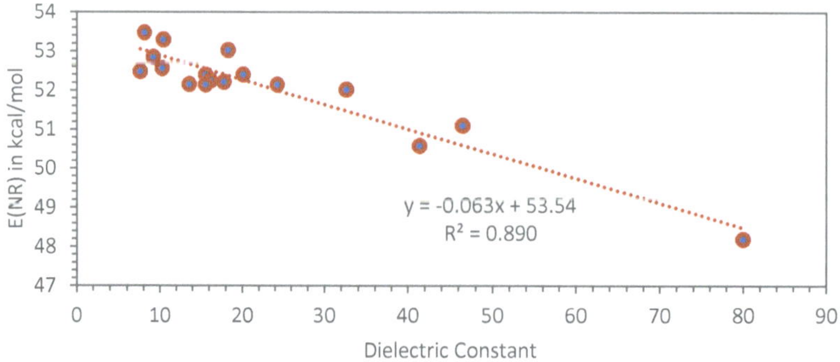

Figure 7. Correlation between the E(NR) values and the dielectric constant (or relative permittivity) of all the alcohols listed in Table 2, including water.

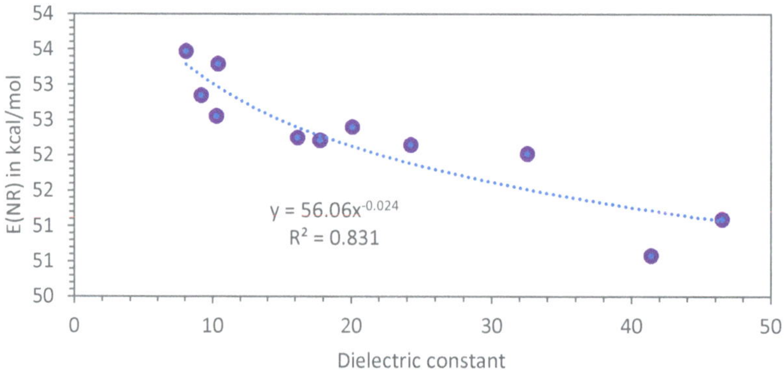

Figure 8. Correlation between the E(NR) values and the dielectric constant (or relative permittivity) of selected "linear" alcohols listed in Table 2 (those without three asterisks).

3.3.3. Plasticizers Evaluated with the Nile Red Polarity Scale E(NR)

A series of new plasticizers were synthesized according to the procedure detailed in ref. [21]. The new plasticizers are essentially the esters of the dimer or trimer acids and hence named "dimerates" or "trimerates", respectively, and shown in bold in Table 3. The other plasticizers and esters reported in Table 3 are shown for reference to see how the dimerates and trimerates are positioned in the E(NR) scale. Furthermore, Table 3 shows another two "new" plasticizers, which are the fatty acid esters of glycerol formal (see Scheme 5), an alcohol derived from renewable sources as the resulting esters of fatty acids [30,31]. Unfortunately, it is not possible to try to make a correlation between the E(NR) values of the plasticizers reported in Table 3 and the corresponding dielectric constant of these liquids. The reason is that for many of the plasticizers reported in Table 3, the ε values are not available from the literature and, moreover, the dimerate and trimerate esters are little known. A similar comment applies also to the fatty acid esters of glycerol formal.

Table 3. Polarity of selected plasticizers measured with Nile red dye.

Plasticizers	λ (nm)	E(NR) kcal/mol	References or Notes on the E(NR) Values
T-DAE (treated distillate aromatic extract)	--	56.20	Ref. [1]
Oleyl oleate	508.1	56.27	Ref. [1]
Ethyl oleate	521.3	54.85	Ref. [1]
MES (mild extract solvate)	521.5	54.83	Ref. [1]
Methyl esters of fatty acids (brassica)	522.0	54.77	Ref. [1]
Methyl esters of fatty acids (cocco)	522.7	54.70	Ref. [1]
Dioctyl adipate (diethylhexyl adipate)	523.0	54.67	Ref. [1]
Soybean oil	524.1	54.55	Ref. [1]
NYTEX Bio (Naphthenic oil and vegetable oil mixture)	525.4	54.42	Ref. [1]
Lauryl dimerate	526.0	54.36	This work
Nonyl dimerate	526.2	54.34	This work
Lauryl trimerate	526.2	54.34	This work
Amyl dimerate	526.6	54.29	This work
Ethylhexyl dimerate	526.7	54.28	This work
Glycerolformal esters of brassica fatty acids	526.8	54.27	This work
Isoamyl dimerate	527.1	54.24	This work
Butyl dimerate	527.1	54.24	This work
Ethyl dimerate	527.7	54.18	This work
Sunflower oil (high oleic)	528.2	54.13	Ref. [1]
Ethyl trimerate	528.4	54.11	This work
Diisododecyl adipate	528.5	54.10	Ref. [1]
Dimer acids	528.7	54.08	This work
Hydrogenated dimer acids	528.7	54.08	This work
Diethyl azelate	528.7	54.08	Ref. [1]
Methyl undecenoate	529.6	53.99	Ref. [1]
PEG dioleate	531.5	53.79	Ref. [1]
Tetrahydrofurfuryl dimerate	531.9	53.75	This work
PEG monoleate	533.4	53.60	Ref. [1]
Ethyl levulinate	536.4	53.30	Ref. [1]
Dimer alcohol	536.5	53.29	This work
Di(ethylhexyl)phthalate	537.0	53.24	Ref. [1]
Epoxidized glycerolformal ester brassica fatty acids	538.2	53.12	This work

Thus, a way to follow the change in polarity of the dimerates and trimerates is shown in Figure 9, where it is possible to see that the polarity of this family of esters is reduced by the length of the aliphatic chain of the alcohol used for esterification with a general equation:

$$E(NR) = 0.022\ n + 54.11 \tag{6}$$

and with n the number of carbon atoms in the alcohol. These results are completely intuitive since the maximum polarity (lowest E(NR) values) are observed in free carboxylic acids, dimer acids and hydrogenated dimer acids, with a linear and measurable increase in

E(NR) on passing to ethanol (n = 2), butanol (n = 2) and so on in the resulting esters. With $R^2 = 0.859$, the correlation is reasonably good.

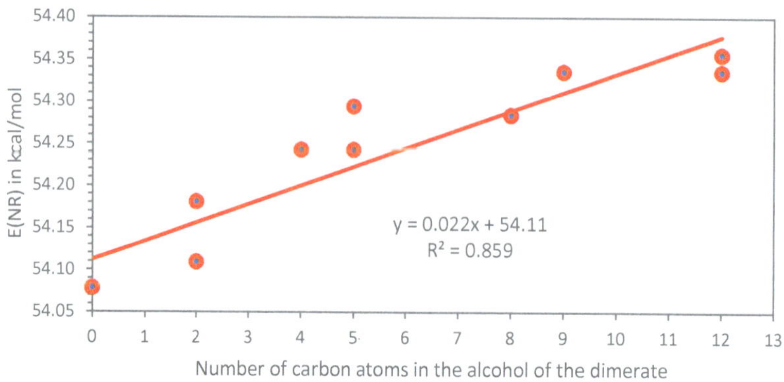

Figure 9. Correlation between the E(NR) values of dimerates and trimerates listed in Table 3 and the number of carbon atoms present in the alcohol of the ester.

The trimerates are considered only in a couple of cases, i.e., as ethyl and as lauryl (C_{12}) esters. In both cases, as well as in the case of the free trimer acids (see Table 1), this class of compound becomes more polar than the corresponding dimerate esters and dimer acids. This experimental fact is completely reasonable since trimer acids have three free carboxylic acid functionalities in contrast with the two functionalities displayed by dimer acids. The same reasoning applies in the case of the corresponding esters, i.e., three ester groups for the trimerates and only two ester groups for the dimerates.

4. Discussion of the Application of Dimerates, Trimerates and Other Esters

Dimerates and trimerates and the fatty acid esters of glycerol formal were developed in order to offer a new series of plasticizers, mainly for the rubber but also for the plastics industries. These new esters were designed to be derived from renewable sources and with the aim of substituting them for the traditional petroleum-based plasticizers used in the rubber industry, like T-DAE (treated distillate aromatic extract) and MES (mild extract solvate) [21,30,31]. Furthermore, the mentioned esters of dimer acids were considered biodegradable, and indeed, they present a reasonably good level of biodegradability [32].

In a previous work [1], the discussion was extensively focused on new plasticizers, which can substitute partially or completely for the traditional petroleum-based plasticizers, and their evaluation through the Reichardt's or Nile red polarity scales. It was shown that a solid correlation between the two scales does exist, and furthermore, Nile red dye is much more suitable than Reichardt's dye $E_T(30)$ for the study of the polarity of ester liquids. Dimerates and trimerates, particularly the former, show polarity levels at the E(NR) scale that make them compatible with the typical polymer matrices used by the rubber and, in particular, the tire industries, i.e., polybutadiene, polyisoprene (or natural rubber) and styrene–butadiene copolymers. Table 4 shows a summary of the E(NR) values of rubber and certain common plastics.

From the data in Table 3, it appears immediately evident that the most suitable dimerate ester candidates to be studied as replacements for petroleum-based MES, in addition to the methyl esters of fatty acids [5], soybean oil and the commercial product Nytex Bio, which is a mixture of naphthenic and vegetable oil, are lauryl (C_{12}), nonyl (C_9) and amyl (C_5) dimerates. Of course, ethylhexyl dimerate is also a candidate, but

2-ethylhexyl alcohol is synthetic and not naturally sourced. On the other hand, lauryl alcohol is derived, for example, from the naturally occurring and renewable palm kernel or coconut oil after fractionation and hydrogenation [33]. 1-Nonanol is industrially produced from the ozonolysis of oleic acid (a widely available and naturally sourced compound), followed by the hydrogenation of the resulting 1-nonanal [34]. Finally, amyl alcohol is naturally occurring from fermentation processes involving corn, molasses, potatoes or other similar substrates and known as fusel oil [35].

The above-mentioned dimerates with E(NR) values closer to those of polybutadiene (BR), polyisoprene (NR) and styrene–butadiene copolymer (S-SBR) rubbers (see Table 4) were indeed selected, synthesized and tested successfully as plasticizers in substitution of both the petroleum-based plasticizers (e.g., MES of Table 3) and vegetable oil plasticizers (Table 3) used in passenger tire tread formulations, with important advantages in winter performance and other tire tread properties [21]. Compatibility between the selected dimerate esters and the rubber matrix was found to be excellent and in agreement with the expectation from the E(NR) scale.

In addition to dimerates, the fatty acid ester of glycerol formal was successfully tested in a tire tread formulation [30,31]. The glycerol formal ester is positioned in the E(NR) scale just between the amyl and ethylhexyl dimerates from one side and the isoamyl and butyl dimerates from the other side (see Table 3). Even in this case, compatibility with the rubber matrix was found to be excellent, and the plasticizing effect and winter performance of the vulcanizate were found to be much better than the results with the reference traditional petroleum-based plasticizers and vegetable oils [30,31].

Furthermore, an epoxidized version of the glycerol formal ester of fatty acids has been developed to provide special compatibility with more polar polymer matrices like nitrile rubber (NBR) and polyvinylchloride (PVC), which have E(NR) values of 53.07 and 52.21 kcal/mol (Table 4) compared with 53.12 kcal/mol for the epoxidized glycerol formal ester of fatty acids (Table 3). Thus, epoxidation is able to shift the glycerol formal ester from 54.27 kcal/mol down to 52.21 kcal/mol in the epoxidized version (Table 3). As expected, the compatibility of the latter, the epoxidized version, with NBR and PVC is simply excellent [30,31]. Another interesting feature to be noticed in Table 3 regards the fact that the E(NR) value of the epoxidized glycerol formal ester (i.e., 53.12 kcal/mol) is equal to or better than the E(NR) value of (diethylhexyl)phthalate (i.e., 53,24 kcal/mol), the most widely used plasticizer in the past but being currently phased out because of concerns regarding its toxicity and its behavior as a xenohormone [36,37].

Table 4. Polarity of selected rubbers and plastics measured with Nile red dye.

Polymers	λ (nm)	E(NR) kcal/mol	References or Notes on the E(NR) Values	Dielectric Constant ε (*)
Polibutadiene (BR)	502.0	56.96	Ref. [1]	2.3
Polyisoprene (IR or NR)	503.2	56.82	Ref. [1]	2.4
Styrene–butadiene copolymer (S-SBR) with styrene 21% & vinyl 50%	509.0	56.17	Ref. [1]	2.5
Polystyrene (PS)	512.0	55.84	Ref. [1]	2.5
Epoxidized natural rubber (ENR-25)	526.4	54.32	Ref. [1]	3.0
Polymethylmethacrylate film (PMMA)	535.0	53.44	Ref. [1]	3.6
Poly(lactic acid) (PLLA)	535.0	53.44	Ref. [1]	3.5 (**)
Nitrile rubber (NBR) with 33% ACN	538.8	53.07	Ref. [1]	3.66
Polyvinylchloride (PVC)	547.6	52.21	This work	3.7

(*) Data from ref. [38]; (**) from ref. [39].

Thus, the epoxidized glycerol formal ester of fatty acids is one of the most interesting and effective potential replacements for (diethylhexyl)phthalate, both in a nitrile-based rubber matrix and in plasticized PVC.

The validity of the liquid polarity measured by Nile red dye and through the resulting E(NR) scale is supported not only from the previous work [1] but also by the correlation found between the E(NR) scale and the dielectric constants of carboxylic acids (Table 1, Figure 6) and alcohols (Table 2, Figures 7 and 8). In the case of the esters, the correlation of their E(NR) values with the length of the aliphatic chains of the alcohols used in the esterification was found (Table 3 and Figure 9).

Because of these exciting results, the correlation between the E(NR) values measured in solid polymer thin films containing Nile red dye and the dielectric constant of the relative polymers was also studied using the data in Table 4. As shown in Figure 10, in this case, a nice, linear correlation was also found linking E(NR) measured in rubber and plastics with their ε values:

$$E(NR) = -2.776\ \varepsilon + 63.13 \tag{7}$$

with the excellent $R^2 = 0.975$. Thus, even in the solid state, Nile red dye remains an excellent polarity probe of the medium in which it is dispersed or dissolved, irrespective of the physical state of the given medium, whether liquid or solid.

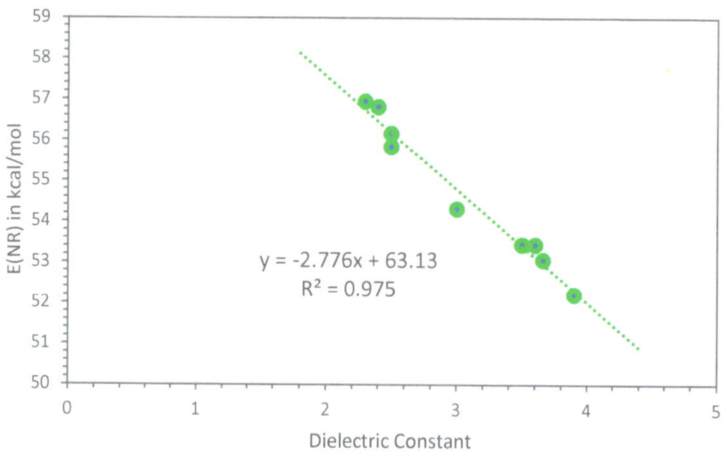

Figure 10. Correlation between the E(NR) values and the dielectric constant (or relative permittivity) of all the rubber and plastics listed in Table 4.

5. Conclusions

Dimer and trimer acids are relatively little-known liquids with relatively complex structures, as discussed in the introduction. Through the study of the electronic absorption spectra of pristine dimer acids and hydrogenated dimer acids, it was possible to conclude that in commercial dimer acids, the aromatic fraction is 35.5%. By contrast, in the case of dimer alcohol, the aromatic fraction was found at a level of 30.1%.

With FT-IR spectroscopy, dimer and trimer acids are practically indistinguishable from the band pattern of a common fatty acid like oleic acid. Of course, dimer alcohols are clearly distinguishable and easily identifiable with respect to either dimer acids or oleic acids.

It was found that Raman spectroscopy is a useful tool for the analysis and easy identification of dimer acids with respect to hydrogenated dimer acids, oleic acids and dimer alcohols. The stretching band of the ethylenic double bonds was found to be a highly diagnostic value in Raman spectroscopy.

Using the solvatochromism of Nile red dye, it was possible to study the polarity of a series of carboxylic acids, including dimer and trimer acids and hydrogenated dimer acids and isostearic acid (all from renewable sources), as shown in Table 1. The validity of this measurement was also confirmed by the correlation found between the E(NR) scale of the carboxylic acids studied and their dielectric constant (or relative permittivity). The experimental data were fitted with a power law, Equation (3).

Because of the success of the E(NR) polarity scale with a series of carboxylic acids, a series of alcohols (including dimer alcohol and glycerol formal, both from renewable sources) was also studied in the same way (see Table 2), and in this case, a fair correlation between the E(NR) scale and the dielectric constant of the alcohols considered was found. A linear correlation was found, as shown by Equation (4). However, when selecting alcohols with linear structures, the correlation of E(NR) with ε assumes a form of a power law of Equation (5).

A series of esters produced from raw materials from renewable sources were synthesized. These esters were made from dimer or trimer acids (both from renewable sources) and alcohols from renewable sources, such as lauryl, 1-nonyl and amyl alcohols. Other alcohols were also considered as dimerate and trimerate esters, as shown in Table 3. Furthermore, the fatty acid ester of glycerol formal (another alcohol from renewable sources) was considered as well as its epoxidized derivative. Table 3 shows the classification of all these plasticizer esters and other plasticizers used for rubber and plastics based on the polarity E(NR) scale using Nile red dye. Unfortunately, it was not possible to study a correlation between the E(NR) values of the plasticizers shown in Table 3 and their dielectric constant for the simple reason that the ε values of most of the plasticizers are not available from the literature data. Instead, a reasonable linear correlation (Equation (6)) was found between the E(NR) values of the dimerate and trimerate esters listed in Table 3 and the number of carbon atoms present in the alcohol chain of the ester.

Finally, a linear correlation (Equation (7)) was found between the E(NR) values measured in a series of polymers in the solid state (as thin solid films) and the relative dielectric constant of the polymers, confirming that Nile red is a powerful polarity probe effective not only in the liquid phase but also in the solid state.

The correlation between the E(NR) values and the dielectric constant of the medium is really intriguing and deserves further investigation.

The typical approach for the selection of the most compatible plasticizer for a given polymer was also shown. In general, it is necessary to select a plasticizer with the same E(NR) value as the polymer:

$$E(NR)_{\text{plasticizer}} \leq E(NR)_{\text{polar polymer}} \quad (8)$$

For polar polymers are also suitable those plasticizers having lower E(NR) values than polymers, i.e., plasticizers which are more polar than polymers.

On the other hand, for a non-polar polymer (e.g., hydrocarbon polymers)

$$E(NR)_{\text{plasticizer}} \geq E(NR)_{\text{non-polar polymer}} \quad (9)$$

are also suitable those plasticizers having higher E(NR) values than that of the polymer, i.e., plasticizer with less polarity than the polymer.

The success of this approach has already been tested in the development of new plasticizers from renewable sources for rubber and plastics.

Funding: This research received no external funding.

Data Availability Statement: This work will be available on the Researchgate website at the author's page. Further inquiries can be directed to the corresponding author.

Conflicts of Interest: The author declares no conflicts of interest.

References

1. Cataldo, F. Conventional and Green Rubber Plasticizers Classified through Nile Red [E(NR)] and Reichardt's Polarity Scale [E_T(30)]. *Liquids* **2024**, *4*, 305–321. [CrossRef]
2. Sandstrom, P.H.; Rodewald, S.; Ramanathan, A. Rubber Composition and Tire with Component Comprised of Polyisoprene Rubber and Soybean Oil. US Patent N°20140135424A1, 15 May 2014.
3. Brunelet, T.; Dinh, M.; Favrot, J.M.; Labrunie, P.; Lopitaux, G.; Royet, J.G. Plasticizing System for Rubber Composition. U.S. Patent N°7834074B2, 16 November 2010.
4. Mohamed, N.R.; Othman, N.; Shuib, R.K. Synergistic effect of sunflower oil and soybean oil as alternative processing oil in the development of greener tyre tread compound. *J. Rubber Res.* **2022**, *25*, 239–249. [CrossRef]
5. Cataldo, F.; Ursini, O.; Angelini, G. Biodiesel as a plasticizer of a SBR-based tire tread formulation. *ISRN Polym. Sci.* **2013**, *2013*, 340426. [CrossRef]
6. Cowan, J.C. Dimer acids. *J. Am. Oil Chem. Soc.* **1962**, *39*, 534–545. [CrossRef]
7. Den Otter, M.J.A.M. The dimerization of oleic acid with a montmorillonite catalyst I: Important process parameters; some main reactions. *Fette Seifen Anstrichm.* **1970**, *72*, 667–673. [CrossRef]
8. Den Otter, M.J.A.M. The Dimerization of Oleic Acid with a Montmorillonite Catalyst II: GLC Analysis of the Monomer; theStructure of the Dimer; a Reaction Model. *Fette Seifen Anstrichm.* **1970**, *72*, 875–883. [CrossRef]
9. Den Otter, M.J.A.M. The Dimerization of Oleic Acid with a Montmorillonite Catalyst III: Test of the Reaction Model. *Fette Seifen Anstrichm.* **1970**, *72*, 1056–1066. [CrossRef]
10. McMahon, D.H.; Crowell, E.P. Characterization of products from clay catalyzed polymerization of tall oil fatty acids. *J. Am. Oil Chem. Soc.* **1974**, *51*, 522–527. [CrossRef]
11. Leonard, E.C. *The Dimer Acids: The Chemical and Physical Properties, Reactions, and Applications of Polymerized Fatty Acids*; Humko Sheffield Chemical: Memphis, TN, USA, 1975.
12. Leonard, E.C. Polymerization-dimer acids. *J. Am. Oil Chem. Soc.* **1979**, *56*, 782A–785A. [CrossRef]
13. Balogh, M.; Laszlo, P. *Organic Chemistry Using Clays*; Springer: Berlin/Heidelberg, Germany, 1993; p. 113.
14. Park, K.J.; Kim, M.; Seok, S.; Kim, Y.W. Quantitative analysis of cyclic dimer fatty acid content in the dimerization product by proton NMR spectroscopy. *Spectrochim. Acta Part A Molec. Biomolec. Spectros.* **2015**, *149*, 402–407. [CrossRef]
15. Zhang, J.; Nuñez, A.; Strahan, G.D.; Ashby, R.; Huang, K.; Moreau, R.A.; Yan, Z.; Chen, L.; Ngo, H. An advanced process for producing structurally selective dimer acids to meet new industrial uses. *Ind. Crop. Prod.* **2020**, *146*, 112132. [CrossRef]
16. Frihart, C.R. Chemistry of Dimer Acid Production from Fatty acids and the Structure–Property Relationships of Polyamides Made from These Dimer Acids. *Polymers* **2023**, *15*, 3345. [CrossRef] [PubMed]
17. Cornils, B.; Lappe, P. Dicarboxylic Acids, Aliphatics. In *Ullmanns Encyclopedia of Industrial Chemistry*, 5th ed.; Gerhartz, W., Ed.; VCH: Weinheim, Germany, 1987; Volume A8, p. 535.
18. Johnson, R.W.; Valdospino, J.M.; Gordon, R.L.; Miller, G.E.; Knight, R.W. Polyamides from fatty acids. In *Encyclopedia of Polymer Science and Engineering*, 2nd ed.; Kroschwitz, J.I., Ed.; Wiley: New York, NY, USA, 1986; Volume 11, pp. 476–489.
19. McAdams, L.V.; Gannon, J.A. Epoxy Resins. In *Encyclopedia of Polymer Science and Engineering*, 2nd ed.; Kroschwitz, J.I., Ed.; Wiley: New York, NY, USA, 1986; Volume 6, p. 352.
20. Bajpai, U.D.N.; Nivedita, M. Dimer Acids-based Polyesters. In *Concise Polymeric Materials Encyclopedia*; Salamone, J.C., Ed.; CRC Press: Boca Raton, FL, USA, 1999; pp. 385–386.
21. Cataldo, F.; Plancq, L.; Genin, F.; Arora, Z. Rubber Composition Comprising Esters from Renewable Sources as Plasticizers. U.S. Patent US11945952B2, 2 April 2020.
22. Santiago, C.; Gandour, R.W.; Houk, K.N.; Nutakul, W.; Cravey, W.E.; Thummel, R.P. Photoelectron and ultraviolet spectra of small-ring fused aromatic molecules as probes of aromatic ring distortions. *J. Am. Chem. Soc.* **1978**, *100*, 3730–3737. [CrossRef]
23. Breuer, T.E. Dimer acids. In *Kirk-Othmer Encyclopedia of Chemical Technology*; Wiley & Sons: New York, NY, USA, 2001. [CrossRef]
24. Perkampus, H.H. *UV-VIS Atlas of Organic Compounds*, 2nd ed.; VCH: Weinheim, Germany, 1992; Spectrum D2/10.
25. Czamara, K.; Majzner, K.; Pacia, M.Z.; Kochan, K.; Kaczor, A.; Baranska, M. Raman spectroscopy of lipids: A review. *J. Raman Spectrosc.* **2015**, *46*, 4–20. [CrossRef]
26. Acree Jr, W.E.; Lang, A.S. Reichardt's Dye-Based Solvent Polarity and Abraham Solvent Parameters: Examining Correlations and Predictive Modeling. *Liquids* **2023**, *3*, 303–313. [CrossRef]

27. Deye, J.F.; Berger, T.A.; Anderson, A.G. Nile Red as a solvatochromic dye for measuring solvent strength in normal liquids and mixtures of normal liquids with supercritical and near critical fluids. *Anal. Chem.* **1990**, *62*, 615–622. [CrossRef]
28. Dean, J.A. (Ed.) *Lange's Handbook of Chemistry*, 15th ed.; McGraw-Hill: New York, NY, USA, 1999; Section 5-105.
29. Weast, R.C. (Ed.) *CRC Handbook of Chemistry and Physics*, 68th ed.; CRC Press: Boca Raton, FL, USA, 1987; Section E-49.
30. Chiaradia, G. Plasticizers for Polymers. *US Patent 2016026 4758A1*, 2016.
31. Chiaradia, G.; Cataldo, F.; Bastiaensen, E.; Colombo, D.; Turesso, C. Esters from Vegetable Oils as Process Oils for the Production of Elastomers. Italian patent Demand 102024000007495, 2024.
32. Randles, S.J. *Synthetic Lubricants and High-Performance Functional Fluids*, 2nd ed.; Rudnick, L.R., Shubkin, R.L., Eds.; CRC Press: Boca Raton, FL, USA, 1999; Chapter 3; p. 69, Table 6.
33. Falbe, J.; Lipps, W.; Mayer, D. Alcohols, Aliphatic. In *Ullmanns Encyclopedia of Industrial Chemistry*, 5th ed.; Gerhartz, W., Ed.; VCH: Weinheim, Germany, 1985; Volume A1, p. 291.
34. Tavassoli-Kafrani, M.H.; Foley, P.; Kharraz, E.; Curtis, J.M. Quantification of nonanal and oleic acid formed during the ozonolysis of vegetable oil free fatty acids or fatty acid methyl esters. *J. Am. Oil Chem. Soc.* **2016**, *93*, 303–310. [CrossRef]
35. Aries, R.S. Alcohol, Industrial. In *Encyclopedia of Chemical Technology*; Kirk, R.E., Othmer, D., Eds.; Interscience Publishers: New York, NY, USA, 1952; Volume 1, pp. 262–264.
36. Davis, B.J.; Maronpot, R.R.; Heindel, J.J. Di-(2-ethylhexyl) phthalate suppresses estradiol and ovulation in cycling rats. *Toxicol. Appl. Pharmacol.* **1994**, *128*, 216–223. [CrossRef]
37. Specht, I.O.; Toft, G.; Hougaard, K.S.; Lindh, C.H.; Lenters, V.; Jönsson, B.A.; Bonde, J.P.E. Associations between serum phthalates and biomarkers of reproductive function in 589 adult men. *Environ. Int.* **2014**, *66*, 146–156. [CrossRef]
38. Brandrup, J.; Immergut, E.H.; Grulke, E.A. (Eds.) *Polymer Handbook*, 4th ed.; Wiley & Sons: New York, NY, USA, 1999; Section V.
39. Boussatour, G.; Cresson, P.Y.; Genestie, B.; Joly, N.; Lasri, T. Dielectric characterization of polylactic acid substrate in the fre quency band 0.5–67 GHz. *IEEE Microw. Wirel. Comp. Lett.* **2018**, *28*, 374–376. [CrossRef]

MDPI AG
Grosspeteranlage 5
4052 Basel
Switzerland
Tel.: +41 61 683 77 34

Liquids Editorial Office
E-mail: liquids@mdpi.com
www.mdpi.com/journal/liquids

www.ingramcontent.com/pod-product-compliance
Lightning Source LLC
Chambersburg PA
CBHW042013260926
51850CB00020B/3346
9783725841912